Jorge Prieto
Jürgen Kiene

Wood Coatings

Chemistry and Practice

VINCENTZ
Providing Professional Information

Cover: Aaron Kohr, AdobeStock

Bibliographische Information der Deutschen Bibliothek
Die Deutsche Bibliothek verzeichnet diese Publikation in der Deutschen Nationalbibliographie; detaillierte bibliographische Daten sind im Internet über http://dnb.ddb.de abrufbar

Prieto, Jorge and Kiene, Jürgen:
Wood Coatings: Chemistry and Practice
Hanover: Vincentz Network 2018
European Coatings Library
ISBN 3-86630-608-3
ISBN 978-3-86630-608-0

Vincentz Network, Plathnerstr. 4c, 30175 Hanover, Germany
T +49 511 9910-033, F +49 511 9910-029, books@european-coatings.com

Layout: Danielsen Mediendesign, Hanover, Germany
Printed by: Grafisches Centrum Cuno GmbH & Co. KG, Calbe, Germany

European Coatings Library

Jorge Prieto
Jürgen Kiene

Wood Coatings

Chemistry and Practice

Translated by Harald-Bernd Schwarz

Foreword

This compact standard reference for the interior and exterior coating of wood and wood-based materials as well as for the decor finish foil coating mainly focuses on the industrial wood coating systems and their coating processes. The significance of the coating of wood and wood-based materials is demonstrated by market data, coating technologies and processes.

10 years after the first edition in German, some trends in the industrial wood coatings are emerging. The implementation of REACH and the VOC Directive has made a significant impact on commodities markets. The processes have become more and more effective. Above all, companies with a corporate structure have profited from this.

The next ten years will also remain exciting. In Germany, the Federal Government wants to initiate the so-called 4th Industrial Revolution with the program 4.0. The industrial production shall be integrated with the modern information and communication technology. Similar developments can also be seen in other countries. In general, the global trend is also evident to apply more environmentally friendly and more efficient coating procedures.

The book gets started with a brief historical overview of wood coatings as well as with an introduction to the technology of wood and wood-based materials. In addition to the classic solvent-borne wood coatings, the modern UV-curing systems and water-borne coating systems as far as powder coatings predominantly are explained with all their advantages and disadvantages. The guide formulations and current procedures for the coating of wood and wood-based materials are depicted in detail. The current window coatings are evaluated on their properties and their applications. The coating of paper foils and their market significance are discussed. Eco-efficient procedures of applications such as "Vacumat" coating, rolling, printing and recovery of coatings are dealt with. The book is rounded off with statements to current trends in the industrial coating of wood and wood-based materials. The practical relevance of this book provides the reader with practical solutions and support for many unanswered questions. By virtue of this highly-complex subject-matter, the authors make no claim to completeness.

This book is aimed at all professionals dealing with the production and coating of wood and wood-based materials: A comprehensive overview on the chemistry and technology of wood coating systems are given to vocational school students, university students, engineers, machine manufacturers, manufacturer of raw materials, coatings, windows, furniture, parquets and doors. A comprehensive reference list is provided to the interested reader in each chapter.

On this occasion the authors would like to thank all the employees and colleagues who contributed to the implementation of the book by means of suggestions, provisions of bibliography and joyous discussions. We would like to thank Ulrich Désor, Rico Emmler, Christiane Swaboda, Mario Beyer, Marcel Prieto, Lars Passauer and Kurt Plöger. Our thanks also go to the co-authors whom we have been able to motivate for this project.

Our special thanks go to our families who not only had to dispense with us on many weekends but have always spurred us on to complete this great project especially in difficult phases.

Senden and Dusseldorf, Germany, January 2018
Jorge Prieto and Jürgen Kiene

We create
chemistry
that makes wood
love coatings.
BASF is your partner for sustainable wood coating solutions based on Laromer® UV resins and dispersions, water-based Joncryl® and Luhydran® dispersions, Basonat® polyisocyanates and Tinuvin® lightstabilizers.
Explore the BASF portfolio with the Resin-Additive-Search on
www.basf.com/resin-additive-search
BASF
We create chemistry

Contents

1 Introduction

The coating technology of wood surfaces significantly has changed over the last 50 years. The need to streamline and to automate processes in order to remain competitive has contributed to new wood-based materials and coating technologies. Driven by the high costs of production in Western Europe, many well-known manufacturers of furniture and wood-based materials have developed new productions in Eastern Europe about 30 years ago. The pioneer was Ingvar Kamprad, the founder of the now world's largest furniture dealer "IKEA" which engaged the first suppliers for furniture manufacturing in Poland already in the year 1961. The purpose of his life is to offer a wide range of well-designed and functional home furnishing products at prices so low that as many people as possible are able to afford the furnishing products. This idea and the emergence of a new consumer group, the so-called "smart shopper" with a vagrant buying behaviour between "Aldi" and champagne, has revolutionized the world of furniture in the manufacturing process as well as in the sale. The social trend to an increased mobility of people has led to a shorter life cycle of furniture than those in the last century. In addition, today's furniture must be compatible to removals and flexible in use within the living environment. If one compares the range of furniture products from renowned furniture manufacturers with each other, a design crystallized out which focuses on lean material thicknesses on the one hand and on very thick top-quality materials on the other hand, or combines these with high contrast. These trends are reflected in the increased use of so-called lightweight panels such as "board-on-frame (BoF)" constructions. The lightweight panels provide the resource-efficient handling of wood as a raw material, because less material is used. Thus, the production costs are reduced drastically. The lower weight of such plates also leads to savings in transportation. Unfortunately, such wood-based materials are developed without the involvement of coating manufacturers. In retrospect, it often becomes apparent that such materials need new modified coating systems and procedures. Also, the switch of pressing processes (switch from hot to cold pressing) in the production of wood-based panels has a significant influence on the coating properties of the carrier material.

For many years, there is a clear trend towards a simpler design of modern furnishings with regard to the coatability in order to save processing costs in the production. Very often flat components are coated with UV coatings in rolling processes.

The most formidable challenge will be the shortage of raw materials and wood in the foreseeable future, both for the paint manufacturers as well as for the manufacturers of furniture, kitchens, doors, parquets and windows. Furthermore, it is expected that the prices for coating raw materials, energy and wood-based materials will increase tremendously. In recent months, for example, the prices for chipboard panels increased by 20 to 30 %, and there is no end in sight. Additionally, the prices for raw materials and wood are strongly influenced by the enhanced demand in the Asia-Pacific region. This makes the application of the above mentioned honeycomb panel structures very interesting. Also, the utilization of waste wood in small amounts in the production of wood-based panels is a new approach to compensate for the shortage of raw materials. Another method of utilizing wood wastes such as wood flour is their use in so-called WPC (wood plastic composites) in combination with polyethylene and polypropylene. In the course of this, these are thermoplastically processable composites with very interesting properties for indoor and outdoor use. For this, new coating concepts on an environmental friendly basis are currently in development. Over the past several years, the scarcity of wood veneers has motivated numerous manufacturers of furniture and laminate flooring to rediscover the indirect gravure printing. In addition, for several years a steady increase in printed and coated paper sheets (decor finish foils) for the manufacturing of furniture and laminate flooring was recorded as low cost alternative options for laminated veneers. In recent years, the increasing use of digital printing in the furniture industry as well as in the flooring industry is to be highlighted. It is assumed that digital printing will become more and more important. In comparison to conventional acid-curing water-borne paints, electron beam curing coating systems have gained market shares in the segment of decor finish foils due to the rational process mode as well as due to significantly proved abrasion resistance and chemical resistance.

In order to keep manufacturers of furniture, kitchens, floorings, windows and other similar products globally competitive, it is essential to deal continuously with the overall optimization of the process of manufacturing and coating of components as many practitioners only look at the purchase price of the paint materials. As it is shown in Figure 1.1, the proportion of costs of the coating materials in the total coating method only amounts between 6 and 12 %. In order to compare the cost of coating materials objectively, the standard unit "cost/ m^2" of coated surface has established. With respect to this, the following important factors such as price per kg of delivered coating, transfer efficiency of the application process, the application rate in g/m^2 and the "solid" (percentage of non-volatile constituents of the coating) have to be taken into account.

Particularly the EU directive "on the limitation of emissions of volatile organic compounds" (VOC[1] Directive 1999/13/EU) as another driving force has contributed to changes of processes and technologies (see Chapter 9). Thus, in the last 10 years a massive exchange of solvent-borne coating systems by UV-curable coating systems and water-borne coatings has

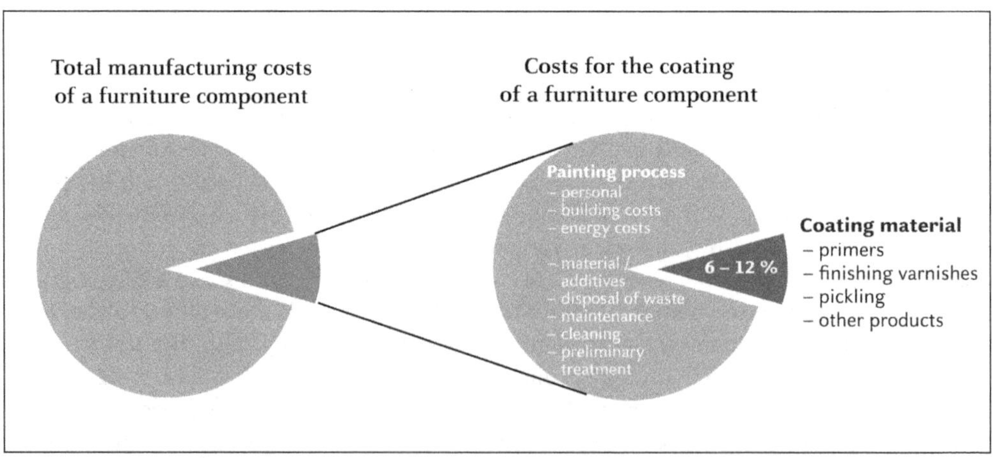

Figure 1.1: Percentage of the coating costs to the total manufacturing process of a furniture component

1 VOC = volatile organic compounds

taken place. The change of technology from strongly solvent-based wood coating systems to more environmental-friendly wood coatings is in full progress, and will accompany us primarily during the next years in Europe. In the meantime, the powder coating (see Chapter 3.1.9) on solid wood and wood materials such as MDF[2] successfully is utilized in the industrial production of furniture in Europe and displaces the classical application with water-borne coatings over there. Completely new developments in the field of wood coating can be expected in the field of nanotechnology. The use of nanoscale particles refers to the improvement of the mechanical and technological properties, such as extremely scratch-resistant coating surfaces. Nanotechnology also is capable to enhance certain functions of the coatings such as an antibacterial effect or photocatalytic properties.

Since around 1989, there is a quite some focus on possible health effects with respect to residual emissions from coated furniture surfaces and wood surfaces [2]. Under the title "Certified Indoor Air", organizations and institutions set themselves the task to use reference values or evaluation schemes (AgBB = German Committee for Health-Related Evaluation of Building Products), respectively, and procedure in the health-related assessment of emissions of volatile organic compounds (VOC and SVOC[3]) from building products. The AgBB scheme, which currently is under discussion, should enable a better evaluation of the health-related assessments of the residual emissions [3].

In the field of outdoor coating, longer warranty periods of at least 5 to 10 years for dimensionally stable components such as windows are required by the consumer. This has led to a continued development of modern aqueous coating systems. At present, in the market of wood windows a UV curing, opaquely pigmented multi-layered construction whose refurbishment intervals amount 8 to 10 years according to the manufacturer's specification is used in addition to water-borne systems. The next few years will decide whether the propagated benefits of UV curing window coating systems confirm in practice. At the same time, the consumption of resistant tropical woods exceeds re-growth of these woods, so that a serious shortage may be experienced in the future. For this reason, one tries to make fast-growing native tree species very durable by chemical reaction (acetylation or "Belmadur" method) or by thermal treatment ("thermally modified wood").

The implementation of the so-called VOC Directive on the limitation of emissions of volatile organic compounds by restricting the marketing of solvent-based paints and coatings influences the further development of window coatings as well as other coating systems for buildings.

1.1 History and development of furniture coatings

Due to the cultural-historical significance of furniture and its influence on the coating technology, this chapter is dedicated to the historical development of the furniture coating. The historical development of other wood-based substrates such as parquet or wood floors and their coating is shortly outlined explicitly in the Chapters 2.5 and 3.

The purpose of coating of wood and wood-based materials, especially furniture, served not only to protect the wood surface, but also provided a representative role up to the art object [4]. The early stages of the decorative coating of furniture are in Europe in the 15th

2 MDF = medium density fibreboard
3 SVOC = semivolatile organic compounds

century, while the first coating methods and coating raw materials are much older. The history of the coating and of the process of coating applications has passed two developments which are initially independent of each other [5]. The first older development was the Chinese paint art (Figure 1.2) whose early stages date back about six to seven thousand years. Only after the landing of the Portuguese in Japan in the year 1543, the paint art from Japan and China has been made accessible to the Europeans.

At that time, various techniques of the paint art for the coating and artistic design of furniture and equipment made of wood and other materials already were maintaining a high technical stand. Among other things, the milky sap from the bark of the Japanese lacquer tree *Rhus verniciflua* also referred to as Rhuslack served as a coating. Other techniques of the Chinese varnish art were relayed by missionaries in the 16th century in Europe. At this time, the seafaring Europeans imported the finished varnish ware from East Asia on a large scale. The increasing demand soon meant that one extensively is employed with the development of coatings in Europe. Since the Asian raw materials were not available, well-known domestic resins were used with great excess by combination with drying oils.

In Europe, the second independent trend was known long before the East Asian coating techniques. It has its origin in the European-Islamic culture. The word "Lakh" and thus the today commonly used word "varnish" comes from the ancient Indian "Sanscrit language" and means "hundred thousand". This is based on the large number of lacquer scale insects whose product of excretion is known to us as shellac. This product arrived Europe via medieval trade routes.

Even before the introduction of shellac in Europe, in ancient times solutions of spruce resin in linseed oil, also known as varnishes, existed which served as protective layers of painted objects. By the year 1000 AD, the monk *Rodgerus of Helmershausen* reported for the first time in his book Schedula Diversarium Artium on the production of coating materials and gives detailed instructions for the design of recipes.

Figure 1.2: Chinese paint art *Source: Verband der deutschen Lackindustrie e.V. (VdL)*

As recently as at the beginning of the 15th century, the colourful design of the furniture by decorative painting became the fashion in Italy. From there, it quickly spread to all countries of Europe. During this period, a craftsman-like direction originated from the carpenter trade including professions such as joiners, cabinet-makers, manufacturers of boxes or carpenters [6].

In the 16th and 17th century, in Europe the East Asian and European-Islamic line of development merged into a common European varnish art. Together with the handicraft varnishing of musical instruments, the decorative design of furniture is a key driver for the development of the coating technology up to the beginning of the 20th century. The varnish application usually was done with a brush which consisted of a mixture of horsehair and plant fibres. As early as 1400, the guild of "Pürsten Pinter" which entirely was devoted to the manufacture of brushes arose. The varnishing of furniture always has been based on the application of paint brushes.

Already for the late Baroque period, the great importance of the coating technique has passed on to the construction of furniture [7]. The coating of wood-based substrates with transparent coatings or varnishes played a subordinate role up to the beginning of the 19th century. However, it was the most important initial point for the development towards modern surface treatment. The recipes known at this time based on materials such as shellac, amber, waxes, mastic, dammar, benzoin, sandarac and copal. Moreover, oxidatively drying oils such as linseed oil which was used not only as a "reactive solvent" were used. Suitable solvents were spirit of wine (ethanol) and turpentine oil. In the course of time one learned to improve the deficiencies of the slow drying of natural resins and oils by simple chemical reactions.

Until about the year 1870, the coaters have produced their own coatings and lacquers. Now many coating factories and oil mills were built which provided the user finished coatings. The esterification of resins and copal with glycerine is known since about 1884.

Other milestones are to be mentioned in the epoch of historicism (approximately 1850 to 1900). During this time, *Michael Thonet* developed a technique for the serial production of chairs and in large numbers. In this era of industrialization (1848 to 1871), the manual work was supported by machines or completely replaced in many companies. For example, in larger fabrics the elaborate polishing of a wooden surface coated with shellac has been widely adopted by polishing machines. Also, the relatively hard cellulose nitrate coating surfaces were very well suited for polishing machines. As from the year 1925, cellulose nitrate lacquers were used in the furniture manufacture.

At the beginning of the 20th century, the furniture industry used bale polishes for the application of coating in the production of high-gloss furniture such as in the "Gelsenkirchener Barock". At that time, attempts were made to work less time-consuming and to minimize the time intervals between the cycles of polishing by using bale polishes with higher concentrations. The necessary resting periods up to the final treatment were achieved by temporary storage for weeks. Only then, the final topcoat was established. Due to enormous progress in the coating technology as well as due to the industrial manufacturing of coatings, partially produced from renewable raw materials and synthetic film-forming agents such as cellulose nitrate (CN) and short oil alkyd resins, the way for today's coating technology was preordained. The redesign and renewal of surface technologies took place by means of the development of the spray technique (approximately in the year 1895). This was done in parallel with the introduction of the above mentioned CN coatings. The technical breakthrough of CN coatings occurred in the 20's of the last century, when the assembly line technology in the automobile industry required a fast drying coating process.

Between the years 1925 and 1935, apart from very few exceptions it can be seen clearly, that the furniture factories still worked according to the principles of handicraft. Only in the early 1930, the spray technology was introduced in the furniture industry gradually.

Today's coating of wood and wood-based materials is less assessed under artistic viewpoints than more from the practical viewpoint. According to DIN 68880, today the term furniture is defined as follows: "Furniture is a fixture for receiving goods, to sit, to lie or for the execution of any work."

The main functions of an appropriate surface treatment are:

- Protection of the surface against mechanical and chemical impacts
- Design (pattern) of the surface
- Increase of the practical value (materially and non-materially)
- Protection of the surfaces in outdoor applications in the form of physical and/or chemical wood preservation.

This point of view arose in Europe, especially in the years after the Second World War. In this reconstruction phase, the German furniture industry was able to secure a large market share in Europe and achieved the leadership in product innovation and production technology [8]. Of course, this also benefits the supplier industry, such as the associated machine manufacturers, tool manufacturers, metal fitting manufacturers, wood coating manufacturers, manufacturers of glue and wood panels. At that time, many of these manufacturers have developed themselves to a world's market leader. In the 1960's of the last century, the furniture industry has been accelerated significantly or modified in their working method by newly developed through-feed machines, chipboards and fast curing wood coatings. The enhanced use of plastics in the manufacture of wood housings in the mid-70's of the 20th century displaced the wood products from this segment. In addition, the significance of the powerful style furniture industries declined dramatically since new furniture trends and tastes had an influence. While at the beginning of the 1980's still a market saturation of mass furniture was recorded, in the mid-1980's a transition from serial to individual production occurred. Many companies optimized their production ("lean production"). But there also were many companies which could not complete the transition to modern times and disappeared from the market. The structural change and the associated adjustments were best managed by the kitchen furniture industry. The turnover of the German kitchen furniture industry exceeded 4.39 billion Euro [9] in the year 2016. Of particular importance is the North Rhine-Westphalian kitchen furniture industry in both the national and the European context. With an annual production volume of 1.5 million built-in kitchens (2014) thus every third to fourth kitchen in Europe was produced in North Rhine-Westphalia [10]. Due to the merger with the new federal states as well as due to the currency exchange of 1:1, the furniture industry of the West experienced an enhanced boom over several years, whereas almost all of the six-large furniture collective combines in the new federal states could not exist in the market. Due to the boom years of the merger, export efforts were further neglected culpably. As in Italy 30 years ago, today the German furniture producers are faced with the question: "Grow or disappear in the insufficiently expandable domestic market". Due to the efforts of the last years, the export of furniture manufactured in Germany is expected to increase from 3 billion Euro in the year 1995 to 6.0 billion Euro (2006) [11]. This corresponds to an assumed export quota of approximately 28 % for the year 2006. The main European competitor of the German furniture industry are the Italian producers. In the tough price competition in Germany, the Italian furniture manufacturers have moved to other export markets and are no longer available as suppliers in the German market in the first place. The Italian domestic market was no longer capable of expansion in the mid-70's, so that Italian furniture manufacturers were forced to open up foreign markets. The Italian furniture industry now is one of the world's leading furniture exporters. The Polish furniture companies have adopted the leading role of the Italian furniture manufacturers and have displaced the Italian furniture manufacturers from the German market. It should be noted that about 60 % of the Polish furniture industry are held by German companies. One of the leading furniture exporting countries worldwide is China which exported furniture with a value of more than 45 billion Euro in the year 2013 [12].

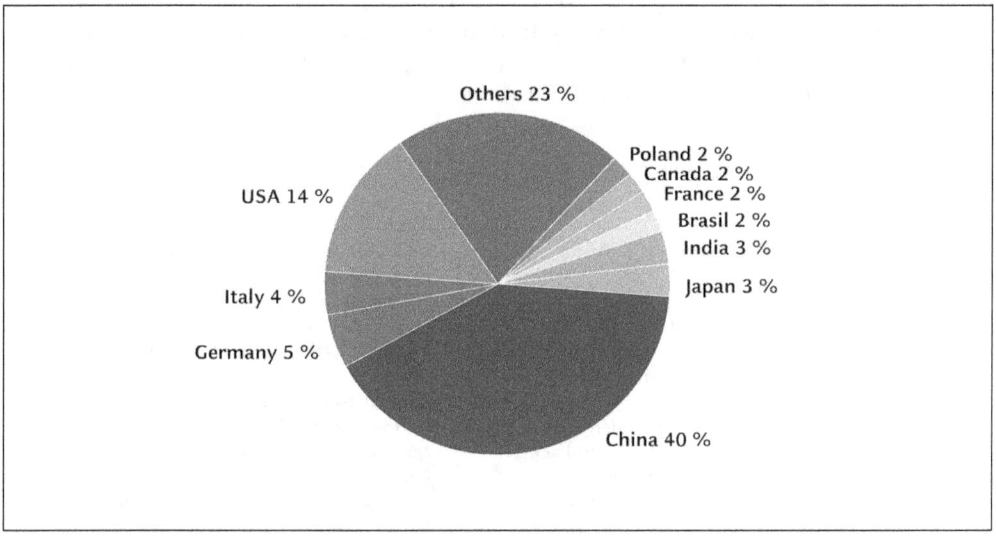

Figure 1.3: Global furniture production 2012 in billion Euro [12, 13]

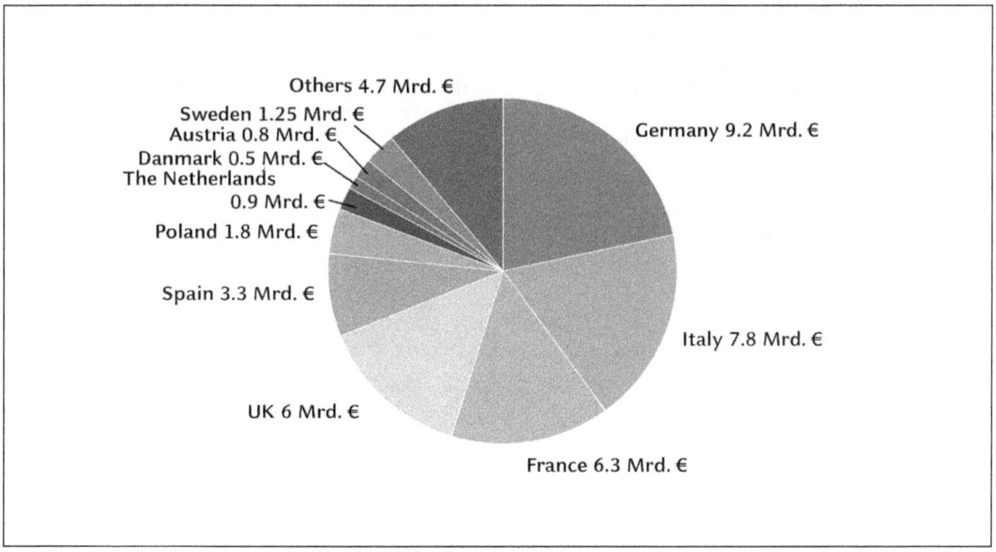

Figure 1.4: Furniture production 2012 (EU 28) in billion Euro [12, 13]

The proportion of personnel expenses in the German furniture industry in 2003 was 27 % for decades, the average size of the company has been 105 employees. In the year 2015, the Germans (400 € per capita) together with the Austrians (362 € per capita) and Sweden (€ 359 per capita) were the European champion in buying furniture [12]. Looking beyond Europe's borders, this trio is even world champion, because potential competitors such as Japan, USA and Canada are behind them.

The EU (European Union) furniture industry consists of 8,800 companies with more than 20 employees each. A total of 600,000 employees are working in these companies. In addition, there are 80,000 companies with fewer than 20 employees (a total of almost 300,000 employees). With a percentage of total EU production of over 20 %, Germany is the most important furniture producer prior to Italy. In foreign markets, furniture "Made in Germany" still is very much appreciated. The worldwide furniture production amounted to approximately 361 billion Euro in the year 2012 (TOP 10). The European furniture production was approximately 42.60 billion Euro in the year 2012 [12, 13].

1.2 References

[1] Goldschmidt, A., Streitberger, H.-J., BASF Handbook Basics of Coating Technology, Vincentz Network, 2007

[2] Fischer, M., Böhm, E., Erkennung und Bewertung von Schadstoffemissionen aus Möbellacken, Erich Schmidt Verlag Berlin, 1994

[3] Wensing, M., Schulz, N., Salthammer, T.: Kontrollierte Innenraumluft, Farbe und Lack 2006, 112. Jahrgang, Nr. 2, S. 61–65

[4] Von Leixner, Othmar: Einführung in die Geschichte des Mobiliars, Konrad Grethleins Verlag, Leipzig, 1909

[5] Sandmann, M.: Die Geschichte der Lackierung und die geschichtliche Entwicklung der Lackauftragsverfahren bis 1940, Semesterarbeit im Studienfach Restaurierung/Konservierung der Fachhochschule Köln, 8. März 1999

[6] Anonymus: Einblicke in die Geschichte der Möbel

[7] Kunze, P.: Mater.+Tech. 4, 40, (1996)

[8] Haas, D.: Die deutsche Möbelindustrie hat eine Zukunft, Holz-Zentralblatt, Nr. 2, 13. Januar 2006, S. 45–46

[9 Märkte + Tendenzen – Küchenmöbel, Gruner + Jahr AG & Co. KG, Nr. 16 Juli 2005

[10] Clusterstudie Forst & Holz NRW, Gesamtbericht Juli 2003, Ministerium für Schule, Wissenschaft und Forschung NRW und Ministerium für Umwelt und Naturschutz, Landwirtschaft und Verbraucherschutz NRW

[11] Bundesagentur für Außenwirtschaft (bfai), Möbelexporte expandieren kräftig, Artikel vom 19. Dezember 2006

[12] Jahrbuch 2005–2006, Die wirtschaftliche Lage der Holz- und Möbelindustrie, herausgegeben von: Hauptverband der deutschen Holz und Kunststoffe verarbeitenden Industrie und verwandter Industriezweige e.V. (HDH) und Verband der deutschen Möbelindustrie e.V. (VDM), November 2005

[13] European Commission: The EU Furniture Market situation and a possible Furniture products initiative, Brussels, November 2014

2 Wood and wood-based materials

2.1 Introduction

In Germany and Europe, until 150 years ago wood was the dominant material apart from natural stone. For Scandinavia, Canada, USA and many other countries wood has retained this position until today. With the beginning of the industrialization in Central Europe, however, wood dramatically lost in importance. In association with a requirement for standardization and top performances in the production with always the same quality, wood could not meet the economic criteria. Thus, wood-based materials have been developed with more homogeneous and new properties which covered all areas of applicability [1-4]. The organic and natural material wood is degraded by environmental factors as well as by organisms, and it is incorporated into the material cycle in terms of its basic components. In order to protect it from the effects of water as well as pollutants and thus from deteriorations, but also to give

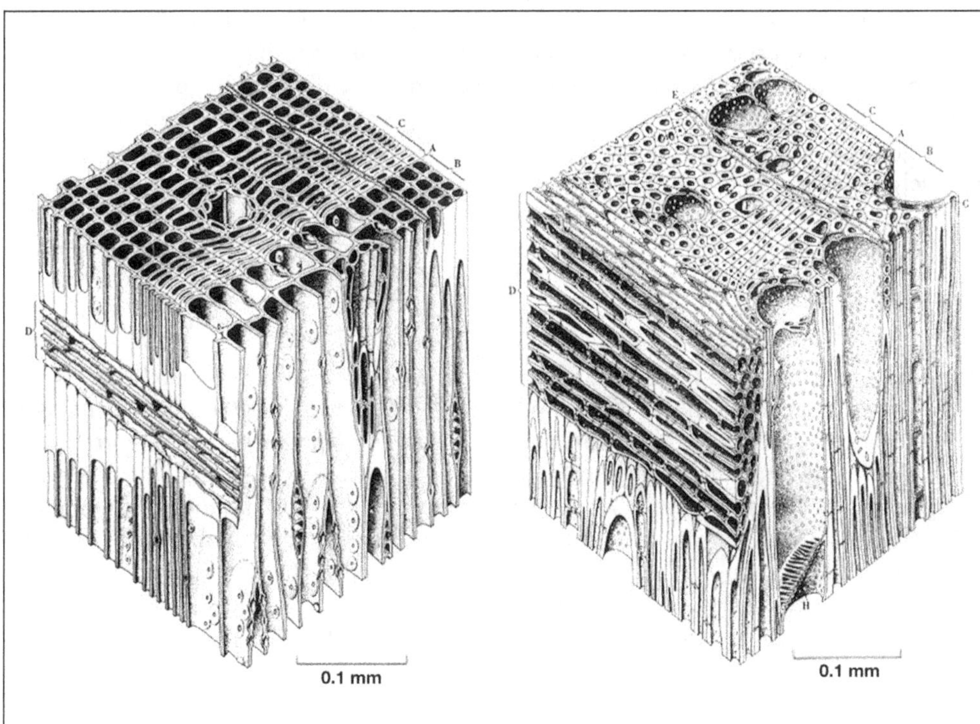

Figure 2.1: Schematic structure of spruce wood (left) and beech wood (right). Meaning: A = growth ring borders, B = early wood, C = late wood, D = medullary rays, F = tracheids, G = wood fibre, H = vessels Source: BASF SE

a decorative appearance to the wood, wood already has been treated early. The surface treatment of wood and wood-based materials still continues to play an important role in the assessment of wood components. However, in contrast to other substrates wood has specific properties that affect the surface finish, or their knowledge in the formulation of wood coatings, establishing procedures etc. is required, respectively [5-8].

2.2 Microscopic wooden construction

Wood is a natural porous composite material. On the one hand, the pore structure ensures a vivid surface as well as a penetration of liquids such as coating materials into the wood surface. On the other hand, the pore structure requires a special consideration such as the coating of coarse-pored wood. With regard to the history of evolution, conifers are older and much simpler structured than hardwoods. Figure 2.1 presents schematically the wood structure as well as the significant structural differences between deciduous woods and coniferous woods.

The porosity is of great importance for many processes of the processing and utilization of wood such as plasticization, coating, bonding and impregnation [9]. Very porous wood only has a low compressive and tensile strength. Thus, wood deforms upon pressing, grinding and buffing and is susceptible to cracking especially where large vessels occur close to the surface. Truncated vessels (pores) which are not filled with coating voluntarily due to their narrow cross-section and due to the air trapped in them cause specific coating defects. With dark wood, the pore wall has to be coloured very strongly. Otherwise the pore wall gleams through silvery grey after painting. Surfaces with exaggerated pores also have an equally unsightly grey shimmer. An air bubble is then located between the coating and wood surface. Obliquely incident light is reflected at the interface coating/air, and the pore appears grey [10].

2.3 Physical and technological properties of wood

Wood can be characterized in a simplified manner as a material which mostly consists of only a few millimetre long fibres and a binder. The well glued wood fibres contribute to an exceptional strength of the wood. With regard to the strength properties, wood corresponds to a modern multi-phase material and also shows its anisotropy. This is caused by the parallel aligned wood fibres oriented toward their height growth. Wood has different properties in the three main axes longitudinal (in the direction of fibres), radial (in the direction of the wood rays) and tangential (in the direction of the growth rings). The cutting directions clearly are visible in Figure 2.2 and described in Table 2.1.

This anisotropy is used in building materials such as planks, beams and boards [11, 12]. An important key figure specific for wood is the bulk density, which of course has a great influence on the strength properties of wood. The bulk density is not only different from wood to wood, but also varies within a wood species.

In temperate latitudes, there primarily exists a difference in the bulk density between early and late wood. The bulk density of wood is between 0.1 kg/m^3 (balsa) and 1.3 kg/m^3 (lignum vitae). It increases with the proportion of cell wall substance and is inversely proportional to the pore space of the wood species. With increasing pore radius of the wood species, also the relative pore volume and thus the relative penetration depth of coatings in the

Table 2.1: Description of the anatomical main cutting direction

Cutting direction	Incision	Synonym
Cross section (Q) the growth rings and the course of the wood rays are visible on the cross-sectional area the wood rays of hardwoods are more developed than those of softwood	Perpendicular to the trunk axis, in the direction of wood rays	Brain cut
Tangential cut (T)	Parallel to the trunk axis, perpendicular to the wood rays	Flat cut tendon cut
Radial cut (R) cut wood rays are referred to as mirror	Through the trunk axis in the direction of the wood rays	Longitudinal cut mirror cut rift cut

pores increases. The pore volume is a parameter for many technological characteristics which determine the performance characteristics of the wood.
Among them are:

- Impregnability of wood
- Degree of swelling and degree of shrinkage
- Wear resistance (hardness, wear resistance)
- Elastomechanical properties

Pores with a macroscopic structure (vessels) have a diameter of 10^{-1} to 10^{-5} cm, while pores with a microscopic structure (wood-cell structure) have a diameter of 10^{-5} to 10^{-7} cm.

2.3.1 Behaviour of wood against humidity

Due to the submicroscopic structure of wood, its material humidity adapts to the ambient air humidity. It is hygroscopic. With decreasing relative humidity, the wood dries while with increasing humidity it is wet. The moisture content of the wood is determined as the **wood humidity** u. The wood humidity not only influences the coatability but also properties such as strength, elasticity or resistance to fungi and insects.

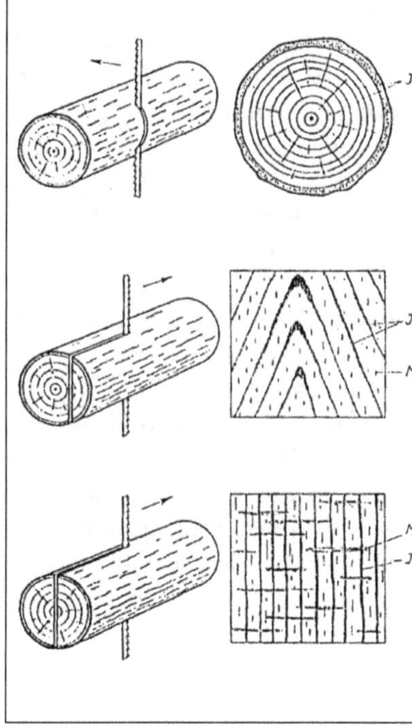

Figure 2.2: Cutting directions of the wood, at the top = radial cut, in the middle = tangential cut, at the bottom = cross-section, M = medullary ray, J = growth ring

Source: Deutsche Gütegemeinschaft Möbel e.V.

Sorption behaviour and capillary water uptake

The system wood/water can be grouped in three limit states:

Table 2.2: Wood equilibrium moisture content of spruce wood at various relative humidities of air (T = 20 °C)

Type of wood	**Relative humidity of air in %**				
	35	50	65	75	85
Spruce	7.0	9.2	11.8	14.4	18

Table 2.3: Recommended values of the mass-based mean wood humidity according to DIN 1052-1 for various mounting conditions

Field of application		**U_m [%]**
All-round closed building	With heating system	6–12
	Without heating system	9–15
Covered open buildings		12–18
Constructions being exposed to the weather conditions on all sides		12–24

- Absolutely dry
  The absorptive area begins at the so-called absolutely dry state with a moisture content of u = 0 %. With this, the state is defined which is established on withdrawal of the humidity at temperatures of 103 to 105 °C up to constant mass by means of technical means.
- Fibre saturation
  When the fibre saturation humidity (uF) is reached, the cell walls are saturated with water, and the cell cavities still contain any dripping liquid water. The fibre saturation point is the area of the maximum level of bound water. Generally, there are the following fibre saturation areas in the different types of wood:
  - European wood species uF = 28 to 35 %
  - Conifer wood general uF = 30 %
  - Deciduous woods with heartwood colouring uF = 22 to 26 %

Upon reaching the fibre saturation point, no further swelling and thus no impact on the strength properties occurs.

- Water saturation
  The micro- and macro-system of the wood is maximally filled with water. Depending on the density of the wood, the wood humidity content is between 77 % (Balsa) and 31 % (lignum vitae). Wood has a large specific internal surface area. For example, spruce has a specific internal surface area of 220 m^2/g. It has the ability to absorb water and steam as well as to release the absorbed moisture again. Depending on the moisture content and depending on the temperature of the surrounding air, a certain equilibrium moisture can be obtained. Depending on the climate zone, wood products retain an average moisture content of 12 to 21 % outdoors, while in living rooms wood products retain an average moisture of around 8 to 10 %. Table 2.2 lists the wood equilibrium moisture contents of spruce at different values of air humidity and a temperature of 20 °C.

The differences in the equilibrium moisture content of domestic wood species are low. One exception are the tropical wood species. The equilibrium moisture content of the wood can be reduced significantly by means of thermal or hydrothermal pre-treatment (for example high temperature drying). A significant reduction in the wood humidity occurs at a temperature of about 200 °C. The heat-pressure treatment reduces the content of hemi-cellulose and thereby the moisture content as well as improved dimensional stability. The equilibrium moisture content and the shrinkage behaviour of wood will be reduced by up to 50 % by means of the thermal treatment at temperatures between 180 and 240 °C. The process of acetylation also induces a substantial reduction in the equilibrium moisture content and improved dimensional stability (see Chapter 2.4.3.1). Below the fibre saturation, the water uptake of the wood occurs via diffusion and/or via capillary forces. The uptake rate of water decreases with the increasing density of wood. Too wet or too dry processed wood induces tensions in the moisture balance to the ambient climate and in the formation of cracks in the wood or in the coating, respectively. In general, the wood humidity should be <12 % for the exterior coating and >10 % for the interior coating. The reference values of wood humidity in the wooden construction should be respected according to DIN 1052-1 (see Table 2.3). The water uptake per unit time in the fibre direction significantly is enhanced in comparison to the water uptake in radial or tangential direction. This statement also applies to the uptake of moisture from the air. For large cross sections (e.g. laminated wood beam), the equilibrium moisture content is achieved over the entire cross-section only after a long storage time. Thus, a full rewetting of dry spruce wood is very time consuming.

2.3.1.1 Swelling and shrinkage of wood

Due to the above-described absorption as well as delivery of water molecules into the molecular structure of the wood, wood swells during moisture absorption and shrinks during loss of heat. The anisotropy of the wood causes different degrees of swelling and shrinking in the 3 main directions of the wood. The following values can be kept in mind as a rule of thumb:
- Tangential approximately 10 % (in the direction of the growth rings)
- Radial approximately 5 % (in the direction of the wood rays)
- Longitudinal approximately 0.1 % (in the direction of the fibres)

The shrinkage and swelling are thereby proportional to the alteration of the moisture content. Table 2.4 presents the differential shrinkage dimensions according to DIN 68100 for selected wood species.

According to DIN 1052-1, calculated values for the degree of shrinkage and swelling also are available (see Table 2.5). For specified types of wood, it is usually calculated with the mean value of tangential and radial values, since the actual development of the growth rings in the sawn wood is not predictable.

The swelling and shrinkage of wood and wood-based materials naturally have a major impact on the coating, because the coating has to survive these dimensional variations of the wood without adhesion losses or cracking.

2.3.2 Hardness of the wood

Another technological key feature of the wood is its hardness. Hardness is defined as the resistance of a body which it opposes to a penetrating harder body under an external force. The hardness primarily is determined by means of the ball indentation hardness test according to Brinell (HB) at a moisture content of u = 12 % (DIN EN 1534). The bulk density, the fibre direction, the relative amounts of early wood and late wood, the contents of lignin and resin as well as the moisture have an impact on the determination of hardness. For

Table 2.4: Differential shrinkage of different wood species according to the standard DIN 68100

Wood species	Differential shrinkage V in % per 1% wood moisture content change		
	Radial	Tangential	Medium
Afromosia	0.18	0.32	0.25
Maple	0.15	0.26	0.21
Bongossi (azobé)	0.31	0.40	0.36
Beech	0.20	0.41	0.31
Douglas	0.15	0.27	0.21
Oak	0.16	0.36	0.26
Ash tree	0.21	0.38	0.30
Spruce	0.19	0.39	0.29
Pine	0.19	0.36	0.28
Larch	0.14	0.30	0.22
Meranti. red	0.11	0.35	0.18
Sapeli mahagoni	0.24	0.32	0.28
Teak	0.16	0.26	0.21

Table 2.5: Calculated values for the shrinkage and swelling according to the standard DIN 1052-1

	Wood species	Values for shrinkage and swelling for the change of wood moisture content remaining 1% beneath the fibre saturation range
1	Spruce, pine, fir, larch, douglas, southern pine, western hemlock, glued laminated wood, oak	0.24[1)]
2	Beech, keruing (yang), angelique, greenheart	0.30[1)]
3	Teak, afzelia, merbau	0.20[1)]
4	Azobé (bongossi)	0.36[1)]

1 Average value of the values tangential and radial to the growth ring or growth zone, respectively

example, the HB-value of spruce is 1.2 perpendicular to the fibre direction, and 3.2 in the fibre direction. The hardness and, thus, the choice of wood plays an important role especially in the coating of floors and parquets. The Brinell hardness demonstratively gives statements about the dangers of shoe heels. It was reported by Emmler[14] that only floors with a Brinell hardness less than <25 MPa have deep indentations. However, laminate floorings do not exhibit indentations.

2.4 Chemistry of the wood, wood components and wood protection

In addition to the morphology, the chemistry of the wood also has a great impact on the coating result. All types of wood mainly consist of cellulose (40 to 50 %) as scaffold material, lignin (20 to 30 %) as a matrix (binder) and wood polyoses (15 to 25 %). Wood components such as resins, oils, waxes, alkaloids, dyes, tannins, inorganic and organic acids as well as salts (see Figure 2.3) occur in amounts between 1 % and 30 %. These wood components affect the properties and usability of the wood in a different manner depending on the type and quantity of their storage.

Lignin is the structure-building component of the lignified plant tissue. It is composed of a high polymer compound consisting of phenyl propane units. The amount of these phenyl propane units varies depending on the type of wood. This amount accounts for 25 to 32 % for conifers and 18 to 32 % for hardwood. The water-insoluble lignin is cleaved photolytic into water-soluble compounds by means of UV radiation. This situation is reflected in the darkening or changes in colour of the wood, in the discoloration under transparent coatings and in the yellowing of so-called wood-containing papers. Cellulose and polyoses also may contribute to the discoloration under UV irradiation. These discolorations are weaker, but can be emphasized in the long term. The photochemical degradation of cellulose also contributes to

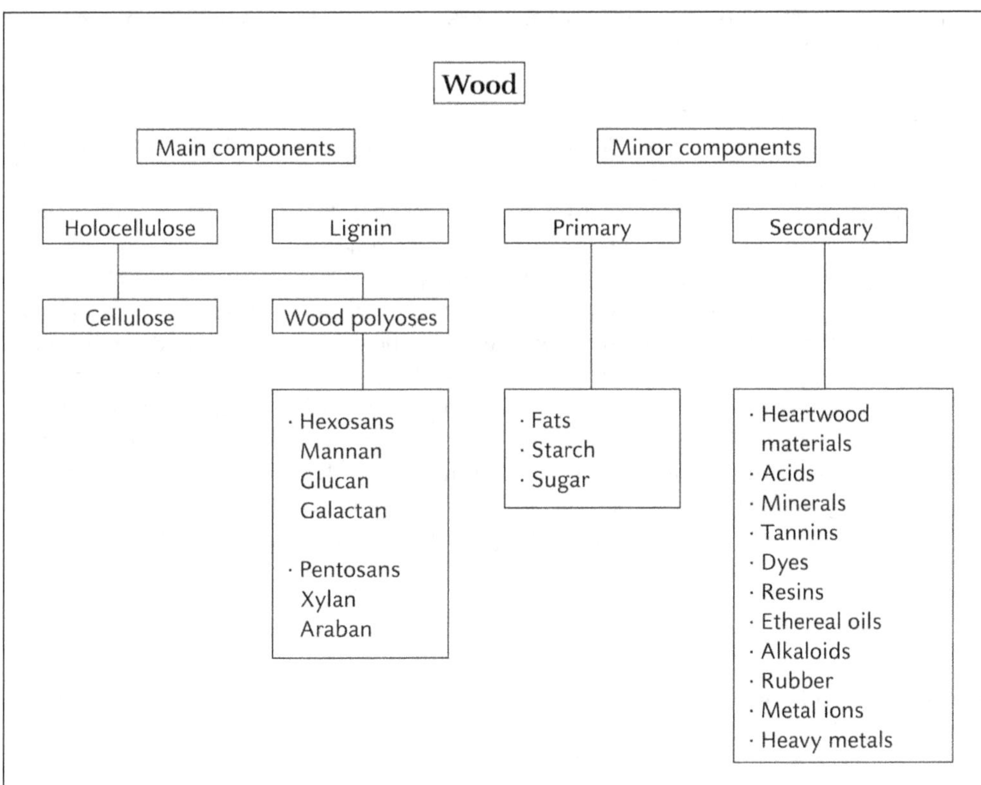

Figure 2.3: Chemical composition of the wood

the reduction of the mechanical strength. Especially in exterior applications, primarily cellulose and polyoses are degraded under weathering.

2.4.1 Wood components

The term 'wood components', also called accessory components, refers to all materials in the wood that do not belong to the structural elements. These components may impact the properties and the utility value of wood in many respects. The wood components are responsible for the smell and colour of the wood, and often the reason for discolorations. The content of these components depends on the type of wood and their location. The warmer the growth area, the more wood components are in the woods. Thus, the resistance to fungi and insects is increased. Furthermore, a higher lignin content, an enhanced hardness and density are found in these woods [15]. According to Gruber [16], wood components are divided into different groups:

- Inorganic components, also referred to as "ash"
- Fats and extractives
- Natural resins

2.4.1.1 Inorganic components

The inorganic constituents especially include calcium, potassium and magnesium compounds as well as trace elements such as cations of iron, manganese, boron, etc. [17]. The content of the inorganic components amounts between 0.2 % and 1 % in moderate zones of growth. It is partly about 5 % in tropical areas.

2.4.1.2 Fats and extractives

These include all low molecular weight, hydrophobic compounds of the wood. Essentially, this involves substances listed in Table 2.6.

The listed fats mainly are glycerol esters of long chain fatty acids with a chain length of 16 to 18 carbon atoms. Other chain lengths are less common. Inhibitory wood ingredients such as polyphenols can delay the curing of unsaturated polyester coatings and UV curable coatings for example [18]. Moreover, the inhibitory wood ingredients may cause discolorations when they react with dyes that are dissolved out from the coating. Tannic acid e.g. from oak wood can cause spotted discolorations in contact with other components such as metals or antifungal agents. This particularly is observable in water-soluble coatings or wood stains as most dyeing wood constituents are water-soluble. For example, blue to blue-black discoloration on oak and other tannin-rich woods are feared. In this regard, wood components are not the trigger for this discoloration. This discoloration is triggered by iron impurities in the form of abrasion of tools (fillings and grinding chips) or by the use of iron-containing water.

2.4.1.3 Natural resins

Natural resins (see Table 2.7) are a mixture of different chemical substances such as resin acids (resinol acid), resin alcohols (resinols), esters of resin acids and resin alcohols, phenyl derivatives with the properties of tanning agents (resino tannol) as well as unsaturated, partially oxygen-containing compounds. Turpentine is the most famous extractive mixture occurring together with resin. As a solvent, turpentine has the task to dilute the incorporated resins and allow it to flow. Since the components of turpentine are volatile, the components

Table 2.6: Extractives of wood according to Gruber [16]

Group	**Chemical structure**
Terpenes	Cyclic diisoprenoids
Terpenoids and resins	Diterpenes
Fats	Glycerides
Waxes	Long-chain esters
Lignans	Degradation products of lignin
Flavenoids	Flavanes, flavones
Tannins (hardwoods)	Polyphenoles
Stilbenes (conifers)	Diphenylene ethene

Table 2.7: Ingredients of natural resins according to Gruber [16]

Resin component	**Enclosed group of substances**
Resin acids (resinol acids)	Abietic acids, pimaric acids, palustrine acids
Resin alcohols (resinols)	Diterpene alcohols
Phenols	Polyhydric phenols, gallic acid ester
Resin esters	Esters from resin acids and resin alcohols
Unsaturated compounds	Monoterpenes, sesquiterpenes and diterpenes
Turpentine	Pinene, 3-carene, camphor
Etherial oils	Alcohols, aldehydes, ketones, ...

reduce with increasing storage time of the woods. Monoterpenes such as 3-carene or camphor are components of natural resin which may cause allergic reactions in sensitive people. Beside α- and β-pinenes and other monoterpenes, these ingredients are detected even in high amounts during the measurement of the remaining emissions of pine containing furniture components (see Chapter 9.2.1.2).

Roffael and Shepherd [19] describe that chipboard panels consisting of fresh pine wood chips after gluing with urea-formaldehyde resins have significantly lower bonding strengths than chipboard boards consisting of embedded wood chips. These justify the higher bonding strength of chipboards consisting of embedded wood chips with a significantly lower content towards the fresh pine wood chips. Resinous wood components can cause precipitations at the lacquer surface. These precipitations mainly are enhanced by high temperatures and dark coatings in the outdoor area or in the coating process. Precipitations cannot be prevented completely by means of coatings. There are special insulating primers, for example for resinous pine trees or tannin containing woods. Heat supply also may mobilize wood components which usually do not cause damages. Thus, often "grease spots" are formed at the Tchitola veneer already at a temperature of 40 °C by exudation of low-melting impurities [20].

2.4.1.4 Influence of the pH value

The pH value also may influence the curing and discoloration of coatings or adhesions. In some glue binders, a low pH accelerates the curing and thus shortens the open time of the glued joint. Other dependencies of the pH value and certain wood extractives on the coating of wood are listed in Table 2.8 [21, 22].

Table 2.8: pH values of exemplary wood species and effects of ingredients on the coating Source: [5]

Wood species	pH value	Nature and impact of the ingredients upon the coating
Dedicuous wood		
Afrormosia	4.5	Retarding the curing, discolouring, corrosive on Fe metals
Afzelia	4.9	Retarding the curing, discolouring
Iroko	6.4	Polyester coating damaging ingredients
Maple	5.1	Risk of discoloration
Birch	4.8	Generation of an inactive surface layer by means of fatty substances
Beech	5.4	Good pickling ability and coatability
Oak	3.8	Tannin-rich, discolouring, blue in the presence of Fe metals
Framire	4.1	Tannin-rich, discolouring, dark colouration in the presence of Fe metals
Merbau	4.3	Discolouring
Mansonia	4.3 to 6.2	Contains ingredients inhibiting the curing of polyester coating
Sipo	6.	Retards the curing, discolouring
Teak	5.1	Often formation of veils
Coniferous wood (softwood)		
Pine	5.1	Rich in resins, risk of resin discharge with increasing temperature
Spruce	5.3	Possibility of minor resin discharge
Larch	4.3	Rich in resins, risk of resin discharge, possibly discolouring
Douglas fir	3.5	Risk of resin discharge, discolouring
Pitch pine	3.5	Rich in resins, risk of resin discharge

DIN 68800: Categorization in hazard classes, EN 335: Categorization in application classes

2.4.2 Wood protection

The term wood protection refers to all measures to prevent or retard a depreciation in value or destruction of wood or wood-based materials. Wood has the following disadvantages:

- Wood can be destroyed and degraded by fungi, insects and microorganisms
- Wood is hygroscopic. Wood may absorb or deliver moisture and swells or shrinks as a result of this
- Wood is flammable

For the reasons mentioned above, wood must be protected during outdoor use if the natural durability is not sufficient. For load bearing and reinforcing components, these measures are summarized in DIN 68800 Part 2. In addition, weather protection by coatings (physical protection of wood) and/or chemical wood protection by paints may be required [23, 24].

Table 2.9: Some commercial wood species and their resistance class according to DIN EN 350-2 (DIN EN 350-2 implies statements on the natural durableness of 120 wood species)

Wood species	Residence class
Oak, european	2
Oak, american	2 – 3
Spruce	4
Framire	2 – 3
Hemlock	2 – 3
Pine, heartwood	3 – 4
Pine, sapwood	5
Larch, heartwood	3 – 4
Niangon	1 – 2
Oregon pine, douglasie	3 – 4
Robinia	1 – 2
Red meranty	2 – 4
Sipo mahogany	2 – 3
Fir	4
Teak	1 – 3*
Western Red Cedar	2 – 3

* Each worse classification for wood from plantations

Table 2.10: Hazard class (HC) according to DIN 68800-3

Hazard class	Use	Exposure by			
		Insects	Fungi	Leaching	Mildew-rottenness
0	Inside installed wood, constantly dry	no [1)]	no	no	no
1	Inside installed wood, constantly dry	yes	no	no	no
2	Wood that is exposed not to the ground contact nor to the weathering or leaching even directly; a temporary humidification is possible	yes	yes	no	no
3	Wood is exposed to weathering or condensation, but not to ground contact	yes	yes	yes	no
4	Wood in permanent ground contact or exposed to continuous strong humidification	yes	yes	yes	yes

2.4.2.1 Natural durability

The natural durability of numerous wood species is summarized in DIN EN 350-2 (natural durability of wood to fungal and insect infestation). The commercial wood classes and their resistance classes are listed in Table 2.9. The indication of the resistance class refers to the heartwoods. The sapwoods have a lower resistance without exception. The durability classes are organized so that the class 5 exhibits the least (not permanent) and class 1 the highest (very durable) natural resistance [25].

Those wood species whose natural resistance does not meet the requirements of the application areas (risk classes according to DIN 68800-3 and EN 335[1]) must be protected by means of suitable wood protection measures.

These measures may consist of structural and chemical measures according to DIN 68800, part 1 and 3. The standards DIN 68800-3 and DIN EN 335-1 consistently classify wood structures in so-called hazard classes (HC) or application classes, as shown in Table 2.10. The higher the hazard class, the greater the biological exposure to which the wood is subjected.

2.4.2.2 Constructive wood protection

The conditions favouring humidification, insect infestation, fungus infestation, etc., have to be avoided in the planning and construction of wooden structures. Constructive (structural) wood protection primarily is a reduction in moisture stress or a quick discharge of water and a prevention of an uncontrolled insect infestation. If the moisture content of the wood is not significantly above 20 % over a long term, a fungal infection is not expected (HC 0). The following incomplete basic principles have to be complied with the constructive wood protection:

- Protection against rainfall and splashing water
- Protection against humidity conduction from water, soil or neighbouring components
- Protection against formation of condensation water in wood constructions
- Humidity protection during the construction
- Appropriate wood quality and avoidance of cracks
- Good ventilation of the total wood construction
- Chamfering of end grain surfaces, still better coverage
- Formation of eaves
- Rounding off of sharp edges
- Protection of horizontal wood components
- Rear ventilation of exterior claddings

2.4.2.3 Chemical wood protection

Chemical wood preservatives and procedures should only be applied if all constructive possibilities of wood protection have been exhausted or are not possible by the design. Chemical wood protection is achieved by the application or impregnation of the wood with chemical wood preservatives. According to DIN 68800 part 3, the chemical wood preservatives have to feature the mark of conformity of the German Institute for Building Technology (DIBt), provided it deals with load bearing wood structures that are subject to building regulations. For all other components, chemical wood preservatives with a mark of conformity of the RAL Quality Association Wood preservatives should be used. The chemical wood protection is mandatory regulated[2]) in DIN 68800. The use of wood preservatives always requires the

1 DIN 68800: Categorization in hazard classes, EN 335: Categorization in application classes

2 The DIBT directory of wood preservatives [27] annually published by DIBt contains a list of all wood preservatives with a Building Inspectorate Approval, all wood preservatives with a RAL quality label and protection agents against blue staining fungi according to the VdL Directive, however, without the text of the approvals, thus without warranty.

Table 2.11: Field of application of wood preservatives according to hazard classes [26]

Type of wood preservation agent				Field of application/HC HC 1	HC 2	HC 3	HC 4
Water-borne	Not fixing			x	x	–	–
	Fixing	Chromate-containing	Without Cu (CF, CFB)	x	x	–	–
			with Cu (CK, CKB, CKF, ZKF)	x	x	x	x
			with arsenic (CKA)	–	–	x	x
		Chromate-free	Without Cu (Quat 1, triazols	x	x	x	–
			with Cu (Cu-Quat 1, Cu-HDO 2, Cu-triazols)	x	x	x	x
		Others	Triazols pyrethroids	x	x	x	–
Solvent borne				x	x	x	–
Tar oils				–	–	x	x

use of biocides, because organisms that attack or destroy wood should be cauterised or repelled by means of chemical wood preservatives. According to § 3b, section 1, no. 2 of the Chemicals Act, biocidal active substances are substances with a general or specific effect on or against harmful organisms that are intended for use in biocidal products. Microorganisms including viruses or fungi having an equivalent effect and purpose are classified as such substances. Wood preservatives have to be reported to and approved by BAuA (Federal Office for Occupational Safety and Health).

The main principles for the use of wood preservatives are presented in a simplified form in Table 2.11 for the different hazard classes according to the standard DIN 68800-3 [26]. Wood preservatives are preparations which generally consist of active ingredients (biocides), binders (for fixation), emulsifiers, pigments, solvents and other auxiliaries.

Chromium compounds in wood preservatives mainly are used to fix the active ingredients in the wood, thereby the leaching can be prevented effectively. The use of chromates is controversial due to their toxicity. The authors just want to refer to TRGS 618 "Substitutes and restrictions for use of chromium (VI) containing wood preservatives". There exists a variety of methods [28] for the chemical wood protection treatment. A distinction especially is made between pressure impregnation, dip impregnation (storage process) and methods for surface treatment. In the pressure impregnation, a complete penetration of the sapwood is achieved with wood preservatives (full protection). In the storage process (long-term immersion) woods have to submerged over a long period of time (24 hrs. up to several days) completely. The penetration depth has to be greater than 10 mm (depth protection); otherwise it is a pure edge protection. Depending on the type of wood, a penetration depth of several microns up to several mm is achieved with this method.

Most heartwoods are difficult to impregnate due to their anatomy. However, it is especially important to soak the less durable sapwood as deep as possible and with this to protect it. Woods with direct ground contact, such as palisades, are a good example for the field of application in which a premature destruction due to an infestation by fungi only can be prevented by the use of naturally durable wood species or by a preventive chemical protection. This usually is achieved with the boiler pressure impregnation method.

The penetration depth of anti-blueing agents in pine tree sapwood has to be greater than 1.5 mm (inspection according to EN 152). Thus, the user cannot achieve a full chemical protection against wood-destroying fungi by means of a brush. If the wood moisture content inside the wood permanently amounts more than 30 %, there exists the danger of putrefaction in the interior of the wood, especially in larger sized wood cross sections despite chemical wood protection (surface treatment processes) and coating. Wood-destroying fungi partially can be active only far above the fibre saturation[3]). Wood-discolouring fungi (blue-stain fungi) do not infest resistant wood species such as sapwood of pine tree species in excess of 18 to 20 %. In such cases, an anti-sapstain preservative, often combined with an agent for mould infestation, is useful. However, it is not a chemical wood preservative according to the standard DIN 68800-3. Moisture protective measures against insects are ineffective. However, smooth and crack-free surfaces can prevent oviposition on non-resistant wood species and thus an infestation of the construction.

2.4.3 Chemical and thermal modifications of wood

The chemical and thermal treatment is used to make wood more dimensional stable and durable possible without the use of wood preservation agents. It is also important that woods of inferior quality are used to obtain more high-grade and more homogeneous woods for the coating by means of a finishing process. Besides improving the durability of the wood, the process of wood modification partly results in an increase in the hardness of the wood as well as in an increase of the following properties of the wood:

- Dimensional stability (reduction swelling-shrinkage)
- Resistance to weathering/UV resistance
- Realisation of thermoplastic properties
- Enhancement of the gluing ability
- Thermal stability

2.4.3.1 Acetylation of wood

In this process, the hydroxyl groups of the cellulose and lignin are chemically converted with acetic anhydride (see Figure 2.4) by means of a technical process. A hydrophobization of the wood occurs by esterification of the hydroxyl groups. The acetylation of the wood induces a

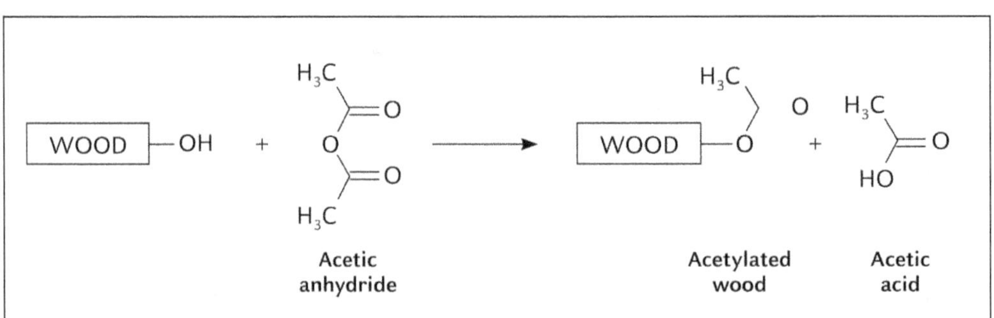

Figure 2.4: Principle of acetylation of wood

3 The dry rot also can be active significantly under fibre saturation

strong improvement of the resistance, dimensional stability as well as UV stability depending on the type of wood and thus to a more resistant wood for outdoor applications (durability of class 1). Depending on the type of wood, also a change in colour of the wood occurs during the process of acetylation. Acetylated poplar wood already has been used successfully in the hydraulic engineering [29]. Low residues of free acetic acid being perceived as smell can be expected in the acetylated wood.

Despite the elaborate, multi-stage manufacturing process under enhanced pressure, an industrial production plant with a capacity of 30,000 m³ per year of acetylated solid wood (pine tree plantations) went into operation in the Netherlands at the company Titan Wood BV. It produces acetylated wood that is available specifically for windows, doors, external claddings and terrace floors. According to the manufacturer's specifications, the acetylated wood is characterized by the following properties [30]:

- Resistance class 1 according to the standard DIN EN 350-2
- Bulk density 500 to 550 kg/m³
- Reduced swelling behaviour and shrinkage behaviour of up to 80 % for untreated wood (excellent dimensional tolerance)
- The moisture content of wood is about 4 to 5 % after acetylation
- An impregnation is not necessary. Only a primer which offers blue stain protection is necessary
- A 10-year warranty on translucent and opaque coating systems is given
- 50 years durability under above ground application (wood)

2.4.3.2 Modification with melamine resins

In chemical modification of wood with melamine resins, the cross-linking mainly occurs due to the reaction with hydroxyl groups of the wood (polycondensation). The coating with melamine resins also induces dimensionally stable and harder woods. The protection from putrefaction also is improved. The woods treated with melamine resins also are referred to as "polymer timber". The disadvantages of this procedure are the currently very high costs of process.

2.4.3.3 Modification with DMDHEU derivatives

The "Belmadur" process is used to withdraw air from the wood in a boiler pressure impregnation plant. Thereafter, an aqueous solution of resin completely penetrates into the wood at a pressure of about 10 bar. By subsequent heating at a temperature of more than 100 °C, the cellulose fibres crosslink with each other. The cross-linking solution contains a modified 1,3-dihydroxymethyl-4,5-dihydroxy-imidazol-2-one (m-DMDHEU) and special catalysts (see Figure 2.5) which polycondensates with the hydroxyl groups of the cellulose.

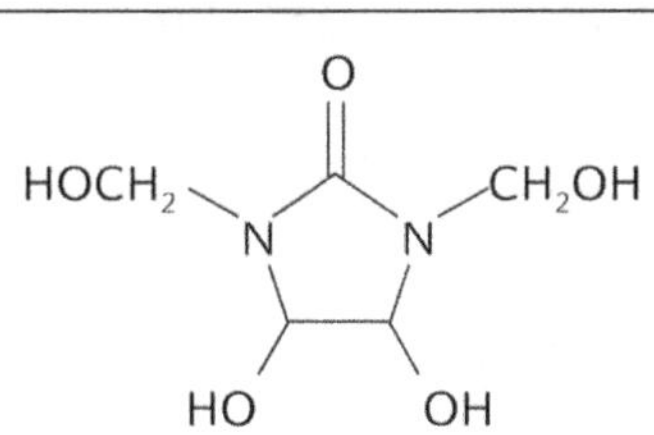

1,3-dihydroxymethyl-4,5-dihydroxy-imidazol-2-one
DMDHEU = Dimethylol-dihydroxy-ethyleneurea

Figure 2.5: Chemical structure of DMDHEU

The process of drying and curing shall be controlled such that no unacceptable emissions of formaldehyde may occur. The "Belmadur" process is protected under patent law [31]. The integration of the modified DMDHEU causes a permanent swelling of the cell walls. Thus, the additional intake of water with humidification significantly is reduced, and the hardness (see Figure 2.6) of the wood considerably is increased. The durability of the coating materials significantly can be extended by reducing the swelling and shrinkage (see Figure 2.7).

2.4.3.4 Thermally modified wood

In recent years, the technology of heat treatment at a reduced oxygen content has been developed significantly for practical maturity in order to achieve dimensionally stable and durable wood. In the heating of wood far above temperatures of 150 to 240 °C, hemi-cellulose, cellulose and lignin are modified in the above order. The modification occurs throughout the wood cross-section. Depending on the process, the heating is carried out in a hot air/water vapour atmosphere, in an oxygen-free nitrogen environment or in an oil bath. *Scheiding* [32] found that the durability especially of the beech tree against brown rot fungi as well as against white rot fungi can be enhanced by the thermal treatment from Class 5 to Class 1, whereas in pine trees only a durability class of 3 can be achieved. The thermal modification does not protect against blue-stain fungi. A well-known shortcoming of the thermal modification of wood is the reduction of the mechanical strength by 10 to 30 %, while the tensile strength even is reduced as much as 30 to 60 % [33]. It has been shown on thermally modified copper beech wood that the wood becomes darker with increasing the intensity of treatment. The density and hardness of the wood decrease with increasing intensity of treatment [34]. The equilibrium moisture content in a standard atmosphere decreases significantly. The report of Scheiding also illustrates that the thermal modification is the only possibility for severely impregnable wood species to increase the durability. Practical investigations in

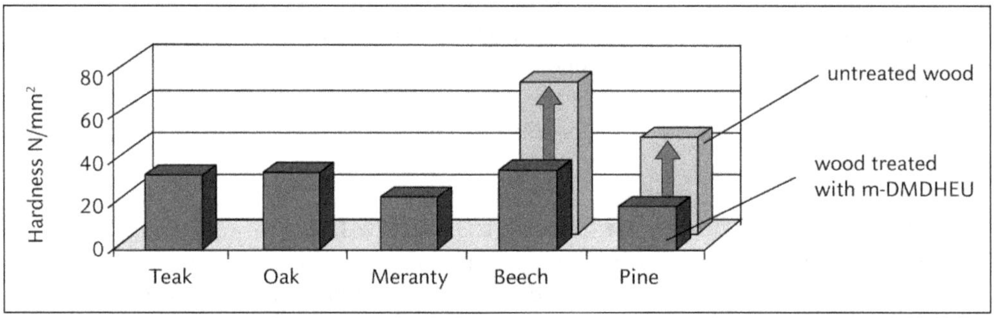

Figure 2.6: Hardness of beech wood and pine wood after cross-linking with modified DMDHEU

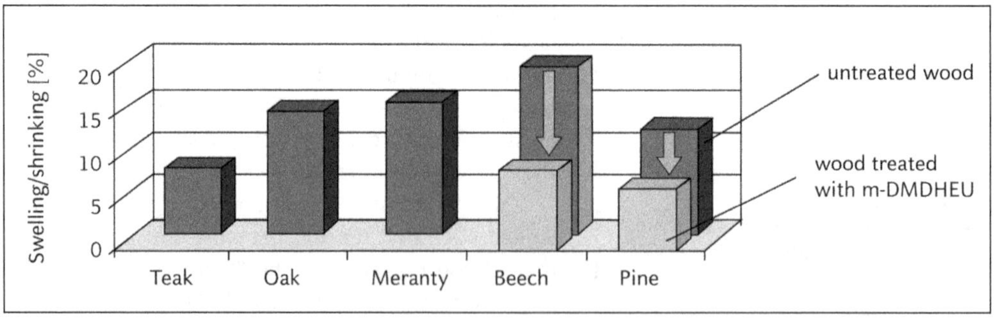

Figure 2.7: Reduced swelling and shrinking of beech wood and pine wood after cross-linking with modified DMDHEU

Sweden have outlined, that joining means made of steel corrode more powerful in contact with thermally modified wood than in contact to untreated wood. It is believed that in some processes of thermal modification of wood, the increased corrosion behaviour due to the formation and whereabouts of formic acid as well as acetic acid is the reason.

2.4.3.5 Wood-Plastic-Composites (WPC)

The use of natural fibre-reinforced plastics currently is a rapidly expanding market with high growth rates. The world leaders in the production and use of WPC are North America and China followed by Europe. In North America, nearly 2.4 Mio. tons of WPC were sold on the market in the year 2012 [35]. In Europe, experts estimate the current WPC production to about 400,000 t in 2015. Wood-plastic composites (WPC) are composite materials which can be processed thermoplastically.

Thermoplastics can be transformed in almost any shapes usually by extrusion. Thus, wood-plastic composites particularly are suitable for spatially designed, self-supporting, dimensionally stable objects [36]. The composition of the currently known WPC materials is formulated as follows:

- The proportion of wood (wood flour) amounts at least 50 % (filler)
- Thermoplastics such as polyethylene, polypropylene, and polyvinyl chloride (matrix)
- Additives in the percent range, such as pigments and light stabilizers

In Europe, currently there exists no standard or clear definition about the material WPC[4]. WPC components currently are applied uncoated. Uncoated or abraded WPC components are not resistant to abrasion, scratch-resistant and become discoloured. So far, WPC components have not found a broad application in furniture [37]. Efforts to equip the surfaces of WPC materials with plastic films, powder coatings or liquid coating materials are in full progress. With increasing of the number of thermoplastics, WPC materials can be coated more difficult than classic wood materials. This is due to the low surface tension of these WPC materials. Adhesion problems often occur. In many cases, a pre-treatment (flame treatment) or the special use of adhesion primers also is required. Since WPC wood materials are absolutely free of splinters, they are particularly suitable for patios and verandas. In addition, garden furniture, window sills, shelves or even veneered skirting boards can be used. In the furniture manufacturing, the first chairs consisting of WPC materials are in use. On the part of the furniture industry, currently reluctance prevails as WPC materials still are relatively new on the market, and many aspects such as weathering resistance, quality standards and coatability still lack practical experience.

2.5 Wood products and wood-based materials

Wood is used as a synonym for a wide variety of wood-based materials and wood products. The forestry sector provides the different types of wood from coniferous forests, deciduous forests and mixed forests to the timber industry. It offers selected assortments of round wood (see Figure 2.8) in different dimensions and qualities. The sawmill industry predominantly processes strong and precious round wood to sawn wood for purposes of construction and

4 However there exists a draft standard for WPC: DIN CEN/TS 15534-2 and a new DIN EN 15534-1

decoration. This massive wood is used, for example, for roof trusses, wooden bridges, claddings, doors, windows, floors, stairs, furniture, and special constructions.

Not all requests of a designer can be met by means of natural wood. This is due to the structure or, in other words, to the naturalness of wood. As a natural material, wood is not free of errors. Inhomogeneities due to the growth such as tree rings, visible branches as well as wood defects such as spiral growth complicate an engineering design of building components as often as the site-specific high variability of the mechanical properties. High safety factors consider these variations in the design of wood components. Hence, significantly larger cross-sections arise for load-bearing components, as these are required by the mechanical properties of the wood. It is the aim of the wood-based industry to eliminate this disadvantage while maintaining the strengthening structures. It produces panel materials from weak and low-value forest wood as well as residues from the sawmill industry, the veneer industry and further processing wood industry and woodworking (wood chips, sawdust, wood shavings) (see Table 2.12).

In the following, only the essential wooden substrates and their specifications relevant to an industrial surface coating are presented. The specifications to wood coatings on various wood products depend on the fields of application of the materials. Hereafter, we will distinguish between the requirements of wood products for outdoor applications and the requirements of wood products for interior applications. Thus, the relevant specifications as well as the standards for wood products for outdoor applications and their surface treatment explicitly are described in Chapter 3.2.

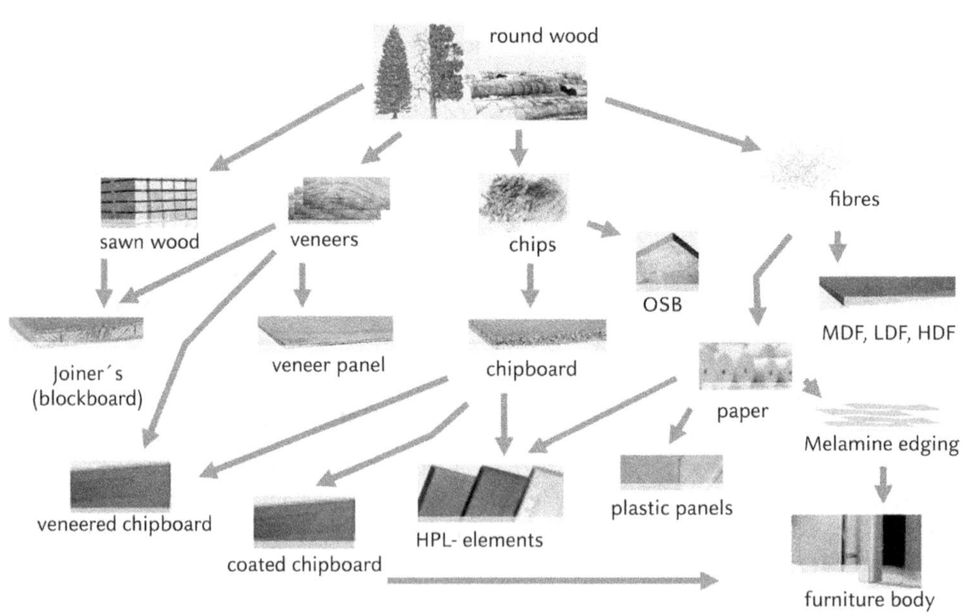

Figure 2.8: Pedigree of the wood-based materials

Source: Mantau, U.: Lecture "Wood markets, subsequent markets, globalization"

Table 2.12: Definition of the wood-based materials regarding to their production (exemplary)

Source: U. Mantau, University of Hamburg

Dismantling	Intermediate product	Assembling	Semi-finished product
Sawing	Bars	Crosswise gluing	Blockboard
Peeling	Veneers	Crosswise gluing cutting and crosswise gluing radial gluing	Veneer plywood laminboards star plywood
Cutting	Wood shavings	Mineral binding and pressing	Wood wool panel
Slicing machining	Chips	Gluing and plugging gluing and pressing gluing and calender pressing	Extrude panel flat press panel calender chipboard
Breaking down	Fraying	Dewatering pressing strong pressing	Porous fibre panel medium hard fibre panel hard fibre panel

Table 2.13: Sawn wood-based products and materials from sawn wood-based products (exemplary)

Structural woos	Sawn wood for carpentry works, furniture manufacture and interior design	Glued sawn wood-based materials
Pales, boards, planks	Wooden scantlings,	Glued laminated wood (BSH)
Bars, squared wood	unfinished profiles	Core-split lumber
Solid structural wood ("KVH")	Wood for floors	Laminated wood beam
Solid wood ("MH")	Wood in stair components	Laminated scantlings
		Solid wood panels
		Multi-layer parket flooring

2.5.1 Wood-based materials consisting of sawn wood

Table 2.13 exemplary presents sawn wood products as well as materials from sawn wood products.

2.5.1.1 Solid wood panels

Solid wood panels consist of planks or lamellae (thickness > 3.5 mm) and are used for the application in dry environment, wet areas as well as in outdoor applications. A distinction is made between single-layer and multi-layer solid wood panels with respect to the standard DIN EN 12775: 2001-04 "Solid wood panels – Classification and terminology". In interior constructions, solid wood panels are used for door leaves, stair steps and floors. The technological origin of the multi-layer solid wood panels actually used in the furniture sector as well as in the building sector lies in the so-called wood core plywood (today: laminboards or blockboards) [38]. In Europe, more than 15 manufacturers also feature a general building inspectorate approval which enables the production of plates with defined attributes for structural purposes.

2.5.1.2 Floors

In Europe, wood-based floors are known since the Middle Ages. Around the 13th century, raw planks and later planed floor boards consisting of softwoods such as fir trees, spruces or pine trees were still placed side by side. From the 16th century, the flooring parquet in its present form emerged from these floor boards in its present form as a representative flooring [39]. Today, a range of different materials are combined under the term wooden floor:

- Parquet (solid and multi-layer parquet)
- Massive floorboards
- Veneered floors
- Wood blocks

The general characteristics of wood floors are specified in the standard DIN EN 13756 "Terminology for wood-based floors and parquet floors". Since 2003, the first European product standards for wood flooring parquet became effective in Germany and other European countries. There are the following standards:

- Wood flooring – Solid wood parquet – Vertical finger, wide finger and module brick; German version EN 14761
- DIN EN 1533, Wood flooring – Determination of bending strength under static load – Test methods; German version EN 1533
- DIN EN 1534, Wood flooring – Determination of resistance to indentation – Test method; German version EN 1534
- DIN EN 1910, Wood flooring and wood panelling and cladding – Determination of dimensional stability; German version EN 1910
- DIN EN 13226, Wood flooring – Solid parquet elements with grooves and/or tongues; German version EN 13226
- DIN EN 13227, Wood flooring – Solid lamparquet products; German version EN 13227
- DIN EN 13228, Wood flooring – Solid wood overlay flooring elements including blocks with an interlocking system; German version EN 13228
- DIN EN 13442, Wood flooring and wood panelling and cladding – Determination of the resistance to chemical agents; German version EN 13442
- DIN EN 13488 Standard, 2003 Wood-based floors – mosaic parquet elements; German version EN 13488
- DIN EN 13489, Wood-flooring and parquet – Multi-layer parquet elements; German version EN 13489
- DIN EN 13647, Wood flooring and wood panelling and cladding – Determination of geometrical characteristics; German version EN 13647
- DIN EN 13696, Wood flooring – Test methods to determine elasticity and resistance to wear and impact resistance; German version EN 13696
- DIN EN 14293, Adhesives – Adhesives for bonding parquet to subfloor – Test methods and minimum requirements; German version EN 14293
- DIN EN 14342, Wood flooring – Characteristics, evaluation of conformity and marking; German version EN 14342

Parquet

During the industrialization of the 19th century (1848 to 1871), the parquet found its breakthrough and way into the houses as well as public buildings of the citizens. The grant of the patent for multilayer parquet floorings in the year 1939 to *Gustav Kähr* is an important milestone in the history of wood-based floorings. This construction made the floor more stable, and in addition used the wood as a raw material in a new and resource efficient manner [40]. The surface treatment of wood with oils has been known since ancient times. However, it was not until the 1950ies that the sealing of soils used today has been perfected and made the floor easy to clean. At the same time, the craft sector became more and more professional; thus, since the year 1970 there exists the training profession of parquet layer in Germany. Today, the natural characteristics of real wood are more in demand as never before – and parquet is estimated as a high-quality flooring for all types of rooms.

Laminate floors

In the year 1930, laminate was produced for the first time by the companies *Römmler* and *Masa*. In the 1950ies and 1960ies, this material mainly was used in the furniture sector. Quality and durability have been continuously improved by means of a continuous development of the techniques. A laminate floor is a multi-layered, rigid flooring with a decorative layer based on the printing technology. In this respect, the actual laminate layer[5] is glued and pressed on a support material in the main HDF (High Density Fibreboard, see Chapter 2.5.4). In general, the following types of laminate flooring element are distinguished [41]:

- PL (High Pressure Laminate) Laminate flooring elements with a surface consisting of decorative high-pressure laminates
- CPL (Continuous Pressure Laminate) Laminate flooring elements with a surface consisting of a decorative continuously pressed laminate
- DPL (Direct Pressure Laminate) Laminate flooring elements with a surface which directly is pressed on a supporting material
- EPL (Electron beam Laminate) Laminate flooring elements with a electron beam cured surface
- PDL (Printed Direct Laminate) Direct pressurized floor (see Chapter 3.1.6.9.2)

The features of the laminate floorings are described in the following standards: DIN EN 13329, EN 14978 and DIN EN 15468.

2.5.2 Veneers

The term 'veneer' was borrowed from the French "fournir" in the 16th century and described the process of overlaying less valuable wood with a fine sheet of wood. Today, the term 'veneer' is standardized and defined as a thin sheet of wood which is separated by peeling, cutting or sawing from the trunk or trunk segment. At the beginning of the 19th century, it succeeded to mechanize the working practices of the production of veneer sheets. The today's veneer industry originated with the first slicing machines in Hamburg (1870). While veneers were the dominating surface material in the 1960ies, in the 1970ies melamine films and decoration foils received a greater emphasis. The veneer lost its dominating significance down

5 A laminate (Latin: lamina = layer) is a multi-layered, duroplastic material which arises from compressing and sticking together of at least two layers of the same or different materials.

Table 2.14: Types of veneer with respect to the manufacturing process

Type of veneer	Advantages	Disadvantages
Rotary cut veneer	Large veneer sheet widths producible rationally producible in large quantities	Large shrinkage of the width, non-recurring, widely grained, little striking texture, right side is rough
Excenter cut veneer	Rationally producible texture is similar to the texture of sliced veneers usable such as sliced veneers	Large shrinkage of the width, sheets of veneer are producible only with a limited width right side is rough
Sliced veneer	Large veneer lengths producible smooth surfaces textures depending upon the cut direction simply (radial cut) or with grain patterns (tangential cut)	Right side exhibits hair-line cracks discolouration by means of vapours is possible veneer sheets may have different widths within a block
Sawn veneer	No pre-treatment is necessary natural wood colour is preserved free of cracks and tensions	High losses resulting from wastes of sawdust surface is rough sawn and often uneven

to the present day. Veneers conceptually are standardized in DIN 68330. According to their thickness, their reference moisture content as well as their permissible deviations, veneers are standardized in DIN 4079. In DIN 68330, veneers additionally are distinguished with respect to the manufacturing process (peeled veneer, sliced veneer, sawn veneer) and their intended application (face veneer, layon, veneer for furniture interiors, etc.).

2.5.2.1 Manufacture of veneers

Veneers are manufactured from round wood which is divided into the desired block length after cutting out of defects. Wood species, growth of the wood, trunk diameter and the desired future veneer figure decide how the individual veneer blocks are separated and processed in each case. Tree trunks from which peeled veneers have to be prepared, usually are steamed at first, cut to length and debarked. Veneer trunks which have to be processed to sliced veneers usually are cut at first, then debarked, shaped and steamed. By means of steaming, the wood received the desired colour whether it be through colour changes such as the reddish colour of beech or the intensification of the colour in walnut tree, mahogany and pear wood. Only a few types of wood (such as maple, birch, lime-tree) are processed without pre-treatment. After steaming the manufacture of veneers is performed by means of various methods, their advantages and disadvantages briefly are described in Table 2.14.

Sawn veneers

Nowadays, sawn veneers are produced in thicknesses of 1 to 4 mm or more only in special cases. Veneers predominantly are produced by non-cutting methods such as slicing or cutting.

Sliced veneers

A distinction is made between flat knives, true quarter cutters, flat quarter cutters and faux quarter cutters. Different veneer patterns are presented in Figure 2.9.

Peeled veneers

For high quality wood species such as birch as well as decorative veined wood species such as walnut, poplar, ash tree and other trees, the block is sawed in its longitudinal direction so that one single sheet is formed at each rotation. Hence, veneers approximately with the same design arise. If the block is not clamped in its central axis exactly but eccentrically into the peeling machine, then during every revolution only that portion of the block is led against the knife which describes the outer circular arc. In this technology referred to as eccentric peeling or semi-circular peeling veneer sheets with a nearly equal width are produced. These veneer sheets have a veneer pattern which exhibits the features of cut veneers. The so-called stay log peeling is a further development of this method.

Fineline veneers

Fineline veneers, also called design veneers, are based on a patent by Williamson (1935). In this method, peeled veneers with required dimensions are manufactured from solid woods. After bleaching and dyeing, in compression mouldings the veneers again are glued to blocks in a predetermined colour sequence (urea-based glue). By means of various processing techniques, the coloured block again is processed to veneer in the desired reproduction or the desired design. In this method, especially bleached wood of poplar, lime tree and obeche is used. Here it is to be noted that fineline veneers based on poplar wood strongly become yellow when exposed to UV light (exposition to sunlight) and have to be coated appropriately with coatings containing light stabilizers which only temporarily counteract colour variations. It is important to note that fineline veneers are much more absorptive than regular wood veneers.

2.5.2.2 Processing of veneers

Due to the manufacturing process, veneers have cutting or peeling cracks that occur on the side adjacent to the cutter. That leaf page which is directed to the core of the stem also is referred to as the "right side". This "right side" is rough and may cause processing problems when staining and coating.

Sliced veneers primarily are processed as face veneers (fine veneers). The gluing of face veneers has the purpose to give a decorative appearance to the visible surfaces. Even thin eccentric veneers or veneers which are peeled regarding to the stay-log method serve as face veneers. The most thicker peeled veneers are used as cross-ply veneers. In order to achieve beautiful and decorative surface designs, the veneers are selected so that a mirror-inverted veneer image occurs after assembling. This effect is achieved by the so-called falling. The unfolding of one or two veneer sheets being superimposed in a stack around a longitudinal or transverse joint is referred to as a simple fall. As a result, always a smooth and a rough side are next to each other. Of course, these sides have a different capacity for absorbing coating materials. This can be countered with an adequate surface

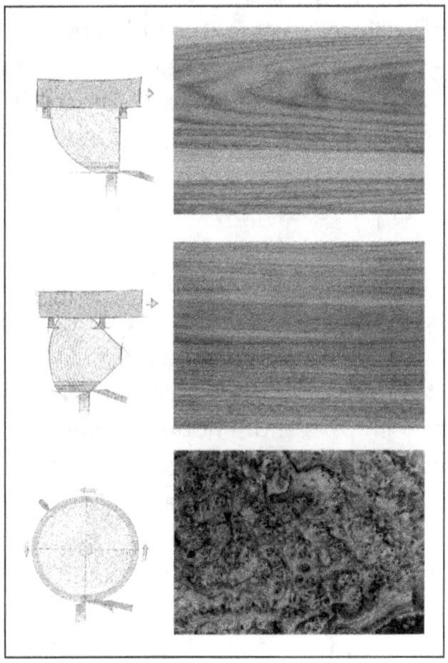

Figure 2.9: Several cross-sectional images of the different techniques. Top: Faux-Quartier-Cutters; Middle: Real-Quartier-Cutters; Bottom: Peeling

Source: Möbel, die nichts zu verbergen haben, Deutsche Gütegemeinschaft Möbel e.V.

grinding. A careful joint bonding is important for the processing, since these bondings may crack due to the swelling behaviour or shrinking behaviour of the veneer sheets glued together.

2.5.3 Wood-based materials based on chipboard panels

Chipboard panel

Wood-based materials based on chipboard panels have prevailed on the market of wood-based materials, worldwide. The chipboard panel has achieved the greatest importance in the furniture construction sector. The industrial production started in the 1940ies in Germany. The utilization rate of trees which amounted to around 40 % at that time should be increased by means of the development of a chipboard panel. In Bremen, the first 10-tons-chipboard panel plant in the world was put into operation. After the Second World War, the success story of the chipboard panel started and continues right up to today. The chipboard panel is produced by compacting wood chips of various sizes with a binder (usually organic adhesives or resins) under application of heat. Chipboard panels are sold either raw or polished or veneered or laminated with melamine resin or film-laminated. The raw material base of chipboard panels consists of small-dimensioned wood, sawmill by-products, industrial wood and used wood. Thus, today the recycling rate already is nearly 80 %. Depending on the functionality and intended purpose, several types of standards are distinguished. These are defined in accordance with the standard DIN EN 312. A total of 7 plates were defined (P1–P7). A distinction is made for boards for general, load-bearing and highly stressable purposes for indoor applications as well as for the moisture range. The different chipboard panels may reach the market either raw or surface-refined whereby the following methods of surface treatment mainly are used:

- Veneering with veneer thicknesses between 0.5 and 2.0 mm
- Plastic coating in accordance with the standard DIN 68765 by compressing the panels with carrier layers consisting of paper impregnated with thermosetting resins
- Foil coating
- Liquid coating with coatings

Surfaces of very fine chips or surface layer of fibre chips with a reduced swelling property preferably are used for the coating process. After the manufacturing process, an extremely fine calibrated finish has to be made on the surface of the panel in order to prevent coarse structures and irregularities that become clearly visible only after coating. By using thin, narrow chips, a more homogeneous surface is achieved which is advantageous for the reasons outlined above. Furthermore, the thin, narrow chips may induce enhanced bulk densities especially in the middle layers of the narrow surfaces of the chipboard panel. Here, the coating cannot penetrate as deeply as in narrow edges of chipboard panels which are manufactured from thicker, shorter chips. In order to achieve a reasonably good water resistance, specific waxes (paraffins) were added in the manufacture of the panel. If too much paraffin is present at the surface of the chipboard panel, this may create interferences in the subsequent coating system. Interferences may appear as dents, gloss points, holes or poor adhesion.

Formaldehyde in wood-based materials

Formaldehyde is a trace substance in wood and a component of the binder in conventional chipboard panels and fibre boards[6]) used as support materials for certain floors, the furniture manufacture as well as in construction applications. It is a strong irritant gas with high toxicity and demonstrated to be carcinogenic to animals in laboratory experiments. According to recent studies, under certain boundary conditions formaldehyde is likely carcinogenic to humans [42]. Since 1986, the current situation in Germany provides a statutory limit of the formaldehyde emission to 0.1 ppm (emission class E1) with wood-based materials and products consisting of wood-based such as furniture. The wood-based materials industry regularly performs a monitoring of the production according to DIN EN ISO 120 or DIN EN 717-1. Currently, some manufacturers voluntarily limit the formaldehyde emissions to 0.05 ppm by environmental and quality designations as well as own obligations. According to Marutzky [42], a value of 0.05 ppm for coated wood-based materials is the state of the art. A further reduction of formaldehyde emissions to an equalization concentration of 0.03 ppm in the test chamber is associated with additional costs up to 20 % by for example longer process times during the manufacturing process as well as higher glue costs. A so-called emission-free[7]) plate would involve an increase in the price of about 40 to 50 %. In the case of wood-based materials, since the year 1977 the emissions of formaldehyde were reduced in Europe by a factor of 30, as can be seen from Figure 2.10.

OSB panels

According to the standard DIN EN 300, OSB panels (oriented beach board) are materials consisting of long, slender wood chips (strands) of a predetermined shape, thickness and

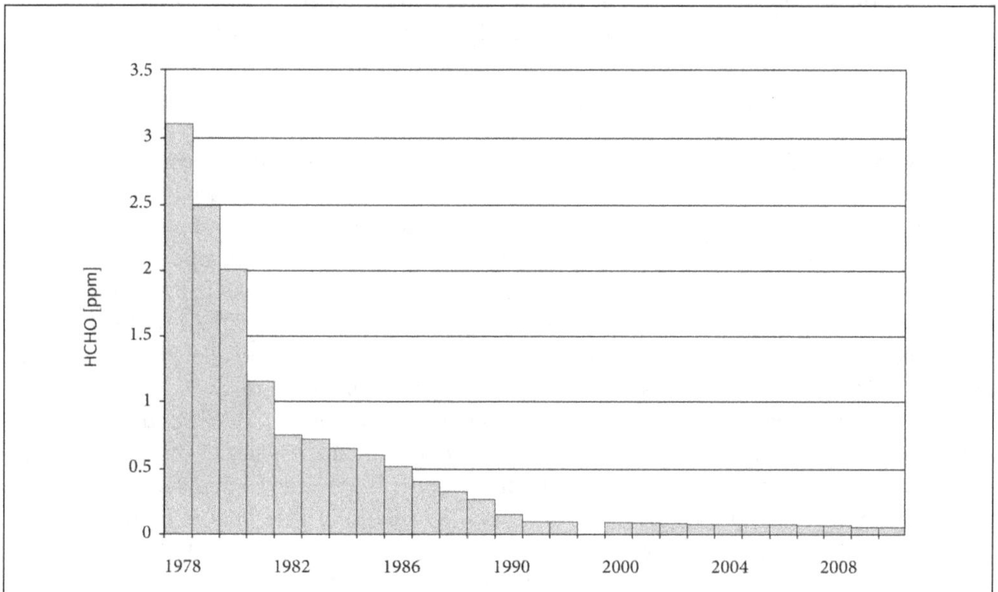

Figure 2.10: Decline of the formaldehyde emissions in Europe [42]

6 Formaldehyde based adhesives (or binder) also are used for the production of solid wood panels, BSH, KVH and other wood-based materials.

7 Emission-free wood panels do not exist, as naturally grown wood also contains and emits formaldehyde. There are manufacturers who produce plates with formaldehyde-free adhesives and designate these plates as "formaldehyde-free – as naturally grown wood", respectively.

Table 2.15: Types of fibre boards based on the dry process according to the standard DIN EN 316

Type of fibre board	Bulk density in kg/m³
ULDF (ultra-light MDF)	> 450 and ≤ 550
LDF (light MDF)	≤ 650
MDF	> 650 and ≤ 800
HDF (high density MDF)	≥ 800

multi-layered boards made with a binder. According to the standard DIN EN 300, four types of OSB panels are distinguished: while OSB/1 and OSB/2 preferably are used for furniture construction and interior finishing, OSB/3 and OSB/4 are used for the construction sector. Several manufacturers also have established particularly beneficial properties on the market according to the General Construction Inspection Approval. The classic OSB panel consists of three layers, wherein the wood chips (strands) of the outer layers are oriented parallel to the length or width of the panel. The chips in the middle layer can be oriented randomly. The perpendicular orientation of the wood chips in the middle layer in comparison to the wood chips in the outer layer has been proven. The good mechanical properties enable the application of the OSB panels in load-bearing structures such as stiffening wall planking and ceiling panelling in wooden constructions. For many years, OSB panels industrially are coated with UV-curable roller coatings for indoor use (fair construction, floors). OSB panels are one of the so-called "engineered wood products" which include other materials such as LSL (laminated lumber beach), PSL (parallel strand lumber) and LVL (laminated veneer lumber).

2.5.4 Wood fibre boards

A distinction is made between medium density fibre boards using the dry process (MDF) and fibre boards using the wet process. Fibre boards using the wet process are not discussed in this book. MDF is a relatively new wood-based material which was placed on the US American market by the company Allied Chemicals in the year 1965. In Europe, from the year 1973 MDF panels were manufactured in the former GDR [43]. In the year 1987, the first MDF factory was put into operation in the Federal Republic of Germany. MDF wood fibre boards with a medium density (MDF - Medium Density Fibreboard) are produced by means of a dry process. MDF wood fibre boards have a very homogeneous structure and a sealed surface. These wood fibre boards consist of very fine wood fibres. In Europe, MDF wood fibre boards mainly are produced from coniferous wood. In Europe, the bulk density of the standard panels is between 600 and 800 kg/m³. A distinction is made between high-density (HDF), medium density (MDF), lightweight (LDF) and ultra-light (ULDF) fibre boards (Table 2.15). In practice, HDF boards have an average raw density between 850 and 1100 kg/m³. While previously approximately 2 to 12 % adhesive resins were used with regard to the dry fibres [44, 45], now up to 18 % glue are used for HDF, while up to 12 % glue were used for MDF.

More than 90 % of all MDF panels are manufactured with urea resins, since urea resins allow a rapid curing within the manufacturing process and have a favourable impact on the price of production. These wood fibre materials have emerged in order to overcome the technological shortcomings of solid wood products which have a relatively large heterogeneity of the material properties even within a wood species. Towards the chipboard panel, one important property of MDF is the better profile design when tight radii are involved [46]. Thus,

MDF substrates can be used to realize three-dimensional furniture objects. This is possible because the MDF panels exhibit a uniform density profile and a homogeneous fibre distribution over the cross-section of the panel in comparison to other panel materials (see Chapter 3.1.9 powder coating materials). Wood fibre materials in the floor area (HDF), furniture manufacture (MDF) and building sector (LDF, ULDF) have the greatest significance. The standardized guidelines for MDF are specified in DIN EN 622-5. Besides the panel type MDF (for application in the furniture sector) there still are, among others, panels for general purposes for use in the moisture range (MDF.H) as well as panels for structural applications in the dry environment (MDF.LA) as well as in the moisture range (MDF.HLS).

Coating of fibre boards with liquid coating materials

Generally, MDF materials are very well suited for the surface coating with liquid coating systems. Unlike chipboard panels, edges (narrow surfaces) of MDF directly can be coated in the multilayer structure. Two-component polyurethane coatings, UV coatings as well as waterborne coatings essentially have established for the coating. In principle, all known wood coating systems are suitable for the coating of MDF surfaces. However, in the coating of milled edges (narrow surface area) and profiles, a preliminary insulation of these highly absorbent areas with suitable materials is recommend in order to avoid cracking [47, 48]. In the coating of MDF panels, reinforced cracking phenomena on narrow surfaces emerged for the first time in the years 1987 to 1991 [49–52]. The former cracking phenomena are due to the following causes:

- Different swellability and related volume fluctuations exposed to humidity at the edges of the MDF panel
- Application of MDF panels with too low transverse tensile strengths (DIN EN 319) <0.8 N/mm^2
- Promotion of the crack formation due to a pronounced density profile. Irregular density distribution over the cross section of the panel (see also Figure 3.107)
- Roughening of MDF edges by means of blunt cutting tools
- Application of not well-coordinated and brittle coating systems
- Looping through of the formerly used two-component polyurethane fillers or unsaturated polyester fillers up to the ground (narrow surfaces)
- No use of insulating coating systems or special insulating primer for the solidification of the wood fibres in the edge region prior to the application of the two-component polyurethane fillers or unsaturated polyester fillers (see Chapter 3.1.4 "2C polyurethane coatings" and 3.1.5 "Unsaturated polyester")

The development of the so-called insulating primers based on 2C polyurethane coatings for the solidification of wood fibres in the region of milled edges and profiles dates back to this time. The applied insulating primers have a very low solid content of < 20 % and a very low viscosity, so that they can penetrate the wood very well. Since the introduction of insulation primers that are applied by spraying, there are no significant complaints with the formation of cracks in the range of narrow surfaces. In order to examine the stability of cracking of coated MDF components prior to their application in the production for the furniture manufacturer, together with paint manufacturers, various test institutes developed new test methods. Due to the elaborated test procedures (for example stress by climate change), today coating systems can be tested accurately with respect to their suitability for different MDF qualities. For several years, modern coating processes for the coating of MDF components use elastic and insulating 2C polyurethane fillers simultaneously.

Thus, it may be dispensed with a preliminary insulation with low viscous materials in many applications. Inspite of these achievements, combinations of an insulating primer with 2C polyurethane fillers or unsaturated polyester fillers still are the safest method (see Chapter 3.1.4 and 3.1.5)

2.5.5 Lightweight construction materials

Due to the increasingly scarce raw material deposits, increasing environmental impacts as well as stringent guidelines, laws and taxes, the lightweight design as a construction principle increasingly will have economic benefits in the future [53, 54]. Thus, also a trend analysis showed that 2/3 of the respondents agreed that wood products significantly become lightweight in the future [55]. An unambiguous definition of the term lightweight construction material does not exist currently. According to a definition by *Thole* [56], lightweight construction materials are isotropic or anisotropic materials whose density is less than the density of the applied materials (raw materials), or where the structure of raw materials specifically is used for mono- or bidirectional solidification without increasing the bulk density. According to Stosch [57], the following classification based on the bulk density (ρ) can be made:

- Lightweight material $\rho \leq 500$ kg/m^3
- Extra lightweight material $\rho \leq 350$ kg/m^3
- Ultra lightweight material $\rho \leq 200$ kg/m^3

According to this classification, also the established wood core plywood and light plywood or the tubular chipboards are included in the definition of this category. Meanwhile, there are more lightweight construction materials [58–63] available. Especially the concept of the honeycomb plate appears [53]. In principle, it is not a new development, since it has been used in the door industry for decades. It has been used already in the 1960ies in combination with an internal circulating frame as a so-called "Board-on-Frame". The kitchen "SieMatic 6006" and the "PC system" of Flötotto should be highlighted as an example. In 1982, the concept of IKEA was rediscovered and again revived with the furniture line "LACK" in the furniture industry. The coating of "Board-on-Frame" materials is described in Chapter 3.1.6.

2.6 References

[1] Albin, F., Dusil, R., Feigl, H. H., Froelich, H. J., Funke: Grundlagen des Möbel- und Innenausbaus, DRW-Verlag 1991

[2] Brickenstein, R.: Holz ein traditioneller Werkstoff mit Zukunft. In: WKI-Bericht Nr. 33 1998, Bauen mit Holz und Holzwerkstoffen, Braunschweig 1998

[3] M. Hölzel, M.: Der Bau- und Werkstoff Holz ist und bleibt im Wettstreit der Rohstoffe konkurrenzfähig, Holz-Zentralblatt No. 79, 1997

[4] Holz – Rohstoff der Zukunft, Informationsdienst Holz, Deutsche Gesellschaft für Holzforschung (DGfH), 2001

[5] Rothkamm, M., Hansemann, W., Böttcher, P.: Lack Handbuch Holz, DRW-Verlag, 2003

[6] Pecina, H., Papryzycki, O.: Lack auf Holz, Hrsgb. Ulrich Zorll, Curt R. Vincentz Verlag, 1995, p.15–22

[7] Böttcher, P.: Oberflächenschutz witterungsexponierter Bauteile. In: WKI-Bericht No. 33 1998, Bauen mit Holz und Holzwerkstoffen, Braunschweig 1998

[8] Böttcher, P.: Beschichtbarkeit von Holz und Holzwerkstoffen. In: Berichtsband des DFO-Lehrgangs Lackieren von Holz- und Holzwerkstoffen, 1999, Braunschweig

[9] R. Wagenführ, R. und Pecina, H.: Untersuchungen an Tropenhölzern über das Eindringen von Polyesterlacken; Holztechnologie 7 (1966) 1; p. 43–47

[10] Düllmann, H., Görk, W.; Möbeloberflächen, Holzkosmetik - vom Furnierschnit bis zum Maserdruck; Hrsgb. Dr. Kurt Herberts & Co., Lackfabriken – Chemische Werke, Wuppertal, 1966

[11] Volker Thole, V.: Neue Holzwerkstoffe, Bauen mit Holz- und Holzwerkstoffen, WKI-Bericht 33, 1998

[12] Volker Thole, Ein weites Feld, HK11, 1998, p. 58–60

[13] Burmester, A.: Formbeständigkeit von Holz gegenüber Feuchtigkeit, Grundlagen und Vergütungsverfahren, BAM-Bericht Nr. 4, Berlin, 1970, p. 179

[14] Emmler, R.: Auf das Trägermaterial kommt es an, Zeitschrift: boden-wand-decke, 5 1999, p. 58–63

[15] Dittrich, H.: Oberflächenschäden und Trocknungsstörungen durch Holzinhaltsstoffe bei der Oberflächenbehandlung von Hölzern, Kompendium Holz, Holz Verlag Mering Kapitel 289

[16] Gruber, E.: Ökologie und Ökonomie der nachwachsenden Rohstoffe, Makromolekulare Chemie, Online-Skript der TU-Darmstadt, 2000, URL: http://www.cellulose-papier.chemie.tu-darmstadt.de/Vorlesungen_Veranstaltungen/ Vorlesungstexte_slides/Vorlesungstexte.html

[17] Sandermann, W., Braun, D., Augustin, H.: Über ungewöhnliche mineralische Einlagerungen in tropischen Hölzern, Sonderdruck aus „Holz als Roh- und Werkstoff", Springer-Verlag, Berlin-Heidelberg-New York, Vol 23,1965, p. 87–96

[18] Sandermann, W., Dietrichs, H.-H., Puth, M.: Über die Trocknungsinhibierung von Lackanstrichen auf Handelhölzern, Sonderdruck aus „Holz als Roh- und Werkstoff", Springer-Verlag, Berlin-Göttingen-Heidelberg, Vol 18,1960, p. 63–75

[19] Roffoel, E., Schäfer, M.: Die Bedeutung der Extraktstoffe des Holzes, Teil 2: Der Einfluss der Extraktstoffe auf die Verleimung, Holzzentralblatt Nr. 115, 1998, p. 1640

[20] Düllmann, H., Görk, W.: Möbeloberflächen – Holzästhetik – vom Furnierschnitt bis zum Maserdruck, Wuppertal: Dr. Kurt Herberts & Co., 1966

[21] Sandermann, W., Rothkamm, M.: pH-Werte von Handelshölzern, 1959, Holz als Roh- und Werkstoff, 17, 433–441, Springer-Verlag Berlin New York

[22] Dietrichs, H.-H.: Holzkunde, 1978, BM Bau- und Möbelschreiner, Sonderheft der chemisch-technologischen Merkblätter, Konradin-Verlag Robert Kohlhammer Stuttgart

[23] Oberflächenbehandlung von Holz im Außenbereich, Merkheft 11, Deutsche Gesellschaft für Holzforschung, München 1998, 2nd ed.

[24] Anforderungen an Beschichtungssysteme von Holzfenstern und Haustüren, VFF-Merblatt HO-03, 1999

[25] Peter Böttchern; Anstriche für Holz und Holzwerkstoffe im Außenbereich, Informationsdienst Holz, 1999, Arbeitsgemeinschaft Holz e.V.

[26] Rapp, A. O., Willeitner, H.: Holzschutz, Vorlesungsmanuskript im Studiengang Holzwirtschaft an der Universität Hamburg, Fachbereich Biologie, Hamburg 2004

[27] URL: http://www.holzfragen.de/seiten/hsm_2001.htm http://www.holzfragen.de/seiten/hsm_2001.htm

[28] Niemz, P.: Holzphysik, Vorlesungsmanuskript, Institut für Baustoffe, Eidgenössische Technische Hochschule Zürich, June 2006

[29] Mair, P.: Modifiziertes Holz – Der Markt der Zukunft?; Zeitschrift Deutscher Gartenbau, 27/2006, p. 10–12

[30] URL: http://www.treffpunkt-fenster.de/16838.html (Sikkens bietet spezielle Oberflächengewährleistung für Accoyaolz), 12. September 2006

[31] Patentschrift WO 2004/033171 A1, BASF, 2004

[32] Kürsten, E.: Möglichkeiten „neu geschaffener" Holzarten, Holz-Zentralblatt, No. 83/2005, p. 1101

[33] Kürsten, E.: Möglichkeiten „neu geschaffener" Holzarten, Holz-Zentralblatt, No. 83/2005, p. 1101

[34] Junghans, K., Niemz, P.: Behandlungs intensität bestimmt Thermoholzqualität, Holz-Zentralblatt, 7. April 2006, No. 14, p. 412

[35] Vogt, D. ect.: Wood-Plastic-Composites (WPC), Märkte in Nordamerika, Japan und Europa mit Schwerpunkt auf Deutschland, Studie vom Nvoa-Institut, Hürth, September 2005

[36] Gahle, C.: WPC mit Thermoverfahren räumlich umformbar, Holz-Zentralblatt, No. 2, 13. January 2006

[37] Gahle, C: Ikea nimmt WPC-Stühle ins Programm, Holz-Zentralblatt, No. 14, 7. April 2006, p. 409

[38] Steffen Tobisch; Methoden zur Beeinflussung ausgewählter Eigenschaften von dreilagigem Massivholzplatten aus Nadelholz, Dissertation, Universität Hamburg, 2006

[39] URL: http://www.parkett.de/parkett/geschichte.html

[40] URL: http://www.kaehrs-parkett.de/uk_EP.php

[41] Merkblatt: Verlegen von Laminatfußbodenelementen, erstellt durch den Verband der europäischen Laminatfußbodenhersteller e.V. (EPLF), deutsche Ausgabe 08-2006

[42] Marutzky, R.: Formaldehyd in der Holz- und Möbelindustrie: Anwendungen, Emissionswerte und aktuelle Entwicklungen, 1. Fachtagung Holztechnologie – Holzwerkstoffe und Beschichtungen, November 2006, Darmstadt, Hanser Verlag

[43] Kehr, E.: Beitrag zur Kurzzeitfestigkeit und zum Dauerstandverhalten von MDF der ersten Generation, Holz als Roh- und Werkstoff 51, p. 229–234, 1993

[44] Deppe, H-J., Ernst, K.: MDF-Mitteldichte Faserplatten, DRW-Verlag, Auflage 1, July 1996

[45] Koster, R. J. A.: Die Entwicklung von Leimen für MDF, Holz-Zentralblatt 116, 1990, p. 104

[46] Deppe, H-J., Ernst, K.: MDF-Mitteldichte Faserplatten, DRW-Verlag, Auflage 1, Juli 1996

[47] Gehrts, E., Kranz, U.: Kantenrisse in lackierten MDF-Platten, Zeitschrift BM, No. 11, 1991, p. 67

[48] Unbekannt: MDF – Der neue Holzwerkstoff (5), Holz- und Kunststoffverarbeitung (HK), Nr. 5, 1988, p. 498–499

[49] Baucks: MDF-Platten und ihre Oberflächenbehandlung, Holz-Kunststoff, No. 2, 1990, p. 20

[50] Unbekannt: Lackieren von MDF, Zeitschrift BM, No. 7, 1988, p. 64–65

[51] Risius, H., Epple, A.: Lackrisse sind vermeidbar, Holz- und Kunststoffverarbeitung (HK), No. 7/8, 1989, p. 900–901

[52] Unbekannt: Arbeiten mit MDF – Oberflächenbehandlung von MDF-Materialien, Der deutsche Schreiner und Tischler, No. 2, 1988, p. 40–41

[53] Thömen, H., Poppensieker, J.: Wabenplatten im Möbelbau, Bundesforschungsanstalt für Forst- und Holzwirtschaft und Universität Hamburg, Arbeitsbericht No. 2, 2005

[54] Thömen, H., Poppensieker, J.: Wabenplatten im Möbelbau sinnvoll einsetzbar, Holz-Zentralblatt, No. 2, January 2006, p. 50–51

[55] Knauf, M.; Frühwald, A.: Trendanalyse Zukunft Holz – Delphistudie zur Entwicklung der deutschen Holzindustrie, Abschlussbericht 2004

[56] Thole, V.: Leichte Werkstoffe aus nachwachsenden Rohstoffen, Vortrag, Braunschweig 2004

[57] Stosch, M.: Design und Konstruktion von Leichtbaumöbeln, Vortrag zum 10 Jowat Symposium, 2004

[58] Stosch, M.: Leichtbau gewinnt an Gewicht, BM Spezial – Leichtbauwerkstoffe, Konradin Verlag Robert Kohlhammer GmbH Stuttgart, 2006, p. 8–12

[59] Leichtbau: keine ganz leichte Angelegenheit, Holz-Zentralblatt, No. 2, January 2006, p. 49 & p. 53

[60] Müller, M., Hultsch, B., Teischinger, A.: Leichtbau – wichtiger Trend bei Holzwerkstoffen, Holz-Zentralblatt, Nr. 33, August 2006, p. 953

[61] Blöttscher, H.: Fix im Maß und Prozess, BM Spezial – Leichtbauwerkstoffe, Konradin Verlag Robert Kohlhammer GmbH Stuttgart, 2006, p. 14–17

[62] Soiné, H.: Ein neuer Holzwerkstoff für die Möbelindustrie?, Holz-Zentralblatt, Januar 2004, p. 62

[63] Grell, R.: Leichte Werkstoffe und Sandwiches à la Carte, BM Spezial – Leichtbauwerkstoffe, Konradin Verlag Robert Kohlhammer GmbH Stuttgart, 2006, p. 34–35

3 Coatings for wood and wood-based materials

Depending on the application, fundamentally a distinction is made between coatings in indoor areas or outdoor applications such as for windows, doors, wooden facades and others. The different requirements for the indoor areas and outdoor areas require a separate treatise. The focus mainly is on industrial coating of wood, wood-based materials and decor finish foils, although the coating of plastic surfaces (such as ABS, PVC, melamine laminates) clearly gained in the industrial coating for example of kitchen and living-room furniture.

3.1 Coatings for indoor applications

Compared with metals, wood is a 'living' material. For this reason, the coating technology has to be selected carefully prior to the surface treatment in order to ensure perfect results from coating. The following non-exhaustive examples greatly may affect the result of painting:
- Type of the wood-based material (suction properties, swelling capability, bulk density)
- Availability of wood components
- Type of glue (PVAC, urea-formaldehyde resin or phenol formaldehyde resin)
- The type of veneer pressing on material supports such as cold grouting or hot grouting
- Fluctuations of the wood moisture content, especially in solid wood
- Wooden sanding: abrasive particle size distribution as well as longitudinal and/or transverse section
- Temperature conditions and air humidity conditions in the coating booth or in the drying process, respectively
- coating parameters such as viscosity, temperature, amount of application, drying conditions as well as curing conditions

For the wood interior coating, various coating technologies are favoured worldwide. Figure 3.1 illustrates the global industrial Wood Coatings Market 2016 according to the coating

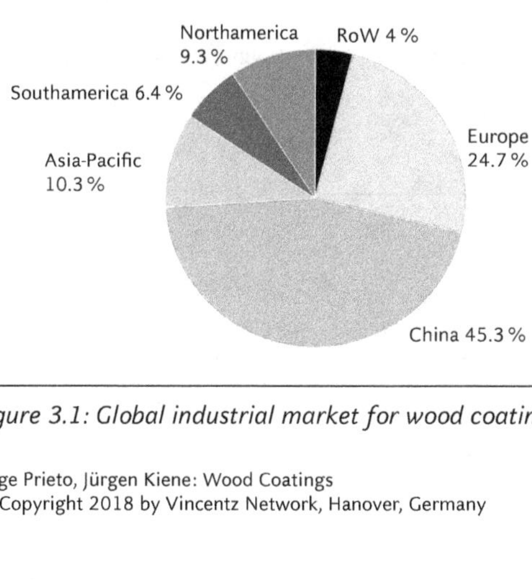

Region	[kt]
Europa	600
China	1100
Asia-Pasific	250
Northamerica	227
Southamerica	155
RoW	98

Figure 3.1: Global industrial market for wood coatings 2016 by different regions

Table 3.1: Percentage distribution of the technologies of wood coatings 2003 for the main Western European countries without outdoor coatings [1]

	CN coatings	Acid-curing coatings	1C and 2C PU coatings	UP coatings	UV/EB coatings	Water-borne coatings	Other products
Italy	4	4	35	18	22	16	1
Spain	13	5	48	10	11	10	3
Germany	26	5	34	4	10	20	1
Great Britain	21	42	10	6	6	14	1
France	25	9	30	11	13	12	–
Sweden	3	50	3	4	16	23	1
Denmark	4	58	3	1	14	18	2
Finland	4	37	11	1	6	40	1
Norway	4	60	6	3	11	15	1

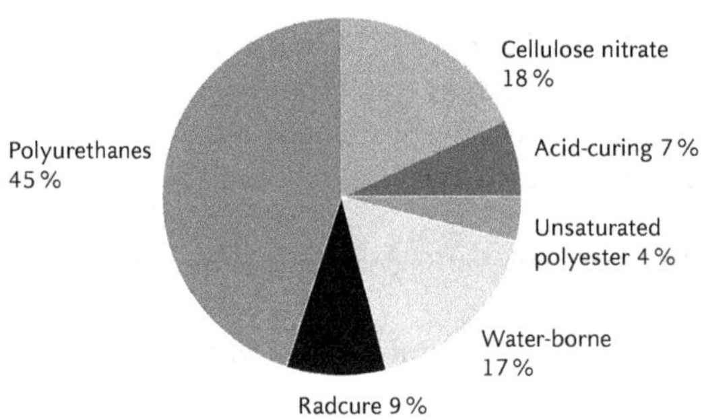

Region	[kt]
1C/2C polyurethanes	1100
Cellulose nitrate	438
Acid-curing	160
Unsaturated Polyester	90
Water-borne	422
Radcure	220

Figure 3.2: Global industrial market for wood coatings 2016 by technologies [1]

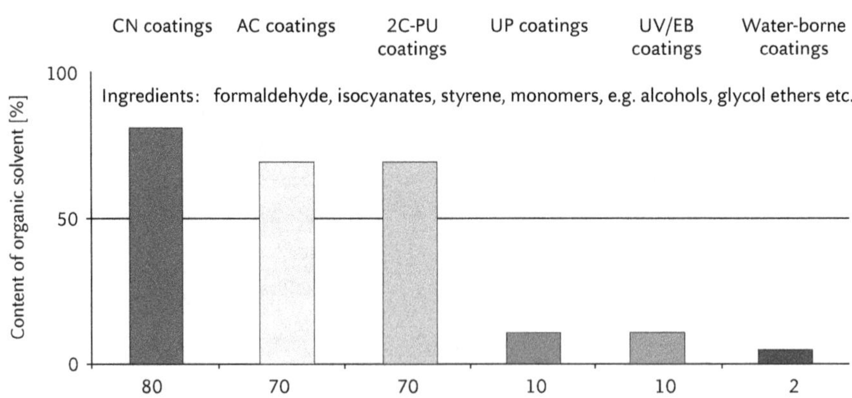

Figure 3.3: Content of organic solvents in different wood coating technologies referring to 100 parts of weight clear coat

technology for indoor coatings with a total tonnage of 2.430 kt [1]. This overview considers the regions Europe, Middle East and Africa (EMEA), North and South America, and Asia-Pacific region. The global and European market data for wood coatings being published in recent years are very different depending on the database used. This is demonstrated by the publications of *Asher, Brenni* and other *sources* [2–5]. Finally, one should know, that the published market data for wood coatings contribute to a better understanding of the technology and needs of the markets and to recognize trends regardless of the tonnage.

The coating systems/technologies nowadays used all over the world are:

Solvent-dilutable

- CN coatings (cellulose nitrate coatings) which are termed as NC coatings
- Acid-curing coatings
- 2C PU coatings (two-component polyurethane coatings)
- UP coatings (unsaturated polyester coatings)
- UV/ESH coatings (UV-curing and electron beam curing coatings)
- Oils and waxes
- Wood stains

Water dilutable

- Water-borne coatings (1C self cross-linking, 2C polyurethane coatings, UV-curing coatings)
- Wood stains

As shown in Table 3.1, predominantly acid-curing coatings are applied in Northern Europe. Compared with this, 2C PU coatings are preferred in Southern Europe (Italy, Spain, France). The wood coating systems being applied in Europe exhibit a different content of organic solvents (Figure 3.3). Depending on the applied technology, formaldehyde or styrene still can be present. Accelerated by the EU-VOC Directive 1999/13 EG, the trend clearly is towards ecological and economical coating systems which do not contain hazardous substances. Thereby,

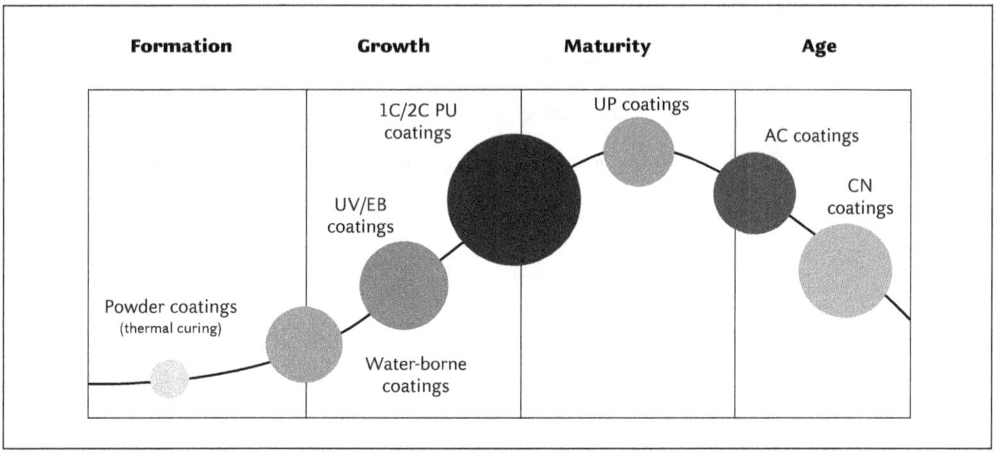

Figure 3.4: Life cycle of the wood coating technologies for the industrial coating

the highly solvent-borne wood coating technologies such as CN coatings, acid-curing coating as well as 1C and 2C PU coatings are replaced gradually by more environmentally friendly systems such as UV coatings, water-borne coatings and modern coating processes. The Figure 3.4 illustrates the life cycle of the wood coating technologies. This figure shows that the UV coatings as well as the water-borne coatings are in a growth phase while CN coatings and AC acid-curing coatings become less important. It cannot be said reliably, whether the powder coatings currently being in development/implementation phase will prevail extensively for the coating of wood materials. According to the information from INFORMATION RESEARCH (IRL), the proportion of UV coatings quantitatively will increase by 4.8 % per year from the year 2003 up to the year 2008, while the proportion of water-borne coatings quantitatively will increase by 6.6 % in the same period of time [1]. According to *Paulus*, the proportion of UV/EB coatings for the wood coating in Europe should increase from approximately 47,300 tons in the year 2004 to 102,000 tons in the year 2015 [6].

The following important properties have to be allowed for the wood coating in the application of the corresponding coating systems [7]:

- Insulation of interfering wood components
- Good wettability of the wood pores
- Low roughening of the wood
- Good elasticity of the coating in order to counteract the formation of cracks
- Favourable mechanical properties, resistant to stresses by abrasion (coating of parquets and wood floorings) as well as a good scratch resistance

Due to the varying requirements and properties of wood coatings in the interior, it is generally classified according to branches and segments of application:

- Household furniture and office furniture
- Kitchen furniture and bathroom furniture
- Chairs
- Doors
- Musical instruments
- Parquet
- Laminate coating floors

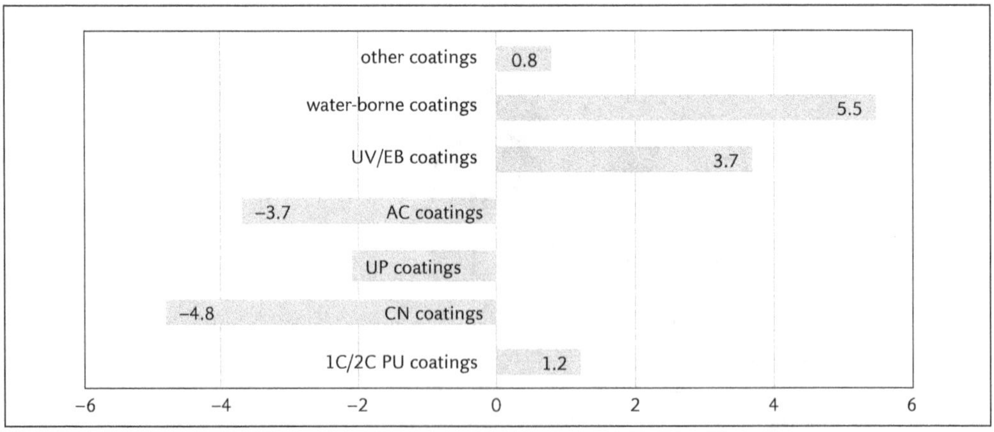

Figure 3.5: Estimated growth rate for coatings technologies

- Panels
- Wooden coffins
- Picture frames
- Other applications

For all of them, the structure, beauty, special character and colouring of the wood develop to their full effect only after an additional surface treatment. In order to achieve different effects in natural coatings for transparent coatings, a distinction can be made between coloured stain/translucent surfaces and fully pigmented coated surfaces.

3.1.1 Wood stains

Staining is defined as the aggressive action of a substance on an object in the sense of an acidic or alkaline impact is understood [8]. Formerly, the staining of wood also was referred to as wood dyeing. The proximity to the textile dyeing resulted in the utilization of approximately the same raw materials and methods. The wood stains have the task of colouring and refining raw wood. The grain of the wood is not blurred, but rather emphasized. Modern wood stains achieve fascinating results from the subtle nuances of the natural colour of the wood to a strong, expressive colouration. Specially adjusted stains may highlight the hard growth rings (positive staining appearance), realize the emphasis of the wood pores (rustic effect) or balance the colour variations (equalizing effect). This chapter mainly deals with the application and issue of wood stains in the industrial production of furnitures, kitchens, parquets and doors. Here, the processing in the craft is not considered, since a large number of publications already exists.

3.1.1.1 History of the industrial wood staining

Even in ancient times, the colouring effect of the animal and vegetable dyes was known. In the Middle Ages, alkalis, iron salts as well as copper salts were discovered as dyes. At least since the 18th century, tannic acids and alkaline solutions (chemical etching) as well as dyes in their bonding to organic matter and metal salts in combination with tannins specifically were used for dyeing of wooden materials [9]. By the end of the 18th century, the loosening of the guild committed relaxations of production in the joinery trade enabled a transition to the beginnings of a small industrial production. The characteristics were collaborative manufacturing processes, implementation of machine work as well as implementation of new materials. At this time, natural dyes mainly were used for the staining of woods. Towards the end of the 19th century, the newly discovered tar dyes established themselves. In this century, the colourful furniture has been replaced by selected veneers of quiet and lively grain. Shiny polishes emphasized the colour and texture of the wood. Mahogany was the dominant veneer wood of this time. The strong demand and naval blockades led to shortages of the popular furniture wood. Mahogany was rare on the market and very expensive. Efforts were made to imitate native woods with special staining formulations and procedures.

3.1.1.2 Composition of wood stains

In classic wood coatings, the film-building agent primarily determines the properties of the formulations. In typical wood stains, the composition of the dye and/or pigment substantially determines the formulation. Figure 3.6 presents the general composition of solvent-dilutable and water-borne wood stains. As can be seen from this figure, a large part of the raw materials used for wood stains is an ingredient of conventional wood coatings.

Dyes

Dyes are chemical compounds which have the property to colour other materials. According to the regulation DIN EN ISO 787-5, dyes are colourants that are soluble in their medium of application. Mainly synthetic organic dyes are used for the industrial wood staining. The dyes used for the wood staining originate from the textile industry. They can be classified to methods pertaining to dyeing technology or according to the chemical structure of the dyes.

The following dye groups are used for the formulation of wood stains:

Acid dyes ("anionic dyes")

Acid dyes, also referred to as anionic dyes, contain carboxyl, hydroxyl and especially sulfonic acid groups, and therefore react as Bronsted acids[1]. Acid dyes mostly are used in the form of their sodium salts. Above all, especially azo compounds, anthraquinone derivatives as well as triphenylmethane derivatives belong to the group of acid dyes.

Metal complex dyes

These are a part of the acid dyes and contain complexed chromium or complexed copper as a central atom. There are metal complex dyes with a different number of acid groups (such as $-SO_3H$). These are soluble in alcohols, glycols, glycol ethers, esters and ketones. The lightfastness often is insufficient for the application in the furniture industry.

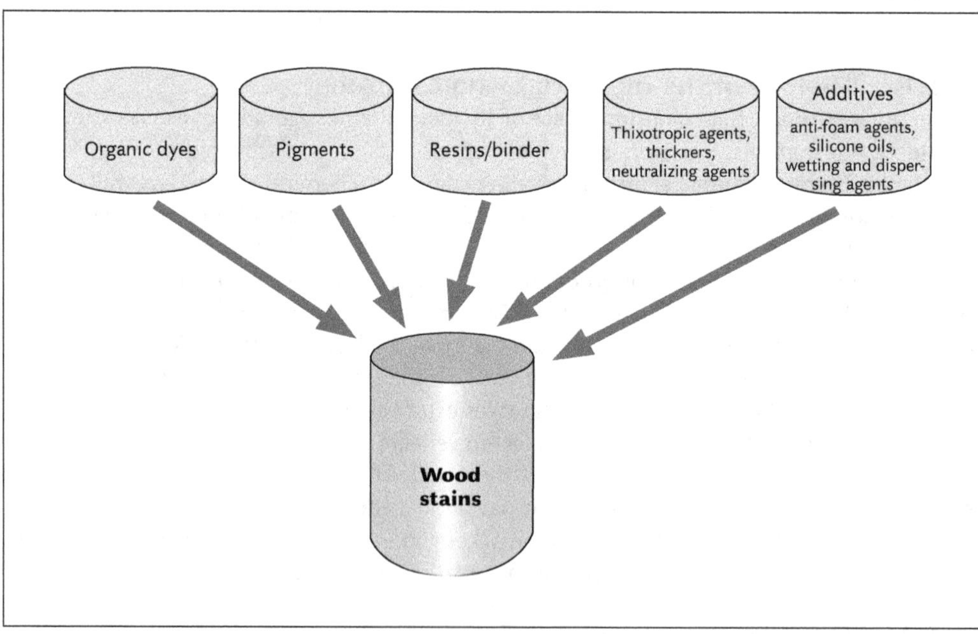

Figure 3.6: Basic raw materials for production of solvent-dilutable and water-borne wood stains

1 An acid can release protons and thus is a proton donator

Substantive dyes

The most important colours in this category also are azo dyes and anthraquinone dyes.

Reactive dyes

In chemical terms, the reactive dyes consist of two components: the chromophoric component (e.g. an azo compound) and, attached to this, a reactive group (with bifunctional reactive dyes also multiple reactive groups) such as a vinyl sulfone group (remazol dyes) or perhaps a triazine group. This reactive component chemically combines with an OH group of the cellulose. These are used in so-called positive stains (coniferous wood stains).

Pigments

Pigments are inorganic or organic, colourful or achromatic colouring agents which are almost insoluble in the application medium in contrast to the colourants. Depending on the application and effect, organic and/or inorganic pigments may be used. Pigment preparations are applied exclusively. In contrast to dyes, pigments are characterized by a better light fastness and chemical resistance. When applying traditional pigments, it is important to note that these are less transparent in comparison to the dyes, and thus cover or disguise, respectively, the wood grain in a stronger matter. For this reason, after many years of pigment preparations with a high level of transparency and simultaneously good lightfastness were developed.

Solvents

The composition of the solvents used depends on the coating process, the desired effect development and the dyes used as well as pigment preparations. Esters, ketones, glycols, glycol ethers, alcohols as well as water were used as solvents.

Resin

All known resins from the wood coating systems such as alkyd resins, acrylic resins, cellulose nitrate, cellulose acetobutyrate as well as UV-curing and water-borne dispersions are employed.
The resins have different functions in wood stains such as:
- Stabilisation of pigment preparations against sedimentation in containers
- Manipulation of the processing properties (such as transfer behaviour of the rolling machine)
- Effect development (for example, rustic effect)
- Improvement of the performance characteristics (such as surface resistance, adhesion together with wood coatings)

Additives

The additives used are selected according to the same criteria as for wood coatings. These are very often defoamers, wetting and dispersing agents, among others.

3.1.1.3 Classification of wood stains

There is no unified system for the classification of wood stains. There is a wide range of common names for the different wood stain systems. Most of them are product designations of the different coating producers. For industrial use, the classification by pure solvent-borne or water-borne wood stains is advisable. For special applications, water-borne combination stains exist which may contain up to 60 % water-miscible, organic solvents. Furthermore, different application methods (such as spraying, dipping or rolling) are distinguished. Additionally, certain wood stains can be specified with respect to their surface effect. Examples of meaningful designations are:

- Solvent-borne stains with rustic effect for rolling
- 1C positive stains, water-borne[2] for spraying

3.1.1.4 Fundamentals for the formulation of wood stains

Before starting with the formulation of wood stains, different criteria, such as processing parameters as well as desired surface effects, have to be considered. Special formulations only can be created under consideration of the following criteria:

- Type of surface effects
- Colouring
- Wood species and pretreatment of wood (e.g. wooden sanding)
- Type of application (such as rolling, with or without distributor brushes) and process parameters (e.g. drying conditions, temperature)
- Coating technology used for the priming and topcoat of wood stain (e.g. UV coatings, UP coatings or water-borne coatings, etc.)

Type of surface effects

The most important surface effects are the positive effect for coniferous woods and the rustic effect in coarsely porous types of woods.

Positive effect for coniferous woods

The term 'positive effect' refers to the effect, that the early wood zones remain bright and dark late wood zones are coloured dark. This is achieved by a combination of specific silicas and/or waxes with reactive dyes.

Rustic effect

The term 'rustic effect' refers to an intense colouring of the wood pores in comparison to the surface. The effect on coarse-pored woods such as oak and ash tree greatly is appreciated and can be influenced by the formulation. In fine-pored woods (maple, birch or beech) the rustic effect is not desired as a general rule.

Colouring

Mainly pigment preparations are used for the formulation of light colours in order to achieve the highest possible light fastness as well as to compensate wooden irregularities. In contrast, darker shades are made with a combination of organic dyes and pigment preparations in order to achieve a more intense colouration and brilliant colour effects. Probably one of the most important properties of an optimal wood stain is the resistance to UV light, the so-called light fastness. The selection of dyes/pigments used in wood stains has to be done very carefully. Depending on the composition and colour shade, wood stains simultaneously have the property to reduce drastically or to prevent the early wood discolouration of untreated wood, respectively.

2 Water-borne wood stains often are called 'hydro stains' or 'aqua stains'.

Wood species and pretreatment of wood

The selection of formulation components also has to consider the wood species and their ingredients (see Section 2.4). The ingredients may not only cause a discolouration, but also influence the wetting behaviour of the wood stains. The frequently prescribed wooden sanding among the users has to be considered in the design of the formulation. The wooden sanding has a strong impact on the uniformity and intensity of the wood staining.

Type of application and process parameters

The process of application and the drying conditions significantly determine the selection of the resin, the solvent as well as the additives. For example, the Shore hardness of the rubber roller influences the amount of application and thus the colour effect within the rolling process. Therefore, it is necessary that the paint manufacturer gets more information about the use of roller prior to the drafting of the wood stain. The amount of application can be controlled by a suitable adjustment of the viscosity.

Applied coating technologies

Apart from the wax stains, after application all wood stains are coated or protected with a clear coat for drying. The coating systems applied in the wood coating (see Chapter 3) affect the visual impression of the colour tone and surface effect. One has to know that there are coating systems which 'light' or 'brighten' the underground. Water-borne coating are more likely to brightening and solvent-borne coating systems. This always has to be considered in the formulation of wood stains. The rejection of the wood stain always has to be performed with the coating systems used by the customer. Furthermore, it has to be taken care that the application of wood stains below polyester coating layers based on unsaturated polyester, styrene, peroxides as well as beneath UV-cured coating systems requires a special adjustment of the stain with regard to the colour stability and adhesive strength.

3.1.1.5 Industrial processing of wood stains

Water-borne wood stains for the immersion

In the furniture industry, the application of wood stains in the immersion method mainly is used for parts of a chair, legs of a table, and sometimes entire chairs. Beech is the mainly applied type of wood, but also oak and ash tree are applied. This usually involves aqueous stains. In the past, pure dye stains without an addition of binders mainly were used. Modern immersion stains still are based on water-borne systems and are formulated with low contents of dyes to a low extent. These immersion stains contain water-borne dispersions, whereby good absorbable resins are combined with resins remaining at the surface. Thus, the level of intensity of the rustic effect is controlled. The discharge behaviour is affected and has to take countermeasures to different suction capabilities. Different additives such as wetting agents and small amounts of organic solvents also are used. After drying the stain, a subsequent priming with water-borne paints by means of an immersion procedure can be performed without pollution by colouring the running coating. The application amount of the topcoats can be reduced due to the amount of resins in the stain. The industrial wood staining by means of immersion processes is performed manually as well as mechanically. Depending on the colour and effect, the residence time in the immersion pool amounts between 30 and 60 seconds. The subsequent drying is made either at ambient conditions (23 °C/50 % relative humidity) for several hours or by means of forced drying (15 to 60 minutes at a temperature of 30 to 50 °C) applying a circulating air dryer.

Solvent-borne spray stains

In the industry, the application of solvent-borne spray stains frequently still is performed manually or by surface spraying machines and/or strip spraying machines. In the pickling of fine-pored wood (such as maple, beech, birch), the wood stains are applied 'dry' or 'fogged'. This is achieved by using very fast evaporating solvents in the formulations in order to achieve a regular stain appearance not accentuated with pores (rustic effect) on fine-pored woods. By adding small amounts of resins and slowly evaporating solvents, a cloudy surface is avoided during the application by means of surface spraying machines. The formulation of stains for large-pored woods contains a large amount of slowly evaporating organic solvents in order to achieve the wetting of the wood pores. The pickling of large-pored woods such as oak or ash tree occurs with higher wet application amounts ('excess') in comparison to fine-pored woods. Most commonly, the wood stains are dispelled after the process of coating.

Water-borne spray stains

The classical spray pickling for the chair industry consists of pure dye stains without adding film forming agents. The applied wood largely consists of massive beech or beech plywood. The thus formulated stains can be applied with an excess and easily dispelled if required.

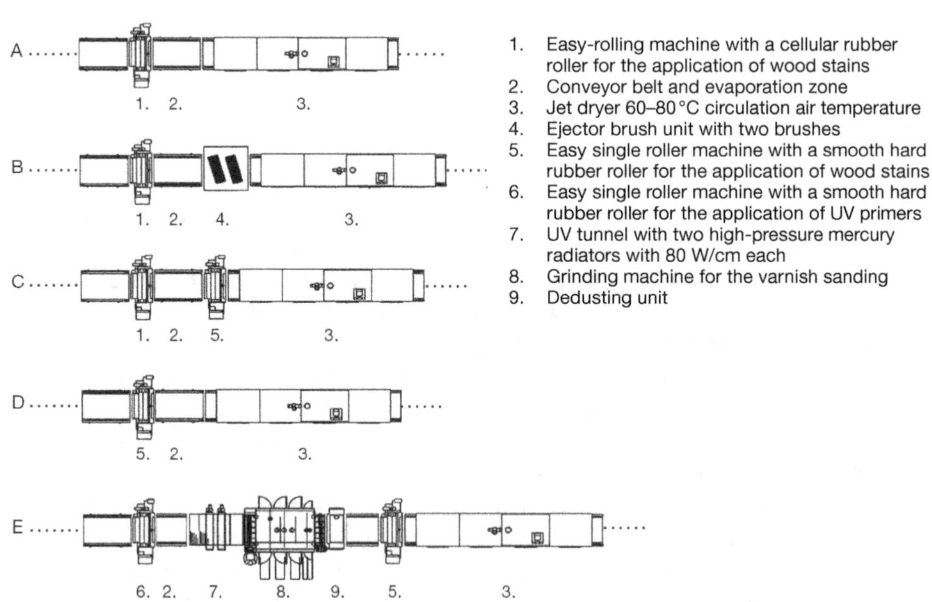

Figure 3.7: Schematic illustration of the significant methods of roller application for the application of wood stains

After drying (2 hours on ambient drying at a temperature of 23 °C and a relative humidity of 50 % or about 10 to 15 minutes on forced drying at a temperature of 30 up to 50 °C), the chairs stained with this are primed with solvent-borne or water-borne lacquer systems and topcoated. Modern water-borne wood stains have to cover the full range of requirements for industrial processing (such as colour-tone effects and process parameters). Along with conventional dyes, also pigment preparations, resins, organic solvents as well as additives are used on this occasion. In the processing of wood stains on softwoods such as spruce and pine with 'positive effect', the compliance with the wet application amount of about 60 to 80 g/m² and the drying parameters for the development of colour effects has to be guaranteed. In the event of a forced drying (30 to 50 °C), no strong movement of air has to be present in the drying tunnel, as it may result in a turbulence in the wet stain as well as to troubled stain appearance, otherwise.

Roller pickling

In industrial production facilities, flat components are applied by means of roller assemblies. Probably, the first roller pickling machine was delivered by the company *Bürkle* in Freudenstadt, Germany [9]. For this purpose, depending on the effect the stain is applied with cellular rubber rollers and, if necessary, equalized in the direction of the fibre by means of so-called ejector brushes. In practice, different cellular rubber rollers are used which vary in their porosity. Cellular rubber rollers of the pores 1, 2 and 3 mainly are used.

With increasing number, the adsorption capacity of the rubber coating and therefore also the application amount on the workpiece increases. Especially for the dark stain effect, a combination of cellular rubber roller and smooth roller is used in practice. Normally, a wet application amount of about 10 to 40 g/m² is applied. In case of combination stains (water/organic solvent) which are applied with a cellular rubber roller to oak or ash tree, the amount of the solvent ranges up to 60 % in order to achieve a good wetting of the pores. The proportion of water roughens the stained surfaces easily and thus ensures that, especially in dark shades of colour, the adhesive strength is improved. The application of wood stains with a smooth hard rubber roller (Shore hardness 40 to 60) is very suitable for colouring of fine-pored wood. Figure 3.7 presents the essential methods of roller application for the application of wood stains. Depending on the shade of colour, the application amount and the drying parameter, the wood stains dry over a period of 20 to 60 seconds (at a circulating air temperature of about 60 to 70 °C). Normally, jet dryers with airflow velocities of nearly 15 to 20 m/s are applied. In order to obtain particularly transparent colour-tone effects, special wood stains are applied to sanded UV primers (see Figure 3.6, variant E).

3.1.2 Cellulose nitrate coatings (CN coatings)

Cellulose nitrate coatings still play an important role in the coating of wood and wood-based materials. Even in the year 2014, with 556,000 tons the market share of cellulose nitrate coatings amounted approximately 24 % of the worldwide produced wood coating systems (2.4 million tonnes) for industrial applications.

The global market for cellulose nitrate coatings is growing by 2 to 3 %. Upon closer examination, however, this development differs regionally with respect to the production volumes. Currently, we observe the further delocalisation of the furniture industry towards Eastern European and Asian countries. Therefore, in China alone a growth of more than 4 % was recorded for cellulose nitrate coatings. In Europe, the trend is reversed. While the market share of cellulose nitrate coatings amounted 17 % in the year 2000, the market share is still by 18 percentage points in the year 2016 (see Figure 3.1, Chapter 3.1). Thus, with the EU VOC directive 1999/13 EU (see Chapter 9.1) the European Union pushes the application of ecological coating systems with a much lower content of solvent than this exists in traditional cellulose nitrate coating formulations. In addition, innovations in water-borne coatings as well as UV coating systems will help to contribute to decrease the market share of cellulose nitrate coatings in Europe in the future. Cellulose nitrate as film forming component is applied not only in classic cellulose nitrate coatings based on short oil alkyd resins, but also as a resin in two-component polyurethane coatings.

3.1.2.1 Historical developments of CN coatings

The development of cellulose nitrate as a resin for coatings closely is connected with tests for the production of explosives made of cellulose and nitric acid just before the mid of the 19th century [12]. The first coating technical applications of cellulose nitrate were made after it was realized in 1880, that castor oil was suitable as a plasticizer for cellulose nitrate coating formulations. After further progress, with the end of World War 1, 1918, the real triumphal procession of cellulose nitrate varnishes started. The availability of large quantities of no longer required gun cotton pushed a rational utilization of this war waste. Further impetus came from the production of low viscous collodion wool[3], which were far more suitable for coating formulations than the previously produced cellulose nitrate resins. The availability of new solvents in the chemical industry also is closely connected with the development of cellulose nitrate lacquers. Thus, in the twenties of the last century cellulose nitrate coatings could replace oil coatings as first automobile series coatings used up to then in the United States. Due to a fast drying these coating systems laid the foundations for series production of automobiles on an assembly line.

At that time, CN coatings dried in almost the same number of hours as an oil coating which needs several days to dry. General Motor Corporation was the first automobile factory which began with the application of CN coatings in the year 1923. A contemporary work study stated that the time required for the car body coating decreases from 336 hours to 15 hours by means of CN coatings. Furthermore, the number of handles has been reduced from 2,400 to 600. The CN coatings were simple to produce and could be used after production immediately. These were conditions for a rapid development of cellulose nitrate coatings. In the year 1950, already with cellulose nitrate coatings coated furniture were produced in the amount of about 500 million dollars in the USA. A combination with oil-modified alkyd resins as well as resins from natural fatty oils and with polyester resins facilitates new applications and improved properties of cellulose nitrate coatings. Today, many other combinations with other types of resins are available which allow a wide range of application of CN coatings.

3.1.2.2 Components of CN coatings

Cellulose nitrate coatings can be formulated with cellulose nitrate as sole resins and in combination with other resins. The essential components of cellulose nitrate coating formulations are illustrated in Figure 3.8.

3 Cellulose nitrate accumulates during the manufacturing process as a flaky solid which colloquially is referred to as collodion wool or only as wool.

Cellulose nitrate

The raw material base for cellulose nitrate is the polysaccharide cellulose[4], the structural component of the wood (see Section 2.4) and almost all plants. Cellulose is a 1,4-linked poly-ß-D-glucopyranoside[5] [17] which consists of approximately 3,000 monomer units and has a molecular weight of approximately 500,000. The monomer units (glucose molecules) mainly are arranged linearly within the cellulose molecule (see Figure 3.9). The individual cellulose strands tend to align parallel to each other and are connected by a large number of hydrogen bonds. Cellulose has a very rigid structure due to these numerous hydrogen bonds. Natural cellulose is not soluble in solvents and not film-forming.

Nowadays, cellulose from coniferous wood mainly is applied for the production of cellulose nitrate. This cellulose is implemented to cellulose nitrate by means of a so-called nitrating acid[6] consisting of sulfuric acid and nitric acid in varying proportions. This is an esterification reaction of nitric acid with the free hydroxy groups of cellulose. This involves the

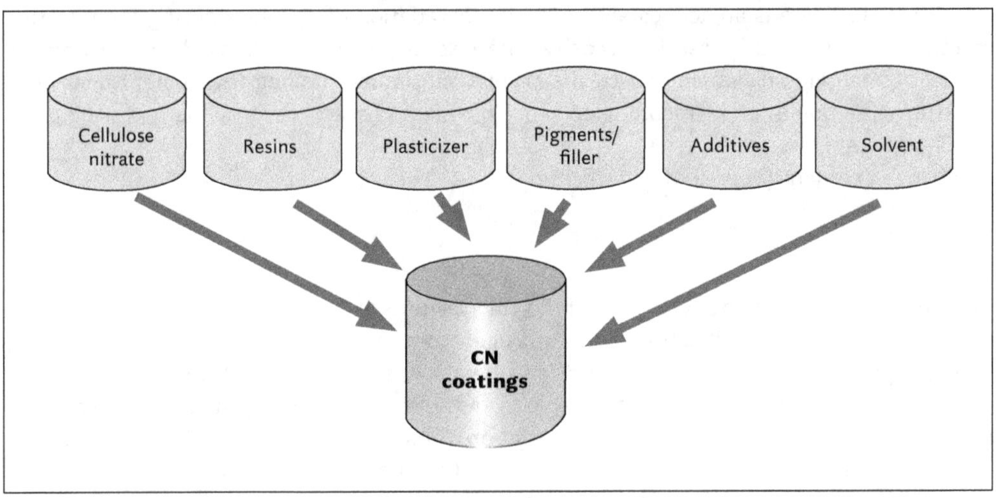

Figure 3.8: General composition of classical cellulose nitrate coating formulations

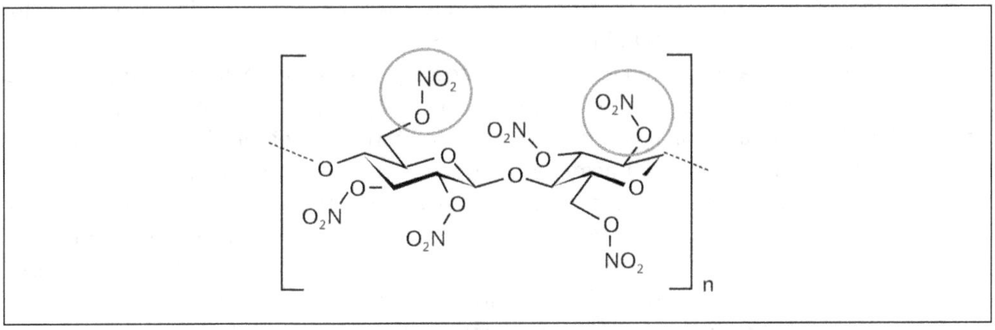

Figure 3.9: Molecular structure of cellulose nitrate

4 Saccharide = sugar, polysaccharides are the polymers of monosaccharides

5 1,4-linked poly-ß-D-glucopyranoside: These are glucose molecules which are linked together via oxygen bridges at the positions 1 and 4 and form a polymer.

6 This nitrating acid also is referred to as mixed acid

Table 3.2: Nitrogen content and degree of esterification of the different types of cellulose nitrate

Type of cellulose nitrate	Nitrogen content	Ester groups per structural unit
E	10.9–11.3 %	11.80–12.30
M	11.3–11.8 %	2.06–2.20
A	11.8–12.3 %	2.20–2.33

formation of a nitric acid ester which also is referred to as nitrate. Cellulose nitrate erroneously is referred to as nitrocellulose, although it does not contain any nitro groups, but ester groups[7]. Cellulose nitrate mainly consists of cellulose and is one of the first "biopolymers" in the industrial development of coatings.

The degree of esterification (conversion factor) can be controlled by means of the content of water. The higher the content of water in the nitrating acid, the lower the degree of esterification. There is an acid-catalyzed chain degradation of cellulose during the nitration reaction and thus a decrease in viscosity. The viscosity can be decreased furthermore by means of a targeted treatment, i.e. by means of a subsequent boiling with water under pressure (pressure boiling). The resulting cellulose nitrate is characterized by the following properties or parameters:

- Nitrogen content (degree of esterification)
- Solubility
- Viscosity

Theoretically, all free hydroxyl groups of the glucose units in the cellulose can be nitrated. The complete degree of esterification is at a nitrogen content of 14.14 %. However, practically only a degree of esterification of 13.6 % can be achieved. Cellulose nitrate with a nitrogen content of more than 12.6 % only is suitable as an explosive. The nitrogen content of cellulose nitrate for use in coatings and printing inks approximately is between 10.7 % and 12.2 %.

The degree of esterification has a great impact on the solubility of cellulose nitrate. With decreasing degree of esterification (nitrogen content), the solubility of cellulose nitrate in alcohols increases, while the solubility in solvents with a lower polarity such as esters increases with an increasing degree of esterification. Generally, there are the following types of cellulose nitrate regarding to their solubility.

The E-types are the standard types of wood coatings as well as leather coatings and are soluble in esters, ketones and glycol ethers. These are well compatible with various other resins and therefore suitable for the so-called "nitro-combination coatings". In comparison to other types, these coatings are characterized by development of hard films, a fast solvent delivery, good blending with alcohols, aliphatic hydrocarbons, aromatic hydrocarbons and by achieving good mechanical properties. The A types[8] with a low degree of esterification are soluble in alcohols, esters, ketones and glycol ethers and are mainly used for heat-sealable coatings on foil (such as aluminium foil) due to their thermoplastic behaviour. The solubility of AM types is between the solubility of the other two types, and thus are not described in this chapter.

7 Nitro compounds have a nitrogen carbon bond as in nitromethane $H_3C\text{-}NO_2$

8 The type name A is based on the solubility in alcohol (ethanol); the type name E is based on the solubility in esters; the type name AM signifies alcohol medium soluble

Table 3.3: Blending capability of genuine solvents in the case of cellulose nitrate

Solvent	**Blend**	
	Aromatic hydrocarbons such as toluene	Aliphatic hydrocarbons
Acetone	4.4	0.8
Methyl ethyl ketone (MEK)	4.3	0.9
Methylisobutylketone (MIBK)	3.6	1.0
Ethyl acetate (EE)	3.3	1.3
Butyl acetate (BA)	2.6	1.5
MEK/MIBK 1:1	3.8	–
MEK/MIBK 20:80	3.6	–
EE/BA 65:35	3.0	–

Solvent

There are existing three types of solvent with regard to the solubility of cellulose nitrate [12]:

- Real solvents completely dissolve cellulose nitrate without the addition of other solvents. Among them these include esters, ketones, glycol ethers, derivatives of glycol ether.
- Latent solvents require the addition of a genuine solvent in order to dissolve cellulose nitrate. Latent solvents increase the dissolving power of a genuine solvent. These include higher alcohols. Methanol[9] and ethanol belong to the genuine solvents (see A-type) and ethers.
- Non-solvents slightly reduce the dissolving power of genuine solvents. Aromatic and aliphatic hydrocarbons belong to this category.

The solvents are combined technically, environmentally as well as economically balanced and perform an important function. This starts with the production of cellulose nitrate coating, accompanies the application and ends with the evaporation from the liquid film. A solvent is a "good solver" when a solution has a low viscosity at a certain concentration of cellulose nitrate.

A good efficacy of a real solver is recognisable in which relationship it can be blended with non-solvers. The blending ability is defined as:

Blending ability = amount of the added blend/amount of the real solver.

The figures in Table 3.3 imply that for example 1 part of acetone (real solver) can be blended with 4.4 parts of aromatic hydrocarbons (here: toluene), but with only 0.8 parts of aliphatic hydrocarbons before gelation or flocculation of cellulose nitrate succeeds. The solvents also play an important role for the film forming of the cellulose nitrate chains and impact the mechanical properties of the coating film. The aim is to select a solvent so that the cellulose ntitrate molecules can be deagglomerated and thus clearly can be used for film forming. As with other coating systems, the proportion of solvents also has to be adapted by means of the evaporation behaviour.

The viscosity of cellulose nitrate coatings is an important measure for their formulation and processing. Of course, the viscosity is influenced by the type of cellulose nitrate

9 Methanol is no more used due to its toxicity.

Table 3.4: Commercial types and designations of commercial cellulose nitrate [18]

ISO 14446 (DIN 53179)	Percentage of the measuring solution	Manufacturer Hagemann	Manufacturer Hercules	Manufacturer Wolff
4E	4.0	H 4	RS 1300''	E 1440
9E	9.0	H 9	RS 15''	E 959
25E	25.0	–	–	E 510
27E	27.0	H 27	RS''	–
28E	28.0	–	–	E 400

Table 3.5: Mechanical properties of cellulose nitrate film [16]

Type	Elongation in %	Tensile strength in N/mm²
E 1440	24–30	98–103
E 1160	23–28	98–103
E 950	23–28	88–98
E 840	20–25	88–98
E 620	12–18	78–88
E 510	8–12	69–78
E 400	5–10	59–69
E 375	< 5	39–49
E 330	< 3	29–49

depending on the chain length of the molecule. The average degree of polymerisation (DP) which is the average number of glucose units per chain molecule is used as a measure for the length of the chain. The DP value varies within wide limits with technical cellulose nitrates, but the DP value is the decisive factor for the mechanical strength of the coating film obtained from solutions. During decomposition, the mechanical strength of cellulose nitrate considerably decreases below the DP value of 250 (molecular weight 70,000).

In practice, the main types of cellulose nitrate are specified on the basis of their solution viscosity, whereby there are different methods of determining the viscosity.

European manufacturers have committed themselves to a classification according to ISO 14446 (formerly DIN 53179). The concentration of cellulose nitrate (non-volatile and pre-dried) being necessary to obtain a solution viscosity of 375 to 425 mPas is specified. Acetone with a moisture content of 5 % is applied as a solvent. The viscosity is determined with the Höppler falling-ball viscometer. Additional measurement techniques can be found in the literature [16]. Common commercial types of cellulose nitrate are listed in Table 3.4.

As shown in Table 3.4, some manufacturers of cellulose nitrate join to this definition while other manufacturers choose viscosity numbers of defined solutions. Apart from the viscosity, also the solid body as well as the mechanical properties of the dried coating film are affected by the type of cellulose nitrate as shown in Table 3.5.

The highly viscous types result in flexible films while the low-viscosity types yield very brittle films. A mixture of highly viscous types with low-viscosity types for the production of a cellulose nitrate film with a medium viscosity is not recommended because coating films of such a mixture mostly have more inferior mechanical properties than coating films from a single type of cellulose nitrate. The medium viscosity types are most commonly applied in the field of wood coatings. By adding plasticizers, the mechanical properties of low molecular weight CN types can be improved as far as these achieve the mechanical elasticity and tensile strength of the medium viscosity type. Cellulose nitrate is explosive and thus is subject to certain safety regulations (transport, storage). Collodion wool is moistened only in alcohols (ethanol, n-butanol, usually 2:1 with cellulose nitrate) or dealt and stored in combination with plasticizers in the form of chips.

Plasticizers

As this term implies, plasticizers serve to plasticise the film and to keep the film elastic. Plasticizers behave similarly to solvents. However, plasticizers are characterised by a much lower vapour pressure compared to the solvent. Thus, plasticizers will remain in the coating film. A distinction is made between gelatinising and non-gelatinising plasticizers:

Gelatinising plasticizers form solid solutions with cellulose nitrate films, thus creating a uniform elasticised system. A blooming or a withdrawal of the softener does not occur.

Non-gelatinising softeners form an inhomogeneous system. The plasticizer tends to remove from the film when heating (exudation). Thus, the film loses its elasticity.

In order to avoid an exudation, non-gelatinising plasticizers often are combined with gelatinising plasticizers. The plasticizers have to meet the following criteria:

- Physiological harmlessness
- Good compatibility with resins and solvents
- Low colour number
- Resistance to light
- Resistance to migration
- Odourless
- Good plasticizing effect

Depending on the chemical origin, plasticizers can be classified as follows [6]:

- ester plasticizer
- Diester of phthalic acid, adipic acid, sebacic acid with longer-chained mono-alcohols
- Triester of citric acid
- Phosphoric acid ester of longer-chained, aliphatic alcohols
- Dibenzoic acid ester of aliphatic diols
- Esters of fatty acids with aliphatic alcohols
- Diesters of polyethylene glycol ethers
- Plasticizers based on epoxidized fatty acid esters or epoxidized oils
- Hydrocarbon plasticizers
- Chlorinated paraffins
- Others

Whereas at former times products such as dibutyl phthalate or other phthalic acid diesters of low molecular aliphatic alcohols were applied, these substances as well as different halogenated aromatic hydrocarbons and certain types of phosphoric acid esters extensively are disappeared from the market due to their toxicological properties. In the year 2002, the EU has reclassified and labelled dibutyl phtalate (DBP). Coatings containing more than 0.5 % of

DBP have to be labelled as poisonous. Due to its possible hormonal effect, the plasticizer is considered as critical for a long time. Epoxidized soybean oil, adipic acid esters (adipates) as well as special citric acid esters even with a FDA[10] approval are added. Ester plasticizers often are applied as gelatinising plasticizers in formulations of cellulose nitrate coatings. Ester plasticizers are combined with cellulose nitrate up to a ratio of 1:2. Most ester plasticizer have a gelatinising effect on cellulose nitrate and mostly have a good resistance to gasoline. Non-polar plasticizers (derivatives of fatty acids and hydrocarbons) do not exhibit the effect of gelatinization.

Resins

Only quite brittle, inelastic films will be received by using cellulose nitrate as the sole resin. Thus, these are combined either with plasticizers as described above or with resins. For use in wood coatings, cellulose nitrate is used especially in combination with alkyd resins (so-called "nitro combination coatings"). Natural resins such as dammar, shellac and rosin are rarely used today in cellulose nitrate coating formulations. Short to medium oily alkyd resins based on saturated fatty acids such as synthetic fatty acids or coconut fatty acids have a plasticizing effect on cellulose nitrate. This also applies to alkyd resins of oleic acid-rich fatty acids (for example peanut oil) or alkyd resins based on ricinoleic fatty acids as well as castor oil fatty acids. Also, hard resins are used as a combination partner in order to improve the gloss level and the abundance and to lift the application solid. A cellulose nitrate alkyd resin system can be constructed extremely diverse. In some cases, a formulation consisting of cellulose nitrate and alkyd resin in dilution with a mixture of solvents is sufficient. A plasticizer is dispensable, if a highly viscous type of cellulose nitrate is used, and if the alkyd resin has a sufficient plastification.

Pigments, fillers, additives

In principle, all organic and inorganic pigments which are suitable for indoor application can be used. All known fillers such as calcium carbonate, barium sulphate, kaolin, talc, mica, aluminium silicates, etc. are applicable. In particular, matting agents, waxes, anti-foaming agents, grinding aids and levelling agents are used as additives.

3.1.2.3 Examples of CN coating formulations

For decades, cellulose nitrate coating is the classical coating in the handcrafted and industrial furniture production. The variety of applications ranges from coating structures for frame furniture, living room furniture, bedroom furniture and children's room furniture to small components. The broad compatibility with other resins facilitates a wide variety of applications. CN coatings are characterized by a simple processing in the spray process or curtain coater process, rapid physical drying and excellent wood firing. Elastic, protective and decorative paint films can be obtained from cellulose nitrate in combination with other appropriate coating raw materials. Due to their low application solids (non-volatile content) of 15 to 25 % (CN clear coat), CN coatings are optimally suitable for the coating of open-pore oak furnitures. The CN coatings fulfil the requirements for chemical resistance of furniture surfaces (such as the resistance to alcohol) in accordance with DIN 68861 Part 1C. Due to the

10 The Food and Drug Administration (FDA) is the State Food and Drug Administration of the USA and is subjected to the Ministry of Health.

Table 3.6: Exemplary formulation for a grinding primer

Component	Feature	Parts by weight
Collodion wool E 510; 30 % in isopropanol	Cellulose nitrate – resin	15.00
Mixture of solvents consisting of esters, ketones, alcohols, aliphatic and/or aromatic hydrocarbons	Adjustment of the viscosity and evaporation behaviour	69.00
Alkyd resin 75 % dissolved in ester (such as peanut oil modified)	Resin	12.00
(2 parts adipic acid ester/1 part butyl stearate)	Plasticizer combination	3.00
Zinc stearate	Abrasive additive	1.00
Sum		100.00

Non-volatile amount: 23.5 % efflux time 4 mm-cup: approximately 25 s

simplicity, CN coatings are applied as CN multilayer coatings which are processed as a base coat as well as a topcoat. Due to their good dissolving characteristics by organic solvents, CN coating films without coating sanding can be repaired easily or overcoated.

A considerable disadvantage of the CN systems is, of course, the high proportion of solvents in the formulations, especially in the clear coat systems. Furthermore, CN systems tend to yellow under the influence of UV radiation, if sufficient UV protection is not incorporated, which protects the degradation of the cellulose nitrate chains as well as the wood substrate from changes in colours (see Chapter 8). A common problem with CN lacquer coatings is the migration of plasticizers. The migration of plasticizers occurs e.g. often with PVC plastics, which are used as door sealing tapes for example and thereby have contact with CN paint films.

Cellulose nitrate coating primers

CN primers are for example fine grinding primer, inlet primer, filling primer and brightener. The primers directly are applied on the wood or on stains. The primers fill and wet the surface, dry quickly and are good grindable.

Cellulose nitrate hard coating in different settings of the gloss level

This relates to base coat, multi-layer coatings and topcoats.

3.1.2.3 Example of application

The coating of European oak (veneer) as a combination process of UV primer and CN topcoat in the rolling process is schematically illustrated in Table 3.13.

3.1.2.4 Important tips when dealing with cellulose nitrate

Cellulose nitrate and oxidatively curing coatings may never be processed together in a coating booth since arising peroxides may react violently with cellulose nitrate. This also applies to unsaturated polyester coatings.

Do not leave any residual matter in used barrels as the phlegmatizer evaporates, and thus the dry cellulose nitrate wool is highly flammable. During the storage of the CN drums, direct irradiation of sunlight should be avoided. Also, the barrels have to be turned regularly in

Table 3.7: Exemplary formulation for an inlet primer

Component	Feature	Parts by weight
Collodion wool E 400; 30 % in isopropanol	Cellulose nitrate – resin	18.00
Mixture of solvents consisting of esters, ketones, alcohols, aliphatic and / or aromatic hydrocarbons	Adjustment of the viscosity and evaporation behaviour	67.00
Alkyd resin 75 % dissolved in ester (such as peanut oil modified)	Resin	8.00
Maleic resin	Resin for the enhancement of the plenty and hardness	2.00
(0.6 parts castor oil + 1.4 parts adipic acid ester + 2 parts butyl stearate)	Plasticizer combination	4.00
Zinc stearate	Abrasive additive	1.00
Sum		100.00

Non-volatile amount: 24.5 % efflux time 4 mm-cup: approximately 22 s

Table 3.8: Exemplary formulation for a filling primer

Component	Feature	Parts by weight
Collodion wool E 400; 30 % in isopropanol	Cellulose nitrate – resin	18.00
Mixture of solvents consisting of esters, ketones, alcohols, aliphatic and/or aromatic hydrocarbons	Adjustment of the viscosity and evaporation behaviour	57.50
Alkyd resin 75 % dissolved in ester (such as peanut oil modified)	Resin	16.00
Maleic resin	Resin for the enhancement of the plenty and hardness	2.00
(0.8 parts castor oil + 1.7 parts adipic acid ester + 2.5 parts butyl stearate)	Plasticizer combination	5.00
Zinc stearate	Abrasive additive	1.50
Sum		100.00

Non-volatile amount: 33.0 % efflux time 4 mm-cup: approximately 70 s

Table 3.9: Exemplary formulation for a brightener primer

Component	Feature	Parts by weight
Collodion wool E 620; 30 % in isopropanol	Cellulose nitrate – resin	12.00
Mixture of solvents consisting of esters, ketones, alcohols, aliphatic and/or aromatic hydrocarbons	Adjustment of the viscosity and evaporation behaviour	75.50
Alkyd resin 75 % dissolved in ester (such as peanut oil modified)	Resin	7.00
Cellulose acetobutyrate	Resin for the incompatibility, serves as a brightener	2.00
(1 part adipic acid ester + 2 parts butyl stearate)	Plasticizer combination	3.00
Zinc stearate	Grindability	0.50
Sum		100.00

Non-volatile amount: 19.1 % efflux time 4 mm-cup: approximately 30 s

Table 3.10: Exemplary formulation for a mat cellulose nitrate topcoat

Component	Feature	Parts by weight
Collodion wool E 400; 30 % in isopropanol	Cellulose nitrate – resin	15.00
Mixture of solvents consisting of esters, ketones, alcohols, aliphatic and/or aromatic hydrocarbons	Adjustment of the viscosity and evaporation behaviour	71.40
Alkyd resin 75 % dissolved in ester (such as peanut oil modified)	Resin	8.00
Alkyd resin	Resin for the enhancement of the plenty and hardness	1.50
Adipic acid ester	Plasticizer combination	3.00
Polyethylene wax	Surface smoothness, slip effect	0.50
Silica, organic chemically treated	Matting agent	0.50
Solution of silicone oil, 10 % in butylacetate	Levelling aids	0.10
Sum		100.00

Non-volatile amount: 22 % efflux time 4 mm-cup: approximately 30 s

Table 3.11: Exemplary formulation for a cellulose nitrate filler

Component	Feature	Parts by weight
Collodion wool E 400; 30 % in isopropanol	Cellulose nitrate – resin	8.00
Mixture of solvents consisting of esters, ketones, alcohols, aliphatic and/or aromatic hydrocarbons	Adjustment of the viscosity and evaporation behaviour	39.00
Alkyd resin 75 % dissolved in ester (such as peanut oil modified)	Resin	12.00
Titanium dioxide	Pigment, colouring	12.00
Adipic acid ester	Plasticizer	3.00
Soya lecithin	Dispersing agent	0.50
Bentone	Anti-settling agent, thixotropization	0.50
Silica	Matting agent	0.50
Barium sulfate	Filler	8.00
Kaolin, organic chemically treated	Filler	8.00
Zinc stearate	Grindability	0.50
Talc	Filler	8.00
Sum		100.00

Non-volatile amount: 54 % efflux time 4 mm-cup: approximately 50 s

Table 3.12: Exemplary formulation for a silk mat cellulose nitrate primer coat

Component	**Feature**	**Parts by weight**
Collodion wool E 400; 30 % in isopropanol	Cellulose nitrate – resin	12.00
Mixture of solvents consisting of esters, ketones, alcohols, aliphatic and/or aromatic hydrocarbons	Adjustment of the viscosity and evaporation behaviour	42.00
Alkyd resin 75 % dissolved in ester (such as peanut oil modified)	Resin	15.00
Titanium dioxide	Pigment, colouring	18.00
Adipic acid ester	Plasticizer	3.00
Soya lecithin	Dispersing agent	0.30
Barium sulphate	Filler	8.00
Bentone	Anti-settling agent, thixotropy agent	0.30
Pyrogenic silica	Anti-settling agent	0.30
Silica, organic chemically treated	Matting agent	0.50
Polyethylene wax	Surface smoothness, slip effect	0.50
Silicone oil solution 10 % dissolved in butyl acetate	Filler	0.10
Sum		100.00

Non-volatile amount (solid): 50.5 % efflux time 4 mm-cup: approximately 50 s

Table 3.13: Coating system on European oak by rolling process

Support	**European oak**
Feed	20 m/min
Wooden sanding	120/150/180 abrasive grain size
Dedusting	Brushing out of pores
Wood stain	Rolling process, apply cellular rubber roller with excess
Ejection	Diagonal alignment of 3 ejector brushes
Drying	Approximately 60 s at 80 °C, 20 m/s air velocity, jet dryer
Coating base coat	Roller application (double roller or 2 single rollers)
Drying	80 Watt/cm mercury high pressure lamp (2 lamps)
Coating sanding	320/400 abrasive grain size
Wood stain	Rolling process, cellular rubber roller (re-staining for blackening of pores of the oak
Ejection	Diagonal alignment of 3 ejector brushes
Drying	Approximately 30 s at 80 °C, 20 m/s air velocity, jet dryer
Finish coating	CN coating silky-mat (approximately 45 % solid), transparent, in the rolling process, grooved rubber roller, approximately 30 g/m^2 (wet application)
Drying	Approximately 6 minutes at 20 up to 50 °C in circulating air/jet dryer and subsequent cooling compartment 1 min/25 °C at 20 m/s air velocity

order to distribute the phlegmatizer uniformly. Cellulose nitrate coatings include solvents with low flashpoints and high evaporation numbers. These cause a severe cooling of the produced droplets by the spray application so that water may condense at the droplets. This results in a white, blunt and unsightly coating film (whitening). In order to avoid this effect, some butanol and/or glycol ether is given into the coating system. Water forms a so-called azeotrope water/mixture of solvents[11] of these substances, and thus is easily removed from the system by evaporating this azeotropic mixture [19].

3.1.3 Acid-curing coatings

The acid-curing coatings are a successful system of coatings in the English, North American and Scandinavian furniture industry for many years [20]. Even about 25 years ago, due to the excellent coating-technical characteristics of the acid-curing paints were used in large quantities for the coating of furniture, doors, panels, floorings, parquet, small parts and in the interior design, the estimated market share was 20 %. However, in Europe, the use of acid-curing paints declines. While the acid-curing paints still had a market share of 12.3 % in the year 2003, the market share in the coming years is likely to decrease. Despite the good price-performance ratio of acid-curing coatings – the acid-curing systems are priced between CN and 2C PU coating systems – these coatings have any significance for wood coating in Germany as well as in Central and Southern Europe. In Germany, the application is restricted to just a few special areas, such as the aqueous decor finish foils [21]. Thereby, these coating systems offer many advantages such as:
- Long pot life
- Fast drying
- Early stackability
- Enhanced hardness
- Good resistance to chemicals
- High scratch resistance and abrasion resistance

However, an essential disadvantage of acid-curing coatings increasingly has resulted in a loss of significance since the mid-80ies of the last century. Acid-curing coatings split off formaldehyde during the cross-linking. With regarding to a newspaper article in the German's weekly news magazine *'Der Spiegel'* in the year 1992, a large Scandinavian furniture manufacturer did not use acid-curing coatings any more as of the year 1993 [22]. Beside the fact that formaldehyde may be split off from finished furnitures in the event of non-compliance with curing parameters and other influencing factors, a high proportion of solvents of the classical acid-curing coating systems as well as a tendency towards brittleness with potential crack formation at layers are reasons for the decline in the use of acid-curing coatings. In the furniture industry, efforts are made to substitute the existing amounts of acid-hardening coating systems by more environmentally compatible coatings. In the United States, Scandinavia and with smaller market shares in England acid-curing coatings still are in use.

11 An azeotrope mixture occurs when a distillative separation of the components of the mixture is no longer possible. An example of this is the system ethanol/water with a water content of 4 % which boils at a temperature of 78 °C without altering the composition.

3.1.3.1 Components of acid-curing coatings

Resins

Acid-curing clear coat systems as well as pigmented coating systems usually contain approximately 50 to 80 % of organic solvents in the formulations. A classical formulation of clear coat contains a combination of alkyd resins/urea resins and/or melamine resins, physically drying resins such as cellulose nitrate wool, organic solvents, additives and corresponding silica (matting agent) in order to control the gloss level. As a rule, these are cured chemically as a 2C system by addition of an acid (*para*-toluene sulfonic acid) (polycondensation). The essential components of the formulations of acid-curing coatings are shown schematically in Figure 3.10.

Amino resins

Acid-curing coatings mainly contain urea resins as base resins. These refer to curable oligomers and polymers of urea and aldehydes, preferably formaldehyde. These amino features a coating-technical significance only in a partially or fully etherified form and mostly using primary alcohols such as n-butanol or isobutanol in the etherification reaction. The low molecular, butyl etherified types of amino resins mostly have a hexahydrotriazine structure as a structure type (see Figure 3.11). The first generation of acid-curing wood coating formulations primarily only contained these non-plasticized urea[12] as a resin. It was realized relatively quickly, that pure urea resin formulations result in cracking and spalling phenomena due to their brittleness of the coating film by addition of plasticizers in the formulation. It was tried to reduce the brittleness of the coating film by addition of plasticizers in the formulation. Plasticizers which are not embedded in the coating film migrated from the film over the time and resulted in a temporarily delayed cracking. Only the use of plasticized urea resins and/or the addition of alkyd resins as external plasticers led to good or satisfactory results. The compatibility with alkyd resins decreases with a higher degree of condensation of urea resins [23]. Brittle urea resins are more compatible with many alkyd resins than the elasticised urea resins. Therefore, non-plasticized urea resins mostly are used in acid-curing coatings. An addition of 10 to 20 % melamine resin with regard to urea resin extends the pot life of acid-curing coating formulations. Furthermore, thereby fullness and shine can be enhanced and the drying time can be prolonged. The reactivity of acid-curing urea coatings considerably is higher than the reactivity of melamine resins [24]. Due to the higher price level of the melamine resins, acid-curing melamine resins only are used in special fields.

Alkyd resins

Alkyd resins with semi-drying and non-drying oils or fatty acids primarily are used as a combination resin in acid-curing coatings. Preferably, coconut alkyds, ricinene alkyds and peanut oil alkyds with an oil length of not more than 55 % and an acid number not less than 25 and not more than 34 are used. The mixing ratio of urea resin for alkyd resin may range from 7:3 up to 3:7 varies, depending on the desired properties. A smaller proportion of alkyd resins involves a brittle coating film susceptible to cracking. A higher proportion for alkyd resins soften this coating. A significant improvement of many properties such as stackability, drying time, filling power, resistance to chemicals as well as resistance to water is achieved

12 In the wood-processing industry, non-plasticized urea resins widely are used as adhesive resins.

by chemical reaction of urea resins with alkyd resins. Such pre-reacted resin also is designated as pre-condensate.

Cellulose nitrates

Fast drying coatings with an enhanced filling power can be received by combining cellulose nitrate with plasticising agents. The higher the proportion of cellulose nitrate in relation to the urea/alkyd resin combination, the better the mattability, the faster the drying, and the more pronounced the pore appearance of the wood. However, the resistances of the films decrease with increasing proportion of cellulose nitrate, and the coatings contain an enhanced proportion of the solvents. If the ratio of the urea/alkyd resin combination to cellulose nitrate approaches a value of 1:1, acid-hardening 1C coatings can be formulated. Mono butyl phosphoric acids are used as acid catalyst for these systems.

Acetal resins

In search of resins of acid-curing coatings which do not separate formaldehyde, various resins have been developed in the 1990s. One possibility is the formulation of acid-curing systems without amino resins. This can be done by means of acetal resins. A model-like structure of the acetal resin is illustrated in Figure 3.1-12.

In Scandinavia, these binders still play a role, because the E1 criteria on furniture surfaces can be met (see also Chapter 2.5).

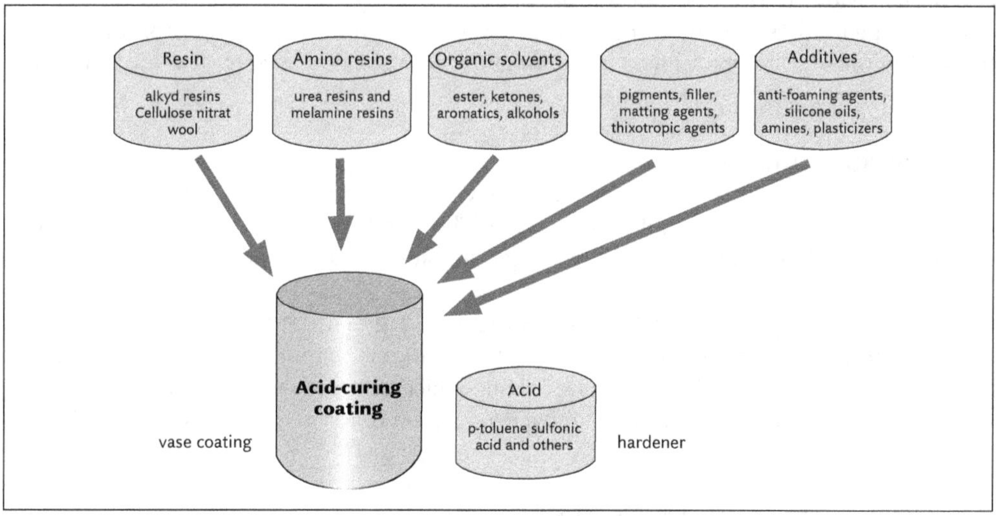

Figure 3.10: General composition of formulations of acid-curing coatings

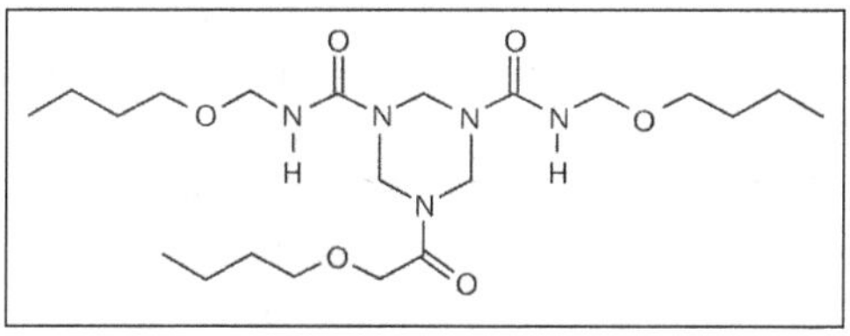

Figure 3.11: Model of butanol etherified low-molecular urea resin

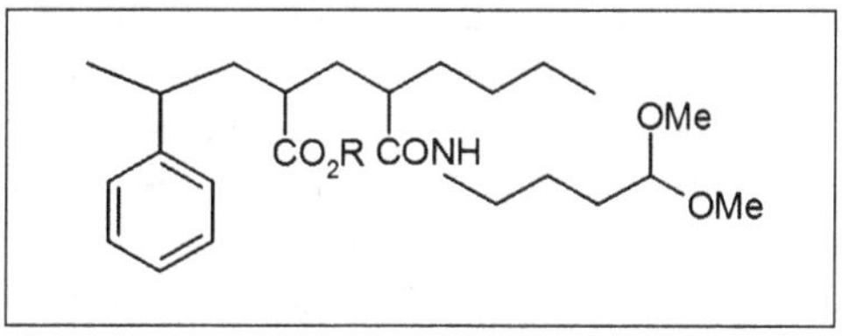

Figure 3.12: Model of an acetal resin [21, 25]

Hardeners

The majority of acid-curing coatings are typical 2C coatings. Mineral acids such as hydrochloric acid or phosphoric acid and their esters as well as organic acids such as para-toluene sulfonic acid (abbreviated p-TS or p-TSS) can be used as a hardener component. In the past, the use of hydrochloric acid has resulted in discolouration of some types of wood such as beech or pine. Another disadvantage of hydrochloric acid is its corrosive behaviour. Thus, para-toluene sulfonic acid has prevailed for acid-curing coatings. The quantity of application usually amounts 4 to 6 % of an approximately 10 to 20 % solution of p-TS based on the base coat. The pot lives amount approximately 3 to 7 days for a typical clear coat after adding 4 % p-TS. By doubling the concentration of p-TS a halving of the pot life can be achieved. To high concentrations of p-TS should be avoided, since p-TS can sweat out due to incompatibilities with the base coat from the dried film. Also, 1C systems can be formulated with phosphoric acid and their esters, because the pot life of these systems may amount between 2 to 12 months. In the dry paint and in connection with air humidity, an overdosage of the hardener contributes to a hydrolysis of urea resins and thus to a reduction of the quality of the painted surfaces. Furthermore, to high dosages of hardeners result in embrittlement's of the coating films.

Solvents

The non-volatile ingredients ('solids') of solvent-borne acid-curing coatings is approximately 30 to 50 %. The correct selection of the solvent depends on their dissolving power and evaporation behaviour during the film formation. Aromatic hydrocarbons, alcohols and glycol ethers can be used in systems without the use of cellulose nitrate. A higher proportion of esters and ketones have to be applied in combination with cellulose nitrate. The substitution of aromatic hydrocarbons by primary alcohols extends to pot life [26].

Further components

All raw materials used in acid-curing coatings have to be acid-resistant. Carbonate fillers and acid-sensitive pigments must not be processed. The grindability of acid-curing primers can be improved by 2 % amide waxes based on the coating. Silica is used as a matting agent.

3.1.3.2 Examples of recipes for acid-curing coatings

As an example, two examples of formulation of acid-curable coatings are listed here. The first example of formulation (Table 3.14) is a coating which can be used wet-in-wet especially in the curtain coating process. A system with a low content of formaldehyde is described in the Table 3.15.

3.1.3.3 Example of application

As a 2C system, acid-curing coatings have to be processed within a certain time before the viscosity increases and the materials begins to gelatinize (pot life). Fundamentally, the acid-curing coating systems are characterized by a rapid physical film formation, hard and abrasion-resistant films as well as good chemical resistance. The cured coating films meet the requirements of DIN 68861 Part 1B with respect to the chemical resistance and are mainly applied by spraying or curtain coater. Figure 3.13 illustrates the industrial coating of veneered chipboards with UV rolling priming and an acid-curing curtain coater with approximately 80 to 100 g/m^2 as a topcoat. The evaporation of the solvents as well as the chemical curing take place in a convection dryer (circulatory air operation, about 3 minutes, 70 to 100 °C) and by

Table 3.14: Acid-curing clear coat for furniture surface

Component	Function	Parts by weight
Alkyd resin[1)]	Resin, plastifying	39.00
Urea resin, non-plastified[2)]	Resin	25.50
Polyethylene wax[3)] AM-6 wax, 10 % in xylene		4.00
Anti-foaming agent[4)]		0.05
Silicone anti-foaming agent[5)]		0.30
Silica	Matting agent	3.00
Xylene	Solvent	11.30
Butanol	Solvent	16.85
Sum		100.00
p-TSS, 25 % in butanol	Acid hardener	2.60
Sum		102.60

Pot life: 4 days, efflux time 4 mm-cup: without hardener approximately 37 s, with hardener approximately 35 s

1) Alftalat AR 300 (Cytec)
2) Plastopal (BASF SE)
3) AM-6 wax, 10 % in xylene (BASF SE)
4) Byk 052 (Byk-Chemie GmbH)
5) Byk 141 (Byk-Chemie GmbH)

Tabelle 3.15: Acid-curing clear coat low content of formaldehyde

Component	Function	Parts by weight
Cellulose nitrate with softening agent[1)]	Resin, plastification, acceleration of the drying process	11.00
Alkyd resin[2)]	Resin, plastification	21.00
Urea resin non-plastified[3)]	Resin, content of formaldehyde < 0.3%	5.50
Butyl acetate	Solvent	20.00
N-butanol	Solvent	5.50
Iso-butanol	Solvent	5.00
Ethyl acetate	Solvent	10.00
Toluene	Solvent	11.50
Ethanol	Solvent	5.00
Xylene	Solvent	4.00
Polyethylene resin[4)]		0.05
Silica	Matting agent	1.00
Sum		100.00
P-TSS, 25 % in butanol	Acid hardener	2.60
Sum		102.60

Pot life: > 3 months, efflux time 4 mm cup: without hardener approximately 75 s, with hardener 70 s

1) NC E 560 +18 % DBP (Wolff Walsrode AG)
2) "Worleekyd" T 36 (Worlée Chemie GmbH)
3) "Plastopal" H73IB (BASF SE)
4) "AM-6"-Wax (BASF SE)

infrared drying at a maximum object temperature of 120 °C for about 1 1/2 to 2 minutes. Subsequently, the surfaces are cooled for a time period of about 2 minutes at a temperature of 30 °C and subsequently stacked.

3.1.4 Two-component polyurethane coating (2C PU coating)

In recent years, the two-component polyurethane coatings (2C PU coatings[13]) have replaced the acid-curing coatings and cellulose nitrate lacquers in many areas of the coating of wood as well as wood-based materials. It is characteristic for all two-component systems that two reactive components also known as base coat and hardener are mixed just prior to their processing. Generally, in 2C PU coatings hydroxyl group-containing resins (base coats) react with hardeners based on polyisocyanates under formation of polyurethane structure units. It is a polyaddition reaction which creates a three-dimensional network at a high functionality of the reaction components. Figure 3.14 illustrates the main principle of this polyaddition reaction using dioles and diisocyanates.

In the year 1937, the main principle of the chemical reaction was detected and patented [27] by Otto Bayer (founder of the company Bayer AG). The technical use of polyurethane chemistry in the coating industry began in the 1950s. Since that time, 2C PU coatings have become one of the most important technologies worldwide. The often used name "DD

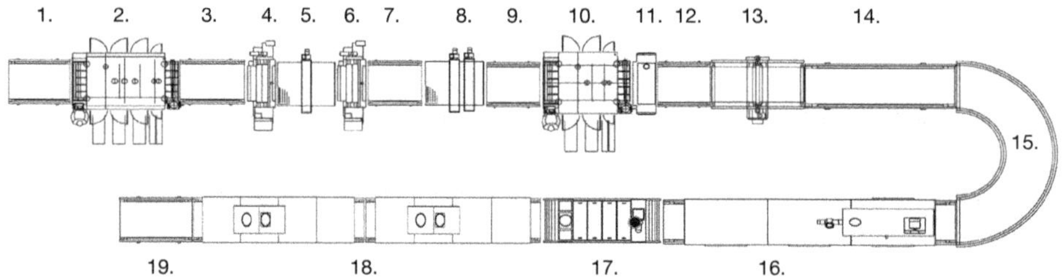

1. Support belt
2. Grinding machine for the wooden sanding
3. Conveyor belt for the control line
4. Single roller for the UV base coat 12 g/m² wet application UV dryer with a
5. UV lamp Hg 80 W/cm for the "UV gelation"
6. Single roller for the UV base coat approx. 12 g/m² wet application
7. conveyor belt
8. UV dryer with two UV lamps, Hg with each 80 W/cm
9. Conveyor belt
10. Grinding machine for the coating sanding
11. Dedusting
12. Conveyor belt
13. Curtain coater machine
14. Evaporation zone 25–30 °C, approx. 60–80 s
15. Evaporation zone curve
16. Circulating air dryer 70–90 °C, approx. 180 s
17. IR lamp approx. 100–120 °C
18. Object temperature, approx. 90 s
19. Cooling compartment approx. 30 °C, approx. 180 s

Figure 3.13: Coating line with UV rolling primer and acid-curing curtain coating

13 Today, polyurethane coatings are referred to as PU coatings according to DIN 55950 while the name PU coating still often is found in the Anglo-Saxon use of language [28].

coating" is a trademark of the Covestro AG and is derived from the commercial products (components) "Desmophen" for the base coat component (resin) and "Desmodur" for the hardener component (polyisocyanate). Meanwhile, there are a variety of different raw materials producers that produce resin and polyisocyanates for the 2C PU coatings. With regard to the respective use, tailor-made macromolecules can be created by means of the above-mentioned procedure of polyaddition.

Due to the fact that base coat and hardener of 2C PU coating mostly react at ambient temperature with each other, these mixtures only have a limited processing time, the so-called pot life. The determining factors of the pot life are described below.

3.1.4.1 Components of 2C PU coatings

The components of 2C PU coating are listed in the Figure 3.15. Apart from the hydroxyl group-containing resins (polyols), the base coat also contains solvents, pigments, fillers, additives and possibly combination resins as well as catalysts.

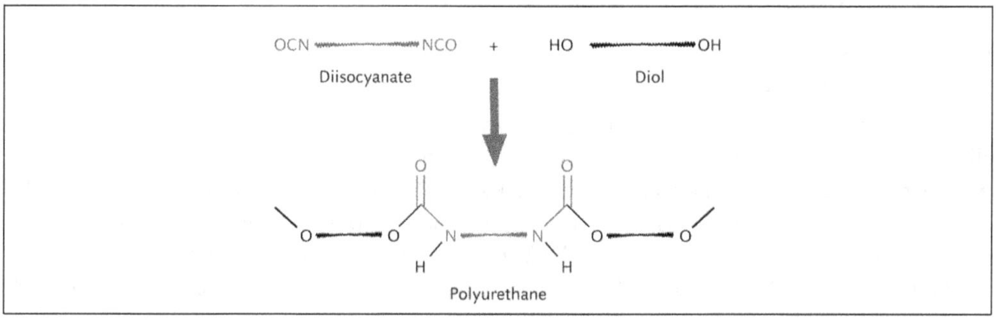

Figure 3.14: Principle of polyaddition for the production of polyurethane using diol components and diisocyanate components

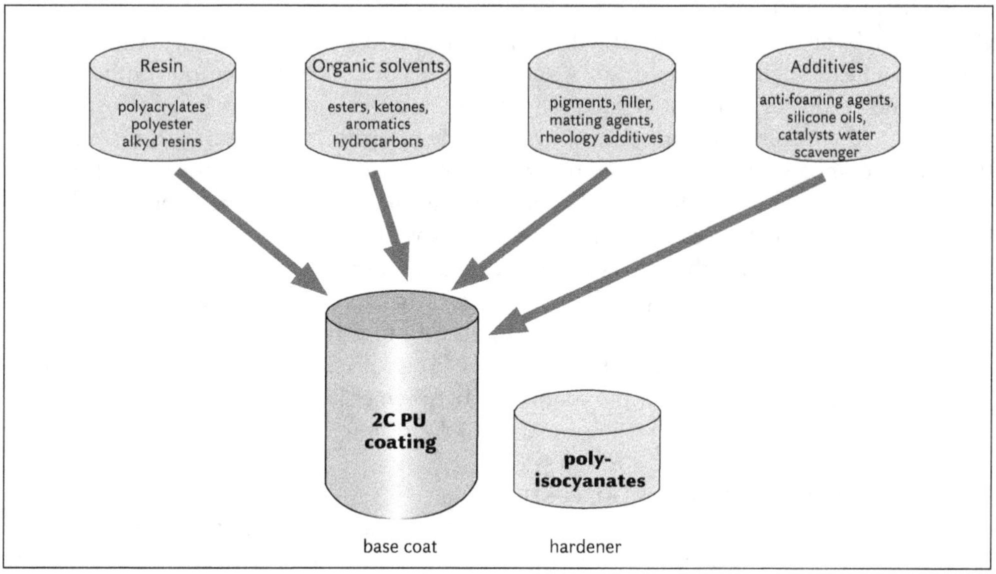

Figure 3.15: Components of 2C PU coatings

Resin

In principle, a variety of hydroxyl group-containing resins can be used for cross-linking with polyisocyanates [29]:

- Saturated polyesters
- Alkyd resins
- Polyacrylate resins
- Polyether
- Epoxy resins as well as epoxy esters
- PVC copolymers
- Polyvinyl acetale

Depending on the country, different resin systems are used for the industrial coating of wood. In southern countries such as France, Greece, Italy, Spain and Turkey predominantly hydroxyl group-containing alkyd resins or combinations of alkyd resins with polyesters are selected, whereas in Germany mainly hydroxyl group-containing polyacrylate resins are used as a component of base coats.

Polyacrylate resins

Polyacrylic resins containing hydroxyl groups (polyacrylate polyole) are very suitable for chemical and weather resistant 2C PU coating formulations, especially in combination with aliphatic polyisocyanates. These are produced by copolymerisation of (meth)acrylic acid, (meth)acrylic acid esters as well as hydroxy alkyl(meth)acrylates and are styrene-free for the coating of wood and wood-based materials. The thus produced polyacrylate polymers are hard resins with molecular weight usually between 2,000 to 4,000. These polyacrylates are placed on the market as 40 to 70 % solutions with a hydroxyl group content of 1 to 5 % [30].

Polyester resins

Hydroxyl group-containing polyesters (polyester polyols) are the most important resins for 2C PU coatings. A variety of aliphatic diol components or polyol components such as diethyleneglycol, butanediol, glycerin, trimethylolpropane etc. are used as alcoholic starting components and esterified with dicarboxylic acids such as adipic acid and various phthalic acids or their anhydrids, respectively. The linearly constructed and lightly branched polyesters usually are liquid at room temperature while the highly branched types feature a resin character. Polyurethanes made from heavily branched polyester polyols are characterized by high hardness, good solvent resistance and chemical resistance. The hardness and flexibility of the films almost arbitrarily can be controlled by means of a versatile combination of the different polyester polyols with various polyisocyanates. The proportion of hydroxyl groups in the commercial products roughly amounts between 1 to 8 % [30]. Due to the higher content of hydroxyl groups with respect to polyacrylate polyoles, higher densities of cross-linking are achievable with polyester polyols. Thus, these are especially suitable for use in chemical-resistant coatings. A comparison of the properties between polyester polyols and polyacrylate polyols for use in 2C PU coating is exemplary listed in Table 3.16.

Alkyd resins (oil modified polyesters)

Alkyd resins for 2C PU coating are short oily types of saturated fatty acids with relatively high numbers of hydroxyl groups and low acid numbers. Various resins of the type castor oil, coconut oil and soya oil/ricinene oil are suitable. Due to their good pigment properties

Table 3.16: Comparison of the properties between polyester resins and polyacrylate resins for 2C PU coatings

Comparison polyester resins/polyacrylate resins for 2C PU coatings

2C PU coating based on a hydroxyl-group-containing polyesters	2C PU coating based on a hydroxyl-group containing polyacrylate
Molecular weight: 1,000–2,000 moderate physical drying	Molecular weight: 2,000–4,000 excellent physical drying
Good compatibility with polyisocyanates	Limited compatibility with polyisocyanates
Good compatibility with resins such as CAB and cellulose nitrate wool	Limited compatibility with resins
The hardness, solvent resistance, chemical resistance increase with increasing branching of the resin structure	Good flexibility and adhesion despite the high hardness, good resistance to light
	Elastification with polyesters (linear)
Application solids (solids) > 30 % are possible	Application solids > 30 % (clear coats) are difficult
Very good coating fullness	Risk of orange peel effect low coating fullness

and substrate wetting properties these resins are suitable for pigmented topcoats with regard to high gloss. Alkyd resins do not achieve the cross-linking densities which are achievable with polyester polyols. However, alkyd resins are characterized by a good price performance ratio.

Combination resins

Further resins can be combined with polyol components in order to improve the gradation and to control the physical drying of 2C PU coatings. Thus, cellulose acetobutyrate at a concentration of 0.2 to 10 % as well as cellulose nitrate in a similar concentration can be incorporated into the base coat. Cellulose nitrate also promotes the wetting or stabilisation of pigments. Otherwise, cellulose nitrate cannot be combined with aromatic polyisocyanates because it then results in a reinforced yellowing especially for clear coats. Also, polyvinyl acetate, polyvinyl butyral, mixing polymers of PVC/PVAC may improve the gradation and the adhesive strength. Their quantities of application are between 0.5 and 3 % relative to 100 parts by weight of the total formulation.

Hardeners

A variety of diisocyanates and polyisocyanates are available for the cross-linking of the hydroxyl group-containing resins whereby only certain products could prevail on the basis of technical availability and profitability. The different polyisocyanates are accessible from diisocyanates by means of different reaction pathways. An example here are some derivatives such as toluene diisocyanate (TDI), methylenediphenyl diisocyanate (MDI), hexamethylene diisocyanate (HDI) and isophorone diisocyanate (IPDI). In the scope of this book, the authors want to waive a detailed discription of the different diisocyanate or polyisocyanates and refer to the secondary literature [28, 29, 30]. In particular, a distinction is made between aromatic and aliphatic diisocyanates or polyisocyanates. The aromatic types are not only much more reactive than aliphatic derivatives, but also become yellow under light exposure just in clear coating systems and are less weather resistant. In addition to the mentioned and desired reaction of the isocyanate group with the hydroxyl group, other side reactions are possible [28]. These include secondary reactions of isocyanates with each other (such as formation of allophanate and isocyanurate)

$$\underset{\textit{Urethane}}{\text{R-NH-COOR}} + \text{R-NCO} \longrightarrow \underset{\textit{Allophanate}}{\text{R-N(CO-NH-R)COOR}}$$

and primary reactions with other functional groups such as water. The reaction of the isocyanate group with water results in an elimination of carbon dioxide by forming an amine which may react with a further isocyanate group under formation of an urea derivative.

$$\text{R-NCO} + \text{H}_2\text{O} \longrightarrow \text{R-NH}_2 + \text{CO}_2 \quad \text{R-NH}_2 + \text{R'-NCO} \longrightarrow \underset{\textit{Urea}}{\text{R-NH-CO-NH-R'}}$$

This urea derivative may react with another isocyanate group under formation of a biuret.

$$\text{R-NH-CO-NH-R} + \text{R-NCO} \longrightarrow \underset{\textit{Biuret}}{\text{R-N(CO-NH-R)}_2}$$

Some of these primary and secondary reactions also are applied for a selectively production of polyisocyanate hardeners.

Diisocyanates

With the exception of the MDI, the above-mentioned aromatic as well as aliphatic diisocyanates have a low vapour pressures and are classified as toxic. These components are not used as a hardener for 2C PU coatings.

Polyisocyanates

However, an oligomerization of the diisocyanate generates polyisocyanates with an enhanced molar mass and a vapour pressure which is lower than the vapour pressure of the diisocyanate and thereby providing a prerequisite for safe handling from the industrial hygiene point of view. However, a contact with the skin and the mucous membranes should be avoided. The oligomerization increases the functionality of the isocyanate at relatively low viscosities. This leads to a high crosslinking density in the coatings. The following possibilities of oligomerization are implemented technically:

- Trimerization of diisocyanates to isocyanurates
- Conversion of an excess of diisocyanate with triol to urethane
- Trimerization of diisocyanates to biuretes
- Implementation of two diisocyanates with an aliphatic alcohol to form an allophanate

For the coating of wood and wood-based materials, the following polyisocyanates have prevailed as hardener components in 2C PU coatings:

- TDI trimethylolpropane adduct (such as "Desmodur L" by Covestro, "Basonat PLR 8525" by BASF)

- Trimerised HDI as biuret (such as "Desmodur N" by Covestro, "Tolonate HDB" Rhodia, "Basonat HI" by BASF and "Luxate HDB" by Lyondell), see Figure 3.16
- Mixed trimerisate of TDI and HDI (such as "Desmodur HL" by Covestro)

Although the polyisocyanates already may react with hydroxyl group-containing resins to form polyurethanes at ambient temperature, however this reaction is too slow for a sufficiently rapid curing of the coating especially when using aliphatic polyisocyanates. Thus, catalysts are used to accelerate the curing process. In particular, tertiary amines and various metal salts are the most effective catalysts. The effect of catalysts is based on complex formation with the reaction partners (isocyanate group and hydroxyl group) and lowering the activation energy of the addition reaction. 1,4-Diaza-bicyclo-(2,2,2)-octane (DABCO) has a high catalytic efficiency within the amine catalysts. However, the metal salts of organic and inorganic acids have a substantially stronger effect than tertiary amines. Dibutyl tin dilaurate (DBTL) is the most effective and most popular catalyst within this group.

In combination, DABCO and DBTL have a reinforcing (synergistic) effect. Due to the toxicity of organotin compounds, one is still looking for suitable alternatives to DBTL. Currently, no equivalent substitute is available. The catalysts will be incorporated into the base coat. The normal concentrations for metal salts are 0.02 to 1.00 % and for the amines 0.01 to 0.50 % based on the base coat. These generally are incorporated as a 10 % solution into the formulation. An overdosage has to be avoided since otherwise the technological properties of the coating film may change negatively.

Solvent, pigments, fillers and additives

Especially esters and ketones are used as a solvent. Aromatic hydrocarbons such as toluene, xylene or the higher boiling aromatic hydrocarbons and petrol can be used as blend solvent. The solvent or the mixture of solvents, respectively, have to be free of water as far as possible and must not exceed a level of 0.05 % water.

It has to be ensured that the pigments and fillers do not contain too much adherent moisture or take part in unwanted side reactions. For example, zinc oxide as well as some carbon black pigments may result for example in a shortening of the service life of the coatings (base coat and hardener). Suitable "water scavengers" are added to the base coat or to the hardener solution for the stabilisation against humidity. These are highly reactive mono-isocyanates (such as "additive TI" from OMG Borchers) or further orthoester of formic acid (such as "additive OF" from OMG Borchers). p-Toluenesulfonyl isocyanate (p-tosyl isocyanate "TI additives") is considered to be one of the most effective substances. Figure 3.17 illustrates the reaction of p-toluenesulfonyl isocyanate with water. The reactivity of the carbon of the isocyanate group is increased strongly by the sulfonyl group.

Furthermore, additives such as matting agents, deaerating agents or agents for viscosity control are added.

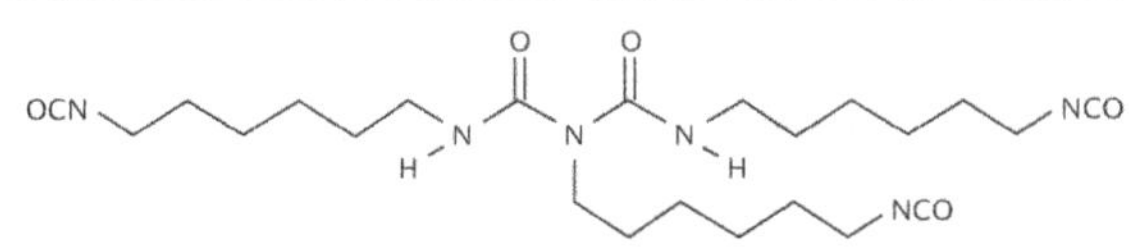

Figure 3.16: Trimerisized HDI as biuret

3.1.4.2 Application of 2C PU coatings

2C PU coatings are characterized by a limited pot life. The pot life specifies in which time the coating still can be processed without occurrence of coating defects. It has become established to determine the pot life by means of the application viscosity. The adjusted processing viscosity should be doubled maximally upon reaching the pot life. Ideally, the pot life should amount 8 hours (1 work shift). The components base coat and hardener are mixed together only prior to the application. The blend ratio base coat: hardener may vary between 100:5 and 100:100. Analytical studies have confirmed that after about 7 days (23 °C and a relative humidity of approximately 50 %) the chemical reaction between the base coat and polyisocyanate in the most 2C PU coatings completely takes place [31]. In practice, it sometimes seems that the addition of the hardener does not occur accurately or not at all. In extreme cases, this may result in formation of cracks or losses in adhesive strength, respectively, in the case of polyacrylate containing base coats being cross-linked with polyisocyanates due to the missing the flexibilising properties of the hardener component (such as HDI biuret).

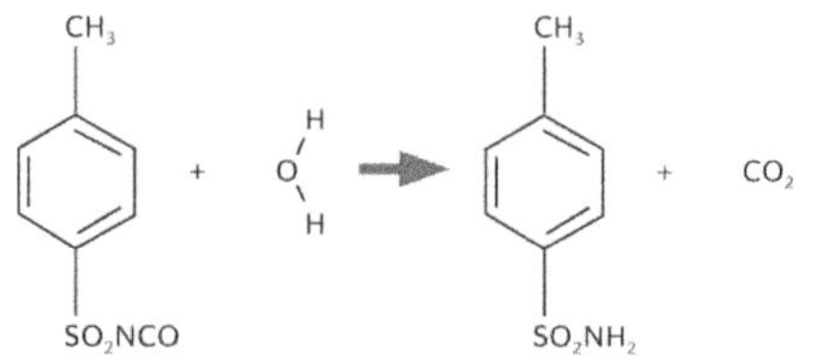

Figure 3.17: Reaction of p-tosyl isocyanate with water

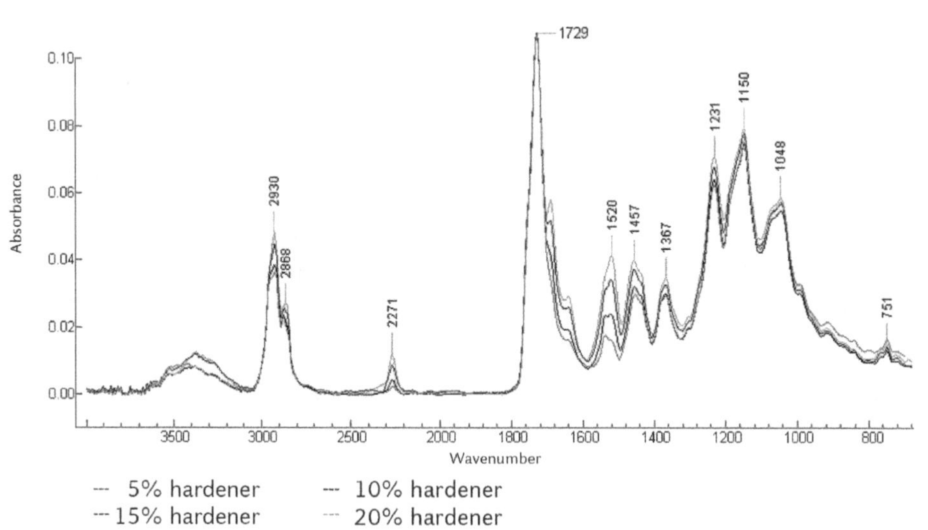

Figure 3.18: ATR spectrum of 2C PU coating depending on the concentration of polyisocyanate

14 Only a thin surface layer is investigated by means of infrared spectroscopy using ATR (attenuated total reflection). For this, the proportion of non-reflected wavelengths (absorbed wavelengths) specifies the functionality of individual chemical compounds and their proportion in the surface compared with a standard.

Table 3.17: Test standards for the furniture surfaces

Test standard no.	Title of the standard
DIN 68861 part 1	Furniture surfaces – behaviour with chemical stress. The standard was withdrawn with the release of EN 12720. However, the standard DIN 68861 part 1 was introduced once more after a short time because the European standard does not contain a classification.
DIN 68861 part 2	Furniture surfaces – behaviour with stresses by abrasion
DIN 68861 part 4	Furniture surfaces – behaviour with scratching strain
DIN 68861 part 6	Furniture surfaces – behaviour with burning cigarette ash
DIN 68861 part 7	Furniture surfaces – behaviour with dry heat The standard has been withdrawn with the publication of EN 12722. However, thereafter the DIN 68861 part 7 was reintroduced because the European Standard does not contain a classification.
DIN 68861 part 8	Furniture surfaces – behaviour with dry heat The standard has been withdrawn with the publication of EN 12722. However, thereafter the DIN 68861 part 7 was reintroduced because the European Standard does not contain a classification
DIN EN 12720	Furniture – evaluation of the resistance of surfaces to cold surfaces
DIN EN 12721	Furniture – evaluation of the resistance of surfaces to humid heat
DIN EN 12722	Furniture – evaluation of the resistance of surfaces to dry heat
DIN 4102	Fire behaviour/low flammability With regard to DIN 4102 part 1, low flammability coatings have to be deployed on carrier plates (approved according to DIN 4102-B1) in order to fulfil the test criteria
IOS-MAT-0066[1)]	2. Physical properties 2.1 Test method-IOS-T-0002/1.1-recommended class P1–R7

1) Specification of Ikea

A lower dosage or overdosage of the hardener component can be analysed by means of ATR spectroscopy[14] in comparison to reference patterns (see Figure 3.18).

Depending on the binder system, there exist solids between 20 and 50 % for transparent or high gloss clear coat systems, respectively. 2C PU coatings are used where enhanced demands on the mechanical and chemical resistance exist. The two-component polyurethane coatings are characterized by the following points [32]:

- 2C PU coatings are universally applicable. They have an excellent barrier effect against bleeding of wood constituents (teak, rosewood, pine and oak).
- The chemical resistances are excellent and comply with the furniture and kitchen standard DIN 68861 Part 1A and 1B. Additionally, these systems also comply with the high requirements of the IOS-MAT-0066 of the stress group R2. Especially the often criticized alcohol test (1 hour 48 % alcohol solution in water) is fulfilled even in dull mat two-component polyurethane systems without any signs of gleam.
- Depending on the composition of coatings and hardeners, the 2C PU coatings are characterized by a very good resistance to light.
- Depending on the formulation, hard and yet elastic surfaces are received.
- PU lacquers usually are flame retardant adjustable.

Table 3.18: Maximal scratching strain [N] of coated wood surfaces depending on the substrate

Substrate	**Testing in accordance with DIN 68861 part 4 behaviour with scratching strain**
Chipboard panel, cherry tree veneered	1.5–2
Chipboard panel, beech veneered	2.0
Chipboard panel, anegre veneered	1.5–2
Colid alder	1.0
Glass	3.0

- These are soft PVC-stable.
- High abrasion resistance, scratch resistance and impact strength also are adjustable via the coating formulation.
- The 2C PU coatings are particularly appropriate for highly stressed steps or wet areas with strongly fluctuating relative humidity and temperature.

The 2C PU coatings mainly will be applied in the injection procedure. All common methods such as compressed air nozzles, airless with and without air support and electrostatic coatings are used. 2C spraying units are used for the processing of 2C PU coatings with a very short processing time (pot life). The base coating and the hardener directly are mixed in the spray head of the application pistol. For years, the 2C PU coatings are applied for high-quality kitchen fronts and furniture fronts. The coatings are used as base and topcoat (preferably in the northern countries). In the southern countries of Europe, but also in South America a separate application of a 2C PU primer and a corresponding 2C PU topcoat are preferred. The 2C PU primers are specially designed for high filling capacity of the subsoil as well as for a very good sandability. The topcoat is made with a topcoat system which is responsible for the gloss level and elegance of the surface. The temporal distance between the application of the primer and the topcoat should be less than 24 hours after primer sanding in order to ensure a good adhesion. Still, it is especially important to remove the sanding dust completely in order to avoid effects of obfuscation and grey wood pores. The essential testing standards according to which furniture surfaces are tested are listed in Table 3.17. It often happens that furniture manufacturers have developed their own house standards, or list them in the requirement profile.

An example of application of high-quality furniture fronts with a silk-mat 2C PU clear coat as a base coat and topcoat (blend ratio base coat/hardener 10:1) is:

- Application of a graduated wooden sanding in case of rough wood surfaces (grit of sandpaper 120/150/180)
- Dedusting
- Hand spraying with bucket gun, nozzle 1.5 to 1.8 mm, pressure 2.5 to 3.5 bar
- 2C PU coating as a primer coating, approximately 100 to 120 g/m²
- Drying/curing at 23 °C and 50 % relative humidity, approximately 3 to 6 hours
- Coating sanding/smoothing with 320/400 grit of sandpaper
- 2C PU coating as a topcoat, approximately 100 to 120 g/m²
- Drying/curing at 23 °C and 50 % relative humidity, 1 to 2 hours
- Subsequently, 4 to 5 hours of forced circulating air drying at approximately 40 to 45 °C

Table 3.19: High-gloss finishing based on 2C UP or 2C PU filler white with pigmented 2C PU finishing coats in spraying process

Optionally: 2C PU insulation	**Application for insulation and solidification of the milled edges and interior profile**
2C UP filler white (containing styrene)	300–500 g/m² wet application in several work stages
2C PU filler white as an alternative to 2C UP coating filler white	150–500 g/m² 1–3 work stages
Drying/chemical curing at ambient temperature	23 °C and 50 % relative humidity
Coating sanding	400–500 grit
High-gloss topcoat	2C PU topcoat in the respective colour shade depending on the requirements, 1–3 coating applications with an interim drying for a period of 15–30 min at 23 °C and 15 % relative humidity
Drying/chemical curing	Minimal 24–48 hours at 23 °C and 50 % relative humidity
After-treatment	Depending on the demand for quality, the surface can be grinded, wobbled and polished to a high gloss. The grinding and polishing works is referred to as buffing[1].

1) Buffing is a dismantling polishing procedure based off. For this, at first a fully curing coating layer with different sandpaper grits from coarse to fine is grinded. subsequently, the grinding is performed by addition of polishing paste or buffing wax, respectively, on a buffing wheel. This is known as buffing. The buffing is performed by means of a felt or pile band on belt grinders, angle grinders with packs of linen disks, nettle disks or Molton disks, buffing bucks or on special buffing automates in order to achieve high-gloss surfaces.

- Cooling to 20 to 25 °C
- Stacking

Coating properties:
- Behaviour with chemical stress according to DIN 68861 part 1B is fulfilled
- Behaviour with scratching strain according to DIN 68861 part 4 4C
- Microscopically measured thickness of the dry film 45 to 50 μm

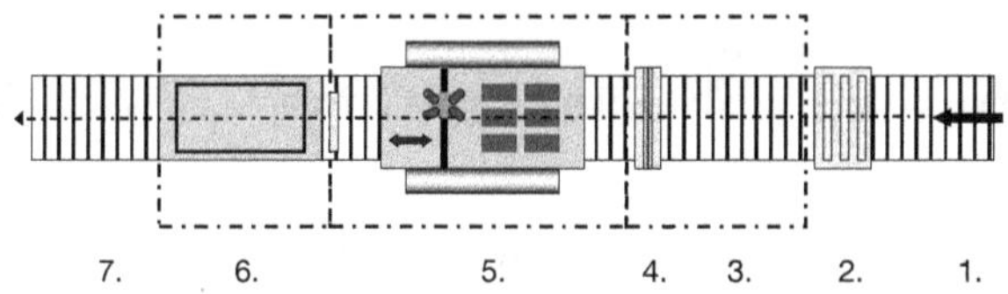

1. Conveyor belt (occupancy)
2. Dedusting scrubber (3 pieces)
3. Conveyor belt (encapsulated chamber)
4. compressed air dedusting
5. Spraying robot (encapsulated cabinet) with 4 spraying guns (inline machine)
6. Rack dryer (6 racks) approximately 25–30 °C residence time approximately 45 minutes
7. Conveyor belt (withdrawal)

Figure 3.19: Coating line for the high-gloss finishing with pigmented 2C PU coatings

The behaviour with scratching strain highly depends on the underground. The correlation between the substrate and the result of the scratching strain is illustrated in Table 3.18.

2C PU coatings both in transparent and in highly pigmented systems have been proven for the production of high-gloss coating surfaces for high-quality kitchen cabinets and furniture fronts. MDF boards have been established as a carrier in this area of high-gloss finish. At present, there are two proven methods for building a PU high-gloss finish on MDF boards. In the first method, the MDF boards (surface + edge) is coated with styrene-containing 2C UP coating filler (pigmented white). In order to prevent a collapse of the subsequent coating with the high-gloss coating based on a 2C PU coating system and thus in order to prevent a restless appearance of the glossy surface, 2C UP fillers 300 to 500 g/m^2 are applied in several steps.

Gloss-coated kitchen cabinets which are coated on the basis of styrene-containing 2C UP fillers are characterized by an in-depth mirror effect. In recent years, the 2C UP coating fillers were a subject of critical discussion since styrene used as a reactive diluent is not chemically incorporated to 100 %. In the journal *Ökotest* (issue 9/2003) more than 1,000 µg/m^3 air of free styrene were found in kitchen furniture fronts. For many years, the 2C PU fillers are used as an alternative styrene-free filler. Depending on the quality and composition of the MDF board (medium density fibreboard, see Chapter 2.5), the 2C PU filler is applied 1 to 2 times at application rates of 150 to 200 g/m^2. It often occurs that the milled areas of a MDF board (edge or inner profiles) are coated with a special 2C PU clear coat insulation. This isolation has the task to stick together the wood fibres previously raised by the milling process in order to prevent a formation of cracks due to moisture penetration in the area of the edge and inner profile. Moreover, the use of a 2C PU insulation prior to the application of the filler increases the impact resistance of the overall coating at the edges. Table 3.19 describes an example of a high-gloss finish.

Figure 3.1-19 illustrates a coating line as it is used for high-gloss coating with pigmented 2C PU topcoat. In order to avoid dust ingress during the high-gloss finishing, the MDF fronts filler coated in advance are grinded in a separate room. In position 2 and 4 of this coating line, at first the dust is removed by means of brushes and compressed air. After that, the coating is applied in a batch operation. Normally, at first the coating of the edges by means of airmix spray guns is performed using a coating program set in advance. Subsequently, the topcoat of the surface by means of airmix spray guns is applied. The coating cycle takes about 2.5 to 3 minutes depending on the number of structural components.

3.1.5 Unsaturated polyester coatings (UP coatings)

3.1.5.1 Introduction to the chemistry and technology of UP coatings

The unsaturated polyester coatings, also referred to as UP coatings and used in wood coatings, primarily consist of unsaturated polyesters dissolved in styrene. These are converted chemically in the presence of initiators (organic peroxides) and accelerators (such as cobalt salts). The chemical reaction with styrene under cleavage of the double bonds of the double bonds of the unsaturated polyesters is referred to as copolymerisation. The fundamental work on the exploration of the chemistry of unsaturated polyester resins was made at the end of the 1930s (German patents IG Farben) [33]. At the beginning of the 1950s, the coating industry worked intensively on the application of unsaturated polyester resins. In the 1960s, the UP coatings mainly are used in the production of transparent high-gloss surfaces in all sectors of the furniture industry. At that time, the new products (paraffin-containing UP coatings) rapidly established themselves in the housing industry as well as in the manufacture of pianos. The subsequent wave of pigmented furnitures ('white wave') required the use of pigmented paraffin-containing UP finish effect coatings [34]. The further development of unsaturated polyester resins resulted in the introduction of pigmented, paraffin-free UP coatings. Today, the unsaturated polyester coatings only are used for special applications such as the coating of pianos in Europe and Asia (see Figure 3.20) as well as for exclusive musical instruments, pieces of furniture, the interior design of luxury yachts and aircraft [35]. Also, dashboards of noble wooden interior parts in automobiles are coated with UP coatings [36]. In the southern European supplier industry for kitchen furniture, still the advantages of a fast curing process and a good price-performance ratio of white UP fillers are very appreciated. However, the residual amounts of non-converted styrene that are detectable by extraction [37] are detrimental.

3.1.5.2 Components of UP coatings

The unsaturated polyester coatings consist of approximately 100 % of non-volatile components (solids). Thus, these unsaturated polyester coatings rank among the first coating systems with a low content of VOC for the coating of wood furniture. The constituents of UP coatings are illustrated in Figure 3.21.

Unsaturated polyester resins

The unsaturated polyester coatings primarily consist of unsaturated polyester resins (UP resins) consisting of saturated and unsaturated dicarboxylic acids and diols. These predominantly are linear, soluble products of a polycondensation reaction. The main chain of the

Figure 3.20: Grand piano Company Fazioli F 278

resins contains olefinic double bonds which can copolymerize as a reactive diluent in combination with styrene. The radical polymerisation is strongly or completely inhibited by atmospheric oxygen [38]. For this reason, the first unsaturated polyester resins ('paraffin polyester') in the coating formulation contain paraffins as an additive acting as an inhibitor against the atmospheric oxygen. Further developments result in air-drying UP resins, also referred to gloss polyester, containing an autoxidable group which can react with atmospheric oxygen. In the polymerization process, the unsaturated polyester based on maleic anhydride, phthalic anhydride, glycoles and styrene are inhibited during the curing process by the ubiquitous atmospheric oxygen.

The polymerisation inhibition manifests in thin layers of soft coating films and in thick layers of sticky surfaces and sensitivity of the solvent [41, 42]. The inhibition of the free-radical polymerisation occurs by accumulation by oxygen (active species of O_2 diradicals) on radicals which are formed during the polymerisation process. This will cause a premature termination of the free-radical polymerisation. The elimination of the inhibiting effect of the atmospheric oxygen is possible by means of the addition of paraffins in small quantities (0.001 to 1 %). With the beginning of the polymerisation reaction, as a result of the density difference the paraffin which initially is soluble in the liquid coating continuously is transported to the surface to form a barrier against oxygen.

Paraffin consuming polyester resins

The unsaturated polyester resins still employed today are produced by means of polycondensation (melting condensation) from polyols and dicarboxylic acids at temperatures between approximately 150 and 200 °C in order to avoid a premature gelation. The chemical reaction of the formation of esters is illustrated in Figure 3.22.

Within the process of manufacturing, it is absolutely necessary to work without the atmospheric oxygen since even trace amounts of oxygen may cause a gelation of the resin. In addition, unwanted discolouration may occur. Low amounts of so-called inhibitors/stabilisers – usually hydroquinone or p-(tertiary butyl) catechol – are used in order to prevent the process of gelation throughout the process of production. It is known from literature that a further addition of small amounts of copper (0.0005 to 0.01 % Cu) in the form of its salts such as copper naphthenate among other things synergistically increase the effect of stabilisation of the hydroquinone [39]. The water which is formed in the reaction of esterification is either directly or azeotropically distilled off in a closed loop by means of a suitable solvent such as toluene or xylene. the polymerizable double bonds are obtained from the maleic acid esters or from their esters such as fumar acid esters which are firmed from the maleic acid esters at higher temperatures. The pronounced property of copolymerization of unsaturated polyester directly is connected with the formation of the fumaric acid ester groups (trans-configuration) [41]. Today, the degree of isomerisation is determined by means of the NMR spectroscopy and Raman spectroscopy. The proportion of phthalic acid in the formulation of the resins is very important for the compatibility of the resins towards paraffin and styrene. Furthermore, the hard resin character of the UP resins increases with the increasing proportion of phthalic acid. It is known from the literature, that the compatibility towards styrene also is improved by addition of low amounts of 1,3-butane diol [38].

Furthermore, it is known that the structural design of the UP resins influences their performance characteristics. It is reported that unsaturated polyester resins with double bonds of the fumaric acid/maleic acid at the end of the polymer chain exhibit significantly better physical properties after the copolymerisation with styrene in comparison to unsaturated

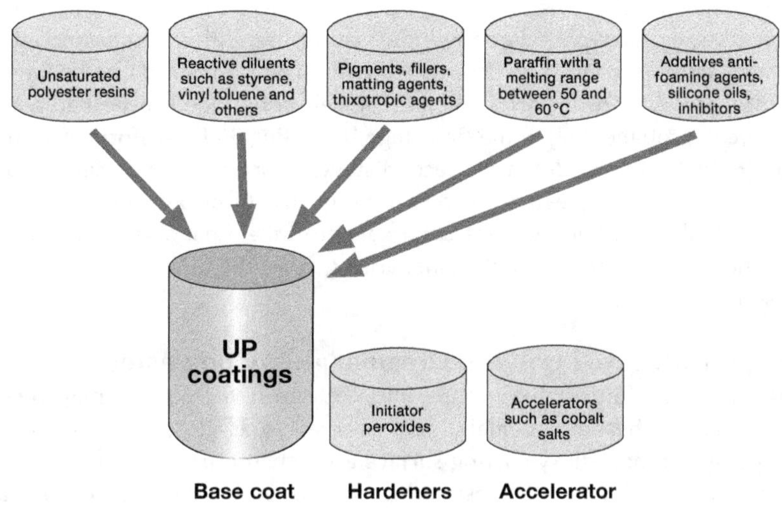

Figure 3.21: Composition of UP coatings

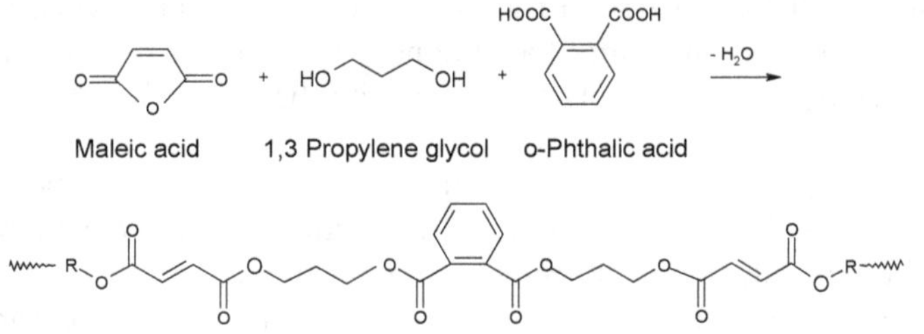

Figure 3.22: Principle of polycondensation for the production of UP resins

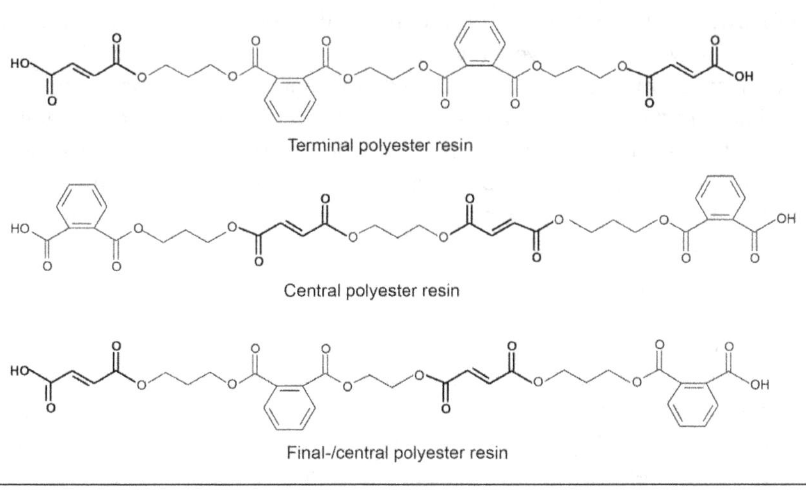

Figure 3.23: Effect of the production process on the structural properties of unsaturated polyester resins

polyester resins with double bonds in the middle of the polymer chain. The so-called tails polyester resins are synthesized via a two-stage polycondensation stage. in the first stage, the saturated dicarboxylic acid (phthalic anhydride) is esterified with the total amount of the polyol and subsequently (second stage) converted with maleic anhydride by boiling. However, the pure centre polyester resins arise when the maleic anhydride is esterified with the total amount of glycol (propane diol) in the first stage by boiling and transformed with the dicarboxylic acid (phthalic anhydride) in the second stage. In practice, the unsaturated polyester resins are produced in a single-stage process. Interim structures between the tail and centre structures are developed whereby the structure of one of the other outweighs depends on the procedure. The different structural elements arising from the different manufacturing processes are summarized in Figure 3.23.

Air-drying unsaturated polyester resins (glossy polyester)

Due to the fitting of autoxidable groups into the resin matrix, one may renounce the addition of paraffin. These resins mainly are referred to as glossy polyesters since these have left a glossy and non-sticky coating surface after chemical curing. In the area of wood coating those modifications of the resin which are based on the use of mono-allyl ether or diallyl ether of the trimethylolpropane have proved themselves. The resins are used technically since about the year 1955. The curing mechanism of the glossy polyester has been investigated systematically by *Traenckner* and *Pohl* [40]. The aliphatic CH bonding in α-position to the allyl double bond can be cleaved simply homolytically due to its low binding energy. Figure 3.24 illustrates the options for reaction of the allyl ether compounds with atmospheric oxygen and the polymerisation with the double bonds of the fumaric acid polyester.

An allyl radical (I) is formed by adding a radical to an allylic double bond. The allyl radical can be consumed either by H abstraction (II) or addition (IV) to the double bond of a fumaric acid group. The allyl radical formed by the reaction (II) reacts with the atmospheric oxygen to hydroperoxides (III). The hydroperoxides decompose in the presence of siccatives (for example cobalt naphthenate) and thus initiate more chain reactions.

With this mechanism, the polyester chains may crosslink freely and produce non-sticky coating surfaces. The curing behaviour was examined intensively in a wide variety of studies [40]. The following curing behaviour was found by IR spectroscopic analysis for allyl ether polyester (glossy polyester) dissolved in styrene [38]:

The normal copolymerisation between the double bonds of the fumaric acid and the styrene predominates in deeper layers. The allylic double bonds are attached to the resin matrix only to a small degree. The oxidative curing of the double bonds of the allyl ether predominates at the coating surface. An exemplary recipe for an allyl ether containing resin is mentioned in Table 3.20. Trimethylol propane diallyl ether as a component of modification is adjusted to the polyester by the existing hydroxyl group.

The partial exchange of phthalic acid by tetrahydrophthalic acid also cause non-sticky and rigid coating surface. Table 3.21 presents two literature examples of recipe based on tetrahydrophthalic acid [41].

IR spectroscopic investigations by Demmler [44] have shown that the loosely bound hydrogen atoms in the 3-position of the tetrahydro phthalic acid ring can easily be abstracted by a starter radical (peroxide) resulting in a carbon radical. This radical position can react with atmospheric oxygen. This creates a peroxy radical which can react to the hydro peroxide. It is believed, that the hydro peroxide may form a peroxy radical or an oxy radical both with the divalent or the trivalent cobalt in a redox reaction. This peroxy radical or oxy radical is stabilized by separation of a hydrogen radical under formation of a ketone (see Figure 3.25). Frequently, glycols of the diethylene glycol type are used since it is assumed that these are less susceptible to atmospheric oxygen. This is due to the ability of the glycols to form peroxides.

Monomers/reactive diluents

The monomers used also are referred to as reactive diluents since the monomers chemically copolymerise with the unsaturated polyester and thus are incorporated into the matrix of the interconnected system beside regulating the viscosity of the UP coatings.

To date, styrene is the most important reactive diluent for unsaturated polyester resins. Approximately 35 million tons of styrene per year are produced in estimated 1,500 plants all over the world. Styrene is classified as a substance with carcinogenic and genotoxic effects. The effective strength of styrene is considered to be in compliance with the 'AGW' value so that a significant amount to the cancer risk to humans is expected as very low. In Germany,

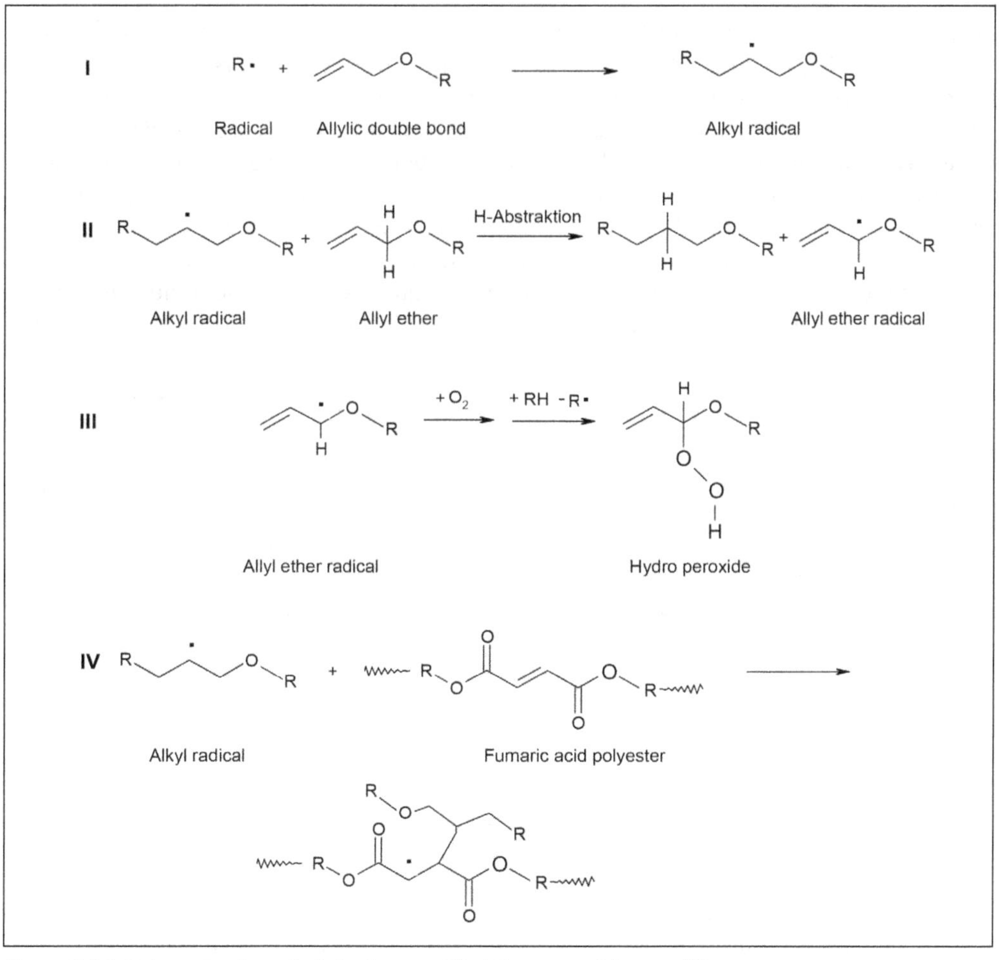

Figure 3.24: Polymerisation of allyl ether modified fumaric acid esters [40]

Table 3.20: Example recipes of a diallyl ether containing unsaturated polyester resin [41]

Raw materials	Examle [mole]
Phthalic anhydride	1.0
Maleic anhydride	1.0
Propylene glycol	1.8
Trimethylolpropane diallyl ether	0.4

Table 3.21: Recipe suggestions for tetrahydrophthalic acid containing unsaturated polyesters [41]

Raw materials	Example 1 [mole-%]	Example 2 [mole-%]
Diethylene glycol	89.25	105.00
Glycerine	10.30	-
Tetrahydrophthalic acid	50.00	25.00
Fumaric acid	50.00	75.00

the occupational exposure limit (AGW in Germany) for styrene is fixed at 20 ppm (86 mg/m³) since 1/2006, according TRGS 900 in Germany. Nevertheless, styrene as a reactant in the copolymerisation reaction still is important for unsaturated polyester coatings. Several attempts have been made to find alternative reactive diluents to styrene. Beside styrene, vinyl toluene, vinyl ether or methacrylate such as 1,4-butylene ether glycol methacrylate are used as reactive diluents or as a substitute in the coating of wood. In the market, the unsaturated polyester resins are offered as a solution in organic solvents or in styrene. In the delivery form, the content of styrene usually amounts 30 to 35 %. Styrene and fumaric acid esters tend to form copolymers with an alternating sequence of monomer units. The reactivity of the monomers firstly was described quantitatively by *Alfrey* and *Price* [50, 51]. Alfrey and

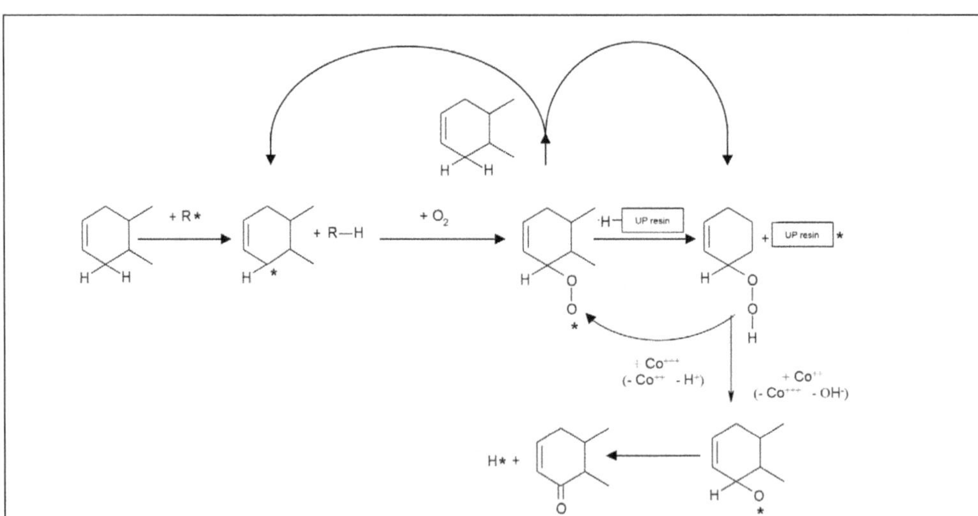

Figure 3.25: Suggested reaction mechanism of tetrahydro phthalic polyesters with atmospheric oxygen (schematic) [46]

Price implemented the two physical quantities e and Q as a characteristic value for monomers. Thus, the rate constant k_{ij} can be described as follows:

$$k_{ij} = P_i \cdot Q_j \cdot \exp(-e_i e_j)$$

For simplicity, monomers with different Q values and similar e-values result in an ideally azeotropic copolymerisation while monomers with different Q values and widely diverging e values result in an alternating copolymerisation. Numerous practical investigations by *Demmler*[46] and *Hamann*[42] have shown that the ratio of polyester/styrene as well as the conditions of the curing process significantly influence the ratio of polyester/styrene within incorporation. Hamann et al. could demonstrate approximately 2 mol of styrene per 1 mol of double bond of the polyester for hardened UP coatings. Further investigations by Demmler indicate that styrene is incorporated into the polymer matrix more irregularly since the conditions for an azeotropic copolymerisation are not respected often. Thus, the ratio polyester/styrene changes during the free-radical chain-growth polymerisation as styrene evaporates more or less depending on the conditions.

Today, the unsaturated polyesters primarily are used in applications in the wood sector in South Europe, Eastern Europe and South America. These are formulations of wood coatings which may contain so-called reactive diluents such as dipropylene glycol diacrylate or organic solvents in addition to styrene. Styrene-free formulations also are in use. The curing is performed by a fast alternating copolymerisation of styrene with the unsaturated polyester resin. Figure 3.26 illustrates the property profiles of paraffin-free as well as paraffin-containing unsaturated polyester coatings.

Hardener and accelerator

For the initiation of a polymerisation reaction, generally organic peroxides are used as initiators which decompose into radicals and thus initiate the radical chain reaction.

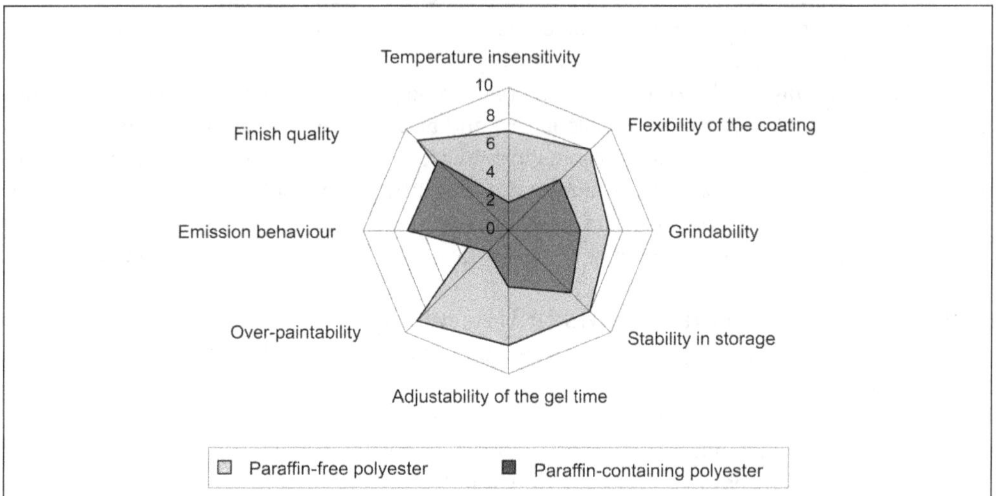

Figure 3.26: Comparison of property profile of paraffin-free and paraffin-containing unsaturated polyester coatings

Cyclohexanone peroxide (CHPO), methyl ethyl ketone peroxide (MEKP), benzoyl peroxide as well as cumene hydroperoxide have been proven in practice. For many years, the application of cyclohexanone peroxide in UP coatings is declining since between 50 and 150 ppm cyclohexanone have been found at residual emission measurements of automotive interior components after 28 days. Today, methyl ethyl ketone peroxide has the greatest importance in the wood coating. The cleavage of the peroxide into radicals only takes place after exceeding a certain energy barrier, the so-called activation energy. The activation energy is so high, that a curing would have a duration of days or weeks only with usage of peroxides. The activation energy can be reduced by means of heat (hot curing) or accelerators (cold curing). Common accelerators particularly are metal salts such as cobalt or manganese salts which can be used alone or in combination with tertiary amines. Cobalt soaps (such as cobalt octoate) preferably are used for the coating of wood and wood-based materials. These cobalt soaps are used as a siccative in oxidatively curing alkyd resins. The mechanism of the degradation of peroxides accelerated by cobalt salts is described in Chapter 3.2.1.4. The combination of metal salts and amines is not used for the application of UP coatings in the coating of wood and wood-based materials. At temperatures below 15 °C, the reactivity of cobalt accelerated systems rapidly decreases. The curing layer remains soft and not grindable in pigmented formulations for a long time. In filled systems, the danger of an adsorption of cobalt on the surface of fillers or pigments exists. The catalytic effect of the cobalt accelerator may be remained by adding a wetting agent based on calcium octoate for example on the surface of the fillers and pigments. With respect to publications in the spring of 1996 (Ökotest and corresponding TV shows), a carcinogenic potential by cobalt siccatives is discussed [59]. The Federal Environmental Agency denied the award of the eco-label (Blue Angel) for coatings containing cobalt siccatives. A referendum of the European Printing Ink Association (EuPIA) of March 2004 proposes a waiving of the application of the use of cobalt siccatives as soon as possible [58]. The toxicological properties of cobalt (II) containing compounds which are definitely not clarified yet and, coupled with this, the threatening reclassification of cobalt siccatives as carcinogenic according to EU category Carc. Cat. 1 A, Carc. 1 B (TRGS 2017) are the reason for this measure. However, the replacement of cobalt siccatives in UP coatings is very difficult. In comparison to cobalt octoates, possible alternatives such as potassium octoate did not supply results of equivalently quality [56].

In the commonly used formulations, the respective peroxides are mixed with an inert liquid (for example, 50 % peroxide in plasticizer) and added in an amount of about 5 to 10 % based on the UP resin including styrene shortly before application. In recent years, efforts were made to replace peroxides containing dibutyl phthalate by substitutes in order to prevent residual emissions from the cured UP coatings. Cobalt salts are added with a concentration of 0.2 to 2 % whereas enhanced concentrations are applied in pigmented systems.

Other formulation components of UP coatings

Apart from the compatibility with peroxides, the pigments and fillers used in UP coatings have to meet the following requirements:

- The storage capability of UP systems does not have to be affected
- The accelerators may not be absorbed by the pigments and fillers
- The fillers and pigment may not cause any additional crosslinking

Calcium carbonate, barium sulphate, talc and kaolin mainly are used as fillers. Especially the volume shrinkage within the polymerisation of the UP resins with monomers can be re-

Table 3.22: Workflow for the coating of fine veneer surfaces with UP coatings in high gloss procedure

1st day	**Rough grinding and pickling of wood** At first, the rough grinding of the wood is performed. The desired colouring is achieved by the application of aqueous wood stains using the spraying process.
2nd day	**Insulation of the wood components and patination** The insulation of the wooden components is performed with 2C polyurethane coatings. The subsequent processing of the insulation has to be performed after 2–3 hours after soonest or maximally after 20 hours in order to prevent problems with the adhesive strength.
3rd day	**Application of the UP coatings** Depending on the procedure, between 3 and 10 individual layers of an unsaturated polyester are applied by means of the spraying process. Up to 200 g/m² of polyester are applied wet per layer. After drying, the total coating thickness amounts approximately 700 up to 800 µm [23].
4th/5th day	**Drying phase/curing phase** The coated surfaces are stored for 24 up to 48 hours under ambient conditions (23 °C/50 % relative humidity of the air). If the curing phase is broken, then this may result in a sagging of coatings layers within the subsequent buffing process.
6th day	**Grinding to the polyester coating** The grinding of the polyester coating consists of several graduated grinding processes in order to receive a plane surface. Furthermore, small disturbances as well as dirt entrapments are eliminated.
7th day	**Buffing and polishing** The production of high-glossy surfaces requires a buffing and polishing of the coatings surface subsequent to the grinding process.

duced by using fillers. The selection of suitable paraffins for paraffin challenging UP coatings is made according to their melting behaviour since the paraffin should not crystallize during the cooling process subsequently to the polymerisation reaction. Reference products have a melting temperature in the range from 51 up to 53 °C. In the winter, paraffins with a melting range between 46 and 48 °C preferably are used, while in the summer paraffins with a melting temperature in the range from 56 up to 58 °C are applied. Since the applied paraffins may be subject to fluctuations from batch to batch, larger amounts of paraffins are purchased and used over years. Thus, a constant delivery quality of UP coatings can be guaranteed. Furthermore, the added paraffins also prevent the evaporation of mono-styrene almost entirely. Depending on the melting range of the paraffin, during the curing process the loss of styrene can be reduced to values below 5 mass-%. Without addition of paraffins, the loss of styrene amounts 60 to 80 % after 90 minutes [42]. In order to improve the storage stability and to extend the processing time, inhibitors such as hydroquinone and p-tert.-butyl catechol in the ppm-range are added.

3.1.5.3 Applications of UV coatings

UP coatings are applied by spraying or by using curtain coater. The spraying process requires an application of coatings by a bucket gun or a 2C spraying machine since the short pot life of several minutes up to one hour require a rapid processing. In the 2C spraying process, the two components separately are fed into the bucket gun and only mixed during the actual spraying process at the exit of the nozzle. The curtain coater process is another proven type of the application of UP coatings. Here, there exist some process variants such as contact curtain coater process, double-head curtain coater process as well as the sandwich process which is described in more detail in Chapter 5.5.

Almost all species of wood are suitable for the UP coating. Exceptions are woods whose ingredients prevent the chemical curing of the unsaturated polyester. These include Iroko wood, rosewood, macassar wood, mansonia wood, myrtle grain wood, teak wood, vavona wood et cetera. Such woods should be insulated before coating with UP coatings. The insulation is achieved by 2C polyurethane coatings with a very low viscosity. Depending on the wood species, it is necessary to apply the insulating coating several times with intermediate drying. Figure 3.22 illustrates the working method for the coating of decorative components (dashboard) of precious wood for the automotive industry and exclusive furniture elements.

The coated precious woods are applied for the manufacturing of dashboards, central consoles, steering wheels, gearshift grips and others. The carrier material may be a layer-glued, form pressed and fine veneered components such as burl wood. The components are coated multiple times in order to achieve the desired depth of the high-gloss surface. Subsequent to the appropriate curing phase, the components are grinded by means of graduated abrasive paper grit. The best conditions for polishing exist if the coated and polished surface has a surface roughness of approximately 1.5 µm [52]. For this, usually a two-stage grinding is necessary in which an abrasive paper with a 600 grit is used in the second grinding phase. In the last grinding process, one should grind across the future direction of polishing. The subsequent polishing can be performed faster and with less pressure. The high-gloss polishing is carried out by means of a polishing machine equipped with a felt belt. For this purpose, special polishing pastes and waxes are used. The wax residues can be removed by means of a polishing agent. Depending on the heat resistance, coatings generally should be polished with a cutting rate of 6 to 24 m/s [52]. An excessive heating of the coating layer should be avoided when polishing since subsequently it may result in a troubled, orange peel. In practice, surface temperatures of 80 to 100 °C are measured in the buffing process. Generally, the following instructions should be considered when applying UP coatings:

- Optimal processing temperatures are between 22 and 28 °C
- Warm up the freezed material to a temperature of approximately 30 °C and cool subsequently on 22 to 23 °C, process the material only subsequently
- The crystallized components in the paraffinic solution have to be dissolved completely by stirring and heating. The heating may occur in a water bath (30 to 45 °C).
- The mixing ratio between UP coating and hardener or activator has to be met absolutely.
- The overpaintability (gelation time[15]) between the individual steps of the coating can be extended up to an hour by adding of retarders. High temperatures and/or low humidity of the air accelerate the gelation time.
- Compliance with the prescribed wet application amounts. Excessive application amounts in a spraying application may cause air inclusions and a grey discolouration of the coating surface, if necessary. Due to the rapid physical drying, too low application amounts result in short gelation times and prevent from seeping out of paraffin in paraffin-containing UP coatings.

15 Gelation/gelation time/gel point: In practice, the gelation time is the time which remains to the processor to process the UP coating. At the molecular level, according to Flory a resin system reaches the gel point when an infinite network is formed for the first time [60]. This means that the molecules from the one end of a resin continuously are linked to the molecules on the other end; It does not mean that all molecules participate in this network.

Table 3.23: Comparison of the RIM process with the conventional UP coating technology for the coating of components with real wood surfaces [54]

	Spray painting with UP clear coatings	**ClearRIM technology**[1)] **with 2C PU clear coatings system**
Procedural principle	Multi-layered spray painting	Curtain coater or recasting, respectively, of a carrier component with a thin 2C PU clear coatings layer
Curing time of the coating	Approximately 72 hours	2.5 up to 4 minutes
Proportion of post-treatment	Very high, grinding, polishing	Marginal repolishing
Residual emission	90 % higher in comparison to the ClearRIM process	90 % lower in comparison to the UP coatings coating
Current expenses	Very high due to elaborate manual work and long process throughout times. Loss of material due to coatings over-spraying and additional disposal costs	Low. Minimal manual effort and drastically reduced process throughput times
Investment costs	Low, but very high current expenses	Higher, costs for tool and reaction moulding machine

1) ClearRIM Technology is the label of the marketing of Hennecke GmbH, Polyurethane Technology in Sankt Augustin

- Curtain coating on warm components results in surface disturbances in the paraffin layer (orange peel, rupturing, sticking)
- The high gloss buffing only should be cancelled after a curing time of 48 to 72 hours of the UP coating in order to achieve optimum surface effects.

RIM process

For several years one is engaged in alternating coating systems in order to substitute the multilayered and very time-consuming coating systems. In addition, for years the automotive industry is trying to reduce the residual organic emissions from the interior [53]. Since May 2001, the RIM process (Reaction Injection *Molding*)[16] for the coating of precious wood veneer components with 2C polyurethane systems is performed in the automotive industry [54]. In the RIM process, the building component which is coated with precious wood veneer on its visible side is inserted into a tool. The remaining gap of approximately 0.8 mm between the veneer surface and the tool cavity is poured out with a 2C PU clear coat system upon a mixing head in a high pressure process and cured within a few minutes at temperatures between 100 and 160 °C [54, 56]. So-called release agends based on silicone or wax have to be applied in the removal of the components from the tool. In recent years, one fundamentally tries to avoid the release agents. Table 3.23 illustrates the advantages and disadvantages of the RIM process in comparison to the conventional coating with unsaturated polyester clear coats.

16 RIM = Reaction Injection Molding. Two or more reactive liquids are mixed under high pressure, poured into the tool, and cured to duromers. This is the only plastic processing method in which the chemical reaction is performed in the tool.

3.1.6 Radiation curing coating systems

In a narrower sense, radiation curing is the curing of coating materials by means of electron beams (EB) or ultraviolet light (UV). In industrial practice mainly the radical UV technology is applied for the coating of wood-based materials. Floors, doors, furniture, wall coverings, decor finish foils as well as corc successfully are coated for years. The EB technology only is used in special applications. The cationic polymerisation is not relevant or an application in the wood coating. Thus, the following explanations will focus on the radical UV-curing.

3.1.6.1 Historical review

Most likely the first UV-induced chain polymerisation already was performed in ancient times in Egyptian mummies [61].

In the year 1955, Chester M. McCloskey and John Bond exemplary described numerous attempts to cure unsaturated polyester resins with UV radiation [62]. They found, that unsaturated polyester resins chemically harden by the addition of photoinitiators such as halogen-containing naphthalene derivatives α-halogen ketones under UV radiation. They also systematically examined the influence of the photoinitiator on the rate of polymerisation depending on the amount added. It was recognized that the type and intensity of the UV radiation affects the rate of polymerisation. The curing of styrene-free unsaturated polyesters was systematically investigated by *Charlesby* et al. *They found that such systems crosslink at a relatively low degree of polymerisation of about 5 basic units under the impact of ultraviolet radiation* [63].

Already at the beginning of the 1960ies, in Germany high solids containing (up to 100 % volatile content) unsaturated 2C polyester coatings have been processed on furniture coating lines industrially. For the first time, polyester coatings enabled an efficient coating of the furnishings in the assembly-line work. The most important products at this time were polishing coatings which could be stacked after a drying time of approx. 20 minutes. Subsequently to the 20 hour post-curing in the stack of plates, the grinding and polishing of the polyester surface is performed. In order to meet the increasing demand for furniture and the pressure of rationalization in the furniture industry, it was attempted to improve the coating capacity by increasing the speed of the conveyor belt of the equipment. The target could be exploited to the limit of what is feasible technically by means of the further development of more reactive unsaturated polyester resins, by increasing the concentration of the accelerator (cobalt salts) and peroxide, by means of promoters (β-diketones) as well as by increasing the drying temperatur [64, 65].

Foremost at the beginning of the middle of the 1960ies, in Europe the UV curing of unsaturated polyesters was picked up by the coating industries and raw materials industries. Some patent applications originated during this period. These patent applications used the patents of the company Du Pont [66] from the year 1946 dealing with the application of benzoin ether of primary alcohols as photoinitiators for unsaturated acrylates. The application of these photoinitiators in unsaturated polyester resins was investigated [67 68].

At the end of the year 1967, the curing of unsaturated polyester coatings by means of UV radiation was suggested in the German technical press [69]. The advantage of this new process consisted of an enormous shortening of the curing time for the coating material of unsaturated polyesters most commonly applied at that time. Moreover, the benefits in the effective shortening of process through time and the space savings by means of more compact coating plants were recognized.

Due to the existing knowledge on raw materials regarding to unsaturated polyester resins, the implementation of the first commercially available photoinitiators and by providing the first usable UV lamps on the part of the machine industry, in Europe the UV-curable polyester coatings could be used in an industrial application for the first time. Bayer AG was one of the main pioneers in the initiation of the UV curing process in the year 1967 and introduced the UV-PE method on the market [70]. In the year 1967, the first known types of resin were introduced by the company Bayer under the name designation "Roskydal UV 10" as well as by the company BASF SE under the designation "Ludopal 8275" on the market [71]. Despite the pronounced tendency of yellowing, in the early days of UV-curing benzoin ethers (benzoin butyl ether and benzoin isopropyl ether) were used as photoinitiators in coatings. The initial hysteria in the application of UV-curing putty was inhibited by a veritable paint crack wave on the chipboard. The cracks were induced by improper combination in the UV putty formulations [72]. The first super actinic fluorescent lamps, also referred to as low pressure mercury discharge lamps, had a power output of 0.5 to 1 W/cm. This resulted in a UV curing time of approx. 5 minutes. In the year 1970, high-pressure mercury vapour lamps with an output of 20 to 30 W/cm were introduced into the market resulting in an additional boost to the further market penetration of UV coatings.

The first applications were UV putty and UV primers which were applied with rolling machines. Already in the year 1969, companies such as Greco, Moralt, Novopan, Paidi, Schlingmann and Triangel applied UV primers for the coating of chipboards [69]. For aesthetic and qualitative reasons, the finish coating was carried out by means of a curtain coater machine using acid-curing coatings. The feed rates of the coating plants amounted to approx. 2 to 4 m/min as the rate of polymerisation of the polyester systems dissolved in styrene during the UV curing is very low. In the United States, the first UV plants (flat road) for wooden substrates was taken into account in the year 1971 [73].

During the process of UV coating, the unsaturated polyester resins being dissolved in styrene are disturbed by the atmospheric oxygen in the polymerisation. In the early stages, this was noticable due to the sticky surfaces. In order to avoid this, low amounts of paraffins have been added to the polyester resin. The irradiation (pre-gelation) of a polyester coating with super actinic fluorescent lamps (mercury low-pressure lamps) initiates a much slower polymerisation due to the low specific power of the lamps. A perfect paraffin mirror is formed on the surface after approx. 60 to 90 seconds. If a mercury high-pressure UV lamp is applied directly, the paint would polymerise so rapidly that the mercury mirror as a protection against atmospheric oxygen is not to be formed. Prior to the UV-curing, a variety of fluorescent lamps with a low intensity of radiation were used in a UV gelling station in order to pre-gel a paraffin-containing UV coating up to the formation of a paraffin mirror [74]. The paraffin mirror on the surface of the coating simultaneously reduces the evaporation of styrene and blocks against the disturbing influence of the atmospheric oxygen in the subsequent UV-curing. Only upon the pre-gelation of the coating film by means of fluorescent lamps, the coating was cured for about 30 seconds with high-pressure mercury lamps with a specific power output of approx. 30 W/cm.

In comparison to currently known methods, the long UV exposure time of at least 30 seconds is long. Modern UV processes require only a few seconds of UV irradiation until curing depending on the speed of the conveyor belt and performance of the UV lamps.

At the end of the 1970ies, the photoinitiator 2,2-dimethoxy-2-phenyl acetophenone (benzil dimethyl ketal) with improved features such as enhanced reactivity, good storage stability, less tendency to yellowing and an optimal price performance ratio in comparison to benzoin ethers was established. Until today, the benzyl dimethyl ketals have maintained an important role in the formulation of polyester resin systems dissolved in styrene.

A rapid industrial application of the radiation curing occurred at this time in the European furniture industry and panels industries as compared to the classically cured, unsaturated polyester clear coats the procedure was cheaper as well as more rational. Thus, one entirely could dispense with the peroxide hardener [75]. In the year 1975, approx. 5,000 metric tons per year of polyester resin have been processed already in Europe. This mainly resulted in about 8,000 to 10,000 tonnes per year of UV putty with an enhanced content of filler. At that time, more than 100 UV curing plants were in operation in Europe. In the year 1976, there were 75 plants for the coating of wood as well as 15 plants for the coating of wooden doors in the USA [76]. The end of the aera of UV crapers was at 1976/1977, when the pigmented wave in Germany was gone. The new developments in the Northern European region concerned itself with transparent UV-rolling coatings for open-pored wood effects. Already after a short time, it was realized that the new developments based on styrene-containing unsaturated polyester resins only were operational to a limited extend since the low UV reactivity which is reflected in low feed rates of the coating plant (approx. 2 to 3 m/min) was an inhibiting factor for the productivity. In addition, the high proportion of styrene in the UV coatings led to a swelling and destruction of the rubberized coating rollers [77]. For this reason, these systems only were used to a limited range of applications such as in the panel industry and door industry. The target to make the UV-curable process faster and more efficiently as well as the improvements of the properties of the coatings such as adhesion, flexibility, reduction of the film yellowing and optimization of the storage stability of the finished formulations of the coatings resulted in the acryl functional epoxy resins and polyester resins. In 1974/1975, the first commercial resins have been offered in the commodities market. In parallel, more powerful UV lamps with a power output of 80 W/cm and with reflectors were introduced at the end of the 1970ies.

The next steps dealt with the replacement of styrene by monomer containing acrylic resins. It was in the early days, when skin irritations occurred due to the applied monomers when handling these coating systems. At the beginning of the 1980ies, the first UV coating systems without low-molecular monomers (reactive diluents) were established in the door industry. However, the applied topcoat systems contained 15 to 25 % organic solvents in order to control the application viscosity. The breakthrough in the furniture industry only was possible with the supply of appropriate pickling dyes/formulations which do not result in the inhibition of the polymerisation process and dealumination of the UV coatings under UV rolling coatings [78]. The feed rate of the coating lines amounted to approx. 15 to 20 m/min. At that time, generally the front surfaces of the furnitures additionally were coated with cellulose nitrate coatings (CN coatings) or 2C PU coatings after the UV coating. The corpus parts were coated with UV primers and UV topcoats. The launching of the photoinitiators hydroxy cyclohexyl phenyl ketone (HCPK) and 2-hydroxy-2-methyl phenyl propanone (HMPP) at the beginning of the 1990ies enabled the formulation of low-yellowing UV coatings with very good UV reactivities [79]. Under the brand name "Irgacure 184" and "Darocur 1173", these photoinitiators performed their triumphal procession against benzil dimethyl ketal (BDMK) and acetophenone diethyl ketal (ADEK).

Pigmented coating systems

The euphoria of the launching of UV curing polyester coatings in the year 1967 later was marred by the fact that the UV curing of pigmented polyester coating systems was not possible since the photoiniators and UV lamps existing at that time do not allow a UV curing in deeper layers. In the early 1970ies, The European raw materials industry as well as the major paint manufacturers tried to produce white pigmented UV coatings with the specific incompatibilities [80, 81]. Both wanted to avoid the application of UV-absorbing pigments such as titanium dioxide since both do not allow an UV curing. The patent specification of the company Vianova Kunstharze AG describes how photoinitiator containing unsaturated polyester resin solutions result in white covering coatings [82] due to the specific incompatibilities by combining with cellulose acetobutyrate or cellulose nitrate, copolymers of vinyl chloride, vinyl acetate, vinyl alcohol, highly dispersive silica and fillers such as talc, dolomite, kaolin. More patents dealt with the application of metal sulphides (molybdenum disulphide) which only weakly absorb in the UV range, certainly weaker than the applied photoinitiators, and thus enable a curing [83].

Until the late of 1970ies, titanium dioxide containing coatings successfully could be cured in a mono-cure system by means of the application of new UV lamps with higher intensity of the UV radiation and by means of derivatives of thioxanthone in combination with tertiary amines (synergist) as photoinitiators [84]. Methyl thioxanthone with methyl diethanolamine as synergist was used as a photoinitiator. Methyl thioxanthone absorbs UV light in the longwave range (max = 384 nm) and thus may absorb enough UV radiation in the transmission range of the titanium dioxide. These initiators only are effective in unsaturated acrylate systems dissolved in reactive diluents. The disadvantage is the necessary acceleration of amine resulting in a significant yellow colouring of these systems which also tends to progressive yellowing due to exposure to light [85]. Additionally, unsaturated polyester resins dominated the furniture industry at that time, and these coating systems could not be cured by means of thioxanthones. At the beginning of the 1980ies, the focus lied on the application of magnesium titanate and zinc sulphate as white pigments. In parallel, new UV lamps with an enhancement of the spotlight power output to 120 to 200 W/cm were developed. Despite the fact that both pigments have a limited absorption in the longwave UV range in comparison to the titanium dioxides anatase and rutile, magnesium titanate and zinc sulphide are applied restrictedly due to their insufficient properties [86].

Only the implementation of a new photoinitiator class, the acyl phosphine oxides (2,4,6-trimethyl benzoyl diphenyl phosphine oxide) brought a significant progress in UV curing of fully pigmented white UV curing systems [86, 87]. In styrene-containing unsaturated polyester resins as well as in unsaturated acrylic resins, the TMPO containing wood coatings are characterized by a good UV reactivity, low tendency to yellowing and good storage stability of the formulations. At the beginning of the 1980ies, the photoinitiator mainly has been applied in the Italian furniture industry. Despite of some contradictionary statements, until the late of the 1980ies the TMPO mainly was applied in the so-called dual cure coating process (combination of peroxide hardening and UV curing) for fully pigmented wood coatings [88]. In the year 1988 it was assumed that three quarters of more than 50 coating systems in Italy were operated with peroxide coating while one quarter of these coating systems were operated with UV coating (mono-cure systems) [85, 89]. Also, combinations of the polyaddition (polyisocyanates) with UV curing were interesting during the curing of fully pigmented wood coatings. In the year 1986, a further boost was achieved by introducing gallium doped mercury high-pressure bulbs with an emission at wavelengths up to 420 nm. In such a system, a coating roller as well as a curtain coater machine in an excess pressure cabin were available in order to prevent dust entrance. A belt band dryer for a dust-free evaporation or pre-gelation, respectively, is applied to the evaporation and pre-gelation [90].

Table 3.24: Important milestones in the development of raw materials, coatings and processes of UV coatable wood coatings

1967	First unsaturated polyesters for the application in radiation curing are established in Europe ("Roskydal UV-10" from Bayer AG, "Ludopal 8275" from BASF SE).
1968	The company Hildebrand Maschinenbau GmbH introduced the first industrial UV plant under the "Triangel" project name for the levelling of chipboards into the market.
1968	Non-pigmented wood furnishing coatings with paraffin containing, styrene-type unsaturated polyester resins
1969	Establishment of transparent UV rolling fillers for the chipboard industry. These firstly caused crack formations in the coating due to inappropriate formulations of the coating (composition of the filler).
1970	Pigmented primer coatings were pre-gelled in the dual core-process (cobalt/peroxide) and subsequently and commonly UV cured with transparent mono cure-UV topcoats.
1972	UV gloss polyester of styrene type for high-gloss and mat thick film systems with low content of allyl ether as a classical gloss polyester ("Roskydal UV 300")
1975	Establishment of the first industrial prepolymers based on acrylate (epoxy acrylate and polyester acrylate).
1977	Urethane acrylates based on TDI ("Roskydal KL5-2442")
1979	Aliphatic urethane acrylates for abrasion resistant parquet coatings. Water emulsifiable unsaturated polyester resin with good options of recycling
1982/1983	Establishment of a new low yellowing and UV reactive class of photoinitiators of α-hydroxyl alkyl phenones ("Darocur 1173" und "Irgacure 184").
1983	Photoinitiator TMPO (2,4,6-trimethyl benzoyl diphenyl phosphine oxide) for the curing of fully pigmented white UV mono cure coatings without yellowness features ("Lucerin LR 8728", the present "Lucerin TPO").
1985	Large-scale breakthrough in the application of dual core curing (peroxide/UV curing) based on unsaturated polyester resins in Italy
1987	Establishment of a Vacumat process for the UV coating of strips, round rods, panels, profiles, edges and ribs.

UV curing was performed in a subsequent UV dryer. In such or similar coating machines, more than 1 million m² furniture surfaces were coated with opaque pigmented UV wood coating systems in Italy in the year 1986 [85].

At the beginning of the 1990ies, in Europe pigmented combination constructions in the rolling process increasingly were applied in the panel industry, furniture industry and door industry. Different coating technologies (water-borne pigmented 1C rolling primers, pigmented UV rolling primers, transparent UV topcoat) were applied in a correct combination with each other in order to achieve economically white opaque coatings with enhanced chemical resistances. Furthermore, bis-acyl phosphine oxides were established in the coating industry in the middle of the 1990ies.

At the same time, one was intensively engaged in the electron beam curing in the furniture industry. In the years 1960 and 1962, the theoretical physical fundamentals already have been described extensively by *Charlesby* and *Chapiro* [70]. The fundamental details on the electron beam curing of coatings are described by Hoffmann in the year 1966, by Meyer-Jungnick in the years 1967/1968 as well as by Tawn in the year 1968 in numerous reports [91–95]. It became clear very quickly that electron beams can harden fully pigmented

Continuation Table 3.24

1987	Establishment of gallium-doped UV lamps. First amine modified polyether acrylate.
1992/1993	Dispersions of acrylated polyurethanes for water-borne wood coatings.
1993	Market launch of bis-acyl phosphine oxides for an efficient curing of pigmented UV coatings under the brand name "Irgacure 1700" and later "Irgacure 819".
1996	Low viscous urethane acrylates, free from reactive thinners
1999	Dual cure hardener with isocyanate groups and acrylic groups in a molecule
2000	Establishment of the first low-maintenance and mat UV finishing coatings with a solid content of approx. 100 %, free from reactive thinners and without application of organic solvents for the roller application
2001	Enhanced application of grooved rubber rollers for the topcoat application of UV roller coatings (transparent/pigmented) in order to obtain coatings with an optical appearance of curtain coatings.
2002/2003	Application of fully pigmented and opaque coatings in a straight UV process with grooved rollers. Mainly white, black and particular colour shades.
2004	Renaissance of IPL laminate flooring (IPL = Indirect printed Laminate) in a rolling process with UV-curing coating systems
2008	Enhanced utilisation of digital printing in the furniture industry and in the laminate industry with UV-curing inks and UV rolling coatings
2011	Worldwide launching of the UV-Hot Coatings-process for a direct coating of melamine resin coated MDF plates with high-gloss UV coatings in the rolling process
2012	Implementation of UV-LED in the industrial UV-curing of wood-based materials for the deep curing of pigmented UV primers
2014	High-End-high gloss and „Super mat" UV coating surface by Calandar Coating Inert Technology of Hymmen for the coating of melamine resin coated MDF plates
2015	Implementation of the UV-Coating-Melacoat-technology (O. Nolte GmbH) for the direct coating of melamine resin coated decors consisting of wood-based materials
2017	Presentation of the Digital Lacquer Embossing-procedure (DLE) for the generation of three-dimensional wood surfaces with UV-curing coating systems (Fa. Hymmen)

coatings [96]. This promoted the parallel continued development of electron beam curing as well as UV curing systems. However, in the year 1968 the industrial application of electron beam curing was blocked by high investment costs of 125,000 to 140,000 Euro [96]. In addition, there were very diverse technical problems which enabled an industrial application only a few years later. The titanium foil as an emission window for electrons to the curing substrate/coating also is worth mentioning which had to be replaced after at least 30 minutes [96].

It took several years until the first coating line for the electron beam curing of fully pigmented coatings in the door industry could be taken in operation on an industrial scale. In the year 1973, the company Svedex in Varsseveld/NL with its door brand (Dextüra) converted from acid-curing paint systems to electron beam coatings and is one of the pioneers in the field of electron beam curing up until today (see Chapter 7.5). The procedure advantages that were drawn up at that days are valid up today [97]:

- The space requirement of a conventional drying for acid-curing coatings could be reduced by 80 % by means of EB;
- The consumption of coatings could be reduced by 30 %.

- Any problems of emission with organic solvents, since EB coatings consist of non-volatile components by approx. 100 %.
- The production quantities of coated doors could be tripled per unit of time.
- The repair quota could be reduced by approx. 50 % by means of a more scratch-resistant and harder coating surface in the in-house transport.

As the second company in Europe, Theuma in Bekkevoort/Belgium applied the advantages of fully pigmented EB coatings. This company mainly produces white, fully opaque doors until today [97]. In the USA, in the year 1977 the first EB plant for wood coating was taken in operation by the company PPG. The electron beam curing could not assert oneself until now in the curing of wood substrates due to the higher investment costs of equipment and due to the still existing technical restrictions such as inertisation, adjustment of low degree of gloss and enhanced investment costs. Important milestones in the development of raw materials, coatings and processes of UV coatable wood coatings are listed in Table 3.24.

3.1.6.2 Chemistry and technology

Radiation curing wood coatings have a structure very similar to conventional coatings. The main difference is that the resin contains functional groups which polymerise under exposure to ultraviolet light (UV) or ionizing radiation (EB). Thus, a three-dimensional insoluble network is established.

As a special feature on the UV curing, so-called photoinitiators are required as special additives. A distinction is made between the radical and ionic polymerisation for UV initiated systems. The radical polymerisation based coating systems exclusively are applied in the coating of wood. The process of electron beam curing does not require initiators because the high radiant energy generates a sufficient number of radicals for the spontaneous process of

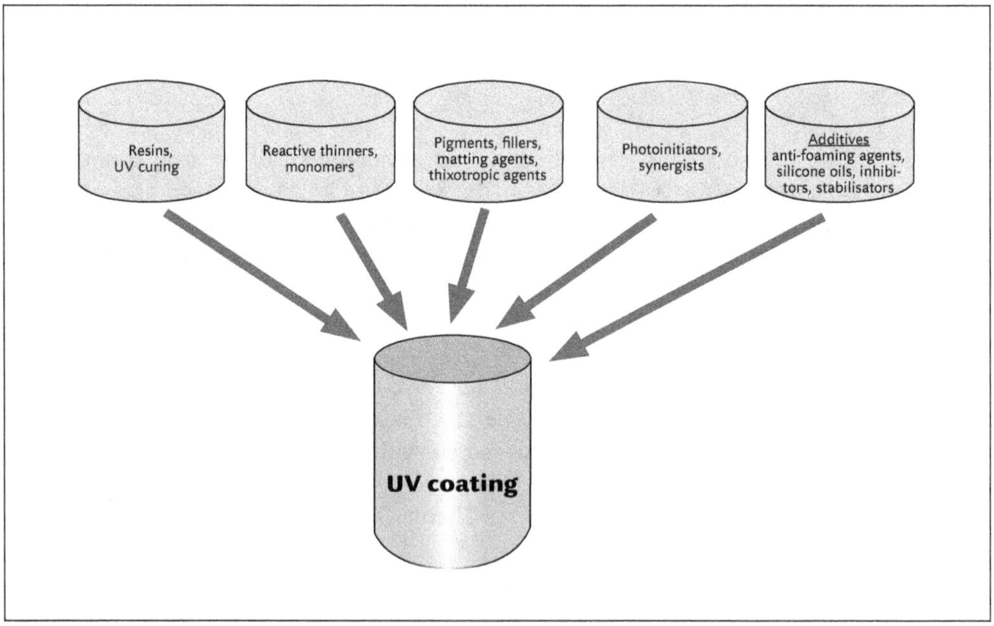

Figure 3.27: Composition of radiation curing coating systems [101]

polymerisation [98, 99]. Further differences between EB system and UV systems can be found for example in the excitation mechanism and the energy input [100].

3.1.6.3 Components of UV curing coating formulations

UV curing coating systems that are applied in the wood industry and furniture industry for example contain several of the shown components in specific percental parts.

3.1.6.4 Polymerisable resins

The essential element of radiation curing coatings is the resin[17]. These are resins which contain built-in C=C double bonds. Essentially, the resins determine the fundamental properties of the cured coating film such as abrasion resistance and chemical resistance, flexibility, hardness, tensile strength and flexural strength [100]. The molecular weights (M_n) usually are between 500 and 2500 g/mol. Further enhancements may result in an increase in viscosity which normally is undesirable for the processing.

Especially, low molecular weight polycondensates and polyaddition products are suitable for the manufacture of reactive resin [101]. The most important UV resins for the radical polymerisation in the field of wood coating are:

- Unsaturated polyesters
- Epoxy acrylates
- Polyester acrylates
- Polyether acrylates
- Amine-modified polyether acrylates
- Urethane acrylates

Apart from the acrylic acid-free unsaturated polyesters, the UV resins shown are classified as eurymeric acrylates (derived from Greek 'euru' equal size and 'meris' equal part) [102]. These

Esterification with acrylic acid to bulk ester

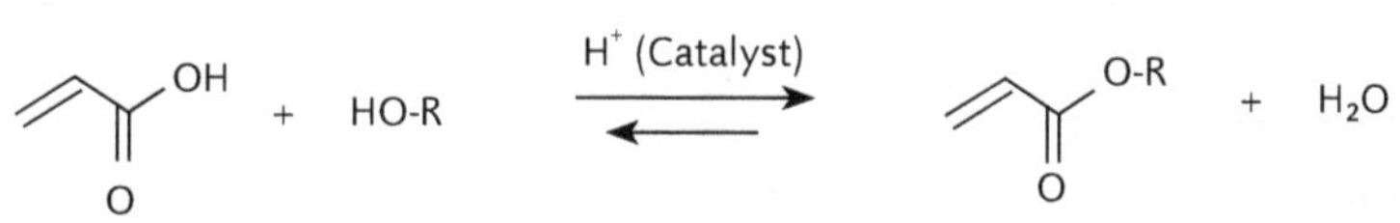

Equilibrium reaction	Yields strongly dependent on reaction conditions
Catalysis	Sulfuric acid, p-toluene sulfonic acid et cetero (1 %)
Stabilisation	Phenothiazine, hydroquinone, methyl hydroquinone, tert.-butyl cresol
Solvent (entrainer)	Toluene, cyclohexane
Manufacturing method	Azeotropic distillation (separation of water)
Temperature range	< 130 °C
Bulk ester contains:	Entrainer, acrylic ester, acrylic acid, alcohol, incompletely esterified acrylates among polyoles, secondary products such as diacrylic acid products

Figure 3.28: Process parameter for the esterification with acrylic acid [103]

17 In practice, the resins for UV curing coatings often are refered to oligomers or prepolymers, respectively

are resins with a high molecular weight and a broad molecular weight distribution. More than 70 % of acrylic esters applied in wood coatings are manufactured by esterification of alcohols with acrylic acid. The esterification reaction is an equilibrium reaction in favour of the acrylic acid ester. Mainly strong acids such as para-toluene sulfonic acid and sulphuric acid are applied as catalysts. Stabilisers are applied during the manufacturing process as well as in the subsequent reconditioning processes in order to avoid a premature polymerisation. Conventional stabilisers are hydroquinone, methyl hydroquinone, di-tertiary butyl phenol and phenothiazine which are added in low concentrations (ppm range). Entrainers such as cyclohexane and toluene are applied to a better stripping off of the reaction water. The process temperatures amount 110 to 130 °C. Subsequent to the polycondensation reaction, the polycondensates undergo an extensive refinement process in order to eliminate the catalysts and still free acrylic acid. Possible process variants are shown in Figure 3.28. In terms of occupational hygiene, the thus reclaimed acryl esters have a low potential for skin irritations and sensitizations. The resins generally are odour-neutral. The removal or neutralization of catalysts, respectively, reduce possible secondary reactions in the cured coating film such as changes in colour shades.

Unsaturated polyesters

The unsaturated polyesters (UP) based on maleic acid (Chapter 3.1.5) are an important class of resins. These UPs were the first radiation curing resins which found a widespread use in

Reconditioning of bulk ester

Procedural options for the removal of secondary components

a) **Distillation**
solvent, acrylic acid, volatile components, depending on the distillation conditions

b) **Precipitation or neutralisation, respectively, with alkalines**
acids

c) **Washing and distillation**
water soluble components (catalysts, acrylic acid), solvent

d) **Distillation and epoxy addition**
solvent, reaction (polyaddition) of free acrylic acid with epoxy resins

e) **Filtration**
solid contents by means of salts, polymerisates

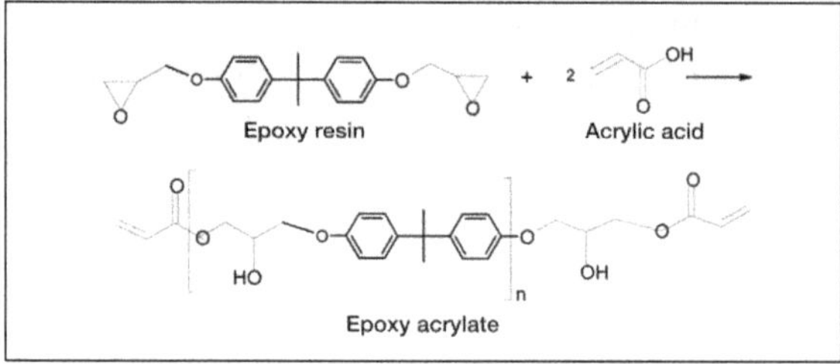

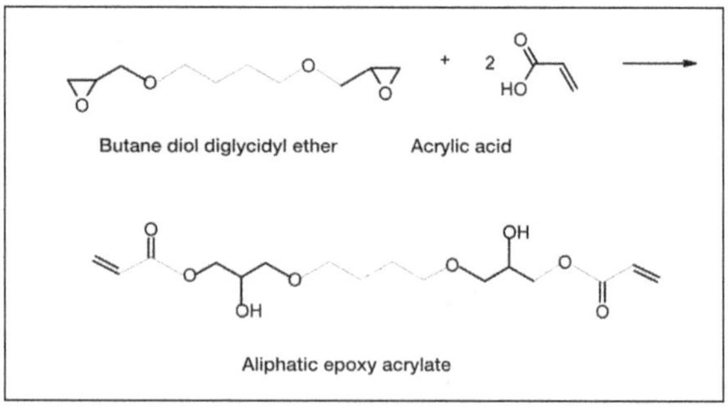

Figure 3.29 (left): Procedural options for the removal of secondary components in the polycondensation process [103]

Figure 3.30 (right above): Synthesis and structure of epoxy acrylates

Figure 3.31 (right below): Aliphatic epoxy acrylate

the industrial application technology on wood-based materials with styrene as a reactive diluent and photoinitiators. The unsaturated polyesters reveal a low reaction rate in UV light limiting their application in modern systems with high throughput rates [104]. This was the result of the stabilisation of radicals on the phenyl ring of styrene during the free-radical chain polymerisation. The stabilisation results in a termination of the radical chain polymerisation since the termination probability of the polymerisation prevails to the chain growth of the polymer (Chapter 3.1.5). Today, the unsaturated polyester primarily are used in applications in the wood sector in South Europe, Eastern Europe and South America. These are wood coating formulations which contain so called reactive diluents such as dipropylene glycol diacrylate or organic solvents in addition to styrene. Styrene-free formulations also are applied. The actually still essential unsaturated polyester resins for UV applications are gloss polyesters which contain a certain ratio of built-in fumaric acid ester groups, trimethylolpropane monoallyl ether groups and trimethylolpropane diallyl ether groups. With this types of resins, hard and non-sticky coating surfaces can be achieved at acceptable speed of the conveyor belt (5 to 10 m/min per UV lamp).

Acrylic resins

The development of highly reactive resins for UV curing led to new prepolymers[18] which are derived from the known classes of resins. With these resins, a double bond is incorporated into the resin matrix by means of acrylic acid or its derivatives.

During the UV curing, the acrylic double bond reacts at least ten times faster than the double bonds in the classic UP systems. Tailor-made resins for the respective requirement profiles can be produced with a special selection of reactants. In the common linguistic usage, these resins are not quite correctly referred to as acrylic resins. These resins contain terminal double bonds which are produced by a conversion of acrylic acid or methacrylic acid with low molecular polycondensates or products from polyaddition reactions. According to the type of these base bodies one subdivides in epoxy acrylates, polyester acrylates, polyether acrylates and urethane acrylate.

Epoxy acrylates

Epoxy acrylates primarily are produced by conversion of liquid epoxy resins based on bisphenol A diglycidyl ether with acrylic acid (see Figure 3.30). The corresponding diacrylate is no longer flowable at ambient temperature.

These are highly reactive resins which result in hard and chemical resistant films. Due to the free hydroxyl groups, the resins exhibit an enhanced wetting behaviour especially for pigments such as titanium dioxides.

In addition, epoxy acrylates are characterized by a very good adhesion strength on various wood substrates and plastic films as well as an excellent chemical resistance. Vicoelastic, hard and highly reactive UV coatings can be formulated [105]. Epoxy acrylates containing an aromatic main structure tend to yellowing ('heat yellowing') when exposed to heat. The high viscosity of the epoxy acrylates is disadvantage. The resin has to be diluted with organic solvents and/or reactive diluents (monomers) for application-technical processing.

In contrast, the aliphatic epoxy acrylates based on butanediol glycidyl ether have a low viscosity. The conversion product of the aliphatic diglycidyl ethers with acrylic acid has a viscosity of approx. 1,000 mPas and can be processed without reactive diluents or solvents [103].

18 In technology, cross-linkable oligomers often are referred as prepolymers.

Table 3.25: Impact of mono-carboxylic acids on the viscosity of epoxy acrylates [105]

Monocarboxylic acid	–	Octanoic acid	Lauric acid	Stearic acid	Ricinoleic acid
Length of the carbon chain	0	8	12	18	18
ICI plate & cone viscosimeter at 50 °C in Pas	> 10	4.2	3.8	3.1	3.3

Table 3.26: OH terminated polyester precursor for the subsequent conversion with acrylic acid [108]

Raw material components	**Weight proportion in [%]**
Adipic acid	25.1
Phthalic acid	25.4
Ethylene glycol	9.4
Trimethylolpropane	17.4
Triethylene glykol	22.7
Acid value	approx. 4
Hydroxyl value (OHZ)	approx. 193
Molecular weight (Mn)	approx. 965
Viscosity 23 °C (70 % in toluene)	4–5 dPas

Due to its hydrophilicity, the resin partially is dilutable with water. Thus environmentally friendly wood coatings with spray viscosity are formulated.

A further possibility for the production of epoxy acrylates with a lower viscosity is the partial replacement of acrylic acid by mono-carboxylic acids such as lauric acid and ricinoleic acid [105]. In addition to the reduction in viscosity due to the incorporation of mono-

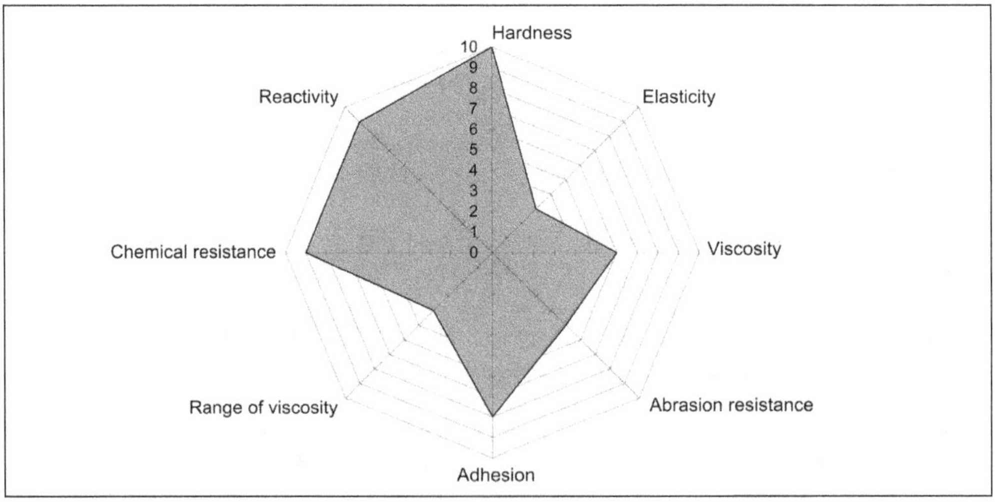

Figure 3.32: Range of properties of aromatic epoxy acrylates [103]

carboxylic acids, the elasticity and adhesion are improved tremendously[106]. Table 3.25 illustrates the influence of different mono-carboxylic acids in the synthesis of epoxy acrylates.

Polyester acrylates

The classical polyester acrylates from polyols and polycarboxylic acids are built up in a two-step process or three-step process[107] (see Figure 3.33). In the first step, a hydroxyl group-containing polyester (polyester polyol) is produced as a precursor. In the second stage, this polyester is reacted with acrylic acid while in the third stage the excess acrylic acid is converted with low molecular weight epoxy resins.

As an example, a recipe for an OH terminated polyester step for the wood coating is illustrated in Table 3.26. The polycondensation will continue as long as an acid number < 4 mg KOH/g and a OH number of 193 mg KOH/g is achieved[108].

In the second reaction stage, the polyester precursor is converted with acrylic acid in the presence of toluene as the entrainer, p-toluene sulfonic acid as a catalyst and hydroquinone as a stabilizer at a temperature of 100 to 110 °C until the acid number is achieved. Subsequently, the esterification catalyst is neutralized with a tertiary amine. An equivalent amount of epoxy resin is added according to the remaining acid number in the third reaction step. The epoxy resin is converted with excess acrylic acid at a temperature of approx. 110 °C until the acid number < 5 mg KOH/g is achieved. Afterwards, the inserted entrainer is distilled off under vacuum. The non-volatile components in the thus produced polyester acrylates exhibit a friction of more than 98 %. Due to the variation possibilities in the selection of raw materials as well as in the selection of manufacturing processes, extremely hard or visco-plastic polyester acrylates can be synthesized. The production of hard and simultaneously elastic resin types is possible only to a limited extent at a solids content of approx. 100 % (non-volatile content) and a low viscosity[103]. In comparison to the aromatic epoxy acrylates and polyether acrylates, the range of viscosities of polyester acrylates available in the market is in the middle level. Polyester acrylates usually require the addition of reactive diluents or solvents in order to adjust the application viscosity. Generally, this class of resins is characterised by an optimum price performance ratio. The polyester acrylates available on the market have a wide range of application properties.

1st reaction step
Synthesis of a hydroxyl-terminated polyester

2nd reaction step
Conversion of the hydroxyl-terminated polyester with acrylic acid

3rd reaction step
Conversion of the residual acrylic acid with low-molecular epoxy resins

Figure 3.33: Three-stage process for the production of polyester acrylates

Due to esterification reactions as well as trans-esterification reactions, acrylic acid esters of low molecular polyols are formed as by-products with the production of polyester acrylates. Depending on the chemical composition, the low molecular esters of acrylic acid also referred to as residual monomers may give rise to occupational hygiene problems in the processing. Moreover, there is a risk that the residual monomers penetrate into the substrate and cannot be cured if the residual monomers directly are applied on highly absorbing wood surfaces. The formation of monomers as by-products from the production of polyester acrylates is due to the equilibrium nature of the esterification reaction [105]. The acid catalysts for the esterification reaction such as p-toluene sulfonic acid also catalyse the transesterification reaction. During the acrylation of a polyester acrylate, the corresponding low molecular weight polyol acrylates fragmentarily can be split out [109]. Thus, the resins contain residual monomers in the order of 3 to 7 % by weight depending on the recipe and manufacturing process.

The formulation of UV coatings containing a low amount of monomers requires a roofing of the manufacturing process for the polyester acrylates. One possibility is the exchange of low-molecular weight polyols by ethoxylated and/or propoxylated polyols with a higher molecular weight. The application of theses polyols does not prevent the side reactions of the polycondensation reaction. The advantage is that the less volatile acrylic acid esters being formed in a side reaction have a higher molecular weight and thus a lower stimulus potential than the low molecular compounds being formed theoretically such as trimethylolpropane mono-acrylate, di-acrylate or tri-acrylate. From one example of the literature it is known that any volatile diol acrylates could be established chromatographically within the application of ethoxylated and/or propoxylated polyols in the manufacturing of polyester acrylates. Acrylates only derived from trimethylolpropane were detected beneath a proportion of 0.8 % by weight [110].

Polyether acrylates

The demand for low-monomer and low-viscosity UV resins which do not need a further addition of monomers or reactive diluents as a viscosity reducer led to the class of polyether acrylates. The polyether acrylates have a broad range of beneficial properties and are increasingly applied in recent years. These polyether acrylates are characterized by their low

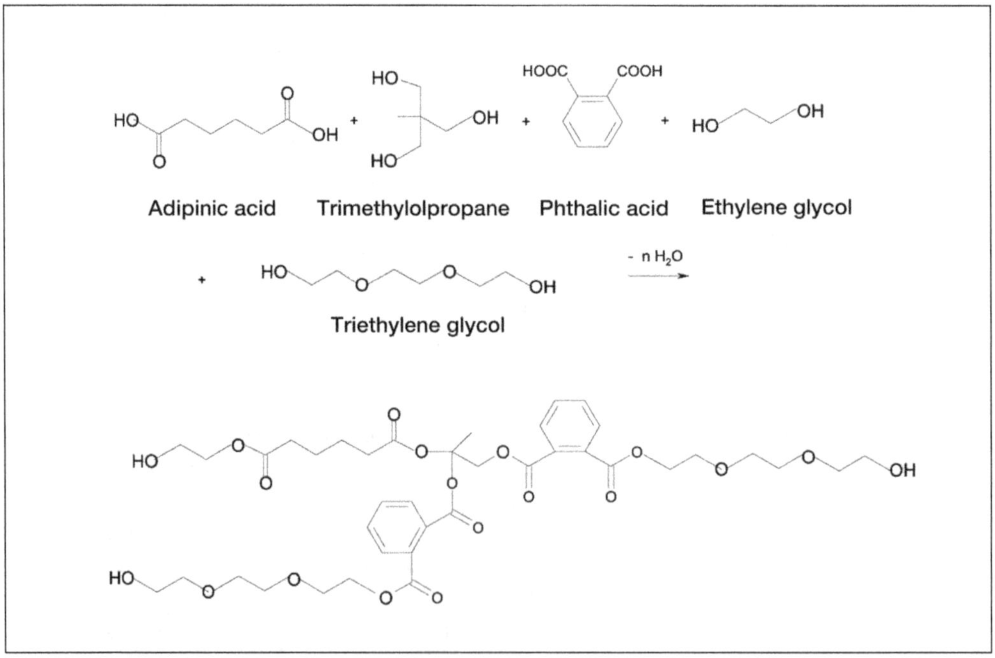

Figure 3.34: Idealised structure of a hydroxyl group terminated polyester as a precursor for the subsequent conversion with acrylic acid

viscosity and good processing characteristics at high UV reactivities. Low-monomer and practically odour-free coatings can be manufactured with these polyether acrylates. They also are applied in the dilution of epoxy acrylates, polyester acrylates and urethane acrylates [111]. In the polyether acrylates, the residual monomers thereby are avoided by application of so-called ethoxylated and/or propoxylated polyols with higher molecular weight instead of low molecular polyols. These polyetherols arise during the conversion of polyols with ethylene oxide and/or propylene oxide [112]. This results in higher molecular polyols whose aryl esters are less volatile, occupational hygienically less critical and still low viscous. In addition, poly-

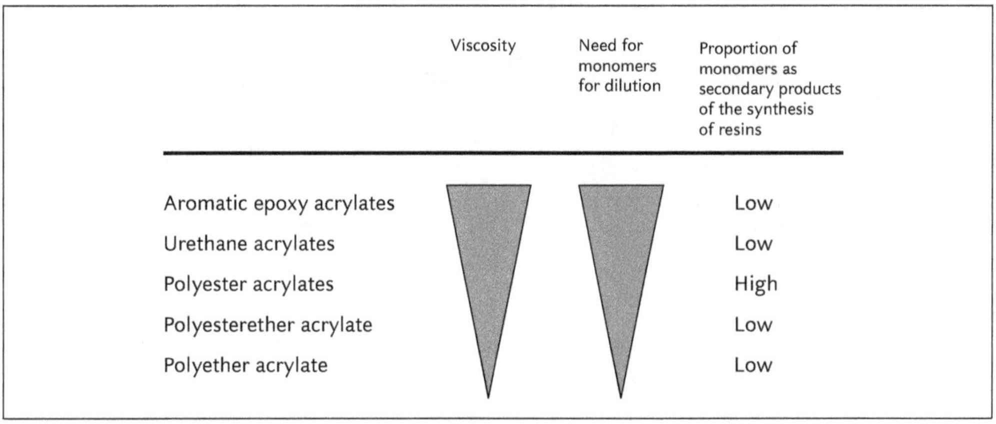

Figure 3.35: Range of viscosities of polyester acrylates in comparison to other resins

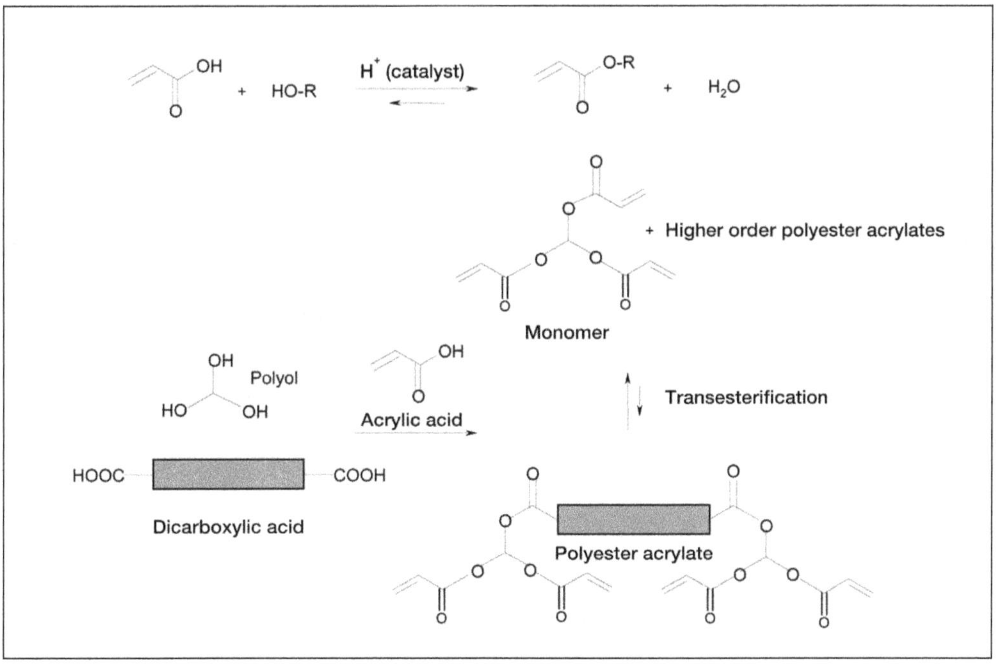

Figure 3.36: Schematic illustration of the development of residual monomers in the synthesis of polyester acrylates [103]

ether acrylates feature a low potential of skin irritations and eye irritations [113]. The polyether groups are stable under the usual conditions of esterification so that polyether acrylates cannot be degradated to monomers. At least, traces of monomers can exist in the finished product, originating from the residues of non-etherified polyols. The polyether acrylates available on the market contain trimethylol propane, glycerine or pentaerythritol as a starter polyol [113]. Difunctional products often are based on ethylene glycol, propylene glycol, hexane diol and neopentyl glycol as a basic polyole. The basic properties such as viscosity, UV reactivity and mechanical properties depend on the degree of ethoxylation and/or propoxylation.

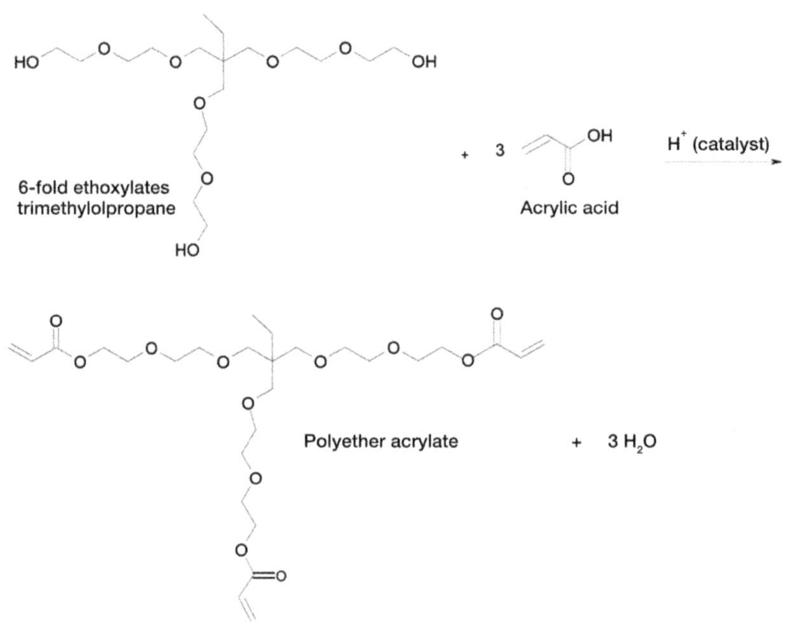

Figure 3.37: Synthesis of polyether acrylates

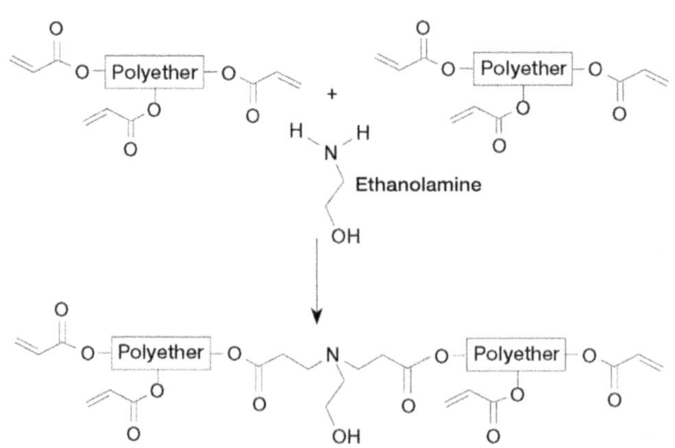

Figure 3.38: Michael adduct of a primary amine with a polyether acrylate

Table 3.27 exemplary formulation for the production of an urethane acrylate[125]

Label of the raw material	Sample weight in [mole]
Isophorone diisocyanate	0.30
Hydroxyethyl acrylate	0.70
Hexamethylene diisocyanate	0.70
Trimethylol propane	0.30
Fourfold ethoxylated trimethylol propane	0.33

Preparation of polyether acrylates

Polyether acrylates are prepared for example by azeotropic esterification of polyether polyols with excess acrylic acid under acid catalysis in the temperature range from 80 to 130 °C.

Conversion with amines

As explained later, some photoinitiators and especially the economically interesting benzophenone only have an effect in conjunction with tertiary hydrogen donors (synergists) such as tertiary amines. The separate addition of amine to the coating formulation leads to disadvantages such as odour nuisance, migration and impairment of the properties of the coating film [114, 115]. For these reasons, functional acrylic resins are converted with primary or secondary amines. The incorporation of amines in the resin matrix is performed via the Michael addition under formation of a secondary or tertiary amine. When using a primary starting amine, the formed secondary amine can continue reacting to the tertiary amine in the same way [116, 117]. The conversion with ordinary amines leads to an increase in the molecular weight as well as in the viscosity and to an increased functionality of the middle double bond per molecule. This has a positive effect on the reactivity and oxygen inhibition. The Michael addition already quantitatively expires without catalysis beneath a temperature of 60 °C [116, 117].

Urethane acrylates

Urethane acrylates often are produced by the reaction of polymeric or monomeric isocyanates, polyols and hydroxy acrylates [118]. The variety of products is very large, since aliphatic or aromatic isocyanates as well as polyols and polyester polyols can be applied [119–122]. This is, why there exist no market-dominating resins, but a large number of speciality chemicals in this class of resins. Highly functional, hard, chemical-resistant resins based on aromatic isocyanates or difunctional, soft and elastic urethane acrylates can be synthesized selectively. An exemplary formulation for the production of urethane acrylates is presented in Table 3.27.

Generally, small traces of dibutyltin dilaurate (0.02 %) as a reaction catalyst and 0.01 % of phenothiazine as a stabilizer are added to the provided reaction mixture [123]. The reaction temperature maximally amounts 60 °C until the isocyanate content of the reaction mixture is below a value of 0.1 %. Urethane acrylates are used very rarely for the normal furniture coating due to the high prices of their raw materials. Urethane acrylates as combination resins for the industrial coating of parquets and veneer rollers are not to be forgotten since urethane acrylates positively affect the flexibility, adhesion and especially the abrasion behaviour. The urethane acrylates known in the market are characterized by their relatively high viscosity. The main reason may be due to formation of hydrogen bonds between the urethane groups [103]. For this reason, the urethane acrylates have to be adjusted with organic solvents, reactive diluents or low viscosity polyether acrylates. Driven by the automotive in-

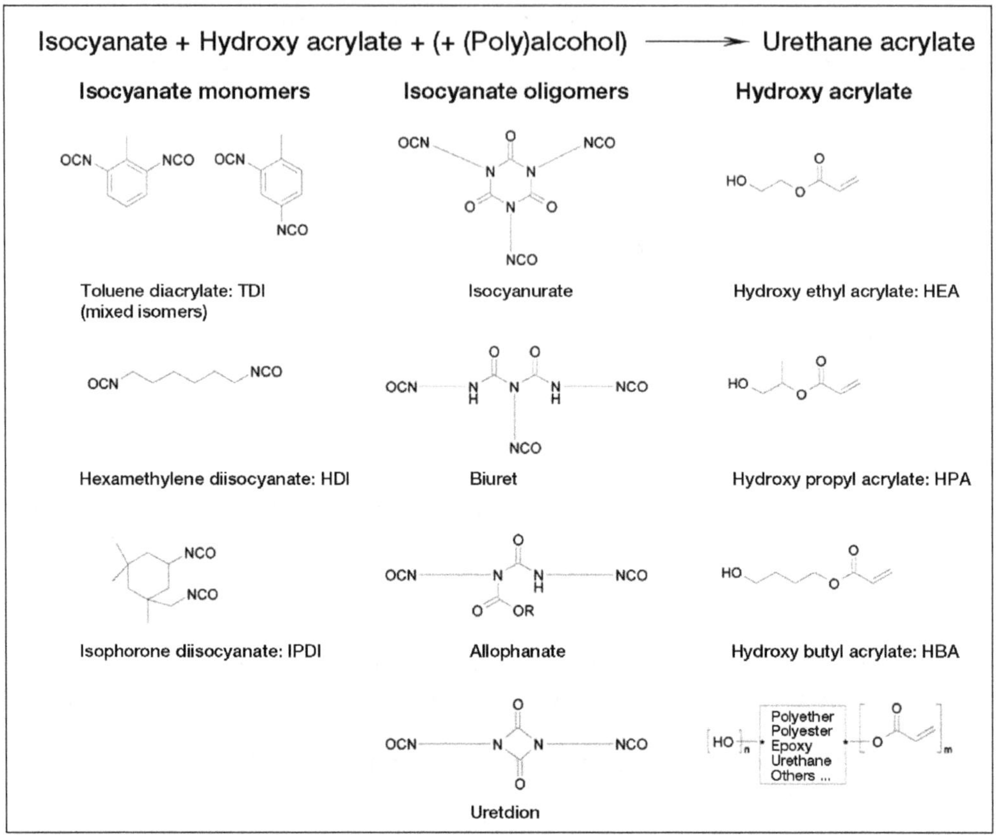

Figure 3.39: Raw material diversity for the production of urethane acrylates

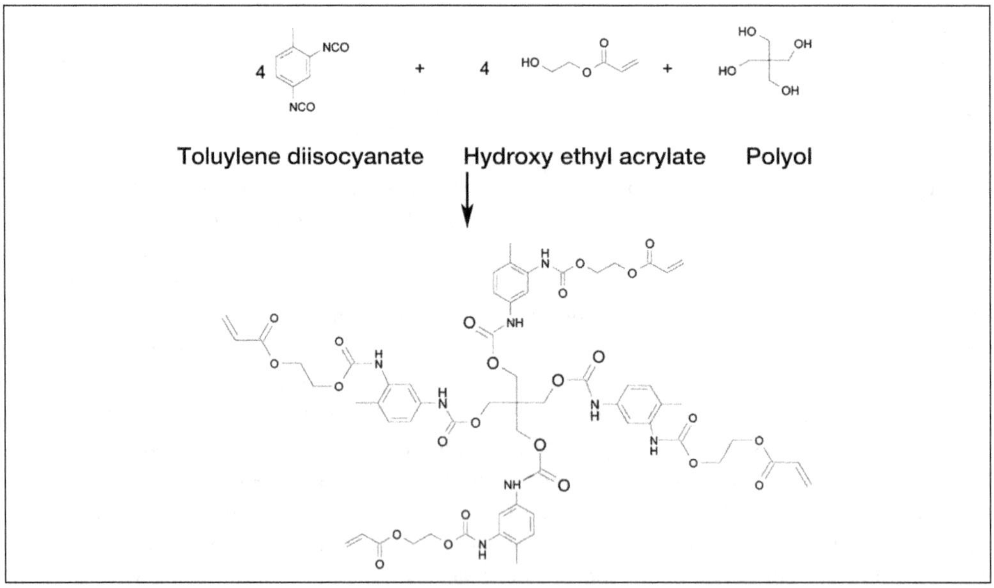

Figure 3.40: Highly functional urethane acrylate based on aromatic isocyanates[125]

dustry, for some time it has been tried to produce low-viscous urethane acrylates free of reactive diluents. Urethane acrylates feature a very low content of residual monomers due to the manufacturing process (polyaddition). It should be considered that urethane acrylates based on aromatic isocyanates are degradated photochemically. The resulting quinoid structures with their conjugated double bonds lead to mild up to very strong discolouration effects [124].

Isocyanato acrylates for the dual cure application

This younger class of resins has arisen from the necessity to combine the advantages of the polyaddition reaction with the advantages of the polymerisation reaction (UV curing) in a resin structure. The application of dual cure systems is known from patent specifications from the years 1977 and 1978 [126, 127]. In recent years, the potential of this application has been rediscovered which again is reflected in numerous patents [128, 130]. The chemical structure of isocyanato acrylates contains free isocyanato groups as well as double bonds. Figure 3.41 illustrates the chemical structure of a commercially available isocyanato acrylate [131]. One speaks of a dual cure system if the curing process is performed in two separate phases while the curing mechanism can be identical in both phases [132, 133]. However, different mechanisms can be applied. In practice, dual cure systems often are compared with hybrid systems incorrectly. Hybrid systems are defined as systems where the curing process is based on two different mechanisms which are performed simultaneously [132]. A simple example of the dual cure application is the pre-gelation of a purely UV curing coating system initially with a low dose of UV irradiation. In a subsequent step, the complete coating layer is cured with a high dose of irradiation. This procedure is applied in order to improve the poor matting of UV coatings specifically [134, 135]. The application of isocyanato acrylates in dual cure systems leads to a variety of combination possibilities in the formulation of wood coatings. Figure 3.42 illustrates that in the direct comparison to the classic 2C polyurethane coatings (hydroxy propyl acrylate and polyisocyanates) the curing of

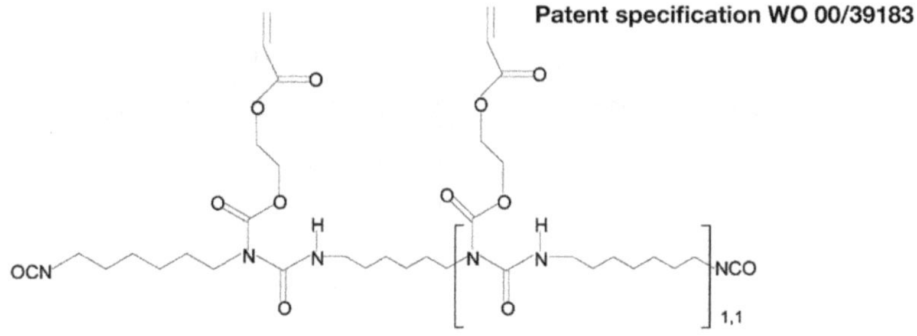

Properties

NCO value:	**15 %**
Content of DB:	**2 mol/kg**
Molecular weight:	**800 g/mol**
Solid (nfA):	**100 %**
Viscosity:	**1200 mPas (23 °C)**
Residual HDI:	**< 0.5 %**
Odour:	**Very low**

Figure 3.41: Chemical structure of isocyanato acrylate ICA 9000 [131]

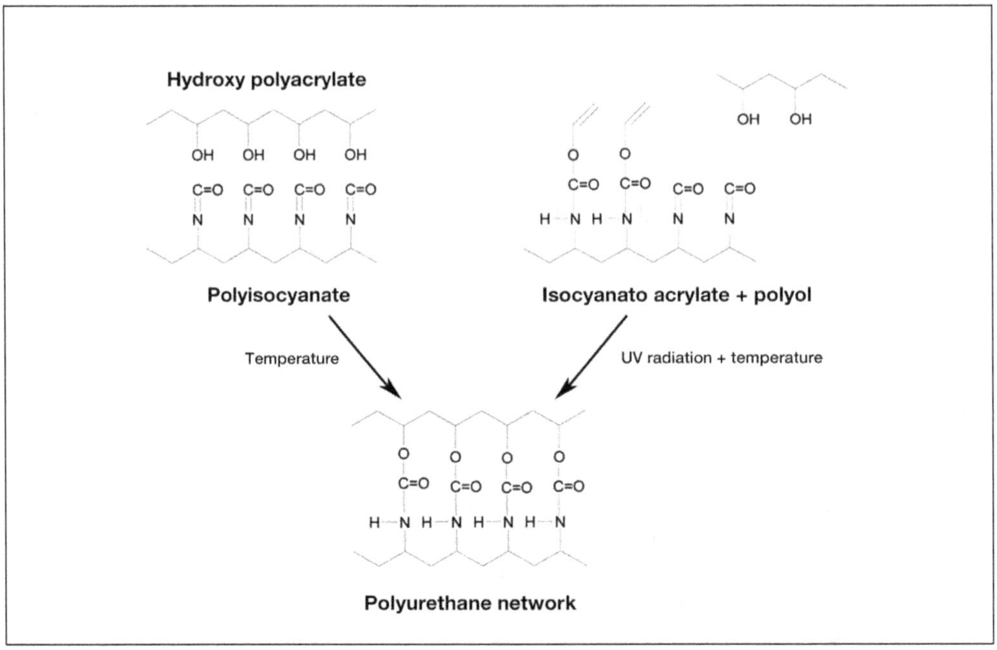

Figure 3.42: Polyurethane networks produced from polyisocyanates and hydroxy polyacrylates or by means of UV curing and polyaddition of isocyanato acrylates with polyols [136]

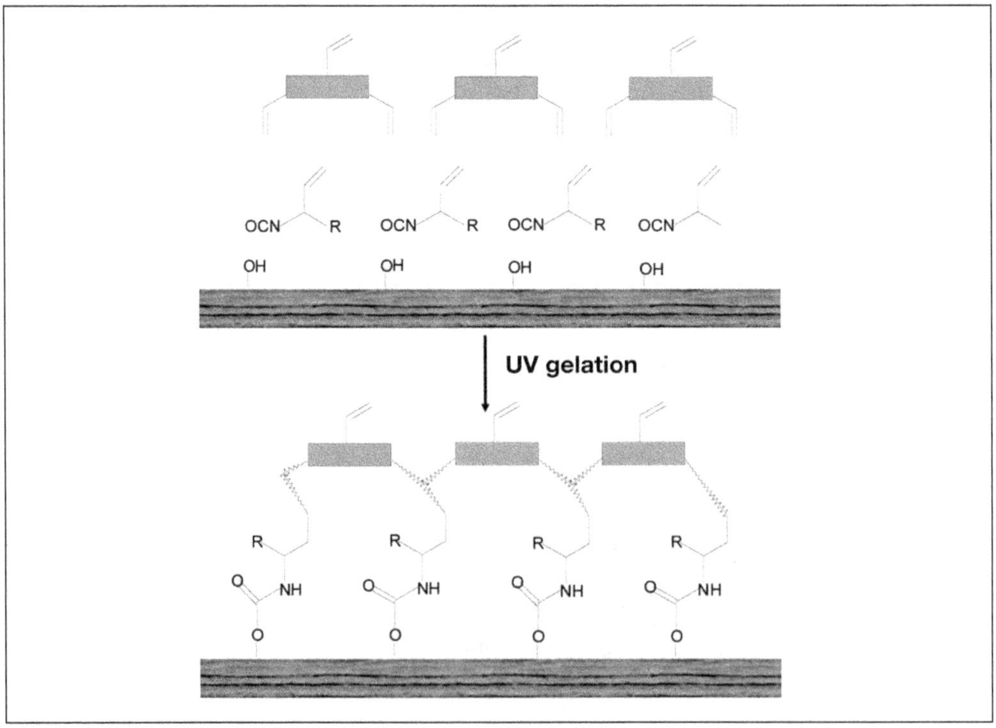

Figure 3.43: Dual cure adhesion promoter

isocyanato acrylates with for example polyols lead to the formation of stable polyurethane networks [136]. Depending on the composition or reactivity, respectively, of the functional groups in the coating formulations appropriate pot-life have to be taken into account.

Wood coating systems of this kind are applied on three-dimensional substrates in such cases in which a complete curing has to be achieved in those shadow zones which cannot be covered by UV lamps. The curability of opaque pigmented coating are further advantages in comparison to only UV curing systems (mono-cure). In addition, isocyanato acrylates are applied as barrier layers and adhesion bridge on sucking or ingredient-rich wood surfaces since the isocyanato acrylates may react with the moisture and the functional groups of the wood components in the depth and in the boundary layer of substrate/coating film. Figure 3.43 illustrates the application of parquets [137]. The barrier effect of the primer against the migration of acyl esters of the subsequently following and monomer-containing topcoats was confirmed by extraction experiments from *Beck* and *Bruchmann* [136]. Figure 3.44 illustrates the extraction results of formulations with and without a barrier primer (dual cure primer) on fine-pored and coarse-pored wood. The extractable components significantly could be reduced below a value of 200 mg/m² (oak) and 5 mg/m² (beech).

Polymerisable monomers

Apart from the polyether acrylates and some polyester acrylates, in general, the most acrylated prepolymers cannot be applied in concentrated form due to their high viscosity. The resins must have to be adjusted on the required processing viscosity by means of suitable

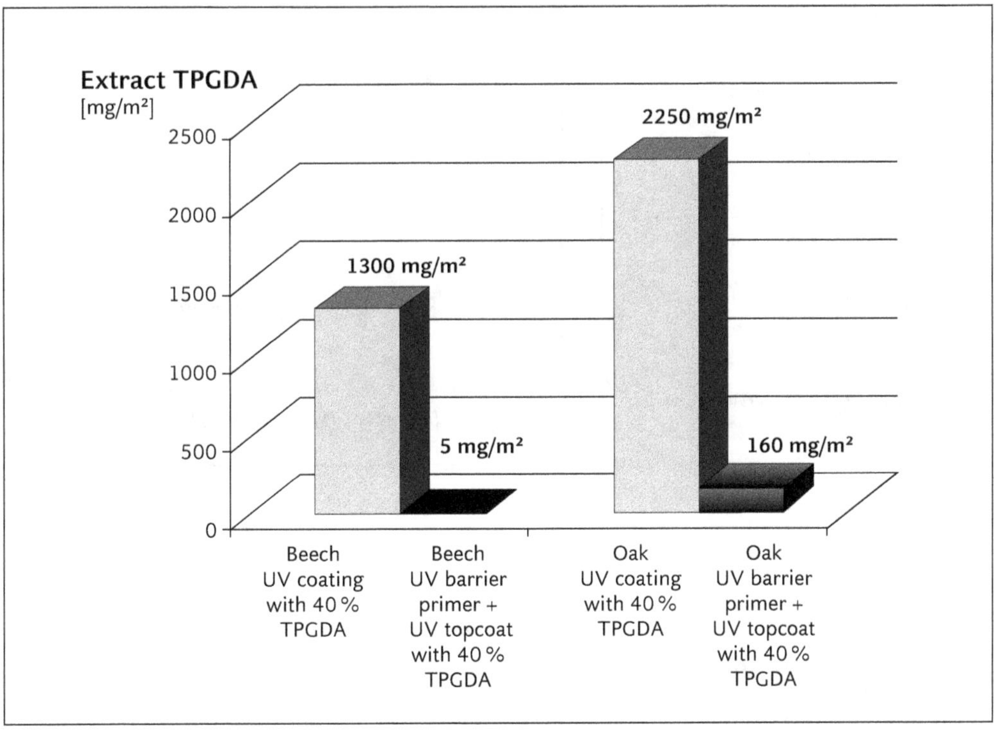

Figure 3.44: Extraction results of the UV roller structures with and without Dual Cure barrier primer in comparison [136]

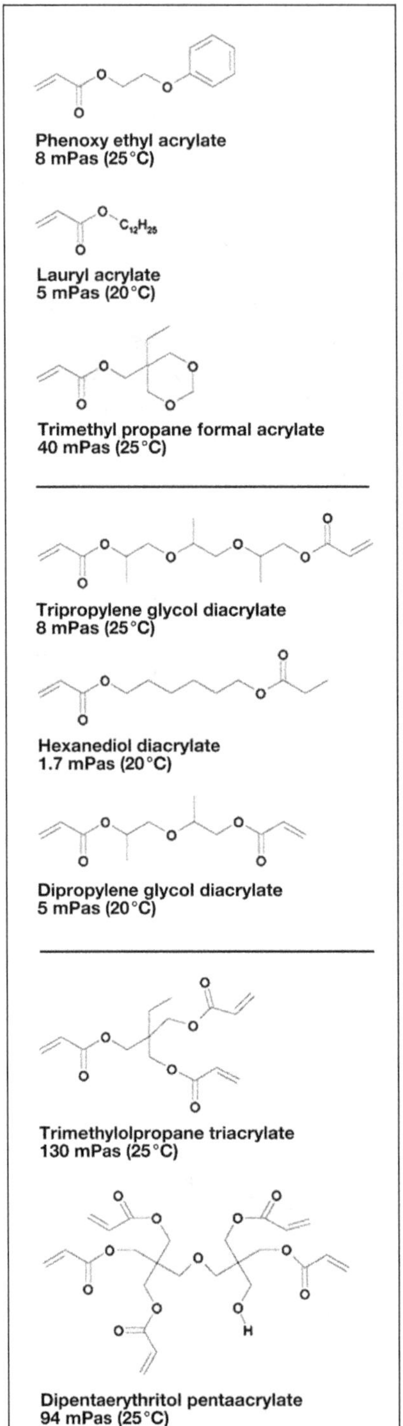

Figure 3.45: Acrylate monomers for wood coatings[125]

Functionality of reactive diluents
mono-, di-, tri-, polyfunctional

Reactivity ——— **increases** ———→

Hardness ——— **increases** ———→

Flexibility ——— **decreases** ———→

Figure 3.46: Impact of the functionality of acrylate monomers on the coating properties[139]

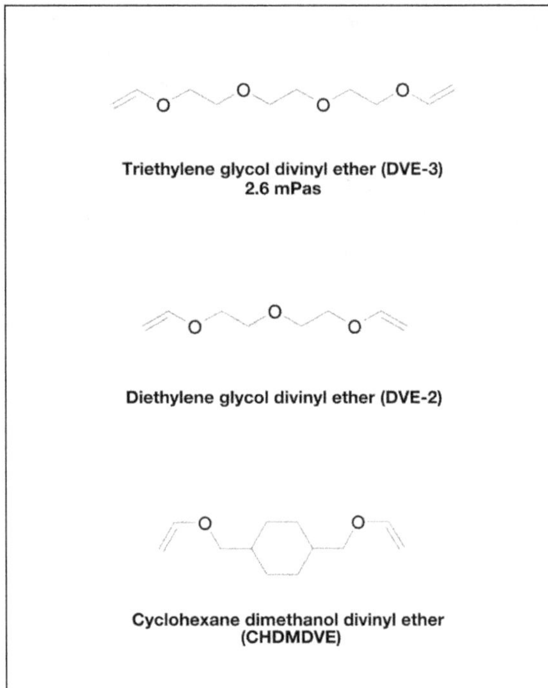

Figure 3.47: Important types of divinyl ether[140]

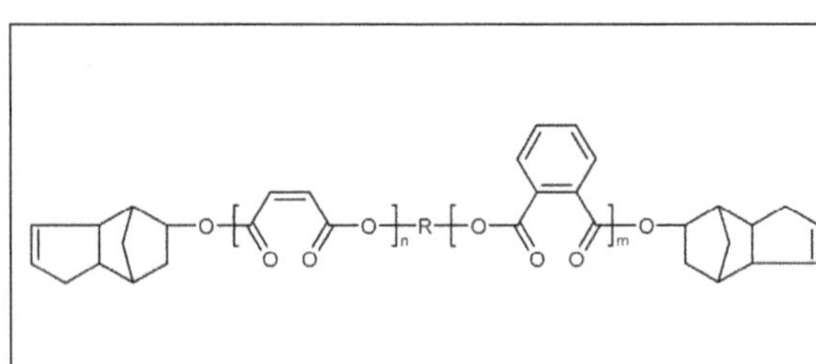

Figure 3.48: Dihydro dicyclopentadienol (DCPD)modified unsaturated polyester[147–149]

substances. These raw materials are to be known as reactive diluents or monomers since these are incorporated in the coating film during the process of UV curing. These are exactly defined substances which are indicated in the EINECS list. These raw materials are classified as stenomeric acrylates (derived from the Greek 'steno' similar to narrow and 'meris' similar portion) since these are exactly defined chemical substances with low molecular weight and a very narrow molecular weight distribution [102]. These substances usually contain less than 1,000 ppm of acrylic acid and less than 2,000 ppm of processing solvents [102]. Mono-functional, di-functional, tri-functional as well as higher functional compounds are available for the adjustment of the viscosity (Figure 3.45).

The technically most important reactive diluents can be found in the class of acrylic acid esters. The following criteria may play a role in the selection:
- High reactivity
- Good dissolving power
- Low viscosity
- Low vapour pressure
- Low toxicity
- Low skin irritation/sensitization
- Functionality

The most important acrylate monomer is tripropylene glycol diacrylate (TPGDA) followed by hexanediole diacrylate (HDDA) and dipropylene glycol diacrylate (DPGDA). The most important trifunctional acrylate is trimethylolpropane triacrylate (TMPTA). Monofunctional compounds are preferred to reduce the viscosity of the coating formulation, to flexibilize or to adjust the gloss level of the coatings. The application of phenoxyethyl acrylate in the targeted reduction of the degree of loss by incompatibilities with unsaturated polyester resins in the range of 3 to 5 % is known from the patent literature [138]. Difunctional or polyfunctional substances lead to higher branched networks. Thus, the resulting coatings are harder and more brittle in the trend (see Figure 3.46).

Wood coatings may contain between 1 and 60 % of reactive diluents. The essential function is to reduce the viscosity of the coating in order to waive the application of organic solvents. Fundamentally, the viscosity of the monomers in particular increases with increasing molecular weight, the polarity, the stiffness of the molecule and the functionality [103]. The dilution effect, the tolerability, the volatility and the odour of the monomers decrease with the increasing molecular weight of the monomers.

Due to the potential of skin irritation as well as due to the labelling obligation of the most acrylic reactive diluents, for several years many trials were done to replace these diluents with non-labelled vinyl products, for example. The reactive diluents are manufacturers by alkaline (alkali hydroxides) catalysed addition of acetylene on alcohol. Subsequently, the vinyl ethers thus received are purified by distillation. This procedure is known as „Reppe vinylation" [140]. Diethylene glycol divinyl ether (DVE-2) as well as triethylene glycol divinyl ether (DVE-3) are the most important vinyl ether for the coating of wood. DVE-3 with a viscosity of 2.8 mPas is low viscous in comparison to many other of the well-known acrylic ester monomers. The viscosity of the tripropylene glycol diacrylate is approx. four times higher (approx. 11 mPas) than the viscosity of diethylene glycol divinyl ether.

In combination with electron-poor maleate/fumarate groups, the electron-rich vinyl ether groups are predistinated for a radical, alternating (1:1) copolymerisation [141]. *Bartlett* and *Nozaki*[142] were the first who made a proposal in the year 1946, that a donator-acceptor-complex could be the possible reason for the alternating structure of copolymers. In the year 1984, Olson et al. [143] demonstrated scientifically by analysing ^{13}C NMR spectra that the mechanism takes place by means of a donor-acceptor complex. The electron-poor maleate/fumarate groups are known as donator monomers. The radical copolymerisation easily occurs between electronical different monomers. The behaviour of the copolymerisation reaction can be explained by means of the Q-e scheme named after *Alfrey* and *Price* [144].

In order to have a better toxicological profile as well as a liberty of labelling, special maleate-containing or fumarate-containing UV resins for alternating copolymerisation with vinyl ethers (e.g. triethylene glycol divinyl ether) were launched into the market by the middle of the 1990ies [145]. These UV resins were marketed under the designation 'non-acrylics'. It turned out rather quickly that the reactivity of these products within the range of the applied quantities for UV roller coatings (standard approx. 5 to 25 g/m^2) has no sufficient UV reactivity or were not even UV curing [146], respectively. Only as from wet application amounts of 80 to 120 g/m^2 well curing coating films could be achieved. The problem of the lower UV reactivity of maleate-containing or fumarate containing polyesters in the copolymerisation reaction with vinyl ethers can be increased by incorporating dihydro dicyclopentadienol (DCPD), whose activated methylene group caused an enhancement of the UV reactivity, also in low wet application quantities of 10 to 20 g/m^2 and a UV reactivity of more than 30 m/min with a mercury high-pressure lamp (80 W/cm) [147–149].

In addition to the maleate-containing or fumarate-containing unsaturated Polyesters, vinyl ethers also can be converted with acrylic esters. Experimentally, the best turnovers of the vinylic double bonds and resistances could be achieved with epoxy acrylates and polyether acrylates with an admixture amount of $\leq$ 10 % DVE-3. Higher values of the admixture amount of DVE-3 resulted in a higher proportion of residual monomers and poor chemical resistances [140]. Further investigations have shown that the less UV reactive unsaturated polyester resins have a higher rate of incorporation of the vinylic groups in comparison to the reactive acryl esters [103]. The complete incorporation of the vinyl ether units depends on its functionality, quantity as well as UV reactivity of the copolymerisation partner [103]. Fundamentally, the formation of acetaldehyde with an acid-catalytic hydrolysis [150] is a disadvantage of the vinyl ether-containing formulations. The formulations on the basis of vinyl ethers should be prescribed acid-free and anhydrous, if possible. The vinyl ethers can undergo a cationic polymerisation in the presence of acids. The coating formulation can be stabilized better in the presence of bases. In addition, vinyl ether-containing formulations tend to yellowing of the coating film.

It is well-known that the low molecular monomers can produce substance-induced skin irritations [104] due to their small molecular size, volatility and their distinctive chemical activity if handled improperly. Also, animal experiments confirmed the potential of skin irritation of a number of these substances, often using the method developed by Draize in animal experiments with albino rabbits [151, 152]. According to the results, the monomers are classified in five groups between non-irritating to highly irritating. Depending on the proportional weight of the UV formulations as well as to the classification, the application of monomers can lead to the labelling obligation of the preparations.

The technically most important monomers are well-known for their toxicological properties and meanwhile also frequently verified. The trend exhibits that the toxicity and the potential of skin irritation decrease with increasing molecular weight. This is first and for most relevant for primary irritations[19], but not for the increase in sensitisation of the skin in response to the immune reflex [113]. In order to reduce the volatility of such substances and therefore any health hazards, the monomers often are ethoxylated or propoxylated. Thereby, the viscosity often increases only marginally [153].

Current tendencies of development of the recent years for the avoidance of low-molecular monomers in UV coatings deal with high-solids (up to 100 %), simultaneously low-monomer coating systems or with the application of water as a diluent [154]. The application of water has been established in many areas of wood industry and furniture industry despite the additionally necessary procedural expenses such as a prior installing an evaporation area and the waste water treatment. The end is still not foreseeable.

Photoinitiators

Since the radiation energy of the industrially applied UV lamps is insufficient for a directly homolytic fission of the carbon-carbon double bond of the applied UV resins and monomers, so-called photoinitiators are required. The binding energy of a carbon-carbon double bond amounts 6.3 eV [98].

According to the Planck formula:

$E = h \cdot \nu$ Eq. (1) with $\nu = c/\lambda$ Eq. (2)

E = Energy (eV)
h = Planck's constant, $h = 6.626 \cdot 10^{-34}$ J s
ν = frequency (1/s)
c = speed of light, $c = 2.998 \cdot 108$ m s^{-1}
λ = wavelength (nm)

This corresponds to a wavelength of approx. 200 nm.

With regard to the Planck's relation, a wavelength of 200 nm is required for the cleavage of a carbon-carbon double bond. The conventional UV lamps emit these wavelengths only to a small degree. In addition, the emitted radiation energy in the range of 200 nm is absorbed by the air, and ozone is formed. The photoinitiators or photosensitizers[20] absorb the light (photons[21]) of the more emission-rich and long-wave area of the UV lamp. Thus, radicals are formed which trigger the radical polymerisation of the reactive components of the UV formulation [155, 156].

The following requirements may be posed to a photoinitiator:

- High reactivity,
- Good dark storage stability,
- Thermal stability,
- Free of yellowing and reasonable priced,
- Good solubility in the coating,
- Odourless and non-toxic.

19 An inflammatory reaction of the skin (accumulation of tissue fluid causing oedema) which is reversible.

20 Photosensitizers are referred to as compounds which transfer energy absorbed in the excited state to other molecules (usually photoinitiators) which in turn form reactive intermediates.

21 A photon is an electromagnetic wave whose wavelength is located in the range of light (photos = light (Greek)) according to the definition. The shorter the wavelength of a photon, the higher is its frequency and the greater the energy of the photon.

Mechanism

The most important principles according to which photoinitiators may deliver radicals can be classified as follows:

- Radical formation by homolytic bond cleavage (intramolecular)
- Radical formation by hydrogen abstraction (intermolecular)

The photoinitiators that trigger the radical polymerisation by intramolecular bond fission (α-fragments) are the benzoin ether, benzil ketale, α-aminoalkyl phenone, hydroxy alkylphenone and the group of phosphine oxides (Figure 3.49). The α-fragments undergo a Norrish I photocleavage[22]. The radicals being formed directly are capable to start the polymerisation (Figure 3.50) [160, 161].

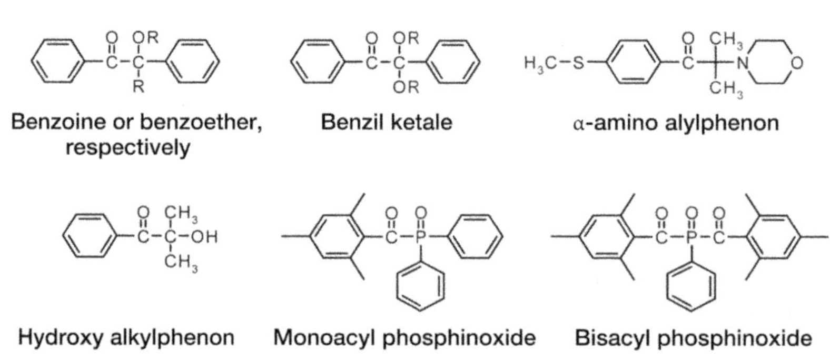

Figure 3.49: Important photoinitiators which initiate a radical polymerisation by means of a homolytic cleavage of the bond

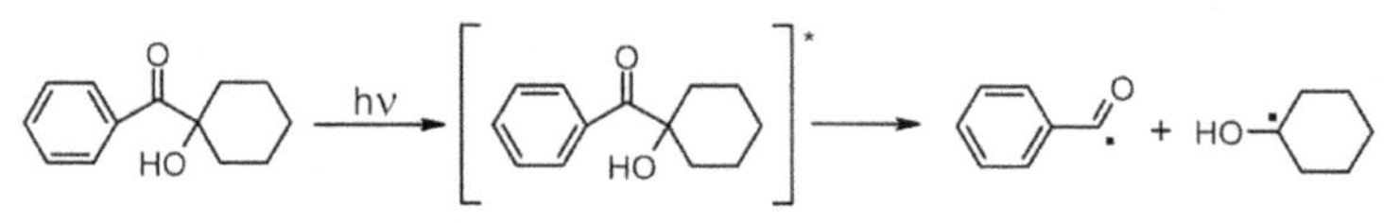

Figure 3.50: Fission exemplified by 1-hydroxy cyclohexylphenylene keton (HCPK) "Irgacure" 184, tradename of the company Ciba

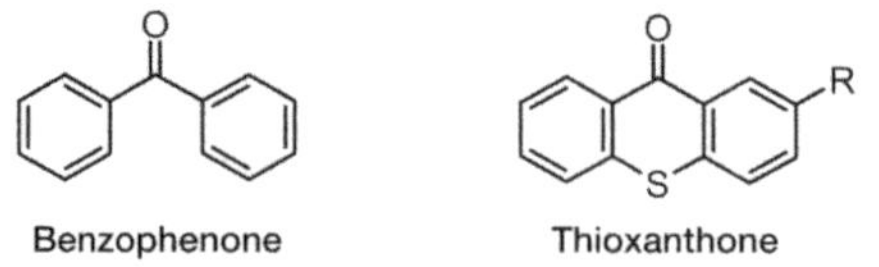

Figure 3.51: Important initiators of the class of hydrogen abstractors [160, 161]

22 Norrish Type I reaction, photochemical decarbonylation with ketones under cleavage of a C-C bond in neighbourhood to a carbonyl group.

The photoinitiators of the hydrogen abstractors include benzophenone, thioxanthone and their derivatives (Figure 3.51). These compounds require co-initiators (synergists[23]) for the formation of radicals with slightly hydrogen atoms such as a hydrogen atom in α-position of alcohols, ethers, amines or mercaptanes.

The mechanism of action of this class of photoinitiators shall be illustrated in the example of benzophenone since benzophenone is the most widely used initiator for UV curing wood coatings due to its low commodity price (Figure 3.52). Benzophenone is transformed into the excited triplet state by means of ultraviolet light. In the presence of amines, photoreduction possibly occurs under formation of exciplexes[24]. The photoreduction proceeds according to an electron transfer mechanism or H-transfer mechanism [156, 162, 144]. The excited benzophenon abstracts an α-hydrogen of the amine applied as a synergist. The formed

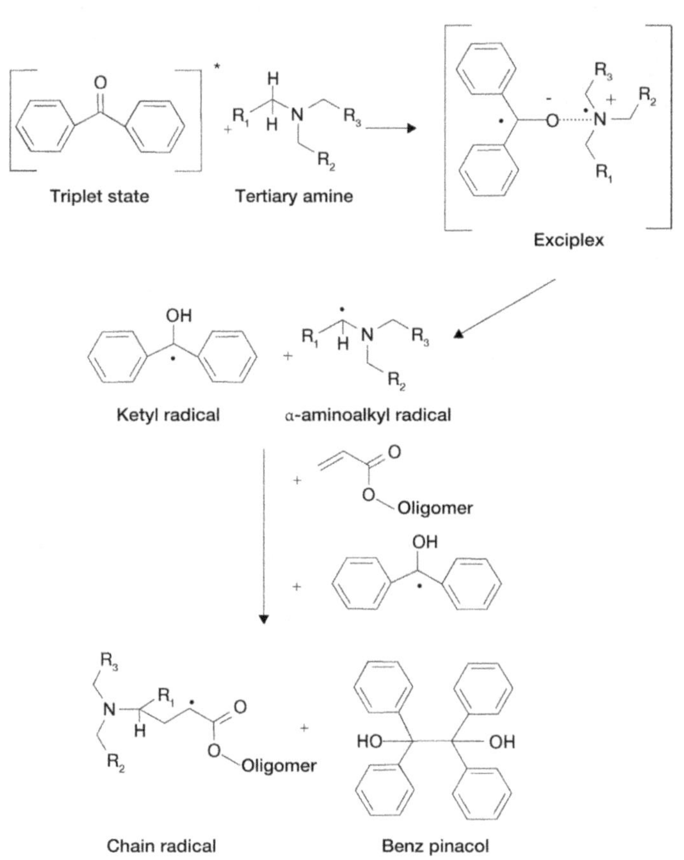

Figure 3.52: Initiation of the radicalic polymerisation exemplified by benzophenon and by a tertiary amine [101

23 Accelerators of the radical photopolymerisation which do not appear as initiators alone.

24 Exciplex ('excited complex'), an electronically excited molecule complex with a non-bonding ground state. This consist of a donor molecule and an acceptor molecule as a collision complex.

α-aminoalkyl radical is responsible for the effective initiation of the free-radical polymerisation under formation of a radical chain. The ketyl radical reacts to pinacol under dimerization, or it terminates the growing of the polymer chain [161, 163].

Photoiniators for pigmented UV coatings

In recent years, the UV curing of pigmented wood coating systems, especially white pigmented wood coating systems enormously becomes more important. Usually, photoinitiators (benzophenone, hydroxyalkylphenone) are applied in non-pigmented UV-cured wood coatings. In white pigmented coating formulations no adequate UV curing is obtained with the above photoinitiators alone, since the white pigment (titanium dioxide) strongly absorbs in the visible range (absorption maxima up to approx. 450 nm), in which the white pigment reflects and scatters the radiation.

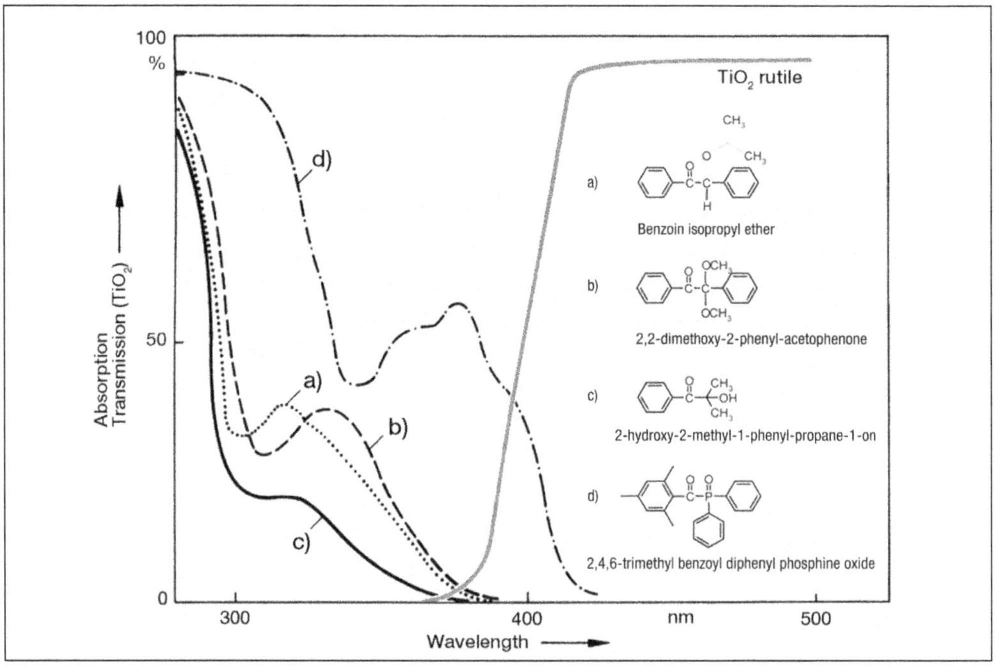

Figure 3.53: UV absorption spectra of different photoinitiators [165]

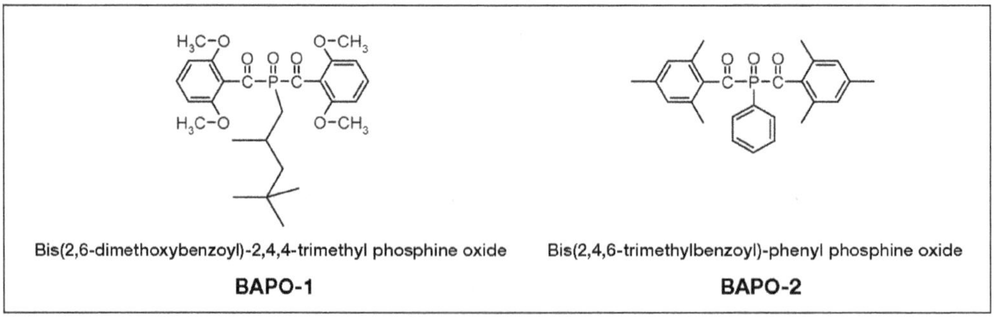

Figure 3.54: Bisacyl phosphinoxide BAPO-1 and BAPO-2 [166, 167]

Thioxanthone derivatives, α-aminoalkyl phenone, monoacyl phosphinoxides as well as bisacyl phosphinoxides are applied for a sufficient curing of white pigmented wood coating formulation with all its advantages and disadvantages. Hydroxyalkyl phenones often are applied as a combination partner. These phenones offer a good UV curing of the coating surface due to their absorption characteristics. Figure 3.53 illustrates an overlapping of the transmission curve of the TiO_2 rutile with the absorption curve of the monoacyl phosphinoxides at a wavelength of about 390 nm, so that a 'spectral window' for the UV curing process remains open [164]. The applied photoinitiators (> 400 nm) partly absorb the blue portion of the visible light and produce a yellow colour impression [164] according to the rules of the subtractive colour mixture.

Currently the most important photoinitiators for the formulation of white pigmented wood coatings are the acyl phosphinoxides (APO). These are characterized by low yellowing and light stability. In the year 1983, the monoacyl phosphinxide (MAPO) 2,4,6-trimethyl benzoyl diphenyl phosphinoxide ('Lucirin TPO') are launched on the market.

About 10 years later, in the year 1993 the photoinitiator class of bisacyl phosphinoxides (BAPO) was introduced commercially. Their absorption is further shifted into the visible spectral range. In direct comparison to the monoacyl phosphinoxides with the same pigment concentration, higher application rates are UV curable by means of bisacyl phosphinoxides. One speaks of a better deep curing process. Monoacyl phosphinoxides may deliver two radicals due to an α-clevage, while bisacyl phosphinoxides may deliver even four radicals (Figure 3.54).

The derivatives of thioxanthane and α-aminoalkyl phenones widely applied in the past exhibit a strong yellowing after UV curing which additionally is enhanced by exposure to daylight. For this reason, the above mentioned photoinitiators do not impact the formulation of white pigmented wood coatings significantly. In addition, the UV coatings containing α-aminoalkyl phenones leave a strong unpleasant smell after UV curing.

Photobleaching effect on the application of bisacyl phosphine oxides

When using bisacyl phosphinoxides it should be considered that the coating surface fades out in the daylight subsequently to the process of UV curing depending on the application concentration. Colourimetrically considered, the yellowness value decreases after exposure to daylight (Δb^*). Figure 3.56 illustrates the systematical investigation of the photobleaching effect of *Wiesing* on the example of monoacyl phosphinoxides and bisacyl phosphinoxides [164]. This process is known as 'photobleaching effect' in the literature. It is expected that the applied photoinitiator may form chromophore groups during the UV curing in the coating process. Following the UV curing step, the planes are posed to natural light, if these are not being stacked on top of the other in the production process. The natural light (UV light) splits the chromophore in fragment layer by layer. These fragments exhibit no absorption in the range of visible light [98]. The paint surfaces appear 'white' to the viewer.

3.1.6.5 Pigments, fillers and matting agents in UV curing wood coatings

Pigments

According to DIN 55943, a pigment is a virtually insoluble inorganic or organic, colourful or achromatic dye in the respective medium [169]. There are white, black and coloured pigments whose task in top coatings include above all to scatter the light, thus to make the coating film opaque and to cover the subsurface. The colour impression results from the reflection or selective or complete absorption of light [168]. Fundamentally, only those pigments are applied for UV curing coatings which do not inhibit or decelerate the photopolymerisation of the total formulation:

- Low UV absorption of the pigments, in particular in the ranges of the absorption bands of the photoinitiators (pigment absorption window) [170, 171],
- No reaction or interaction with the photoinitiator (radical scavenger).

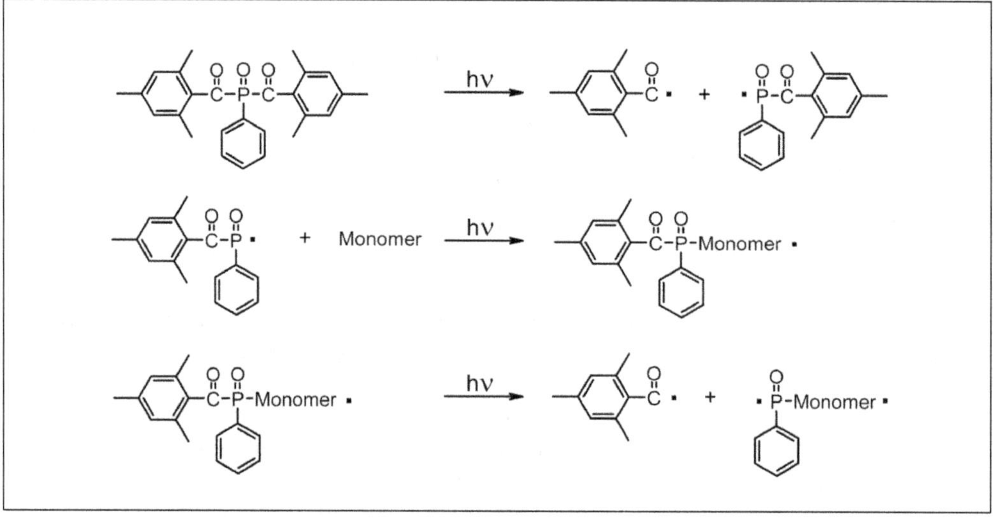

Figure 3.55: Formation of radicals with the example of "Irgacure" 819 (BAPO-2) [166]

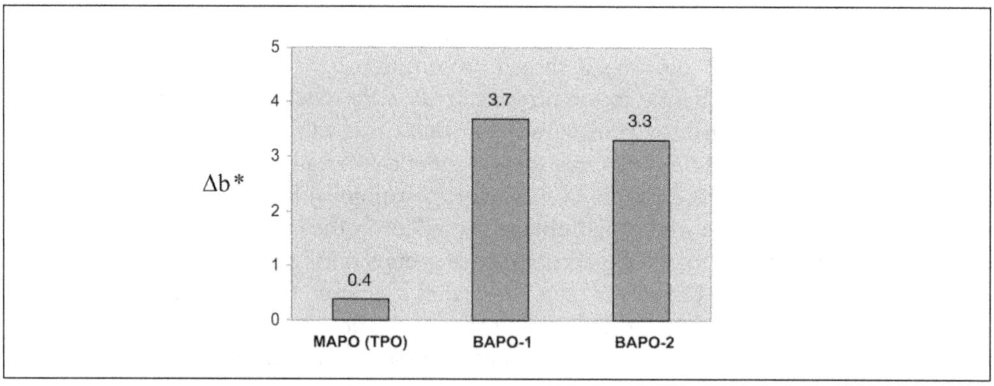

Figure 3.56: Photobleaching of white pigmented and UV cured wood coatings exposure to daylight [164]

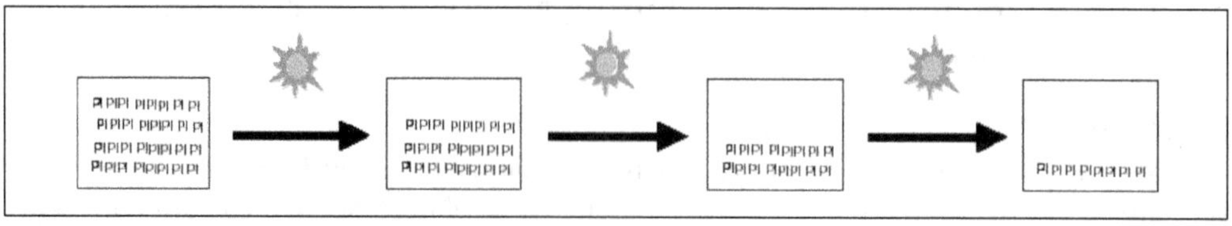

Figure 3.57: Photobleaching effect [168]

Titanium dioxide (rutile structure) is the most commonly applied pigment for UV coatings. It is used to white colouring and coverage of the substrate as well as to brighten all coloured pigments and black pigments. The main factors for the optimal formulation of coating formulations of white pigmented UV curable wood coatings are [172]:

- Type of pigment
- Particle size of the pigment
- Volume concentration of the pigment and
- The type of wavelength distribution of the applied UV light for the curing of the paint film

In comparison to titanium dioxide (rutile structure), alternatives such as magnesium titanate, titanium dioxide (anatase structure) or zinc sulphide have not been able to establish themselves due to the disadvantages in the incorporation and worse covering capacity of the substrate. For example, UV curable coating formulations containing zinc sulphide as a pigment feature strongly tend to yellow (photochemical yellowing) after exposure to natural light (UV light). For example, titanium dioxide (anatase structure) photochemically is more active than the rutile structure. For this structure, the anatase modification has to be coated with aluminium oxide and/or silicon dioxide. The knowledge of the absorption behaviour of the pigments and photoinitiators is fundamental for an optimal photoinitiation of UV curable coating formulations. The processing of pigmented UV coatings always requires the adjustment of pigment and photoinitiators since most of the pigments absorb the UV radiation which is required by the photoinitiator to initiate the free-radical polymerisation.

Figure 3.58 illustrates the absorption characteristics of different white pigments. The figure shows that not only the chemical composition of the pigments (titanium dioxide), but also the crystal modification is responsible for the optical behaviour.

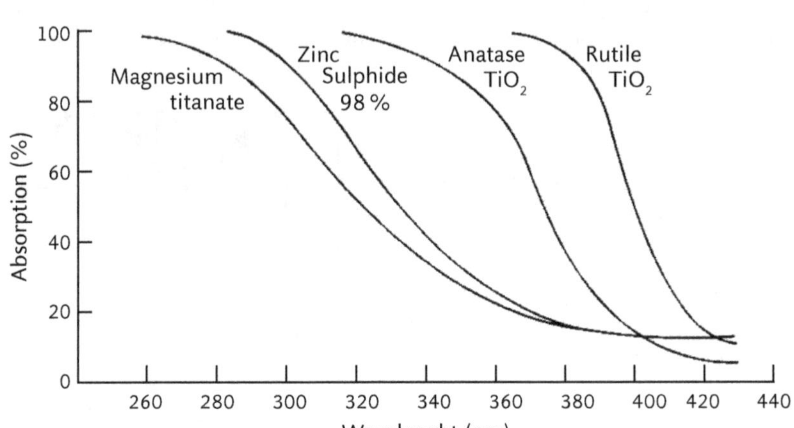

Figure 3.58: UV absorption curves of different white pigments [98, 165]

In the application of organic pigments such as carbon black pigments or phthalocyanine pigments, the storage stability of the wood coating formulations or pigment pastes can strongly be affected depending on the concentration of the pigment. Gelling (thickening) of pigment pastes may occur. In addition, some organic and inorganic pigments may act as free radical scavengers.

Fillers

Common fillers such as carbonates (calcium carbonate), silicon dioxide (fumed and precipitated silica), silicates (talc, kaolin, mica, feldspar), sulphate (precipitated barium sulphate) and organic fillers such as hollow micropheres, and cellulose fibres can be applied in radiation curable systems [170, 169].

Matting agents

The wood and furniture industry is one of the largest consumer of dull matt, matt or silk-matt coating surface. The emphasis here was not the practicality in the foreground, but the appearance. So, the character of the precious woods is best maintained by matt coatings. In contrast to the more obtrusive high-gloss coating surface, a matt coating surface offers a resting point for the eye. Furthermore, it is advantageous, that irregularities in the underground visually can be better compensated by a matt coating than by a glossy coating surface [173]. For over 50 years, synthetic silicas are applied as a matting agent in the coatings industry. Fumed and precipitated silica as well as silica gels are applied. In many cases, an after-treatment of the synthetic silicas with waxes is performed in order to enhance the suspension behaviour and matting efficiency [174].

Matting/reduction of the gloss level

As it is well known, the matt appearance of a coating film surface results from a different form of the visually plan surface. The height, shape and number of surface structures determine the gloss level of the coating [173, 175–177]. The formation of the surface structure is performed during the physical drying or chemical curing. The properties of the resin, especially the drying times and curing times, the drying parameters and the proportion of volatile solvent as well as the viscosity play a decisive role. These parameters determine whether and to

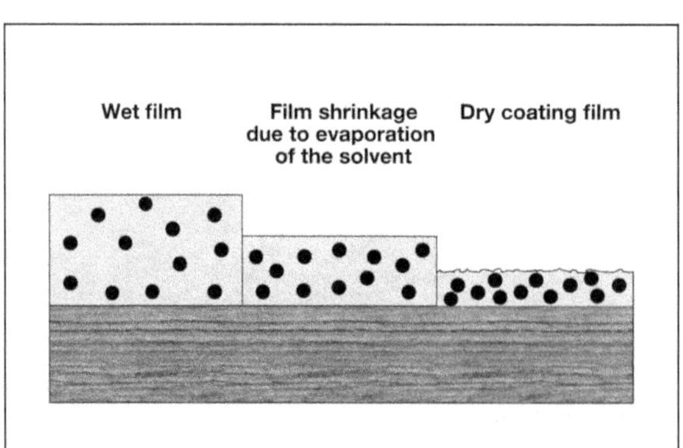

Figure 3.59: Schematic illustration of the formation of a matted coating surface [182]

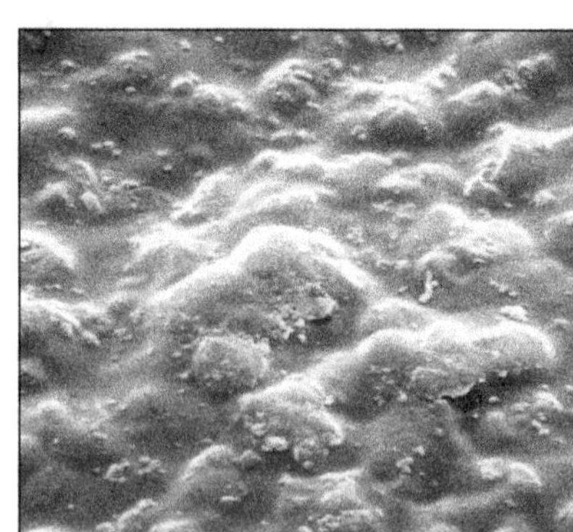

Figure 3.60: Scanning electron microscopy (S.E.M.) picture of a matt coating surface [182]

what extent the incoming light is reflected diffusely in the different incidence angles within the gloss level measurement.

The homogeneous distribution of the matting agent even in the liquid coating film subsequently to the application is the explanation for the emergence of surface defects in conventional lacquer systems. The thickness of the wet film decreases due to the evaporation of the solvents during the physical drying or chemical curing, the lacquer film 'shrinks' (Figure 3.59) [178]. The shrinkage of the film may amount between 25 and 75 %. The result is a more or less rough surface which is not visible to the naked eye (Figure 3.60) [173, 179-181].

The simplest case of matting consists of an increase of the content of pigments or fillers far above the critical pigment volume concentration (CPVC). This method only is applicable in pigmented coatings and often result in a reduction of the mechanical-technological properties of the coating film [173, 183]. The matting of clear coats with pigments or fillers is very limited since turbidity/obfuscation effects have to be considered. In addition, these substances often have a lower matting efficiency [183].

The matting effect can be achieved even without the addition of matting agents. The application of coating components which are not compatible with the components of the formulation such as solvents, resins or plasticizers can be used to achieve mat lacquer surfaces.

Matting/reduction of the gloss of UV curing wood coatings

Radiation curing coatings which generally are fast-curing and on the other solvent-free or solvent-poor and often applied in the curtain coater, Vacumat process or in the rolling process only can be matted difficultly. A film shrinkage for example in the case of solvent-borne coating systems due to the evaporation of solvents [184–186] is hardly possible. Thus, for these systems it is to be expected, that only those matting agent particles which already are near to or on the surface of the wet film effectively influence the matting. Due to the solids content of approx. 100 % high concentrations of matting agents for matting these systems are required. The enhanced concentrations of matting agents in the pump circuit on the rolling machines mean that the stability of the coating formulations cannot be guaranteed. Often it is falsely claimed that the matting agent 'floats' on the thin surface coating layers (5 to 12 µm) and thereby causes a reduction of the gloss level. Due to the application of specifically lightweight hollow glass beads (density of 0.22 g/cm^3), it could be shown in direct comparison with silica-based matting agents that a 'floating' does not occur even after five minutes of evaporation prior to the UV curing [174].

In the rolling process, the processing of mat UV coatings with a content of non-volatile compounds of approx. 100 % (gloss level in the 60 ° measurement angle approx. 5 to 25 units) is not always easy to formulate with the usual combinations of matting agents and waxes. Due to the high quantities of matting agents (up to 15 %), not only the viscosity strongly increases, but also the processing among the users significantly is more complicated or prevented. Combinations of surface-treated silica, polypropylene waxes, polyethylene waxes or polyamide waxes, layered silicates based on magnesium silicate hydrate and organic matting agents with resin character based on urea-formaldehyde are preferred in these coatings [187,188]. Depending on the UV resins system, mixtures which are manufactured from the above mentioned substances can be applied for the production of gloss stable and processable UV mat coatings. Depending on the type of application and in combination with silica, the application of fluorocarbon waxes or polyethylene waxes may result in excess coating problems, i.e. the adhesion to the substrate is no longer guaranteed. For these reasons, fluorocarbon waxes only are applied to a limited extent.

In UV coatings based on acrylates, the photoinitiator system/synergist system has a not too inconsiderable influence on the matting behaviour [189,190]. The economic matting of these coatings requires an optimization of the amine component (amine synergist), since UV curable coatings with an enhanced amount of synergists are extremely difficult to matt. The higher the amount of the amine, the more matting agent has to be added. Figure 3.61 illustrates the correlation between the added amount of amine and the gloss level of the UV topcoat. This example illustrates that the gloss level is in direct connection with the UV reactivity of the coating formulation. The addition of an amine synergist or the application of amine-modified polyether acrylates significantly reduces the inhibition of the polymerisation reaction by means of atmosphere oxygen. Conversely, it can be supposed that in this case the usually disturbing inhibition of the polymerisation reaction by the atmospheric oxygen favours the reduction of the gloss factor.

UV curable wood coatings with an amount of approx. 100 % according to the non-volatiles components may feature a polymerisation shrinkage between 4 to 15 % depending on the composition of the formulation [192–194]. The shrinkage is due to a reduction of the molecular bondings by a radical polymerisation. This volumetric shrinkage may cause adhesion problems especially on smooth substrates such as metal surfaces or plastic surfaces. However, compared to the solvent-borne paint formulations, the volumetric shrinkage has a little influence on the mattability of the coating films.

There has been no lack of attempts to overcome the disadvantage of the poor mattability of radiation-curable systems. The most attempts resulted in impractical methods or require a significant chemical and/or technical effort such as by combining polymerisable commodities with polycondensable components and experiments with different UV radiation intensities which is known as the dual-cure method in the literature [175, 192, 195–199]. In the dual-cure method the gloss level is adjusted by the UV curing process in several steps. The simplest example is the partial curing ('UV gelation') of the coating layer with a long-wave emitting UV lamp (gallium doped mercury high pressure lamp) and the subsequent UV curing ('surface curing') with a mercury high pressure lamp [200]. Thereby, the gloss level seriously is affected by the energy dose/UV intensity. Thus, depending on the UV intensity, a

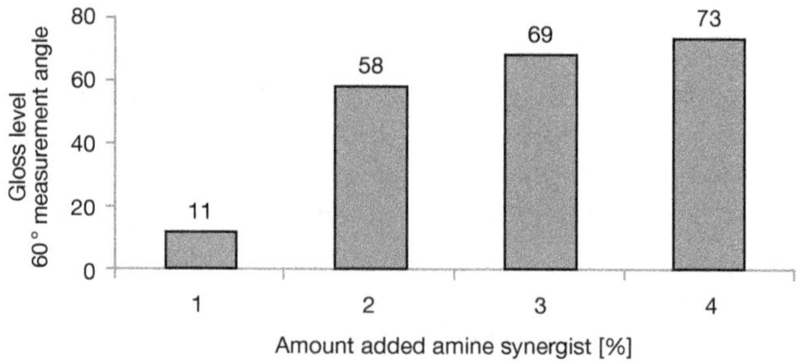

30 % urethane acrylate (aromatic): 70 % tripropylene glycol diacrylate
4 % benzophenon and 1.5 % benzyl dimethyl ketal on UV resin calculated
8 % matting agent, 2x 80 W/cm, 9 m/min, 36 µm wet application

Figure 3.61: Impact of the amine synergist (n-methyl diethanolamine) on the gloss level of an UV finishing coating [184, 190, 191]

pigmented UV topcoat can be adjusted with a wet application amount of approx. 25 g/m² (grooved coating application roller) at gloss levels between 10 and 60 units (60 ° gloss level measurement angle).

Figure 3.62 illustrates systematically that the gloss levels as well as the surface qualities can be influenced by the process in the case of UV rolling finish coatings (100 % share in non-volatile components). In the area of UV topcoat, modern UV rolling flat lines always should have two rolling machines with the possibility of UV gelation between the two rollers in order to adjust the gloss level calibration and quality mere selective. 10 to 15 years ago, in the area of UV topcoat UV flat lines were planned by means of the wet-in-wet process without UV gelation. The experiments documented in the Figures 3.62 and 3.63 present different gloss levels with 100 % UV topcoats depending on the roller process each with a topcoat. The following findings can be achieved by application of different topcoat and multiple examinations of the experiments:

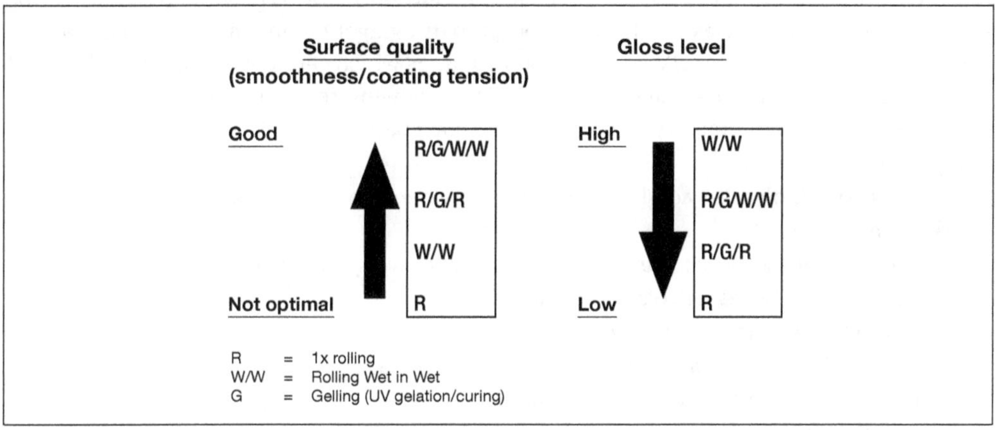

Figure 3.62: Impact of the rolling procedure on the surface quality and gloss level of approx. 100 % UV topcoats[201]

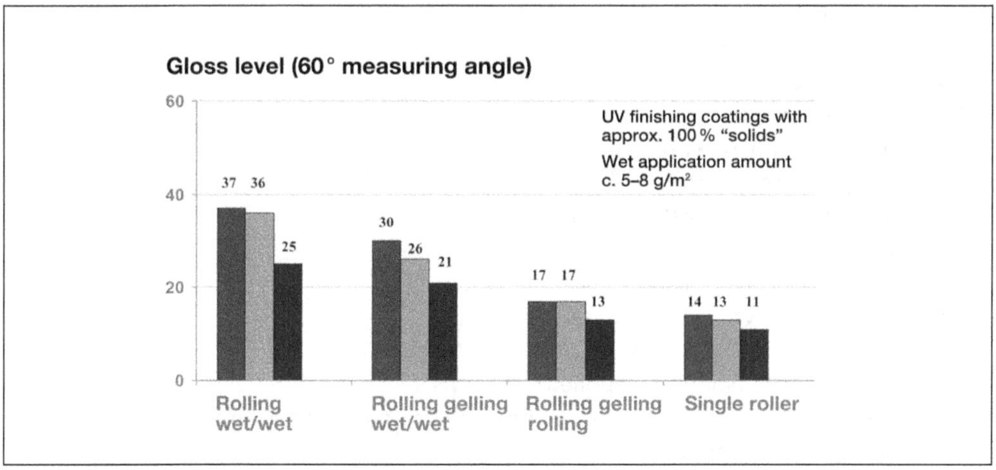

Figure 3.63: Gloss level values in dependence of the rolling process of different 100 % UV topcoats

- A gloss level difference of approx. 50 % always could be established between the wet-in-wet process and the application (roller/UV gelation/roller).
- The gloss level which is achieved in the process rolling/UV gelation/rolling (each roller 5 g/m^2) is equivalent to the gloss level which is achieved with a coating roller (10 g/m^2) with the same coating and application amount. The surface quality and fullness of the procedure (roller/UV gelation/roller) surpasses the surface quality of a simple rolling application.
- In order to achieve a low gloss level as well as an optimal coating surface, the procedure (rolling/UV gelling/rolling) presents itself in comparison to the wet-in-wet procedure and the single rolling procedure. The wet-in-wet procedure is not suitable for very matt wood coatings since the coating formulation requires a very high proportion of matting agents due to the type of process, and since the runtime stability on the coating roller cannot be guaranteed.

The procedure shows that the gloss level can be adjusted more targeted and easier via the energy dose/UV intensity as via the selection of matting agent. Fundamentally, there are no universal matting agents for UV curable wood coatings which can function independently of the UV resins, layer thickness and others [175]. The following recipe ingredients and process parameter affect the mattability of UV curable wood coatings [202, 203]:
- UV resins/monomers
- Type of resins (polyester acrylate, polyether acrylate, epoxy acrylate, urethane acrylate etc.)
- UV reactivity of the resins (functionality, molecular weight distribution)
- Viscosity of the UV resins/coating formulation
- Type of monomers (functionality)
- Blend ratio UV resins/monomers
- Combination of photoinitiators and synergists
- Type of photoinitiators
- Type of synergists (amine base)
- Amounts added of photoinitiators/amine synergists
- Matting agent
- Quantity of application
- Pore volume ml/g
- Particle size distribution
- Surface treatment with waxes
- Procedural/process parameters
- Application amount (layer thickness)
- Substrate temperature and lacquer temperature
- Settling time prior to UV curing (flash off zone)
- Air velocity in the evaporation phase/UV curing
- Energy dose and various intensities of used UV lamps
- Wavelength distribution of the applied UV lamps
- Dual-cure UV curing process in oxygen atmosphere and/inert gas atmosphere (two stage UV curing process)
- Addition of organic solvents
- Physical matting of UV coatings with excimer lamps

Excimer UV lamps for the matting of UV curing coating systems (physical matting)

The physical matting by means of so-called excimer UV lamps is an interesting alternative to the traditional process of matting. The UV curing coating thus is irradiated with excimer UV emitters under inert gas whereby the UV emitters emit photons at a wavelength of 172 nm (Xe_2*) or 222 nm (KrCl*) [196]. Excimer UV lamps are light sources whose principle of op-

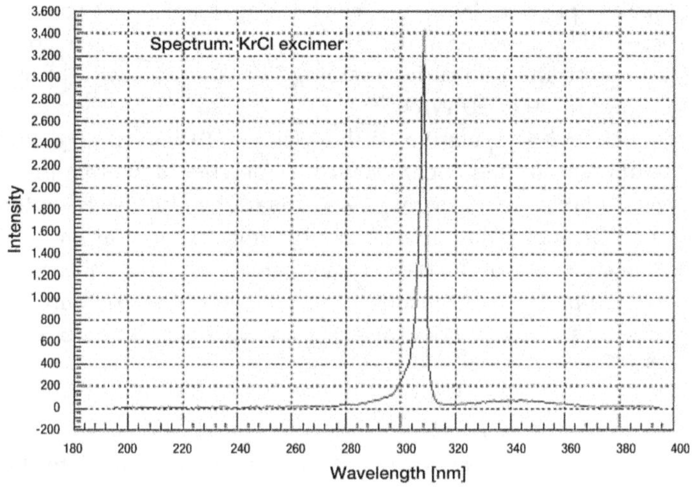

Figure 3.64: Emission spectrum of a krypton chloride lamp (= 222 nm) [204]

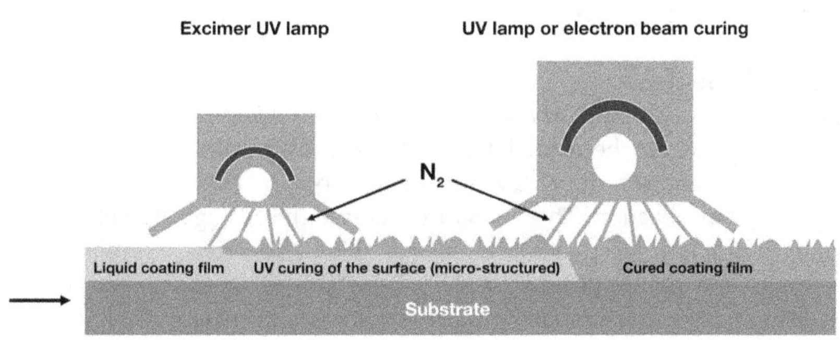

Figure 3.65: Principle of physical matting using an excimer UV lamp [206, 207, 208]

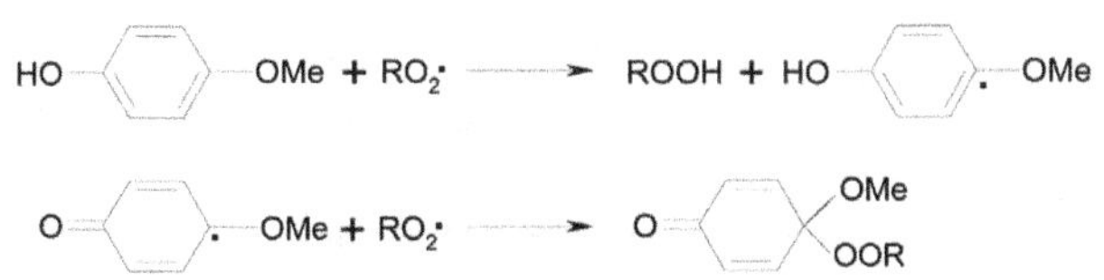

Figure 3.66: Mode of action of hydroquinone methylether as a storage stabilisator [212]

eration is based on the production or decay of excimers. Excimers (excited dimers) are molecules which only exist in an excited state and, therefore, transit back into the (very instable) ground state after a lifetime of few nanoseconds under transmitting of well-defined light quanta [204, 205]. In this ground state, the excimers immediately decompose into their monomers. The occurring volume shrinkage depends on the wavelength, energy input, and film thickness and results in differently developed surface foldings in the micrometre range and millimetre range [196]. A highly deformed film is formed causing a diffuse scattering of light which is perceived as a matt effect. Depending on the power of the lamp, a partial or complete UV curing of the coating surface structure (mattness level) in combination with the UV curing (mercury medium pressure lamp) or electron beam curing for a complete UV curing. This procedure successfully is applied in the area of paper foil finishing in combination with UV curing systems or electron-beam curing systems. The currently available cylindrical excimer lamps have a power of 50 to 90 W/cm [204].

Due to the monochromatic spectrum of the excimer UV lamps, no disturbing proportion of infrared radiation is emitted in comparison to mercury high pressure lamps. The application of excimer UV lamps to temperature-sensitive substrates is very interesting. It has to be considered that the application of excimer UV emitters requires an inertisation in order to prevent oxygen inhibition of the UV curing and particularly the absorption of photons at a wavelength of 172 nm due to atmospheric oxygen. From the literature, tolerable residual oxygen concentrations of up to 9,000 ppm are known which leads to a small consumption of inert gas [196].

Additives

Additives are substances which are added to coating formulations in relatively small quantities in order to improve certain properties or to prevent undesirables properties. Many additives from the conventional range of coatings can be applied in UV systems without restrictions. These are, for example, surface active substances such as defoamers, wetting agents or dispersing agents, levelling agents etc..

Inhibitors/stabilisators

Inhibitors/stabilisators are among the essential additives which are applied for the production of resins as well as for the formulation of UV curable coatings. Inhibitors are applied in the manufacturing and storage of UV resins and also monomers in order to avoid a premature polymerisation reaction. The inhibitors increase the storage stability and ensures a safe handling. In practice, the phenothiazines prevailed as a process stabilisator while hydroquinone monomethylether prevails as a storage stabiliser since the year 1960 [209]. One has to know that hydroquinone monomethylether can fully develop its inhibiting effect only in combination with oxygen and at a temperature of approx. 20 to 100 °C. For this reason, such substances are known as aerobic inhibitors. Thanks to the presence of oxygen, the radicals quickly react to peroxy radicals, since the formation of the peroxy radical significantly is faster than the polymerisation of the acrylic ester [210, 211]. The inhibitor/stabiliser now may react with the formed peroxy radicals. As a first step, a resonance stabilised radical is formed which reacts with a further RO_2 radical. Here, two peroxy radicals are caught per molecule of stabilizer [209].

Phenothiazine has established as a process stabiliser in the production of acrylic ester compounds. Phenothiazine is one of the anaerobic inhibitors/stabilisers. Anaerobic inhibitors have no significance as storage stabiliser or transport stabiliser. The inhibition is implemented by a hydrogen transfer reaction. The radical being formed in the reaction is resonance stabilised and therefore very inert. A further polymerisation is prevented. The peculiarity of phenothiazine compared with hydroquinone monomethylether is that it can react directly with alkyl radicals as well as with peroxy radicals (see Figure 3.67), The oxygen normally required for the conversion with phenothiazine. Phenothiazine is able to intercept the alkyl radicals directly. Thus, phenothiazine is an effective process stabiliser for the synthesis of resin at enhanced temperatures under anaerobic conditions [209]. Phenonthiazine also may decompose peroxides already formed without the occurrence of a free-radical chain reaction by degradation of peroxide. The inhibitor is oxidised by occurring hydroperoxides. Within the oxidation of phenothiazine, coloured degradation products can be formed which may have an impact on the colour number of the resins and monomers [144]. In extreme cases, the colour change of UV curable coating formulations on white surfaces can be explained.

3.1.6.6 Mechanism of the UV initiated radical polymerisation

The UV initiated, free-radical polymerisation transfers the liquid UV resins/monomers into a crosslinked hard coating film even in fractions of a second. Primarily polyfunctional monomers are applied in the wood coating formulations. These polymers form a three-dimensional network within the free-radical polymerisation. The radically induced polymerisation induced by a photoinitiator can be split into three steps [144, 214, 215]:

- Formation of radicals and start of the chain reaction by a chemical initiation
- Chain growth by adding monomers
- Chain termination by radical combination, disproportionation, quenching

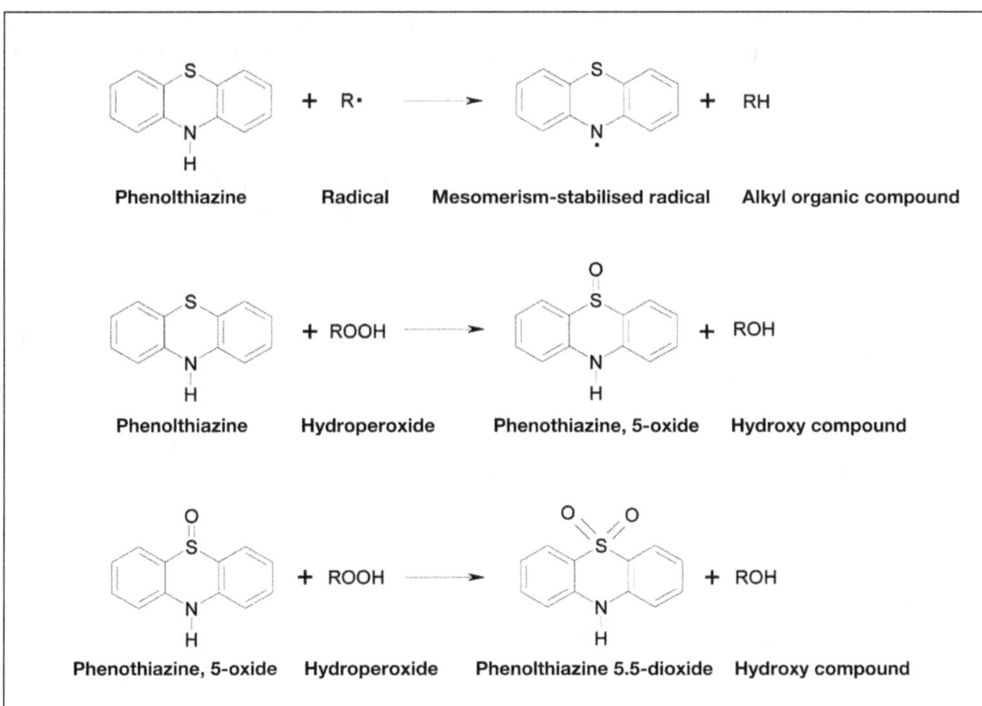

Figure 3.67: Operating principle of phenothiazine (PTZ) as process stabiliser

1. Formation of radicals and start of the chain reaction by chemical initiation
Radicals are necessary in order to start the polymerisation reaction. Initiated by the UV reaction, two radicals are formed from a photoinitiator moleculary by homolysis. The formed primary radicals react with the double bond of a monomer molecule or UV resin under formation of an alkyl radical. Here, the radicals are designated with the symbol R^* where the 'star' represents the unpaired electron.

$$\text{Photoinitiator (R-R)} \xrightarrow{h \cdot \nu} 2\,R^*$$
$$R^* + M \longrightarrow R\text{-}M^*$$

2. Chain growth by addition of monomers
The chain growth reaction is also referred to as 'propagation' which leads to the growth of the polymer chain. In every step of growth, the active radical centre at the end of the polymer chain reacts with a monomer molecule to a polymer radical extended by one unit.

$$R\text{-}M_n^* + M \longrightarrow R\text{-}M_n{+}1^*$$

3. Chain termination by radical combination, disproportionation, quenching[25]
The growth reaction terminates if two polymer radicals react with each other. Both polymer radicals deactivate each other by growing together (combination), or by transferring a hydrogen atom from one chain to the other (disproportionation). This termination reaction leads to a molecule with a saturated end group and to a molecule with an end-capped double bond. The termination reaction by combination and disproportionation of two polymer radicals is also referred to as termination by mutual deactivation. Furthermore, the addition of an initiator radical to a growing polymer chain also leads to a termination of the radical polymerisation.

$$R\text{-}M_n^* + R\text{-}M_m^* \longrightarrow R\text{-}M_n\text{-}M_m\text{-}R \quad \text{(combination)}$$
$$R\text{-}M_n^* + R^* \longrightarrow R\text{-}M_n\text{-}R \quad \text{(addition of an initiator radical)}$$
$$R\text{-}(CH_2)\text{-}CHX^* + R\text{-}(CH_2)\text{-}CHX^* \longrightarrow R\text{-}(CH){=}CHX + R\text{-}(CH_2)\text{-}CH_2X$$

(Disproportionation) or reactions with atmospheric oxygen, radical scavengers, combinations are possible.

Inhibition of the radical polymerisation by atmospheric oxygen

In UV initiated radical polymerisaton reactions, an important and often not desired secondary reaction is the inhibition of the polymerization reaction by atmospheric oxygen. Oxygen is a biradical which thus can bond highly energetic primary radicals within a rapid reaction and thus block further polymerization processes. In addition to the deletion of excitation states, the atmospheric oxygen may react with initiator radicals or growing polymer radicals under formation of stable peroxy radicals. The peroxy radicals are inert due to their stabilisation and further react very slowly [213]. In the presence of atmospheric oxygen, the growth reaction proceeds according to the following scheme [209, 216, 217].

25 Quenching, cancellation, deactivation of electronically excited species by species of the same or other type in the ground state due to the non-radiative process.

$$R^{*} + M \xrightarrow{quick} R\text{-}M^{*}$$

$$R\text{-}M^{*} + {}^{*}O\text{-}O^{*} \xrightarrow{very\ quick} R\text{-}M\text{-}O\text{-}O^{*} \quad \text{(peroxy radical)}$$

$$R\text{-}M\text{-}O\text{-}O^{*} + M \xrightarrow{very\ slow} R\text{-}M\text{-}O\text{-}O\text{-}M$$

This competitive reaction of oxygen to the normal running polymerization is the faster the higher the concentration of oxygen dissolved in the UV resins and monomers, for example, 56 ppm oxygen are dissolved in acrylic acid at air saturation (20 °C). The mechanism of action of oxygen inhibition was described by *Schulz* and *Henrici* [218]. Not only the acrylic monomer, but also the oxygen is incorporated into the polymer within the inhibition reaction. In this case, oxygen functions as a co-monomer. From the literature it is known that the formation of peroxy radicals is 100 times faster than the growth of polymers [213]. In extreme cases, the oxygen inhibition leads to sticky not optimal UV curing coating coatings.

The atmospheric oxygen already is engaged in the cleavage process of the photoinitiator and prevents the formation of primary radicals by deleting excited states (Figure 3.68).

In principle, it should be noted that the diffusion of oxygen depends on the viscosity of the UV coating formulation. The diffusion of oxygen is lower at high-viscosity coating systems [219]. It is known that the content of residual oxygen in UV curing coating films degrades the residual double bonds in the course of time [220].

The thinner the UV coating layer, the stronger the inhibition effect of oxygen. The relatively low analytical method of confocal Raman spectroscopy can be applied to 'look inside' the coating layer and to track the radical polymerisation in different depths during the UV radiation [221]. The intensity of the signal is a measure of the remaining double bonds. The lower the signal intensity, the fewer double bonds still are present and the more complete is the UV curing. The investigations measured with the confocal Raman spectroscopy showed that under the experimental conditions the oxygen inhibition especially is pronounced in the first 8 to 10 µm, and that the signal intensity for the double bonds is very pronounced (Figure 3.71). In deeper layers (> 10 µm), the inhibiting for the double bonds is very pronounced.

Figure 3.72 illustrates the conversion of the double bonds using the real-time IR spectroscopy in dependency of the wet application amount and exposure time. With a wet application amount of approx. 2 to 6 µm, the model formulation based on aromatic epoxy acrylates, trimethylolpropane triacrylate, hexanediol diacrylate, 2,4,6-trimethylbenzoyldiphenyl phosphine oxide and methyldiethanolamine a a co-initiator features a double bond turnover rate

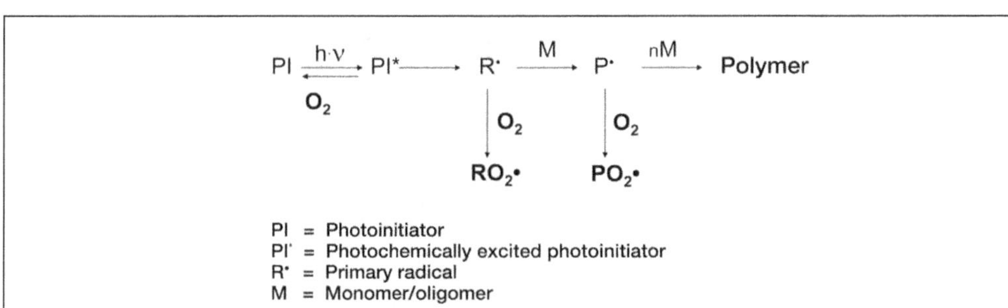

Figure 3.68: Inhibition of oxygen in a photoinitiated and radically polymerising UV coating system [217]

of 80 % according to the double bonds is achieved. The exceptional example is to feature the influence of oxygen inhibition in relation to the wet application amount. The inhibition effect is more pronounced in thinner layers. Such inhibition effects can occur for example at UV topcoat systems with a non-volatile content of approx. 100 % solids applied in the rolling process. In the general case, the wet application amounts account for 4 to 10 g/m² and may decrease heavily due to temperature effects and changes of the process parameters (contact pressure, velocity of the dosing roller and applicator roller, etc.). In this example, the reduction of the application amount can cause curing problems.

The problem of insufficient UV curing due to the influence of atmospheric oxygen can be circumvented in several ways.

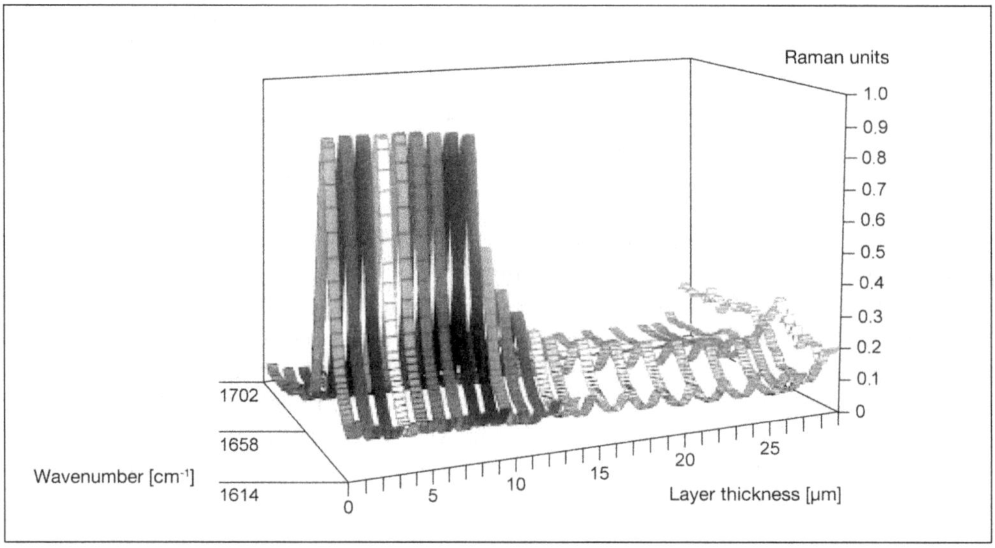

Figure 3.69: Double bond conversion of an UV coating layer depending on the layer thickness detected by confocal Raman spectroscopy [221]

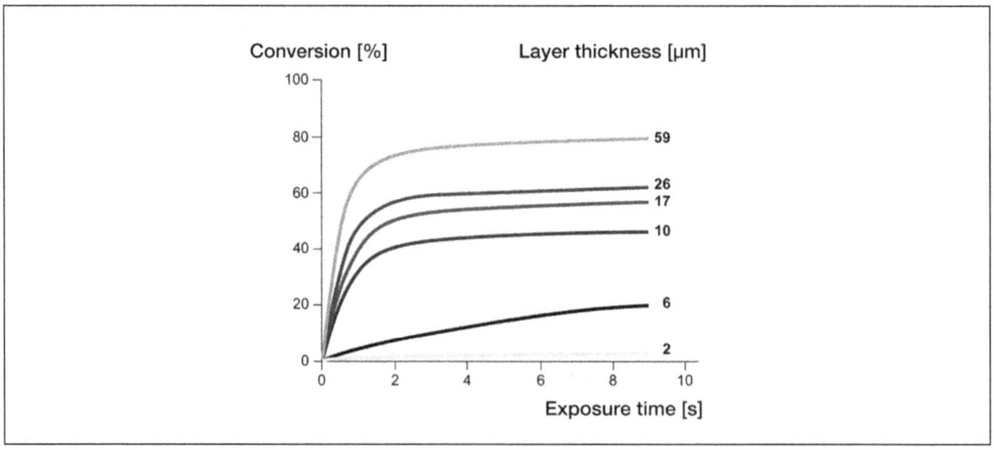

Figure 3.70: Realtime spectroscopic record of the conversion of the double bond depending on the wet application amount [221]

Application of transparent plastic films as barrier materials

Transparent plastic films such as polyester films can be applied in order to prevent the diffusion of oxygen into the coating layer. After applying the UV coating layer, the polyester film is applied on the wet paint film and then UV cured. After removing the foil, a highly scratch-resistant coating surface with a desired gloss level depending on the structure of the film is obtained [222]. The application of these films as a barrier material is procedural complex and only is used in the coating of wood in a few applications such as for special products in the kitchen furniture industry (high-gloss and super matt surfaces).

High concentrations of photoinitiators

Another alternative is to increase the primary radical concentration such that this oxygen in and on the paint film is consumed while the polymerization is faster than the subsequent delivery of oxygen by diffusion. This requires a sufficiently high concentration of photoinitiators in order to ensure the high concentration of primary radicals during the UV curing. This is also referred to as 'knocking down' of the diffusing oxygen [99]. The secondary reactions with the atmospheric oxygen are further reduced by adjusting high UV intensities during the UV curing process [219].

Addition of paraffin waxes

When highly reactive UV resins based on acrylates are applied, the addition of paraffin waxes or similar waxy substances is impractical since a coating film cures faster as a paraffin mirror can be formed in order to protect the surface of the coating film. A paraffin with different melting areas is applied for the UV curing of unsaturated polyesters with styrene since the coating lines feature so-called gelification zones with low pressure lamps of low output enabling a floating of the wax prior to the actual UV curing. The conveyor speeds of such plants range between 8 and 12 m/min (Chapter 3.1.6 unsaturated polyester resins).

Application of oxygen scavengers

Another possibility is the application of amine synergists as active oxygen scavengers (Chapter 3.1.6 Photoinitiators) as a contribution to the elimination of the oxygen inhibition. Preferably tertiary amines are applied which form very reactive α-aminoalkyl radicals by elimination of hydrogen atoms in α-position in combination with benzophenone as hydrogen abstractor for example. The α-aminoalkyl radicals are responsible for the initiation of the radical polymerisation under formation of a radical chain. The amine synergists often are referred to as co-initiators. Depending on the required concentration and application, the application of tertiary amines may cause yellowness features especially for white pigmented UV coating systems or white surfaces.

Furthermore, the application of autoxidable groups in the resin matrix enables an interception of the disturbing atmospheric oxygen within the radical polymerisation. Monoallyl ether or diallyl ether of the trimethylolpropane and tetrahydrophthalic acid have been proven (Chapter 3.1.6 Unsaturated polyester resins). The unsaturated polyester resins modified with these compounds are known as 'gloss polyester'.

UV curing using an inert gas atmosphere

One of the most effective methods to prevent the oxygen inhibition is the exclusion of oxygen within the radical polymerisation by means of an inert gas atmosphere [219, 223–226]. Thus, an UV curing process in an inert gas atmosphere under exclusion of oxygen is possible.

In spite of many resulting advantages for the industrial furniture production, the UV curing in an inert gas atmosphere has not prevailed or established yet, in practice.

- The amount of photoinitiator in the formulation can be reduced significantly [227, 228]
- Less cleavage products of the photoinitiator and less odour problems
- The application of amine synergists can be waived; lower tendency to yellowing [229]
- The application of highly functional monomers is not mandantory
- UV curing dose/UV intensity can be reduced significantly
- No ozone is produced due to the inertisation of the entire UV lamp irradiation area
- No exhaust air is necessary due to the possible cycle operation of the UV dryer with integrated back cooling [230]

This is that the technical effect and the costs for the inertisation of the UV dryer for the furniture industry previously were impractical to implement. In addition, UV coating surfaces (approx. 100 % systems) which are cured under inert gas atmosphere can be matted much worse due to the missing surface inhibition by atmospheric oxygen. This argument has to be taken into account in spite of all disadvantages of the oxygen inhibition. In Europe or around the world, the few plants for the coating of wood-based materials which are operated with nitrogen as inert gas mainly are electron beam facilities (EB) for two-dimensional components. These plants exist for many decades and mainly coat single components such as pigmented interior doors and panel-type materials for the exterior application [231]. The flow conditions have to be considered depending on the component geometry when the inertisation is performed in UV dryer or electron beam systems. A uniform flushing of the materials with nitrogen has to be guaranteed. Otherwise, the UV curing or electron beam curing may result in coating surfaces differing in the behaviour of curing and or the gloss level within a component.

Three-dimensional wood-based materials

Three-dimensional wood-based materials cannot be cured practically with the previous UV curing without the application of inert gases. This is that with the three-dimensional substrates the distance between the lamps and the coating surface must be held as constant as possible by means of a special arrangement of the UV-lamps and the workpiece in order to achieve a uniform irradiation due to the distance-dependent UV intensity. Only like this, the oxygen inhibition can be reduced to a minimum during the process of UV-curing or completely switched off. Due to the large number of components which are applied in the furniture industry, a permanent adjustment of the distances of the lamps to the workpiece is required for the UV curing of the three-dimensional workpieces and for the replacement of the component geometry.

A new, low-cost and simple procedure for UV curing of three-dimensional materials operates under carbon dioxide atmosphere and is of particular interest for industrial application as well as for small batches. In the market, this procedure is well-known as "Larolux" since the year 2001 (Figure 3.71).

The applied carbon dioxide is heavier than air and thus can easily be filled in a plastic container as a 'inert gas sea' and kept on the ground. The supply of dry ice (solid CO_2) is considered as beneficial especially for industrial use. Oxygen contents of lower than 1 % can be adjusted without great effort. The direct comparison between a nitrogen atmosphere and carbon dioxide atmosphere features no significant differences in the UV curing behaviour of a urethane-acrylate formulation as an example (Figure 3.72). The curing process in the carbon dioxide tank or carbon dioxide tunnel with reduced oxygen content enables the reduction of UV radiation and thus the application of simple UV lamps with a broadband spectrum such as an artificial sunlamp.

The coated substrates, for example 3D substrates such as wood chairs, are dipped into the carbon dioxide atmosphere and exposed to light. Reflective surfaces consisting of aluminium are adjusted to the basin in order to illuminate the shaded areas on the substrate.

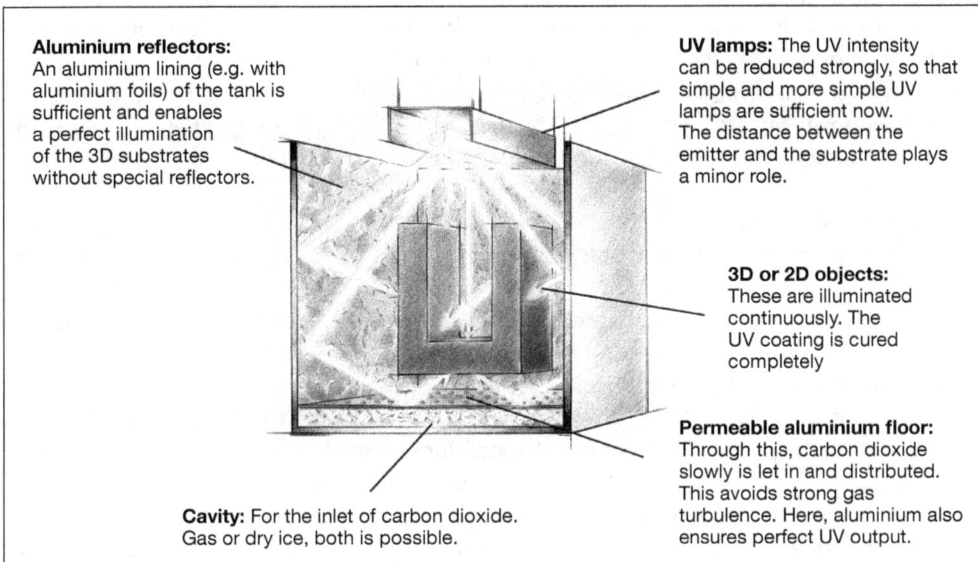

Figure 3.71: UV coating curing in a carbon dioxide atmosphere based on "Larolux"[232]

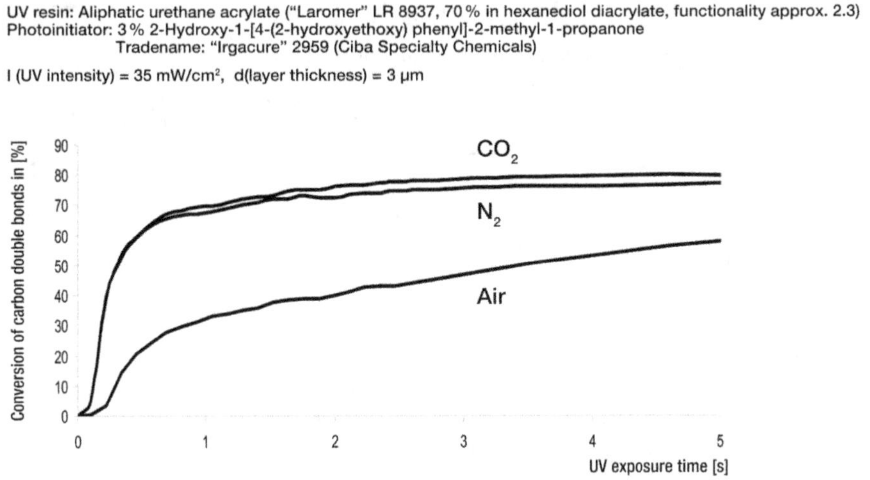

Figure 3.72: Impact of different atmospheres (carbon dioxide atmosphere, nitrogen atmosphere, oxygen atmosphere) on the UV curing behaviour of urethane acrylates[223, 233]

The plant example is very simple, so that it is worthwhile for mall series in handicraft enterprises. In laboratory experiments, a 400-W artificial sunlamp (e.g. "Philips HB 406" with a metal halide lamp with UV-C and UV-B filters for wavelengths > 300 nm) which still emits an energy dose of 1.5 J/cm² in a time period of 30 seconds at a distance of 60 cm successfully was tested [234].

Dangerous UV-C radiation and formation of ozone are avoided by using these UV-lamps. Wood-based materials coated on all sides can completely be cured in this container by irradiation. Further investigation revealed that monofunctional and difunctional monomers can be polymerised with enhanced turnover rates under inert gas conditions (carbon dioxide atmosphere). Photoinitiated hexanediol diacrylate cannot homopolymerise in an oxygen atmosphere. However, in a carbon dioxide atmosphere hexanediol diacrylate can be homopolymerised with turnover rates of more than 80 % (Figure 3.73). The findings open up new paths for the formulation of very flexible and simultaneously hard coatings.

Coating of wood chairs [235, 236]

The first pilot plant for the testing of the "Larolux" process of wood chairs was put into operation in Northern Europe in the autumn of the year 2005. The purpose of this plant is to replace the acid-curing and solvent-based coating systems still applied in Northern Europe by eco-friendly wood coating systems such as one-component solvent-based UV curable waterborne paints. Additionally, there are also economic savings. The plant consists of a compact flow dryer with dimensions of 4 x 2 x 1.5 m (width x height x depth). The interior of the UV dryer is lined with specially treated and shaped aluminium sheets. Three UV lamps with a power output of 2,000 Watt each are arranged behind a quartz glass plate in the UV curing area. One UV lamp is mounted on the ceiling, while two UV lamps are mounted at the sides of the dryer. Preliminary investigations showed that the application of UV lamps in the inside of the dryer result in turbulences of air, which in turn lead to an enhanced consumption

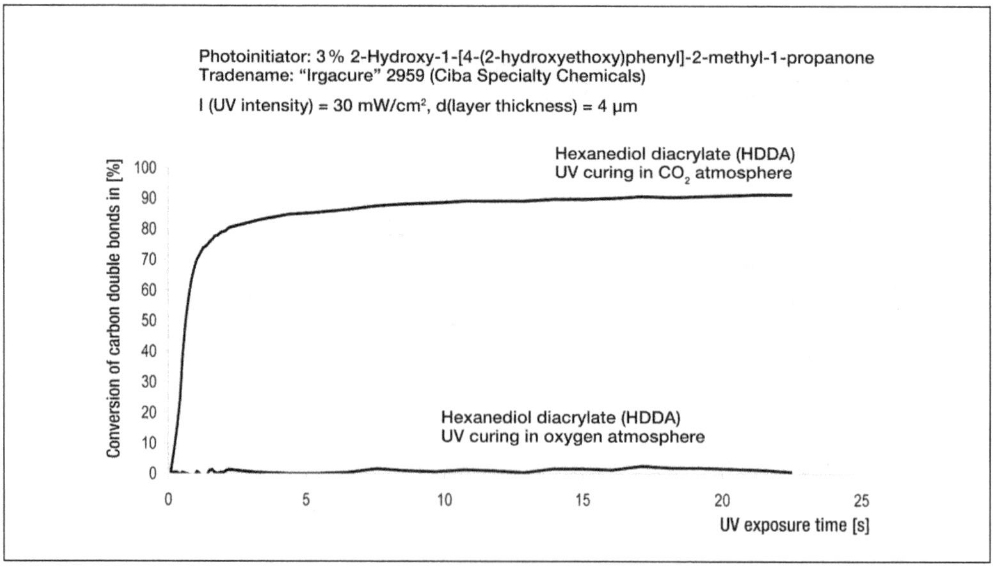

Figure 3.73: Comparison of the UV curing of hexanedioldiacrylate in an oxygen atmosphere and carbon dioxide atmosphere [223, 233]

of carbon dioxide. The turbulences are caused by a thermal convection which is emitted by the UV lamps. The problem could be solved by an external installation of UV lamps and separation from the carbon dioxide pool by using a glass wall. In order to ensure continuously a low oxygen content of 2 to 4 %, a special measurement equipment is applied to measure continuously the content of carbon dioxide. The line speed is approx. 5 m/min.
The UV-curing in a carbon dioxide basin has the following advantages [233]:
- Lower energy costs due to reduced UV performance
- The application of tanning lamps does not supply UV-C radiation and ozone
- UV curing of 3D objects enables greater distances between the UV lamps
- Suitable for both large series and small series
- Reduced amounts of photoinitiators and thus reduced costs of formulation
- Broader formulation capabilities due to the application of acrylates with lower functionality more flexible coatings
- Improvement of the chemical resistance and scratch resistance

Conversion of the double bonds and glass transition temperature during the curing process

During the process of UV curing, generally a non-sticky and handling-resistant coating film is achieved with seconds. This is due to the fact, that the radical polymerisation achieves enhanced degreed of polymerisation even at low degrees of conversion in opposite to the process of polycondensation [237]. Depending on the recipe and process parameters, in practice the degree of conversion of the unsaturated double bonds is due to the vitrification[26] of the coating film during the process of UV curing. Thus, the segment flexibility of the chain molecules is frozen. This happens if the glass temperature[27] exceeds the temperature predominant in the UV curing (depending on the procedure and process). During the curing polymerisation, the liquid coating film passes into the solid state and eventually stops the reaction. At a certain degree of conversion, this curing coating film passes an intermediate state in which the properties of the fluid and the properties of the solid state have the same level. This state is known as the gel point since there exists a network for the first time [238]. Dissolution experiments in different solvents always result in insoluble residues. The degree of conversion at which the gel point is achieved depends on the functionality of all components (resins, monomers), type and concentration of the photoinitiator and the process parameters during the UV curing. The obtained glass transition temperature also depends on the above parameters. The glass transition temperature, heat resistance, resistance to chemicals, hardness of the coating film as well as the internal tensions and the brittleness of the paint film increase with increasing degree of cross-linking [237]. Simultaneously, the elasticity as well as the viscosity of the coating film are reduced.

26 Vitrification is the solidification of a liquid by increasing its viscosity while it is cooled down whereby a crystallization is missing and thus an amorphous material is developed.
27 The glass transition temperature (abbreviation: T_g) is the general term for the temperature at which an amorphous solid merge from the liquid state to a glassy state and vice versa. Above the glass transition temperature, polymers are in a fluid-like state of equilibrium. A free rotation of the carbon-carbon bonds in the main chain is given resulting in changes of the position of the components of the chain molecules (chain segments). When the temperature is below the glass transition temperature, the mobility of these chain segments freezes. The material merges into the glassy state which thermodynamically is a non-equilibrium state characterised by its previous history. This results in a volume relaxation below the glass transition temperature [239, 240].

Further systematic investigations featured that the radical polymerisation is not yet completed after the UV curing and that the hardness of the coating film still increases within the first few days after exposure to the light [239]. This is due to the existence of active radicals and polymerisable double bonds in the coating film immediately after UV curing which react more or less depending on the storage temperature.

Due to the radiation-induced polymerisation within the process of UV curing, UV-curable coating materials suffer under a material-specific film shrinkage which is justified by new covalent bonds between the UV resins and monomers. The intermolecular distances merge into shorter covalent bond gaps. Thus, the density of the coating film increases while its volume decreases. It is important for the manufacturer of the coating films to be able to assess the extent to which film tension effects occur since these may reduce the adhesiveness on smooth substrate surfaces such as melamine resin films.

3.1.6.7 Examples for coating formulations

The examples which are presented here for UV coating formulations illustrate reference formulations from the raw material industry. The aim of this type of presentation is to get an insight into the composition of UV-curable wood coatings.

Basic formulations for the formulation of parquet surfaces

Table 3.28: UV curing adhesion primer for the parquet coating for rolling [241]

Component	Function	Parts by weight
Unsaturated polyester[1)] 70 % in DPGDA	UV resin	75.00
Dipropylene glycol diacrylate (DPGDA)	Reactive diluent, adjustment of the application viscosity	22.00
Bisacylphosphine oxide[2)] (BAPO II)	Photoinitiator for the deep curing process	0.75
2-Hydroxy-2-methyl-1-phenyl-propane-1-one[3)]	Photoinitiator for the surface curing	2.25
Sum	x	100.00

1) "Desmolux" UA VP 2110 (Allnex); allylether modified, acid value approx. 20, viscosity, 23 °C in mPas ca. 17,000
2) "Irgacure" 819 (BASF SE)
3) "Darocure"1173 (BASF SE)

Table 3.29: UV curing adhesion primer, water dilutable for the parquet flooring applied by roller coater [242

Component	Function	Parts by weight
Urethane acrylate[1)] 50 % in water	UV resin	95.00
Methylbenzoyl formate[2)]	Photoinitiator	4.70
Silicone surfactant[3)] (Solution of a polyether modified polydimethylsiloxane)	Subsurface wetting agent	0.30
Sum		100.00

1) "Viaktin" VTE 6155w/50WA (Allnex)
2) "Genocure" MBF (Rahn)
3) "Byk" 346 (Byk-Chemie)

Table 3.30: Corundum containing UV primer for the parquet coating for roller caoting application[243]

Component	Function	Parts by weight
Polyester acrylate[1)]	UV resin	58.00
Urethane acrylate[2)] (aliphatic, 65 % in tripropylene glycol diacrylate)	UV resin (viscoplastic)	18.00
Dipropyleneglycoldiacrylate	Reactive diluent, adjustment of the viscosity	5.00
Al_2O_3 corundum	Improvement of the Taber abrasion	10.00
Polysiloxane/polymere mixture[3)]	Silicone defoamer	1.00
Methyl benzoyl formate[4)]	Photoinitiator	2.00
1-Hydroxy-cyclohexyl-phenyl-ketone[5)]	Photoinitiator	2.00
Sum		100.00

1) "Laromer" PE 44 F
2) "Laromer" UA 19 T
3) "Byk" 088 (Byk-Chemie)
4) "Genocure" MBF (Rahn)
5) "Irgacure" 184 (BASF SE)

Table 3.31: Silky-gloss UV finish coating for the parquet coating for roller coating application[241]

Component	Function	Parts by weight
Aromatic urethane acrylate[1)] (viscous/ flexible)	UV resin	32.60
Polyether acrylate[2)] (amine modified)	UV resin viscosity control	17.10
Dipropylene glycole diacrylate (DPGDA)	Reactive diluent, adjustment of the viscosity	39.50
Silica[3)] (synthetic)	Influence of the gloss level and viscosity	7.40
Pyrogenic silica[4)]	Adjustment of the viscosity	0.80
Benzophenone	Photoinitiator	1.30
1-Hydroxy-cyclohexyl-phenyl-ketone[5)]	Photoinitiator	1.30
Sum		100.00

1) "Desmolux" XP 2614, delivery form 100 %
2) "Desmolux" VP LS 2299, delivery form100 %
3) "Acematt" OK 412 (Evonik)
4) "Acematt" TS 100 (Evonik)
5) "Irgacure" 184 (BASF SE)

Basic formulations for the coating of furniture and doors

Table 3.32: Transparent UV primer coating for the furniture and door industry [243]

Component	Function	Parts by weight
Aromatically modified epoxy acrylate[1)]	UV resin	31.00
Unsaturaed polyester[2)]	UV resin	31.00
Dipropylene glycol diacrylate (DPGDA)	Reactive diluent, adjustment of the viscosity	15.00
Copolymer with acid groups[3)], acid value approx. 53	Wetting additive and dispersing additive	0.50
Talc[4)] (magnesium-silicate-hydrate)	Filler to improve the adhesive strength and grindability	18.00
Benzildimethyl ketale[5)] (alpha, alpha-dimethox-alpha-phenylacetophenone)	Photoinitiator	4.00
Acrylate copolymer[6)]	Improvement of the leveling	0.50
Sum		100.00

Efflux time (6 mm cup, at 23 °C): approx. 70 to 80 seconds, UV reactivity (approx. 25 g/m²): approx. 5 m/min, 1 UV lamp (Hg) 120 W/cm

1) "Laromer" LR 8986, delivery form 100 % (BASF SE)
2) "Laromer" UP 35 D, delivery form 100 % (BASF SE)
3) "Disperbyk" 110 (Byk-Chemie)
4) Talc 10 MO (Imerys Talc)
5) "Irgacure" 651 (BASF SE)
6) "Byk" 361 N (Byk-Chemie)

Table 3.33: Black pigmented UV primer coating for roller coating application of furniture components and doors [243

Component	Function	Parts by weight
Amine modified polyether acrylate[1)]	UV resin with enhanced UV reactivity and good pigment wetting	54.00
Amine modified polyether acrylate[2)] (low viscous)	UV resin adjustment of the viscosity	37.00
Aniline black[3)] pigment Black 1	Pigment, colouring	5.00
2,4,6-trimethylbenzoyl diphenyl phosphine oxide[4)]	Photoinitiator (deep curing process)	1.00
2-Hydroxy-2-methyl-1-phenyl-propane-1-one4)	Photoinitiator (surface curing)	2.00
Cellulose acetobutyrate[5)] 20 % solution in butylacetate	Improvement of the process	1.00
Sum		100.00

Efflux time (4 mm, at 23 °C.): approx. 90 to 120 seconds, UV reactivity (approx. 130 g/m²): approx. 2.0 m/min, 1 minute irradiation with low pressure lamps of the type Philips TL 03, 1 UV lamp (Hg) 120 W/cm

1) "Laromer" PO 84 F, delivery form 100 %
2) Talkum AT 1 (Norwegian Talc AS)
3) "Lucerin" TPO (BASF SE)
4) "Darocur" 1173 (BASF SE)
5) CAB 551-0.01 (Eastman)

Table 3.34: Silky-mat UV finish coating for roller coating application of funiture and doors [242]

Component	Function	Parts by weight
Polyester acrylate[1)]	UV resin with enhanced flexibility, abrasion resistance and very good body	44.70
Tripropylene glycol diacrylate (TPGDA)	Reactive diluent, adjustment of the viscosity	38.00
Benzophenone	Photoinitiator	2.00
2-Hydroxy-2-methyl-1-phenyl-propane-1-one[2)]	Photoinitiator	2.00
Silica[3)] (synthetic, organically post-treated)	Influence of the gloss level and viscosity	7.00
Talc[4)] (magnesium silicate hydrate)	Filler to improve the adhesion and grindability	4.00
Micronised modified HD polyethylene wax[5)]	Improvement of the scratch resistance, surface slip and for the reduction of the gloss level	2.00
Polyester modified polydimethyl siloxane[6)] (10 % solution in xylene)	Slip agent and leveling agent	0.30
Sum		100.00

Viscosity (ISO 3219/23 °C, cone/plate, D 2500 s^{-1}) = approx. 1100 mPas application, amount: approx. 10 to 20 g/m², UV curing: Two UV lamps (Hg) 80 W/cm, 10 m/min.

1) "Viaktin" VTE 6174, delivery form 100 %
2) "Darocur" 1173 (BASF SE)
3) Talkum A10 (Naintsch)
4) "Syloid" ED 50 (Grace)
5) "Ceraflour" 950 (Byk-Chemie)
6) "Baysilon" PL (OMG Borchers)

Table 3.35: Silky-matt, monomer-free UV clear coat for the roller coating application of funiture and doors [243]

Component	Function	Parts by weight
Aromatically modified epoxy acrylate[1)]	UV resin	20.00
Polyether acrylate[2)] (low viscous)	UV resin, adjustment of the application viscosity	53.50
Talc[3)] (magnesium-silicate-hydrate)	Improvement of the adhesion, grindability and adjustment of the gloss level	8.50
Silica[4)] (synthetical)	Influencing of the gloss level and viscosity	11.00
Amine synergist[5)] (reactive tertiary amine)	Enhancement of the UV reactivity	3.00
Benzophenone	Photoinitiator	1.50
1-Hydroxy-cyclohexyl-phenyl-ketone[6)]	Photoinitiator	1.50
Silicone tensid[7)] (solution of a p olyether modified polydimethyl siloxane)	Subsurface aeration agent	0.30
Organic polymer[8)]	Deaerator	0.40
Sum		100.00

Viscosity (Pas at 23 °C): approx. 1.1 Pas, UV reactivity (approx. 10 g/m²): approx. 5 m/min, 1 UV lamp (Hg) 120 W/cm

1) "Laromer" LR 8986, delivery form 100 % (BASF SE)
2) "Laromer" LR 8967, delivery form 100 % (BASF SE)
3) "Mistron" Monomix (Luzenac)
4) "Syloid" 162 C (Grace)
5) "Laromer" LR 8956 (BASF SE)
6) "Irgacure" 184 (BASF SE)
7) "Byk" 346 (Byk-Chemie)
8) "Tego" Airex 920, Delivery form 100 % (Evonik)

Table 3.36: Silky-gloss, monomer-free UV clear coat for spraying application of furniture components and doors [243]

Component	Function	Parts by weight
Polyether acrylate[1)] (low viscous, alkoxylated trimethylolpropane triacrylate)	UV resin	80.00
Talc (magnesium silicate hydrate)[2)]	Improvement of the adhesion, grindability and adjustment of the gloss level	8.00
Influencing of the gloss level and viscosity	Reactive diluent, adjustment of the viscosity	8.00
2-Hydroxy-2-methyl-1-phenyl-propane-1-one[4)]	Photoinitiator	3.60
Acrylate copolymer[5)]	Improvement of the levelling	0.40
Sum		100.00

Viscosity (mPas at 45 °C): approx. 100 mPas, Efflux time (4 mm, at 45 °C): approx. 20 sec
UV reactivity (approx. 75g/m²): approx. 2.5 m/min 1 UV lamp (Hg) 120 W/cm gloss (75 g/m²): approx. 30 E/60°

1) "Laromer" PO 33 F, dekivery form 100 % (BASF SE)
2) Talkum AT 1 (Norwegian Talc AS)
3) "Syloid" ED 80 (OGrace)
4) "Darocur" 1173 (BASF SE)
5) "Byk" 361 N (Byk-Chemie)

Table 3.37: Transparent UV base and topcoat for the "Vacumat" applicaton [242]

Component	Function	Parts by weight
Polyether acrylate[1)] (low viscous)	UV resin (low viscous, excellent UV reactivity)	53.50
Dipropylene glycol diacrylate (DPGDA)	Reactive diluent, adjustment of the viscosity and gloss level	30.60
Methylbenzoyl formate[2)]	Photoinitiator	3.00
Talc[3)] (magnesium-silicate-hydrate)	Improvement of the adhesion, grindability and adjustment of the gloss level	5.50
Silica[4)] (synthetic and organically post-treated)	Influencing of the gloss level and viscosity	5.30
Mikronised modified HD polyethylene wax[5)]	Improvement of the scratch resistance, surface slip and for the reduction of the gloss level	2.10
Sum		100.00

Viscosity (at 23 °C, 25 s^{-1}): approx. 400 mPas, UV reactivity (approx. 10 to 40g/m²): approx. 60 m/min, 1 UV lamp (Hg) 80 W/cm

1) "Viaktin2 VTE 5968, delivery form 100 % (Allnex)
2) "Genocure" MBF (Rahn)
3) Talc A 10 (Naintsch)
4) "Acematt" OK 412 (Evonik)
5) "Ceraflour 950 (Byk-Chemie)

Table 3.38: White pigmented UV base coat and topcoat for the "Vacumat" applicaton [243]

Component	Function	Parts by weight
Modified polyether acrylate[1)] (low viscous)	UV resin (low viscous, good UV reactivity)	17.50
Dipropylene glycol diacrylate	Reactive diluent; adjustment of the application viscosity	43.90
Copolymer with acid groups[2)], acid value approx. 53	Wetting additive and dispersing additive	0.42
Titanium dioxide, rutile type[3)] (post-treated with aluminium compounds and silicon compounds)	White pigment, colouring	21.90
Talc[4)] (magnesium-silicate-hydrate)	Improvement of the adhesion, grindability and adjustment of the gloss level	4.40
Silica[5)] (synthetic and organically post-treated)	Influencing of the gloss level and viscosity	2.20
Micronised modified HD polyethylene wax[6)]	Improvement of the scratch resistance, surface slip and for the reduction of the gloss level	2.20
Amine synergist[7)] (reactive tertiary amine)	Enhancement of the UV reactivity	4.40
Benzophenone	Photoinitiator for the surface curing	0.44
2,4,6-trimethylbenzoyldiphenyl Phosphine oxide[8)]	Photoinitiator for the deep curing	1.80
Bis-acyl phosphine oxide[9)] (BAPO II)	Photoinitiator for the deep curing	0.40
Acrylate copolymer[10)]	Improvement of the leveling	0.44
Sum		100.00

Viskosity (at 23 °C): approx. 20 s, UV reactivity (approx. 25 g/m²): approx. 10 m/min 1 UV lamp (Ga/Hg) 120 W/cm, Gloss level (60 ° measurement angle) at 40g/m², Wet application amount: approx.10

1) "Laromer" LR 8967, Delivery form 100 %, viscosity at 23 °C (DIN EN ISO 3219, velocity gradient D: 250 s^{-1}) = 120 to 190 mPas
2) "Disperbyk" 110 (Byk-Chemie)
3) "Kronos" 2160 (Kronos)
4) Talc 10 MO (Lucenac)
5) "Acematt" OK 607 (Evonik)
6) "Irgacure" 184 (BASF SE)
7) "Lucerin" TPO (BASF SE)
8) "Irgacure" 819 (BASF SE)
9) "Laromer" LR 8956 (BASF SE)
10) "Byk" 361 N (Byk-Chemie)

3.1.6.8 Examples of application for the use of UV curing coating systems

The process for the coating of furniture, kitchens and doors with UV coatings do not differ fundamentally from one another. The examples of application shown in the furniture industry, kitchen industry, parquet industry and door industry present an actual overview of the state of the coating process with UV coatings. The pretreatment of wood, the procedure of application as well as the procedures of drying and curing are discussed in detail in the Chapter 4, 5 and 7.

Table 3.39: Solvent-dilutable UV coatings for the wood coating

Solvent dilutable UV coating systems for the wood coating

Process of application	Types of UV coating	Fraction of non-volatile components of the coating	Fraction of the solvent
Roller coater	UV clear coats		
	UV base coat	approx. 100 %	–
	UV multi-layer coating	approx. 90–100 %	0–15 %
	UV topcoat	approx. 95–100 %	0–5 %
	Pigmented UV coatings		
	UV base coat	approx. 100 %	–
	UV topcoat	95–100 %	0–5 %
Curtain coater	UV clear coats		
	UV topcoat	25–100 %	0–75 %
	Pigmented UV coatings		
	UV topcoat	30–90 %	10–70 %
	UV filler	40–90 %	10–60 %
Spraying	UV clear coats		
	UV primer	25–100 %	0–75 %
	UV multi-layer coating	25–100 %	0–75 %
	Pigmented UV coatings		
	UV filler/primer	30–90 %	10–70 %
	UV multi-layer coating	25–100 %	0–75 %
	UV topcoat	25–100 %	0–75 %
Vacumat	UV clear coats		
	UV base coat	95–100 %	0–5 %
	UV multi-layer coating	95–100 %	0–5 %
	UV topcoat	95–100 %	0–5 %
	Pigmented UV coating		
	UV multi-layer coating	90–100 %	0–10 %
	UV topcoat	95–100 %	0–5 %

Coating of furniture and doors

The UV coatings are applied by curtain coater, spraying, roller coater as well as by the "Vacumat" method in the coating of furnitures and doors. Flat surfaces are coated industrially with almost 100 % 'solids containing' UV primer coatings and UV topcoat by rolling. Depending on the country, different requirements to the coating surface arise. In Northern Europe, furniture surfaces as well as door surfaces often are painted in a thin-layer, so that a protective effect is ensured. However, in Southern Europe the surfaces are painted in a thick layer. The worldwide used solvent-dilutable UV coatings for wood coatings are shown in Table 3.39. Depending on the method of application, UV coatings differ in their solid content [244].

UV coating with UV clear coat

Figure 3.74 illustrates a typical UV rolling line with several single rollers and two coating sanding machines as these are applied in Central and Northern Europe. As it can be seen in Figure 3.75, it can be waived on a second sanding machine for the coating sanding process during the process of planning due to the application of a lightweight levelling machine in the field of priming. Such UV rolling lines are operated with feed rates of 10 to 50 m/min.

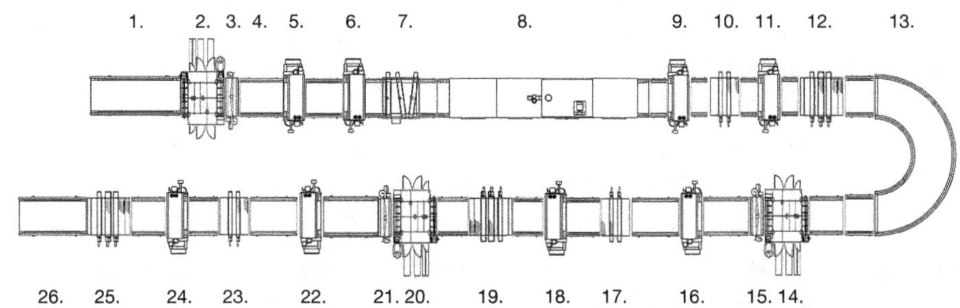

1. Support
2. Grinding machine (grinding belts = transversally/ longitudinally/longitudinally arranged)
3. Dedusting
4. Control line
5. Single roller hard rubber smooth water stain approx. 10–15 g/m²
6. Single roller foam rubber (optionally)
7. Ejector station with 3 brushes (optionally)
8. Jet dryer approx. 60–70 °C circulation air temperature for the drying of the water stain
9. Single roller hard rubber smooth UV rubber coating, transparent approx. 12–15 g/m²
10. UV dryer with 2 mercury high-pressure lamp with 80 W/cm each for the gelling of UV primer
11. Single roller hard rubber smooth UV rubber coating, transparent approx. 10–12 g/m²
12. UV dryer for the UV curing 3 mercury high-pressure lamps with 80 W/cm each
13. Transport curve
14. Grinding machine for the coating sanding
15. Dedusting
16. Single roller hard rubber smooth UV rubber coating, transparent approx. 10–12 g/m²
17. UV dryer with 2 mercury high-pressure lamp with 80 W/cm each for the gelling of UV primer
18. Single roller hard rubber smooth UV rubber coating, transparent approx. 10–12 g/m²
19. UV dryer for the UV curing 3 mercury high-pressure lamps with 80 W/cm each
20. Grinding machine for the coating sanding
21. Dedusting
22. Single roller hard rubber smooth UV rubber coating, transparent approx. 4–5 g/m²
23. UV dryer with 2 mercury high-pressure lamp with 80 W/cm each for the gelling of UV primer
24. Single roller hard rubber smooth UV rubber coating,transparent approx. 4 – 5 g/m²
25. UV dryer with 2 mercury high-pressure lamp with 80 W/cm each
26. Withdrawal of the components

Figure 3.74: Rolling line I for the UV-Coating of veneered and stained furniture components

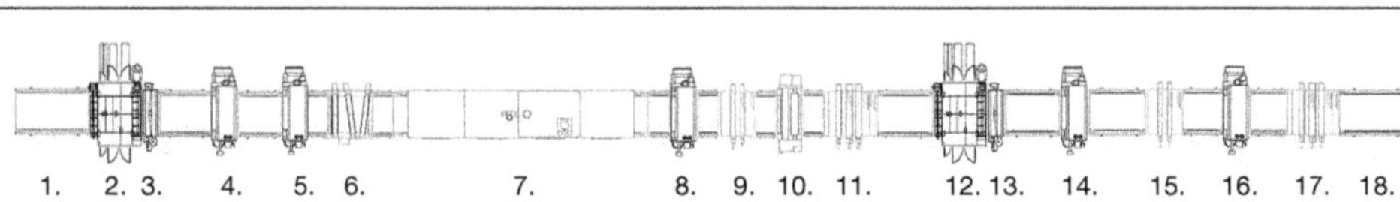

1. Support
2. Grinding machine (grinding belts = transversally/ longitudinally/longitudinally arranged)
3. Dedusting
4. Single roller hard rubber smooth Water stain approx. 10–15 g/m²
5. Single roller foam rubber for dark and intense stained colour shades
6. Ejector station with 3 brushes (optionally)
7. Jet dryer approx. 60–70 °C circulation air temperature for the drying of the water stain
8. Single roller hard rubber smooth UV rubber coating, transparent approx. 12–15 g/m²
9. UV dryer with 2 mercury high-pressure lamp with 80 W/cm each for the gelling of UV primer
10. Light filling compound with levelling roller consisting of steel UV primer approx. 20–25 g/m²
11. UV dryer for the UV curing 3 mercury high-pressure lamps with 80 W/cm each
12. Grinding machine for the coating sanding
13. Dedusting
14. Single roller hard rubber smooth UV rubber coating, transparent approx. 4–5 g/m²
15. UV dryer with 2 mercury high-pressure lamp with 80 W/cm each for the gelling of UV primer
16. Single roller hard rubber smooth UV rubber coating, transparent approx. 4–5 g/m²
17. UV dryer for the UV curing 3 mercury high-pressure lamps with 80 W/cm each
18. Removal of the components

Figure 3.75: Rolling line II for the UV coating of veneered and stained furniture components

There also exist applications where the application of grinding machines for the process of sanding completely can be waived. Instead, a combination of heated calander rolls and brushing units can be applied [245].

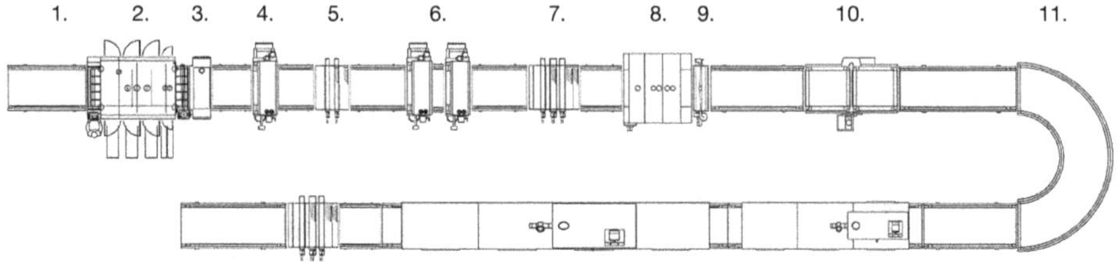

1. Support
2. Grinding machine (grinding belts = transversally/longitudinally/longitudinally arranged)
3. Dedusting
4. Single roller hard rubber smooth water stain approx. 25–30 g/m²
5. UV dryer with 1 mercury high-pressure lamp with 80 W/cm each for the gelling of UV primer
6. Double roller hard rubber smooth (with reverse unit) UV approx. 60 g/m²
7. Jet dryer approx. 60–70 °C circulation air temperature for the drying of the water stain
8. Grinding machine for the coating sanding
9. Dedusting
10. Curtain coater machine UV curtain coater, transparent approx. 120–140 g/m²
11. Transport curve
12. Circulating air-dryer with approx. 25 °C and low airflow rate
13. Dryer fitted with UV low pressure lamp type TL 05/TL 03
14. UV dryer for the UV curing with 3 mercury high-pressure lamps with 80 W/cm
15. Removal of the components

Figure 3.76: Rolling line/curtain coater line for the UV coating of veneered furniture components in South Europe

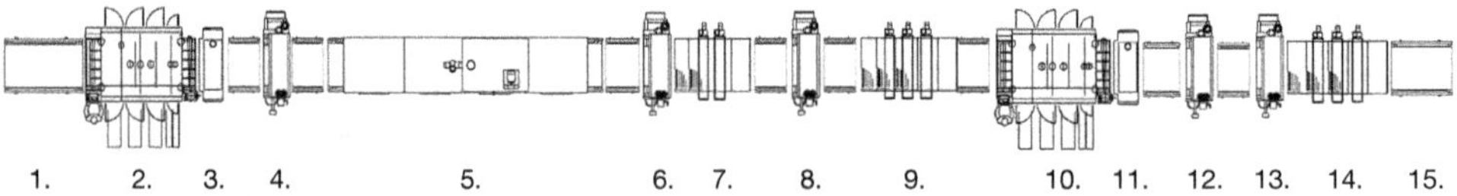

1. Support line
2. Grinding machine
3. Dedusting unit
4. Single roller hard rubber Wood stain approx. 15 g/m²
5. Jet dryer approx. 60–70 °Ccirculation air dryer
6. Single roller hard rubber smooth UV primer coating, transparent approx. 10 g/m²
7. UV dryer with 2 mercury high- pressure lamps with 80 W/cm for the gelling of the UV primer
8. Single roller hard rubber smooth UV primer coating, transparent approx. 10 g/m²
9. UV dryer with 4 mercury high-pressure lamps with 80 W/cm for the curing of the UV primer
10. Grinding machine for the coating sanding
11. Dedusting
12. Single roller hard rubber smooth UV finish coating approx. 25 g/m2 Wet-in-Wet with position 13
13. Single roller with a grooved roller 80 gear, approx. 30 shore
14. UV dryer for the UV curing with 4 mercury high-pressure lamps with 80 W/cm for the curing of the UV primer
15. Removal of the components

Figure 3.77: Rolling line with a grooved rubber roller in the topcoat station for the purpose of topcoat-type surfaces for veneered wood surfaces such as maple, beech, cherry tree and walnut tree

In countries such as Italy, Spain, France and others often closed-pore coating surfaces are preferred. In order to realize this, traditionally furniture elements and door elements are coated by means of a rolling process or curtain coater process (Figure 3.76). A double roller often is used in the field of primering whereby the second roller rotates in opposite direction to the flow direction of the workpiece (Chapter 5.6). This type of machine is referred to as 'reverse system' or 'reverse roll'. As a result, this enables the realisation of enhanced application amounts of coating (approx. 30 to 120 g/m^2) and smooth coated surfaces in a coating application. Subsequently to the UV curing and coating sanding, an application of a still partly solvent containing or styrene containing UV curtain coating with a solids content of 70 to 100 % is made. Such a coating construction provide a very fully and smooth coating surfaces. The feed rates for the rolling/curtain coater lines are maximally 8 to 12 m/min. In recent years, surfaces similar to curtain coatings were procedurally achieved by using grooved rubber rollers for the application of UV topcoats (Figure 3.77).

Profiled furniture components or furniture components with pronounced curvature cannot be coated in the pure UV rolling process. For this reason, such components are coated with UV coatings using the spraying process by means of automatic spray guns. Usually, two-layer or three-layer spraying processes with solvent-borne or water-borne UV primer coatings and UV topcoats are applied. The feed rate of the spraying lines is between 2 to 6/min. As illustrated in the Figure 3.80, the processing time for the overall coating in a spraying line amounts approx. 12 to 20 minutes depending on the process parameters.

Pigmented UV coatings

Even today, in South Europe the application of pigmented UV coatings is performed using so-called rolling/curtain coater lines operating with dual-cure pigmented coatings (Figure 3.78). In a first step, the applied MDF plates are primed by means of a double roller (reverse roller)

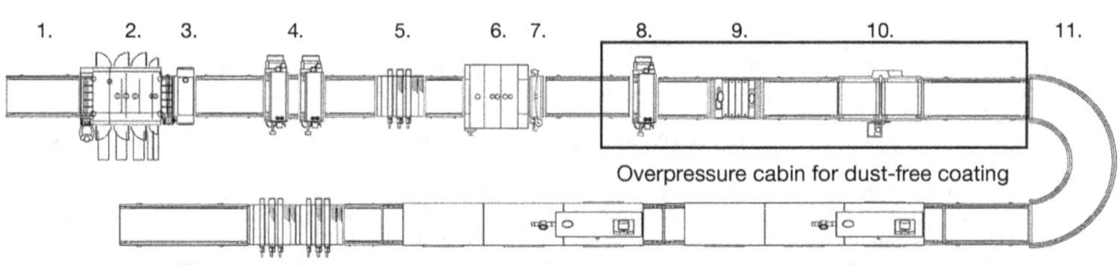

1. Support
2. Grinding machine (grinding belts = transversally/longitudinally/longitudinally arranged)
3. Dedusting
4. Double roller (with reverse unit) approx. 60–70 g/m^2 UV primer coating transparent
5. UV dryer for the UV curing 3 mercury high-pressure lamp with 80 W/cm each
6. Grinding machine for the coating sanding
7. Dedusting
8. Single roller for the application of a peroxide-containing primer ("active ground"), approx. 10–15 g/m^2
9. Infrared drying unit
10. Curtain coater machine Approx. 120–140 g/m^2 UV curtain coater
11. Transport curve, roofed
12. Circulating air-dryer with approx. 25 °C and low airflow rate
13. Dryer fitted with UV low pressure lamp type TL 05/TL 03 for the curing of the coating layer
14. UV dryer for the UV curing with 3 gallium doped mercury high-pressure lamps with 80 W/cm
15. Removal of the components

Figure 3.78: Spraying line for the coating of furniture components for example interior construction of ships

and sealed. Subsequently to the sanding of the coating and to the dedusting, a peroxide-containing priming with the pre-accelerated UV coating (e.g. cobalt salts) as well as due to the impact of the temperature, the radical chain polymerisation is started in the area of the boundary surface. Subsequently to the evaporation phase (approx. 3 to 4 minutes at a temperature of 20 to 25 °C) and gelation (approx. 3 minutes at a temperature of approx. 30 to 40 °C) by means of a variety of low-pressure lamps of the type "Philips TL 05" and "TL 03", the pigmented for the curtain application is UV cured within a few seconds [246, 247].

Such plants are operated with feed rates of about 6 to 10 m/min. White pigmented UV coatings are used most frequently. For several years, modern procedures operate with highly pigmented UV base coatings and UV finish coatings in the rolling process. The in Europe established UV lines for the pigmented coatings of MDF constructions, chipboard constructions and honeycomb constructions operate with feed rates of about 25 to 50 m/min (Figure 3.79). Thus, these UV-lines have a higher profitability in comparison to rolling lines/curtain coater

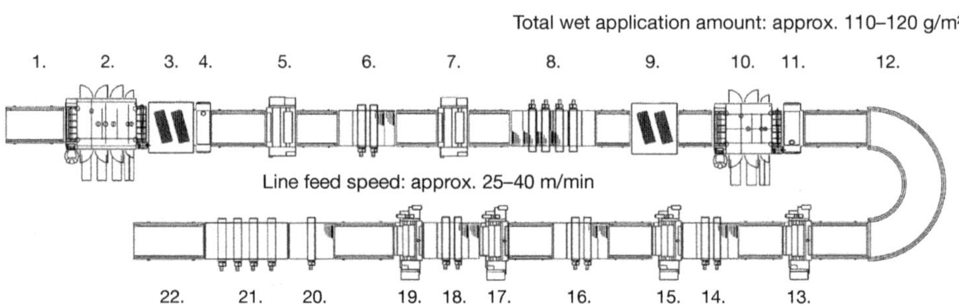

1. Support
2. Grinding machine for the wooden sanding (grinding belts = transversally/longitudinally/ longitudinally with 120–240 sandpaper grain)
3. Brushing device (2 brushes with an inclined arrangement)
4. Dedusting
5. Light filling machine with smooth roller UV primer, transparent approx. 30 g/cm²
6. UV dryer with 2 mercury high-pressure lamp with 120 W/cm each for the gelling of UV primer
7. Light filling machine with smooth roller UV primer, transparent approx. 20 g/cm²
8. UV dryer with 2 mercury high-pressure lamp with 120 W/cm each
9. Brushing device (2 brushes with an inclined arrangement)
10. Grinding machine for the coating sanding (grinding belts = transversally/longitudinally/ longitudinally with 280 sandpaper grain)
11. dedusting
12. ransport curve
13. Single roller hard rubber smooth UV rubber coating, transparent approx. 12–15 g/m²
14. UV dryer with a gallium doped mercury high-pressure lamp and a mercury high-pressure lamp with 80 W/cm each for the gelling
15. Single roller hard rubber smooth UV rubber coating, transparent approx. 12–15 g/m²
16. UV dryer with a gallium doped mercury high-pressure lamp and a mercury high-pressure lamp with 120 W/cm each for the gelling
17. Single roller hard rubber smooth UV rubber coating, transparent approx. 12–15 g/m²
18. UV dryer with a gallium doped mercury high-pressure lamp and a mercury high-pressure lamp with 120 W/cm each for the gelling
19. Single roller with a grooved rubber roller
20. (Shore hardness approx. 25–30, approx. 80 grooves/inch UV topcoat, white pigmented mearly 25 g/m²
20. UV dryer with a gallium doped mercury high-pressure lamp 120 W/cm (height-adjustable for the gelling)
21. UV dryer with a gallium doped mercury high-pressure lamp and a mercury high-pressure lamp with 120 W/cm each for the gelling
22. Destacking of the components

Figure 3.79: Rolling line for the coating of furniture components with pigmented UV coatings

lines with a comparable surface quality. In order to realize surface effects similar to curtains coatings, the pigmented UV finish coating is applied by means of a grooved rubber roller (Shore hardness approx. 25 to 30 and 80 grooves per inch) with a wet application amount of 25 g/m². A height-adjustable, gallium-doped mercury high pressure lamp is applied for the adjustment of the gloss level prior to the actual UV curing.

Rolling process in combination with the indirect gravure printing technology for laminate flooring and furniture components

In the furniture industry as well as in the flooring industry, the application of chipboard panels, hardboard panels and MDF materials considerably increased from the point of view of scarcity as well as the point of view of an economic optimisation of the production processes.

The printing in the indirect gravure printing of wood-based materials is a well-known technology in the furniture production. In the 1960ies, the printing of wood-based materials in indirect gravure printing in the furniture industry was applied for the rear panels of the furniture, interior decors of cupboards and furniture surfaces. One of the pioneers is the company Rauch which started the production of printed surfaces for sleeping rooms in the year [248]. By the midt-1970ies, the printing of wood-based materials enormously was displaced by the fast-growing decorative paper gravure printing [249].

The misleading name 'direct printing' results from the fact that printed designs directly are printed on the wood-based materials and not on paper [250]. In expert jargon, this is referred to as direct printing on wood-based materials by means of the indirect gravure process. Within the indirect gravure, the printing ink is the first to be transferred from the engraved printing roller on a hard rubber roller in order to compensate for irregularities and tolerances for a subsequent printing on wood-based materials. An application cutter which strips off the excess printing ink prior to the transfer to the hard rubber roller is located on the engraved printing roller [251, 252].

Thus, the printing ink remains only in the depressions of the engraved printing roller and then transfers the decorative structure on the hard rubber roller. The printing ink in the depressions are transferred as spots on the roller. For technical reasons, the printed design has to be rasterised in many small cells, so that the wiper scraper on the bar has a uniform supporting surface. Otherwise, the scraper would get out the colour from the cells, if the cells are too large.

When in contact with the wood-based material, the rubber roller transmits the desired decor image on the carrier by applying a corresponding contact pressure. The applied printing cylinders often consist of a nickel-plated steel core with an engraved copper layer which is protected by a thin coat of chromium from mechanical stresses. In addition to the conventional electromechanical engraving, nowadays the laser engraving (direct laser system) increasingly is applied. The laser engraving is able to present decorations still more concisely [254].

In recent years, enormous progress in the electronic control technology of the printing machines have opened up new fields of application for the indirect gravure printing on wood-based for high-quality furniture fronts and laminate flooring [250, 256]. The actually applied printing machines with an operational width of 1,300 to 2,400 mm can be synchronized more exactly via the control technology so that the resulting pressure comes very close to the optical and aesthetic requirements of furniture fronts (Figure 3.82). According to the specifications of the machine manufacturer, the new machines of the new generation with three or four modular cascaded pressure units achieve a fitting accuracy of +/- 0,5 mm [250].

When changing the engraved printing roller (change or design), the machine aggregates automatically are adjusted. The deployed engraved printing rollers have a diameter of 310 to 650 mm. The printing machines of the new generation are easy to clean, and the change of the engraved printing rollers is easier. Today, the change of the engraved printing roller is performed without tools and is performed in a time period of 20 to 30 minutes for two units with an optimized production process. Due to the available change-over units, a switching process takes only 10 minutes [257]. The today's gravure printing rollers, the rubberized applicator rollers and the transportation each are supplied with an infinitely variable drive. Each printing unit has a continuous tape transport (tape transport without belt conveyor). The gravure cylinder scraper is pneumatically adjustable as well as reviewable. A self-cantering conveyor control prevents in-accuraties in the transportation of the workpieces. For years, proven finishing technologies such as impregnated decorative printed papers and PVC foil with a normal layer thickness of approximately 0.15 to 0.25 mm exist for the imitating of wood surfaces. The process of coating for the indirect gravure printing has a total layer thickness of approx. 0.05 to 0.12 mm [161]. The indirect gravure printing is in direct competition to the decorative paper printers which print paper films. The decorative paper printers facing calmly and indifferently the indirect gravure printing previously are started and closely monitor the market activities in this segment [259]. The further development of the market will show to what extent the paper manufacturer and the decorative paper printers register losses of market share.

In practice, a distinction is made between monochrome printing and multi-colour printing. Generally, the monochrome printing is applied for cabinet rear walls and simple panels. The two-colour printing or three-colour printing is applied for high-quality furniture components and laminate flooring. The printing of economically interesting veneers for the imitation of high-quality veneer surfaces also is an application of the indirect gravure printing. Mainly, veneers such as maple, Aningré, birch, beech, Limba or industrially manufactered veneers such as ALPI, the company Alpi SpA Italy, are printed. An important fundamental requirement for an optimum and sharp printing image is a flat surface. Moreover, the quality of the rubberised applicator roller must be controlled regularly since the circumference of the applicator roller may change due to swelling effects or abrasion of the rubber coating [250].

Figure 3.80: Printing cylinder with wood decor design

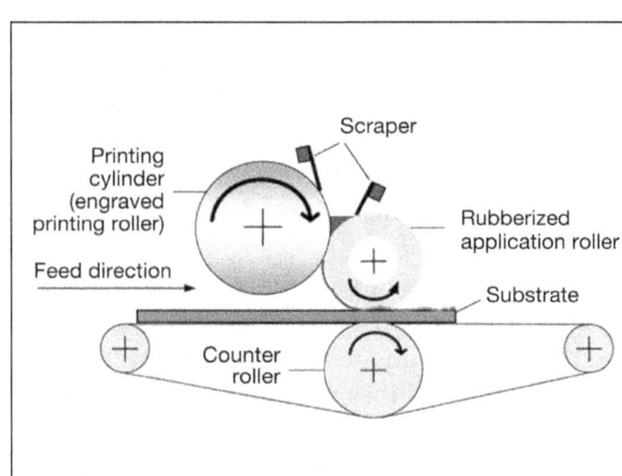

Figure 3.81: Principle of a printing machine

The surface hardness of the rubber roller also should be verified with of a shore hardness test device at regular intervals, since this influence printed design and the applied quantity. In the last few years, thus the printing of wood-based materials has become interesting since the current machine technology economically allow so-called flat lines, feed rates of 80 to 120 m/min in the monochrome printing or two-colour printing as well as 60 to 70 m/min in the three-colour process or four-colour process economically [257, 260]. Mainly water-borne and UV curable printing systems with more than 95 % solids are applied as inks. The advantages and disadvantages of aqueous printing inks are illustrated in Figure 3.85 in comparison to the UV curable printing inks. The economic advantage is approx. 50 to 60 % compared to UV printing inks, calculated on the square meter price [261].

Printing process for laminate floorings and veneer floorings (IPL = Indirect Print Laminate)

The origins of indirect print laminates are due to the year 1997 [261]. The subsequently applied UV coating systems could not yet compete with mechanically technological requirements of a laminate. Also, the former quality of printed designs still was inadequate.

Since 2003/2004, the technique of the indirect gravure printing in combination with newly developed UV curing systems for HDF-based floorings (laminate floorings and veneer floorings) has achieved qualitatively new dimensions which would not have been thought possible until recently [250].

The EPLF Association quantified the global production of laminate floorings and veneer floorings in the indirect printing process in the year 2004 on 20 million m². For the indirect print laminates, this accounts for approx. 3 % of the total world production of 690 million m² in the year 2004. Of such, 63 % were produced in Europe, 19 % in Asia and 8 % in America [258]. In the year 2005, the Association of European Producers of Laminate Flooring (EPLF) officially recognizes the indirect print laminate flooring under the term of laminate flooring or may be distributed as such [257]. The laminate coating is performed multi-layered applying newly developed UV coating systems with almost 100 % solids in the rolling process. This means, that approx. 100 % of the applied wet film remain unchanged as a dry film after UV curing.

Water-dilutable printing inks		**UV curable printing inks**	
Advantages	**Disadvantages**	**Advantages**	**Disadvantages**
Good gradient	Pre-heating of the support plate with IR lamps necessary	Good running stability on the printing machine	Preparation subject to labelling
Very good printed design	Necessity of a regular control of the viscosity	Possibility of high production rates on short plants	Precautionary measures in the handling of UV printing inks
No product labelling			Cleaning of machines with organic solvents
Very broad colour variety			Limited variety of colours
Very good price performance ratio			Enhanced product costs
Possibility to clean machines with tap water			

Figure 3.82: Advantages and disadvantages of water-dilutable and UV curing printing inks [261]

Table 3.40: Selected properties of HDF based floors [258]

	DPL floors (Direct Pressure Laminate)	IPL floors (Indirect Printed Laminate)	Veneer flooring
Quality grade (EN685)	31	31	31
Impact test	8–10 N IC 1 (EN 438-2.11)	8–10 N IC 1 (EN 438-2.11)	> 1200 mm EC 2 (EN 14354 Annex C)
Abeasion resistance	IP > 2,500 revolutions AC 3 (EN 13329)	IP > 2,500 revolutions AC 3 (EN 13329)	IP > 5,000 revolutions WR 2 (EN 14354 Annex D)
Scratch resistance EN 438-2.14 (Scratch Tester, Radial Method)	> 2.5 N	> 2.5 N	not specified
Optical reproduction	+	+	+++
Impression (haptics)	+	+++	+++
Wear of tools	high	high	low
Productivity in the manufacturing process	++	+++	+

In order to further optimize the price per square meter of the laminate flooring, often pigmented aqueous coating systems are applied with the rolling technology in the area of the primer and as a basis for printing primer. In comparison to UV curing and 100 % solids containing systems, the application of aqueous coating systems as a basis for printing primers often has the advantage that the quality of the printed design appears much more concise subsequently to the application of the printing ink. HDF boards often are applied as wood-based materials for the coated laminate floorings. The multi-layered roller design consists of a filling primer (approx. 10 g/m²), a repainting (approx. 20 g/m²), the water-borne basis for printing primer in the corresponding shade of colour (approx. 40 g/m²), a multi-colour printing with water-borne inks (approx. 3 g/m² per printing cylinder),a special, corundum containing UV primer (approx. 90 g/m² for achieving the abrasion class AC 3 according to EN 13329), a UV primer coating (40 g/m²) and finally the sealing with a special UV finish coating (approx. 10 g/m²) [161]. Depending on the profile of requirements, the total applied quantities amount approx. 120 to 200 g/m². The rates of production of the coating lines vary from 40 to 120 m/min depending on the number of printing machines.

Currently, the maximum operational width is 2.45 m. With an operational width of 2.45 m and an average capacity utilisation of approx. 75 %, a coating performance of about 13,200 m²/hour is available [258]. Depending on the requirement profile for laminate floorings (EN 13329), the abrasion behaviour (S42 paper) is achieved by means of the application of a special UV primer of 30 to 90 g/m². This special primer always contains a high percentage (up to 23 %) of corundum (Al_2O_3) which chemically is incorporated with high shear forces and an unspecified process [258]. With respect to unbound corundum particles, the anchoring of the corundum particles in the binder matrix significantly increases the abrasion resistance. In analogy to the modern UV finish coatings for the classical coating of parquets, the UV finish coating contains a silicium dioxide based 'nano modification' in order to increase the scratch resistance and abrasion resistance [258]. The modification also has the advantage that these coating systems can be better repaired subsequently to the grinding process in comparison to the corundum containing UV topcoat systems. Professional circles as well the literature reports on the haptic benefits and the considerable wood-like sound during walking on the coated floors (indirect printed laminates) [256, 262]. Table 3.40 compares selected properties of HDF based floorings (DPL flooring, IPL flooring and veneer flooring) [258]. The table shows that the technical properties of the IPL floorings absolutely are comparable with the traditional laminate floorings (DPL floorings).

Indirect printing process in the furniture industry

In recent years, the furniture industry has rediscovered and greatly pushed ahead the indirect printing process. For years, the manufacturers of quantitatively good mass furniture such as the company IKEA, for example, maintain the coating of visible surface in their supplies such as shelf side panels and shelves, bedroom furniture and even fronts consisting of chipboards in indirect gravure printing. In comparison to the requirements from the field of laminate floorings, the technical requirements and the applied quantities are significantly lower. Currently, chipboards, MDF plates as well as actually frame reinforced honeycomb panels ('Frames on Board') with a HDF layer are applied in the furniture industry.

Mainly two procedures for the printing of wood-based materials exist for sophisticated furniture fronts. The first and long-established procedure operates in the combination with

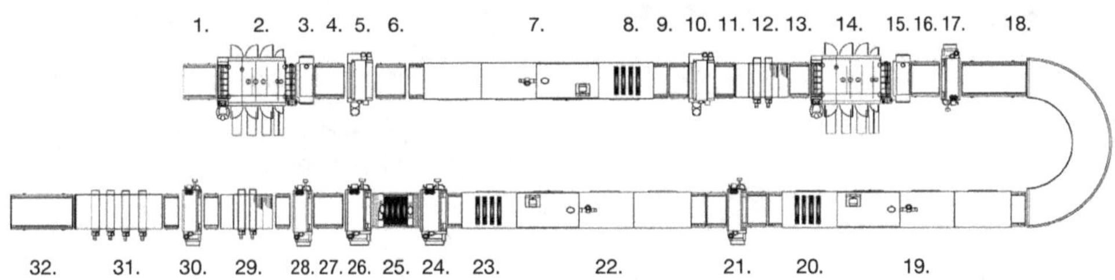

1. Support
2. Grinding machine (grinding belts = transversally/ longitudinally/longitudinally)
3. Dedusting
4. Control line
5. Heavy filling compound approx. 50–70 g/m² water-dilutable putty pigmented
6. Conveyor belt
7. Jet dryer 70–100 °C Circulation air temperature and approx. 20 m/s circulating air velocity
8. Ingfrared lamps (4 pieces)
9. Conveyor belt
10. Heavy filling compound approx. 20–30 g/m² UV coating putty, transparent
11. Conveyor belt
12. UV dryer for UV curing (2 Hg high pressure lamps 80–120 W/cm)
13. Conveyor belt
14. Grinding machine (grinding belts = longitudinally/longitudinally)
15. Dedusting
16. Conveyor belt
17. Single roller 10–12 g/m² water-dilutable printing primer pigmented
18. Conveyor belt/curve
19. Jet dryer 70–100 °C Circulation air temperature and approx. 20 m/s circulating air velocity
20. Infrared lamps (4 pieces)
21. Single roller 10–12 g/m² water-dilutable printing primer pigmented
22. Jet dryer 70–100 °C Circulation air temperature and approx. 20 m/s circulating air velocity
23. Infrared lamps (4 pieces)
24. Printing machine approx. 3 g/m² printing ink
25. Infrared lamps (4 pieces)
26. Printing machine approx. 3 g/m² printing ink
27. Conveyor belt
28. Single roller UV finish coating transparent approx. 5 g/m²
29. UV dryer (gelation), 1 gallium doped and 1 Hg mercury lamp each 80–120 W/cm
30. Single roller UV finish coating, transparent approx. 5 g/m²
31. UV dryer (gelation), 2 gallium doped and 1 Hg mercury lamp each 80–120 W/cm
32. Stacking

Figure 3.83: Plant example for a printing process in combination with water-borne und UV coatings on chipboard panel

pigmented, water-borne roller coatings as well as with 100 % solids containing, pigmented and transparent UV roller coatings. The material mix of water-borne coatings and UV coating systems normally results in a low price per square metre.

Where water-borne coatings are applied, the limiting factor is the drying phase which causes an extremely long planning of the coating systems at high feed rates of > 40 m/min in order to ensure the drying of the water-borne coatings. In order to shorten the drying process or in order to reduce the roughening of the wood-based materials, many coating systems work with a so-called pre-heating of plates consisting of several infrared lamps. Subsequently to the process of pre-heating, the substrate temperature is between 40 and 60 °C. The subsequently rolled-on water-borne coating (spatula) requires a significantly reduced drying time on the pre-heated plate. Empirically, the subsequent drying time can be reduced by approx. 20 to 30 % due to the pre-heated plates. In the last third of the drying zone, the applied nozzle ducts normally are equipped with infrared lamps. Figure 3.83 illustrates an example of an installation for a printing process in the combination process (water-borne coating/UV coating) on a particle board. This plant is designed for a feed rate of approx. 35 to 45 m/min. The printing ink is applied on an aqueous pigmented printing primer.

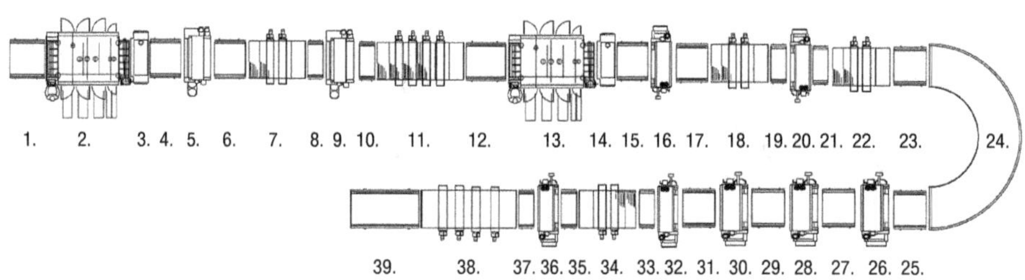

1. Support
2. Grinding machine (grinding belts = transversally/longitudinally/longitudinally)
3. Dedusting
4. Control line
5. Light filling compound UV primer, transparent approx. 20–25 g/m²
6. Conveyor belt
7. UV dryer for gelation (2 Hg high pressure lamps 80–120 W/cm)
8. Conveyor belt
9. Light filling compound UV primer, transparent approx. 20–25 g/m²
10. Conveyor belt
11. UV dryer for gelation (4 Hg high pressure lamps 80–120 W/cm)
12. Conveyor belt
13. Grinding machine (grinding belts = longitudinally/longitudinally)
14. Dedusting
15. Conveyor belt
16. Single roller UV primer, pigmented approx. 10–12 g/m²
17. Conveyor belt
18. UV dryer (gelation/curing) (2 gallium doped lamps, 80–120 W/cm)
19. Conveyor belt
20. Single roller UV primer, pigmented approx. 10–12 g/m²
21. Conveyor belt
22. UV dryer (gelation/curing) (2 Hg high pressure lamps, 80 – 120 W/cm)
23. Conveyor belt
24. Conveyor belt/curve
25. Conveyor belt
26. Printing machine approx. 3 g/m² printing ink
27. Conveyor belt
28. Printing machine approx. 3 g/m² printing ink
29. Conveyor belt
30. Printing machine approx. 3 g/m² printing ink
31. Conveyor belt
32. Single roller hard rubber smooth UV finish coating, transparent approx. 5 g/m²
33. Conveyor belt
34. UV dryer (gelation/curing) (1 gallium doped lamps and 1 Hg mercury lamp each 80–120 W/cm)
35. Conveyor belt
36. Single roller hard rubber smooth UV finish coating, transparent approx. 5 g/m²
37. Conveyor belt
38. UV dryer (UV curing) (2 gallium doped lamps and 2 Hg mercury lamp each 80–120 W/cm)
39. Conveyor belt/destacking

Figure 3.84: Plant example for a printing process with UV curing coating systems

The second and relatively new procedure works only with UV coating systems (approx. 100 % solids). The basic advantages of the pure UV process are due to a shorter plant, since the entire jet dryers for the drying of aqueous rolling coatings can be saved as compared to the combination procedure. Despite the compact design of the facilities, high feed rates can be realised due to the application of UV curing coating systems of > 50 m/min.

Figure 3.84 illustrates a typical plant from the furniture industry for the still-young segment of honeycomb panels based on HDF ('Board of Frame'[28]). An optimal adjustment of the applied ink with a UV curing of the printing primer is very important in order to achieve a clean and concise printed design. By applying a special ink, even a pore structure ('chemical pore') is achieved in combination with the UV topcoat. In this branch, it is also referred to as synchronous pores and universal pores. Depending on the requirements, the total applied quantity of coating materials is between 80 and 100 g/m². The applied products and procedures meet the sophisticated chemical and mechanical/technological requirements of the company IKEA according to IOS-MAT 0066 R 2 and R 4 without any problems.

In order to achieve an optimal printed design and a constant coating quality on chipboard panels and honeycomb panels, a variety of parameters have to be considered:

- To avoid 'frame drawings' or differences in the gloss level between the frame construction and the rest of the panel, respectively, efforts should be made to keep the surface temperature of the substrate as low as possible during the process of coating.
- To avoid chatter marks in the area of the printing ink, it is advisable to verify the position of the doctor blades. Also, an insufficient drying between the application of the first and second printing ink may cause stripes and chatter marks.
- Especially when coating honeycomb panels, it should be ensured that a not too high shore hardness is applied in the rubber coating of the applicator roller.
- The contact pressure of the workpiece should be approx. 0.4 to 0.8 mm during rolling on chipboard panels and approx. 0.8 to 1.2 mm during rolling on frame-of-board constructions.
- Basically, the following applies: the more regular and smoother the substrate surface prior to the application of the printing ink, the more exact the subsequent adjustment of the colour shade and the achievable quality of the printed design.
- When printing on chipboard panels, it should be ensured that the quality of the material and especially the surface density of the panel is crucial for a good result. Due to the manufacturing process of the chipboard panels, there exist qualitative differences between the front and back side of the panel whereby these differences may become apparent during coating procedure. For example, so-called surface defects such as pinholes may occur. The A side ("good side") should always be used. If this is neglected, then a more severe degree of swelling of the wood chips may occur when aqueous products are applied. A preheating of the panels by means of infrared lamps may provide a remedial action.

Coating of round rods, strips, panels, profiles, edges and ribs with the "Vacumat" process

In recent years, in addition to the already established application process such as spraying, curtain and roller coater of the round rods, strips, panels, profiles, edges and ribs, the

28 Board of frames (BoF) refer to the so-called honeycomb panels. These BoF plates are light-weight materials. Lightweight materials are isotropic or anisotropic materials whose density is lower than the density of the applied materials (raw materials), or where the structure of the raw materials specifically can be applied to a mono-directional or bi-directional formation of the structural strength without increasing the bulk density [263].

"Vacumat" application [264–267] is gaining great importance. Chapter 5.7 describes the "Vacumat" process in detail.

UV coating systems for the "Vacumat" application

More environmentally friendly UV coating systems are applied for the coating of wood and wood-based materials. Normally, the UV coating have a solids content of approx. 100 %. In comparison to the still common CN- and 2C PU coatings, the UV "Vacumat" coatings offer the following advantages:

- High feed rates during the coating process (20 to 200 m/min)
- Short drying times and short curing times
- Low amounts of coating application
- Over spray free coating (high application efficiency)
- Very good chemical and mechanical technological properties

As a matter of principle, UV single layer coating systems as well as UV dual layer coating systems are applied for the "Vacumat" coating. Figure 3.85 illustrates a schematic coating line construction for a "Vacumat" process.

UV single layer coating system

The UV "Vacumat" coating is applied both as a primer as well as a finish coating. This has the advantage that a complete coating can be performed by means of a "Vacumat" without changing the coating. The properties of an UV primer and UV topcoat are combined in a system.

UV dual layer coating system

A special UV primer is applied in a dual layer coating process. Subsequently to the UV curing and coating sanding, an UV topcoat in the desired gloss level is applied. In comparison to the UV single layer coating system, the advantage usually is a better grindability of the primer and a good surface smoothness of the finish coating. The feed rate is 25 to 50 m/min depending on the geometry and equipment of the UV curing dryer. The applied quantities are approx. 10 to 40 g/m². The pressure can be maintained at 100 to 280 mbar. The positioning of the UV lamps is essential for an optimal UV curing of the "Vacumat" coatings. The installation of 4 to 5 spotlights is necessary at a feed rate of approx. 30 to 40 m/min. Continuously adjustable mercury high pressure lamps (60/80/100/120 Watt/cm power) in the wavelength range of 280 to 360 nm are applied for transparent and slightly translucent shades. In addition to mercury high pressure, gallium doped UV lamps in the wavelength range of 410 to

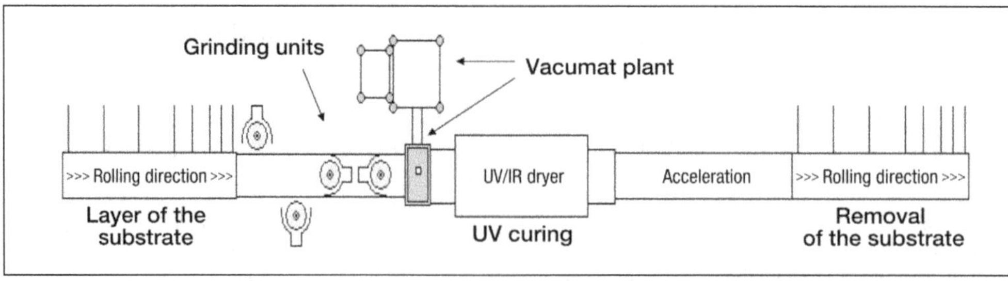

Figure 3.85: Schematic coatings line construction of a "Vacumat" application [268]

Table 3.41: Example of a dual layer process: Open-pored coating of oak transparent in different gloss levels

1. UV primer transparent • UV curing • Intermediate coating sanding	14–18 g/m²
2. UV topcoat transparent • UV curing • Destacking	10–13 g/m²

Table 3.42: Example of a single layer process: Closed-pored coating of beech (transparent)

1. UV primer transparent • UV curing • Intermediate coating sanding	20–25 g/m²
2. UV topcoat transparent • UV curing • Destacking	10–13 g/m²

Table 3.43: Example of a single layer process pigmented: coating of ash tree white

1. UV primer white • UV curing • Intermediate coating sanding	15–28 g/m²
2. UV topcoat white • UV curing • Destacking	10–13 g/m²

420 nm and with a power of 80 to 120 W/cm are applied for the curing of coloured glaze tones and pigmented UV coatings.

Fundamentally, the arrangement of the lamps (e.g. longitudinal or transverse arrangement) as a function of substrate geometry has to be determined in advance with the plant manufacturers and plant operators. Due to innovations in the mechanical engineering as well in the manufacture of coating substrates, operational widths of up to 500 mm are possible today. The Tables 3.41 to 3.43 illustrate examples of single layer processes and dual layer processes.

Industrial coating of parquets with UV roller coatings

Parquet flooring is a floor covering made of wood for interior rooms and still is a popular, exclusive and environmentally friendly flooring with continuous rates of growth. In the year 2002, an extensive life cycle assessment[29] showed in accordance with ISO[30] 14040-14043 that the environmental impacts even counteract the anthrapogenic[31] greenhouse effect during the manufacturing process and application of the wooden floors [269]. In the year 2005, the total production volume of parquet of the FEP countries (FEP – Federation of European

29 Life Cycle Assessment: Compilation and evaluation of input and output flows as well as the potential environmental impacts of a product system throughout its life cycle.
30 ISO: International Standardization Organisation that develops globally applicable standards.
31 Anthropogenic: attributal to human activity

association of the parquet industry) with Czech Republic, Hungary, Poland, Romania exclusive Great Britain and Portugal amounted approx. 96 million square meters (Figure 3.86). Single floors as well as multi-layer floors (the obsolete German terms is prefabricated parquet) which mainly are coated UV curing coatings industrially are applied. With approx. 78 %, the multi-layer parquet is the most frequently applied type of parquet (Figure 3.87). The most commonly applied types of wood were oak with a proportion of approx. 50 %, tropical wood species with approx. 17 % and beech with approx. 9 % of the parquet production in Europe in the year 2005 (Figure 3.88). Modern parquet coatings always are free from organic solvents. Depending on the requirement and depending on the surface effect, the total applied quantity of 50 to 150 g/m² was applied. In order to achieve the highest chemical, mechanical resistances as well as colour effects, wood stains, water-borne primers in combination with UV coatings are applied (Table 3.44). The effects and the adjustment of the colour shade is achieved by means of water-dilutable standard wood stains, still occasionally solvent-based wood stains, high-solid UV stains and translucent pigmented, water-dilutable UV primers.

Wood stains

The so-called UV stains which also contain UV resins as well as the colouring components such as dyes and pigments, predominantly are applied in the United States of America where a rustic stain appearance is desired. The UV stains contain UV curable resins as resin

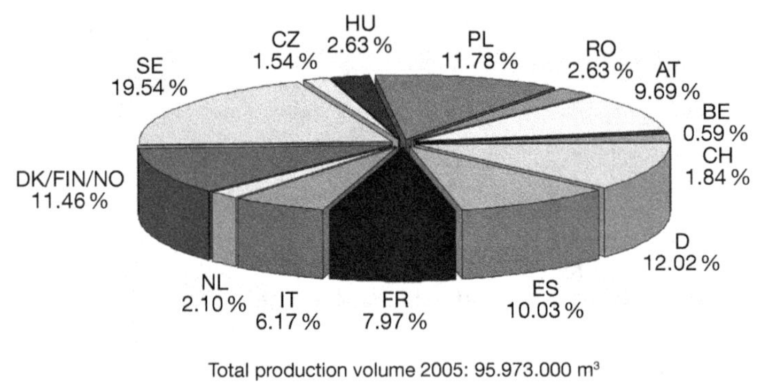

Figure 3.86: Total production volume of wood parquet 2005 of the FEP countries (Federation of European associations of the Parquet industry) with CZ, HU, PL, RO without Great Britain and Portugal[270]

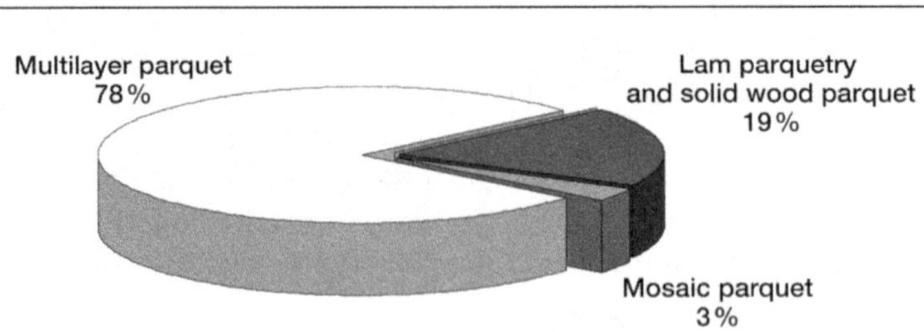

Figure 3.87: Percentage distribution of the wooden parquet types at the total production 2005 of the FEP countries (Federation of the European Parquet Industry) with CZ, HU, PL, RO without Great Britain and Portugal[270]

Table 3.44: UV curing coating systems for the coating of prefabricated parquet in using the rolling process

	Coating system	Function/Properties	Wet application amount [g/m²]	Solids (non-volatile components) [%]
1.	Wood stain, water-dilutable	Effects and colour shading	10–40	5–15
2.	UV base coat, water-dilutable, optionally light translucent pigmented	Effects, colour shading and to achieve a very good adhesion to the support	10–20	20–95
3.	UV putty	To close joints, pores and irregularities on the support in a coating application in order to achieve closed-pore surfaces	25–60	100
4.	UV base coat with addition of corundum ("antiabrasive")	Increase of the abrasion values according to the testing standard EN 438-2. The UV primer is not grindable	10–25	100
5.	UV base coat	Improvement of the filling and application as a grinding primer in order to achieve a smooth coating support	10–20	100
6.	UV base coat for SIS	Achievement of high abrasion values according to SIS 923509 (SIS = Swedish Standard Institute). Known as "Grit Feeder" or "Falling Sand" testing.	10–30	100
7.	UV topcoat antiabrasive adjustment by means of corundum	Scratch-resistant coating surfaces, even resistant to scratching with steel wool. The coating surfaces are non-reparable.	8–10	100
8.	UV topcoat antiscratch coating	Achievement of very scratch-resistant coating surfaces with good ability for repairs	8–10	100

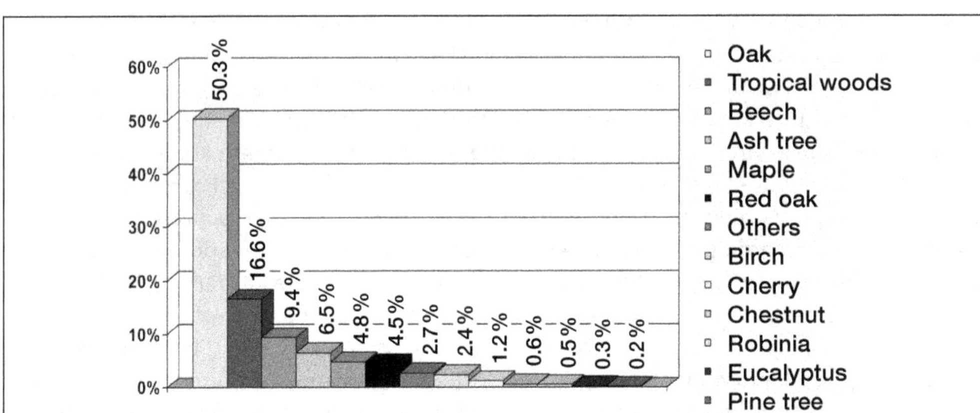

Figure 3.88: Distribution of the wood species applied for parquet floorings 2005 of FEP countries (Federation of the European Parquet Industry) with CZ, HU, PL, RO without Great Britain and Portugal [270]

component and thus have to be cured or crosslinked sufficiently. A very intensive UV radiation may result in a poor intermediate adhesiveness with the subsequent UV coating material. Depending on the suction behaviour of the wooden beam, the UV resins may penetrate into the substrate and are not recognised by the UV radiation. UV stains mainly are applied for large-pore woods such as oak. UV stains are not applied on fine-pored wood since frequently spotted appearances of the stains are achieved. In Europe, mainly aqueous rolling stains are applied which are characterized by a very good association of adhesive strength with the super-adjacent UV coatings. Moreover, the applied pigments and pigment preparations have to be resistant against UV light to a large extent (Chapter 3.1.1 Wood stains).

UV-Primer

For the purpose of achieving a very good adhesion strength, mainly unpigmented or slightly translucent pigmented, aqueous UV primers are applied. The proportion of water ensures a swelling and roughening of the underground and results in a very good overall network for all wood species. UV curing high solid primers with a low water content are applied for special surface effects or for coating plants which do not have a jet dryer for the evaporation of products with a high water content. UV curing high solid primers with a low water content are applied as a first primer directly on the wood support and can be applied in the UV curing without evaporation or coated wet in wet with the subsequent UV primer. Only after this, both coating layers are supplied to the curing. The solids content (non-volatile content) usually is approx. 70 to 95 %.

UV putty

Within a roller application, UV putty perform the function of fill irregularities, pores as well as lamella joints. Closed-pore surfaces are achieved depending on the applied quantity. The effective wet application quantities after the removing/smoothing with the smoothness roller of the putty machine is between 25 and 50 g/m². The UV putty are applied with light and heavy putty machines.

UV base coat with corundum

Corundum-containing UV primers increase the number of revolutions of the "Taber" abraser device up to the abrasion of the total coating in accordance with DIN[32] EN[33] 438-2.6. Depending on the corundum concentration in the UV primer and depending on the application quantity, abrasion values (initial point according to DIN EN 438-2.6) of approx. 150 to 500 rpm relative to 100 µm dry layer can be achieved. Systematic investigations from *Müller* have shown that different types of corundum may result in different abrasion results according to DIN EN 438-2.6 (Figure 3.89) [217]. A very careful selection and operational testing of corundum types is required. Corundum-free UV primers, however, achieve abrasion values (initial point according to DIN EN 438-2.6) of 25 to 150 rpm. By application of corundum-containing UV primers one has to ensure that these are applied as far as possible at the bottom of the total coatings since these are not grindable, or the belts of the grinding machines wear out in a very short time. Normally, the corundum-containing primers are gelled with UV lamps and subsequently overcoated with an oxide-free UV primer. Only then, the actual UV curing and the coating sanding can be done. Excessive application quantities of

32 DIN = German Institute for Standards
33 EN = European Norm

corundum-containing UV primers may cause initial equalising effects or greying on dark wood. Selecting the right type of corundum, storable UV primers resistant to participation can be produced. In order to achieve a high combination with corundum abrasive wear (number of revolutions) as high as possible according to DIN EN 438-2, the UV primers are adjusted 'hard' based on UV resins [272, 273].

UV primer without corundum

Corundum-free UV primers contribute to the further layer structure and usually are good grindable. These also have the task to cover the corundum-containing coating layers in order to prevent the contact of the corundum particles with the abrasive belts of the grinding machine during the subsequent intermediate coating sanding. In addition, the mechanical and chemical properties of the coating layer essentially are affected.

Special UV base coats

These special UV primers are formulated in thus, that these improve the abrasion testing according to SIS 923509 (SIS = Swedish Institute of Standard). The abrasion testing also is known as Grit Feeder or Falling Sand Test. The UV primers are adjusted tough elastically by means of the UV resins. In practice, the UV base coats often are referred to as 'SIS base coats'.

UV topcoat

The UV topcoat is applied as the last coating layer in the coating process. The application generally is done with two coating application rollers in the wet-in-wet process or in the rolling process with UV gelation (rolling/gelling/rolling) between the two coating application rollers prior to the actual curing. The wet application amount is 5 to 10 g/m² per vcoating application

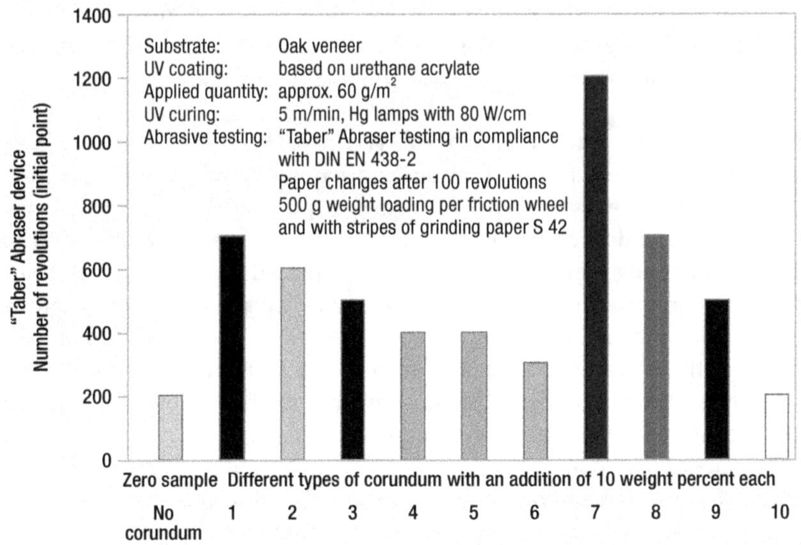

Figure 3.89: Impact of different qualities of corundum on the abrasion behaviour of an UV base coat based on urethane acrylate according to DIN EN 438-2 [217]

roller. Primarily, the UV topcoat has the task to protect the overall structure against chemical and mechanical stress. The UV topcoat for the coating of the parquet is characterized by an enhanced coating hardness and scratch resistance in order to meet the required suitability for use. Two UV coating technologies have proven themselves in practice:

Anti-abrasive adjustment by addition of corundum
The anti-abrasive properties are achieved by addition of specific minerals such as corundum. This micronised corundum powder frequently is applied alone or in combination with other minerals. The scratch resistance of such UV topcoats is so excellent that the coating surface cannot be scratched with steel wool. For this reason, such UV topcoats also are referred to as ceramic UV topcoats. In practical applications, the UV topcoats modified with mineral fillers also have disadvantages such as poor grindability of the coating surface and thus a restricted availability for repairs in the case of prefabricated parquets [274]. The corundum-containing UV topcoats especially are suitable for non-repairable flooring systems such as veneer parquet since the thin veneer layer (approx. 0.6 mm wear layer) would not allow a grinding and thus a repair. In comparison to metallic objects, there is an abrasion which leaves grey markings. Every now and then, also an abrasion towards rubberized soles, plastic rollers of furniture and chairs is observed.

Anti-scratch coatings
The further development of the corundum containing UV topcoat resulted in anti-scratch systems which are free of anti-abrasive minerals. Compared to objects moving on the surface, these UV topcoats do not produce metal abrasions or plastic abrasions. Depending on the formulation components, the anti-scratch coatings are characterized by better scratch strengths such as in corundum-containing UV topcoats. Among other things, the anti-scratch coatings contain nanoparticles. This is why anti-scratch coatings often are referred to as nanocomposite containing UV topcoats. Also, these UV topcoats are not grindable and therefore also restrict repairable. Further developments on the basis of nanoparticles aimed to further improve the hardly visible micro scratches. In addition, repair-friendly UV topcoat systems should be available in the future.

UV curing oils based on renewable raw materials

The desire of the consumers for eco-friendly and natural, matt parquet surfaces ('natural look') has resulted in a steady growth and usage of pure oxidative curing, long-oil alkyd resins and oils. The wood structure is accentuated by oil-based products. The initial still strongly solvent containing products gradually were replaced by products with a higher coating solid. The acceleration of the oxidative curing requires an application of siccatives in small amounts. Metal salts of cobalt or manganese can be used as siccatives. Due to the slow curing mechanism of the oxidative curing, the efficient processing of oil-based products in the process of roller application is not realisable compared to UV curable products. Depending on the type of oil and drying plant, the oil-based products need a circulating air drying of about 3 to 6 minutes at a temperature of about 35 °C or 60 to 90 minutes at a temperature of 23 °C up to dust dryness. Depending on the procedure, the oil surface is brushed from time to time and finally polished. In addition, the danger of spontaneous ignition of oil-impregnated cleaning rags and smouldering fires resulting from grinding of oil-treated is highlighted [189]. For this reason, it has not missed attempts to mix oxidatively curable products with UV resins. At the beginning, the mixes were not compatible and stable when stored. The combination of the two coating technologies results in products that penetrate well into the wood and result in a natural surface. The curing process is accelerated by means of the UV curable amount. Stackable parquet surfaces that do not glue in the stack are obtained. The UV reactivity of oil/UV resin combinations is not sufficient for the application on very high-speed UV rolling lines with feed rates of 20 to 40 m per minute. Thus, since the year 1997, the raw materials industry deals with the so-called UV resins based on oil. It is often about classic UV resins based on multivalent polyols and acrylic acid modified with saturated or monounsaturated fatty acids to a lesser extent. Incorrectly, these UV resins are referred to as 'UV oil' or 'UV cured oil', even though these UV resins contain any oxidatively cured fatty acids with conjugated double bonds. It is vital, that with such modified UV resins oil-like surfaces can be produced that are industrially processed very well. Compared to conventional UV resins, lubricated parquet surfaces are not so resistant chemically and mechanically. The measurable thickness of dry films of lubricated surfaces is up to 25 to 50 % of a conventional UV design for parquet floors. For this reason, lubricated parquet surface.

Miscellaneous UV curing products for the parquet coating

For some year, translucent and fully pigmented repair putty on the basis of 2C systems are applied for the repair of wood defects and holes. There are two systems of repair putty:
- Combination of UV curing with polyaddition by adding polyisocyanates
- Combination of UV curing with polymerisation by adding peroxide-containing curing agent.

In the field of the coating of edges, UV coatings which are applied with a "Vacumat" are used.

Testing of the abrasion resistance of floor surfaces

The abrasion resistance is the dominant feature of quality to characterize the wear resistance of floors [276]. Floor surfaces mainly are abraded or scratched, respectively, due to footwear with adhering dirt. This also is referred as dynamic load. From technical testing of view, this is simulated by means of the so-called 'Taber' abraser test[34]. This test is an internationally acknowledged test method and is mentioned in numerous national and international standards. The stress by abrasion is generated by two rotating rollers which are pressed on the rotating test specimen with a defined force. The defined friction wheels are applied directly or covered with self-adhesive strips of abrasive paper (see test standards DIN 68861 part 2, DIN EN 438-2), or corundum powder (aluminium oxide) is used as the abrasive (see testing standards SS 923509 T1, IOS-T-0507. EN 13696, ASTM[35] F 510-93) which continuously is applied between the friction wheels and the test pieces continuously by means of a metering unit. The resulting wear pattern is a pattern of crossed arches. Thus, the specimen is processed isotropic. Besides the human factor, the result of the examination strongly depends on the choice of the type of the friction rollers, pressure strength of the rollers on the sample surface (each 500 or 1,000 g). Also, fluctuations of the hardness of the rollers (shore hardness) can lead to different results [277]. Depending on the required test standard, there are different evaluation processes which result in significant differences in the number of revolutions.

34 'Taber' abraser test = At the international level, the abrasion is tested based on the "Taber" abraser test, a test procedure developed by the company Taber, USA
35 ASTM = American Society for Testing and Material

Table 3.45: Requirement profile for coating of wooden floorings[280]

Property	Minimum
	1
Fundamental properties:	
Geometric parameters and moisture content according to DIN 13489	Specif. met
Adhesion strength according to DIN EN ISO 2409 [cross-cut rating]	GT0-GT2
Brinell hardness of the wearing surface according to IHD work's specifications 202 [N/mm²]	No specif.
Fouling tendency according to IHG work's specification 427 [step]	1
Chemical resistance: Behaviour at chem. stress according to DIN 68861 Part 1 + water or vintensified with test equipment of DIN EN 13442 (Wooden floorings)	1C; 5 h all PM (except acetone) 5 h, acetone 10 s
Abrasion: Behaviour of abrasive load according to DIN EN V 1396 [u] (abrasive paper procedure)	≥ 50
Ritz hardness according to IHD work's specification 438 in [N/50 µm] and scratch testing/elasticity testing according to Hamburger test instruction. [N]	≥ 0,20 ≥15 N
Impact resistance/elasticity:	
Behaviour at impact load according to IHD work's specification 425 [step] or elasticity testing according to PR DIN EN 13696 [Kegel – No.]	≥ 2 H (8)
Behaviour at resistance to castor wheels with soft castors according to DIN EN 425 (25,000 cycles)	No specif.
Safety-related properties:	
Slip resistance according to DIN EN 13893/DIN 53131 (leather slider) [-]	µ ≥ 0,22
Resistance to cigarette burns according to DIN 68861 Part 6 [class]	No specif.
Resistance to aging:	
Resistance to alternating climate according to IHD work's specifications 424 [cycles]	≥15

Weight loss method

The test specimen is exposed to a specified number of rotation cycles. The abraded portion of the specimen is determined by a differential weighing. The less the number of determinations of the abrasion in mg per number of rotation cycles, the better is the wear resistance against the stress by abrasion.

Visual test method

The number of rotation cycles is determined until a specific criterion for abrasion is achieved. In the test standard DIN EN 438-2, the number of rotation cycles of the test specimen are measured as an initial point (IP) and recorded up to that point at which damages of the specimen occur. The abrasive load will be continued until 95 % of the coating is destroyed. The total number of rotation cycles of the specimen measured up to this point are considered as final abrasion point (final point, FP). The arithmetic mean of the identified number of rotation cycles up to the initial point and the identified number of rotation cycles up to the final point is established by adding both identified numbers of rotation cycles and subsequent

requirement for the performance class:

2	3	4	5	6
Specif. met	Specif. met	Specif. met	Specif. met	Specif. met
GT0-GT2	GT0-GT1	GT0-GT1	GT0-GT1	GT0-GT1
No specif.	> 25	> 25	> 30	> 30
1	0	0	0	0
1C; 16 h all PM (except acetone) 16h, acetone 2 min	1C; 16 h all PM (except acetone) 16 h, acetone 2 min	1B; 16 h all PM (except acetone) 24h, acetone 2 min	1 B; 16 h all PM (except acetone) 24h, acetone 2 min	1 C; 5 h all PM (except acetone) 5 h, acetone 10 s
≥ 80	≥120	≥120	≥180	≥ 200
≥ 0,20 ≥15 N	≥ 0,25 ≥18 N	≥ 0,25 ≥18 N	≥ 0,30 ≥ 20 N	≥ 0,30 ≥ 20 N
≥ 2 H (8)	≥ 3 H (9)	≥ 3 H (9)	≥ 3 I (9)	≥ 3 I (9)
No specif.	no coating	no coating	no coating cracks, no wear appearance	no coating cracks, no wear appearance
μ ≥ 0,22	μ ≥ 0,22	μ ≥ 0,30 (with care products)	μ ≥ 0,30 (with care products)	μ ≥ 0,30 (with care products)
No specif.	No specif.	No specif.	6 D	6 D
≥15	≥15	≥ 20	≥ 20	≥ 20

division by two. When comparing different wear loads of parquet floorings, it always has to be ensured that the initial point as well as the final point are stated as well. In advertising messages, often wear loads are specified where it is known what test standard was applied or where the initial point, final point, or the arithmetic mean of both is meant. In addition, the measured total thickness of the dry film should be specified as the stress by abrasion significantly depends on the total thickness of the dry film.

The international test methods for the determination of the stress by abrasion provide no consistent results for the UV primers applied in the market. Depending on whether the UV primers are set more 'hard' or 'tough elastic' due to the composition of the UV resins or still contain mineral fillers, the results of the respective test method can be influenced. UV primers which are formulated 'hard' and additionally contain corundum, achieve a high number of rotation cycles up to the abrasion of the coating in the so-called 'Taber' abraser test according to the test standards DIN 68861 part 2 and DIN EN 438-2. However, if 'hard' UV primers are tested with regard to the test standards SS 923509 T1, IOS-T-0507, EN 13696 and ASTM F 510-93, a lower number of rotation cycles up to abrasion of the coating is determined and vice versa. Depending on the required test standards by the parquet manufacturer, the coating manufacturers may formulate the UV coatings so that the respective stress by abrasion are met with regard to the test standards DIN EN 438-2 or SS 923509 T1, for example. However, from today's perspective and knowledge a balanced relation between 'hard' and 'elastic/flexible' layers of coating in the overall design for parquet floorings is required [275]. In the meantime, the most parquet manufacturers now have admitted both abrasion tests (DIN EN 438-2 and SIS 923509) in their profile of requirement. In addition to the above mentioned abrasion tests, different test specifications and inspection methods for the determination of the applicability of wooden floors exist. In Table 3.45, *Emmler* illustrates the essential test methods as well as a possible requirement profile for wooden floorings. The stresses of the practice can be divided into certain classes (see Table 3.46). A distinction usually is made between private and public/commercial application, a specified classification for wooden floorings can be imagined in accordance with Table 1 (DIN EN 438-2 or SS 923509 T1). In the application of the test methods it should be not forgotten, that the inhomogeneity of the wood substrate or the wood construction, respectively, affect the results of the determination of the impact strength and indentation [278, 279].

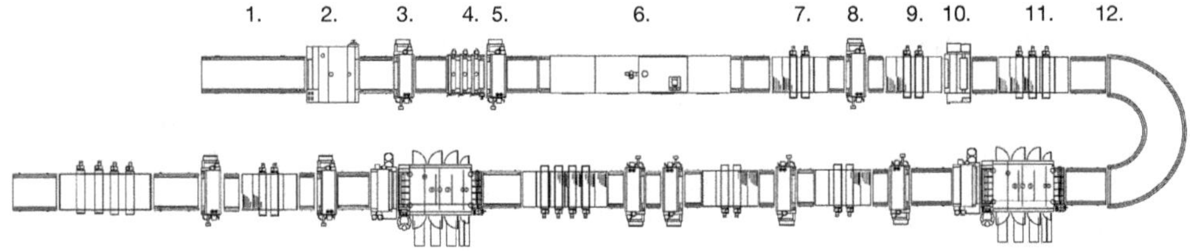

1. Support
2. Grinding machine for the calibration of wood
3. Single roller for the application of wood stains, approx. 10–20 g/m²
4. Ejector brushes for the wood stains
5. Single roller hard rubber smooth
   UV primer, transparent approx. 10–15 g/m²
6. Jet dryer
7. UV curing
   2 mercury high pressure lamps with each 80–120 W/cm
8. Single roller for the application of wood stains, transparent 10 g/m²
9. UV gelation
   2 mercury high pressure lamps with each 80–120 W/cm
10. Levelling machine
    UV putty, transparent approx. 25–30 g/m²
11. UV curing
    2 mercury high pressure lamps with each 80–120 W/cm
12. Transport curve
13. Grinding machine for the coating sanding
14. Dedusting
15. Single roller hard rubber smooth
    UV base coat, transparent approx. 10–15 g/m²
16. UV gelation
    2 mercury high pressure lamps with each 80–120 W/cm
17. Single roller hard rubber smooth
    UV base coat, transparent approx. 10–15 g/m2
18. UV gelation
    2 mercury high pressure lamps with each 80–120 W/cm
19. Single roller hard rubber smooth
    approx. 15–20 g/m² wet-on-wet application with position 20
20. Single roller hard rubber smooth
21. UV curing
    4 mercury high pressure lamps with each 80–120 W/cm
22. Grinding machine for the coating sanding
23. Dedusting
24. Single roller hard rubber smooth
    UV base coat, transparent approx. 5–10 g/m²
25. UV gelation
    2 mercury high pressure lamps with each 80–120 W/cm
26. Single roller hard rubber smooth
    UV primer, transparent approx. 5–10 g/m²
27. UV curing
    4 mercury high pressure lamps with each 80–120 W/cm
28. Withdrawal

Figure 3.90: Rolling line for the coating of parquet floorings

Figure 3.90 depicts a typical UV rolling line for the coating of parquet surface. In this example, the wood calibration (wood pulp) was integrated. However, often the calibration finish is performed on an external line because the maximum feed rate amounts 15 to 20 meter per minute due to mechanical reasons.

There are also other UV flat line concepts at which the wood calibration is set up separately for reasons of capacity and production rate since the calibrated finish only can be performed optimally at a feed rate of < 20 meter per minute. Actual UV roller coating lines for the coating of parquet surfaces are operated with feed rates of 10 up to 50 meter per minute.

3.1.7 Water-borne coatings

Since the 1970ies, water-borne coating systems[36] have expanded into the industrial wood coating and have grown in significance continuously due to a continuous refinement of resins also for high-quality furnitures [281]. With regard to the implementation of the European

VOC directive, many wood processing companies are forced to consider their coating systems, procedures and processes and to adapt to the statutory emission requirements. In the transition or re-planning of plants, respectively, the application of water-borne coatings is one possibility of meeting the legal requirements in Europe and of replacing solvent-borne coating systems. Following the devastating chemical accident in Tianijin (People's Republic of China) on which further accidents followed all over the country, also in emerging markets like China it can be assumed that the chemicals regulation in the People's Republic of China is further tightened. According to the information of the Chinese Coatings Association (CN-CIA), the Ministry for Environment Protection of the People's Republic of Germany will prepare a draft paper how to further reduce the VOCs in the coating industry.

According to DIN EN 971-1, water-borne coating material is defined as a coating material whose resin predominantly is dissolved or dispersed in water. This general definition includes both water-insoluble polymers dispersed in water, water-soluble polymers a well as the transitional region of colloidal solutions. Today, aqueous resins generally can be defined as resin solutions or dispersions whose volatile components predominantly consist of water [282,283].

Depending on the requirement profile of the process and the quality of the coating, transparent, translucent and fully pigmented water coating systems are available nowadays. The most important water-borne coatings for the industrial coating of furnitures and doors are:
- 1-component water-borne coatings (physically drying) self-crosslinking (SC)
- 1-component UV curing water-borne coatings, physically or not physically drying
- 2-component polyurethane water-borne coatings (addition of polyisocyanate as a hardener
- 2-component polyurethane and UV curable water-borne coatings (combination as so-called dual cure systems)

For many years, under solvent-borne coating systems are applied as insulation of knotholes (pine), since the resins of the water-borne coatings are heavier solvable in organic solvents and thus form a barrier for the solvent-containing coating. For a long time, water-borne coatings in solvent-containing wood stains also are applied for the levelling of the staining appearance between the deposed and non-deposed veneers. Pigmented and transparent 1C

36 In practice, water-borne coatings also are referred to as hydro coatings

Table 3.46: Classification of wood-based floorings into application classes [280]

Application class	Field of application	Intensity of application/ characterisation of the strain	Examples of application for wood-based floorings
1	Private residential areas	Weak/occasional application	Bed rooms and guest rooms
2	Private residential areas	Medium/permanent normal application	Living room, dining room, workroom
3	Private residential areas	Strong/intense application	Entrance corridor, children's room, kitchen
4	Application in object area (public/commercial)	Weak/occasional application	Hotel room
5	Object area	Medium/permanent normal application	Boutiques and little stors
6	Object area	Weak/occasional application	Department stores and multipurpose halls

water-borne coatings are applied for a long time on a large scale for the coating of hardboard panels for the back panels for furniture or wooden chairs. In recent years, large amounts of transparent CN coatings have been substituted by self-crosslinking 1C water-borne coatings. Since around the year 2004, 1C curing clear coating systems were processed on industrial lines with regard to the transparent solvent-borne 2C PU coatings. For the high-quality coating of kitchen fronts, for several years pigmented 2C polyurethane water-borne coatings are available which may fulfil the enhanced requirement for chemical resistance as required in DIN 68861 part 1B. Fully pigmented dual cure systems for the kitchen industry are the latest development. A rational coating also for pigmentations which can be difficultly cured by means of UV radiation can be performed by means of a combination of the two cross-linking mechanisms (polymerisation and polyaddition).

3.1.7.1 Components of water-borne coatings

Figure 3.91 illustrates the most important components of water-borne coatings. It has to be taken into account that an immensely large selection of a wide variety of resins and additives for the formulation of water-borne coatings exists.

Resins for the water-borne coatings

Aqueous coating systems mainly are polymer dispersions. The term polymer dispersion is explained in Chapter 3.2.2. There are several possibilities to classify the aqueous resins. The main classification options are as follows:

- According to the size of the present polymer particles in the aqueous phase
- Molecularly dispersed (< 1 nm)
- colloidal dispersed (10 to 100 nm)
- coarsely dispersed (>1.000 nm)

Transition regions between molecularly and colloidally dispersed systems or colloidally and coarsely dispersed systems, respectively, are available for particle sizes between 1 to 10 nm and between 100 to 1,000 nm. Table 3.47 illustrates a finer classification which better considers the transition regions of resins of the applied resins in aqueous systems.

Table 3.47: Classification of organic resins in water[284]

Particle size [nm]	Appearance[1)]	Characterization
> 1.000	Milky white	Coarse polymer dispersions, coarse emulsions
1.000–100	White to bluish	Fine-grain polymer dispersions
100–50	Bluish to transparent	Finest-grain polymer dispersions, dissolved high-polymers[2)], micro emulsions
50–10	Increasingly transparent	Hydrosols of dissolved polymers[3)]
10–1	Transparent (Tyndall effect)	molecular solutions[3)]

1) With substantial difference of the refractive indices of water and coating binder
2) Molecule colloid
3) Coating binder with low molecular weight (e.g. MF resins)

- According to the manufacturing process (such as primary dispersions, secondary dispersions)
- According to the type of resins (such as acrylate dispersions, polyurethane dispersions and others)

Primary dispersions

As it is described in Chapter 3.2.2 for the external coating, the classic acrylic dispersions (emulsion polymers) also are applied for the furniture coating. The fundamental relationships of the production, particle size or particle size distribution, morphology as well as the physical and chemical mechanisms of film formation are described in Chapter 3.2.2.

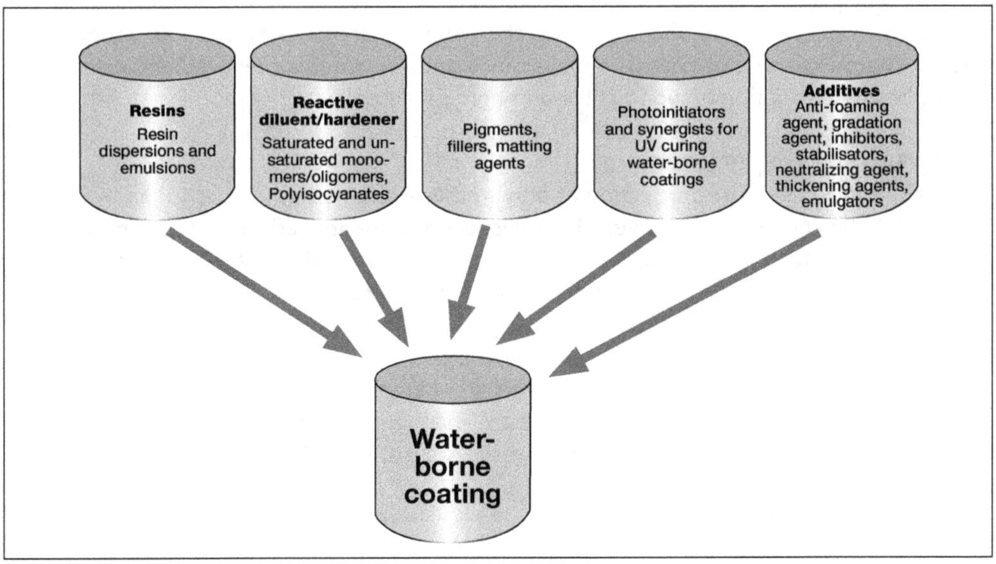

Figure: 3.91: General composition of water-borne coatings

However, the acrylic dispersions for application in the future sector often differ from the acrylic dispersions for the outdoor application in the following properties:
- Enhanced film hardness
- Lower elasticity
- Enhanced temperature of film formation
- Improved chemical resistance
- Different morphology

In the field of furniture coating, non-crosslinking and self-crosslinking acrylic dispersions are applied depending on the required profile. Self-crosslinking acrylic dispersions have prevailed with respect to the chemical resistance of furniture surfaces. In these systems, the chemical cross-linking during the phase of film formation often is achieved by the application of dihydrazides (see Chapter 3.2.2). In recent times, there is quite some focus on new self-crosslinking and UV curable acrylic dispersions with core shell morphology [285].

Emulsions

The term emulsion describes the distribution of two liquid phases[37]. However, the resin emulsions play a marginal role for wood coating compared to acrylate or secondary dispersions. Applications are known from the area of aqueous, UV curable coatings of wood. For example, epoxy acrylates and polyester acrylates are dispersed in water by means of protective colloids (such as n-vinylpyrrolidone co-polymers) or by means of classical emulsifiers. The dispersion in water can be performed by incorporation of non-ionic hydrophilic groups (such as polyethylene glycol units) in unsaturated polyester resins containing allyl ether. The disadvantage of the emulsions is their inclination to the phase separation and the increase in particle size during the storage of the formulations. The release of water from emulsions in the drying process is slower in comparison to acrylate dispersions.

Secondary dispersions

Secondary dispersions are from polymers which are produced in an organic phase or in the melt. Only in a second step of the process, the polymers are transferred in an aqueous dispersion. Mainly classic polyurethane dispersions and UV curable polyurethane dispersions are applied in the coating of wood and wood-borne materials.

Polyurethane dispersions

The main principles of the production of polyurethane dispersions are described in Chapter 3.2.2. Aliphatic and aromatic polyurethane dispersions are applied for the coating of wood and wood-based materials depending on the desired characteristics and economic criteria. Polyurethane dispersions based on aliphatic isocyanates can be applied for lightfast, abrasion resistant and elastic wood coatings. Compared to hydrophobic substances such as fats and plasticizers, polyurethane dispersions have better resistances than traditional acrylic dispersions. In addition, polyurethane dispersions are characterized as traditional acrylic dispersions. In addition, polyurethane dispersions are characterized by a very good pigment wetting, good adhesion and wood firing[38] on various substrates. The application of aromatic

37 Milk is an example of an emulsion. The continuous phase is water containing fine dispersed fat droplets.
38 The term 'wood firing' is defined as a colour deepening which occurs in the wetting of the wood surface by the coating material and permanently highlights the wood grain.

diisocyanate in the manufacturing process results in hard wood coatings which, however, tend to yellowing. Due to the large raw material costs for aliphatic polyurethane dispersions, these polyurethane dispersions often were mixed with acrylic dispersions for the industrial wood coating or not even applied.

UV curing polyurethane dispersions[39]

Water-dispersible polyurethanes can be modified radiation-curable by insertion of polyols containing reactive double bonds [288, 289]. However, isocyanate prepolymers generally are converted with unsaturated hydroxyalkyl acrylates. Terminal double bonds are obtained [290–294]. Figure 3.92 illustrates the structure of the high molecular weight, UV curable polyurethane dispersions with terminal double bonds. Recent results of research indicate that it is better to insert the double bonds along the polyurethane chain and not only to attach acrylic units at the end of the chain. Also, the mixing of acrylic monomers or oligomers with polyurethane dispersions does not result in satisfactory results [295]. The water-borne, UV curable polyurethane dispersions combine the properties of the classical 2C polyurethane systems with the advantage of a quick and economical UV curing. A polyurethane network which additionally is chemically crosslinked by UV radiation already exists in the processable state of the aqueous coating formulations. This process is completed after an evaporation of the water within seconds.

Other polymer dispersions

Carboxyl group containing saturated and amine-neutralised polyester are applied in aqueous 2C polyurethane coatings as well as in decor finish coatings (see Chapter 3.1.10). Subsequently to the neutralisation with amines such as aminomethyl propanol and dimethylethanolamine, these low molecular hydroxy-functional resins form colloidal aqueous solutions which are applied with water-dilutable melamine resins and/or urea resins in stove enamels for paper foils. Resins also applied in aqueous 2C polyurethane coatings in combination with suitable polyisocyanates. Also, carboxyl-functionalised polyacrylates with a molar mass of 50,000 to 150,000 and a typical acid number between 40 and 80 partially are

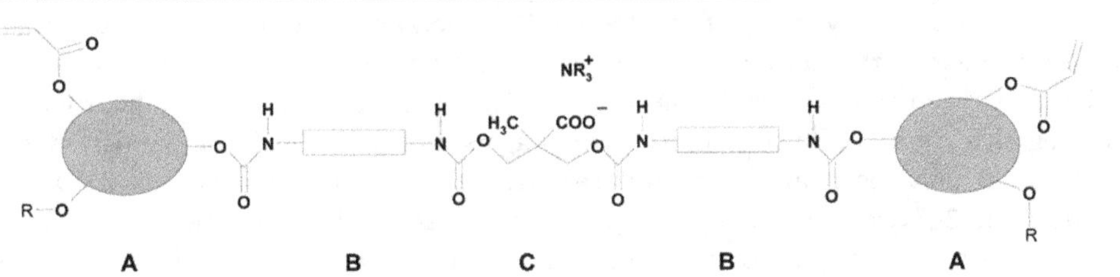

A = polyester acrylate/epoxy acrylate/urethane acrylate/polyester acrylate

B = diisocyanate

C = hydrophilisation (carboxylate groups, ammonium groups and sulphonate groups)

A = polymer or monomer diol

Figure 3.92: Structure of macromolecular, acrylate functionalised polyurethane dispersions [286]

39 Among experts, polyurethane dispersions also are referred to as PUD or PU dispersions

neutralised with amines subsequently to their production. By diluting with water, so-called secondary dispersions with particle sizes between 20 and 200 nm arise [282, 284]. The thus dissolved secondary dispersions can be crosslinked chemically with special polyisocyanates by means of the built-in hydroxyl groups.

Hardeners and reactive thinners

Typical hardeners for the chemical cross-linking of carboxyl- and/or hydroxyl-functionalized dispersions for a worldwide application for the coating of wood and wood-based materials in the interior area are:

- Polyisocyanates
- Polyaziridins
- Polycarbodiimides
- Silanes

Polyisocyanates

Towards the end of the 1980ies, the first processing of aqueous 2C polyurethane coatings on a technical scale succeeded. As already shown in the year 1943 by Otto Bayer, adducts were produced by the chemical reaction of aliphate diisocyanates with sodium bisulfite, whereby these adducts were processed in water [297]. The technical implementation for the coating technology could not be realised for a long time as polyisocyanates strongly react with water under formation of urea and development of carbon dioxide. High quality coatings can be generated only by the development of appropriate hydroxy-functional dispersions which enable the insertion of polyisocyanates into the aqueous phase in a similar way as an emulsifier [296]. Appropriate polyisocyanates also had to be developed. For this, different hardener concepts are available in the market. The types of polyisocyanate are as follows:

- Low viscous and hydrophobic polyisocyanates
- Hydrophilised polyisocyanates
- Sulfonic acid functionalised polyisocyanates

The Figures 3.93 up to 3.95 illustrate different modified polyisocyanate hardeners for the application at aqueous 2C polyurethane coatings. The Figures 3.93 and 3.94 illustrate polyether-modified polyisocyanates of the 1st and 2nd generation. The hydrophilicity remaining after the reaction of the isocyanates with the resins is detrimental.

Particularly, in aqueous, white pigmented 2C polyurethane coatings, the hydrophibicity has a negative impact on the resistance such as against coffee, red wine and mustard. A new generation of polyisocyanates being less sensitive to colouring substances is a HDI trimer modified with 3-(cyclohexylamino)-1-propane sulfonic acid [299]. Furthermore, it can be more easily inserted into the base coating component. Since the property profile of the aqueous polyurethane coatings largely corresponds to the property profile of the conventional systems, the aqueous polyurethane coatings have found a wide application in the practice of painting in a relatively short time. For industrial applications, the aqueous polyurethane coatings are in competition with UV curing coating systems. The disadvantage of 2C PU coatings is the limited processability after addition of the hardener component (pot life).

Polyaziridines

Carboxyl-containing polymers react with polyfunctional aziridines to hydroxyamides. Even these polymers are toxicologically harmless, carboxyl-containing polymers are admitted as a hardener component to aqueous furniture coatings and window coatings. This improves the block resistance (no adhesion in the stack), adhesive strength, alcohol resistance as well as water resistance. From the point of view of the authors, the compounds have no great significance. Due to their toxicological risks, carboxyl-containing polymers should not be applied for the coating of wood and wood-based materials.

Silanes

The addition of functional sealings as a hardener component to aqueous wood coatings initiates cross-linking reactions and selectively improves the chemical resistance. The silane-containing hardeners mainly are applied in North America for the furniture coating. Organofunctional silanes enable a better adhesive strength of coatings as well as adhesives and sealants on the most different surfaces. Thus, organofunctional silanes are applied as an additive in glass coatings. Organofunctional also improve the compatibility and enable the chemical bonding o inorganic fillers at organic resins. Compared to polyisocyanates as a cross-linking component usually results in a lower chemical resistance.

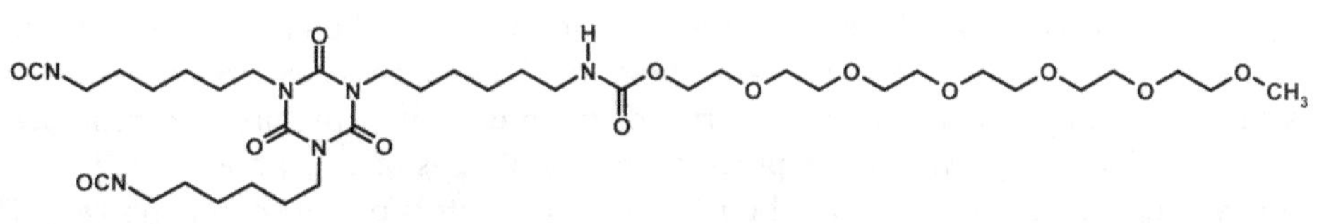

Figure 3.93: Structure example of a polyether modified polyisocyanate (HDI trimer) [298]

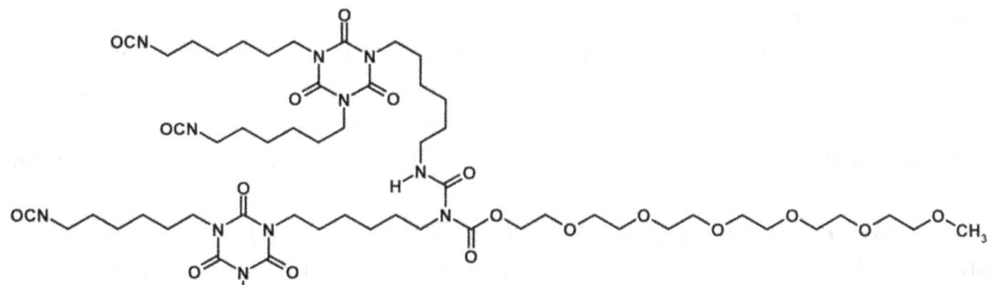

Figure 3.94: Structure example of a polyether modified polyisocyanate (HDI trimer/allophanate type) [298]

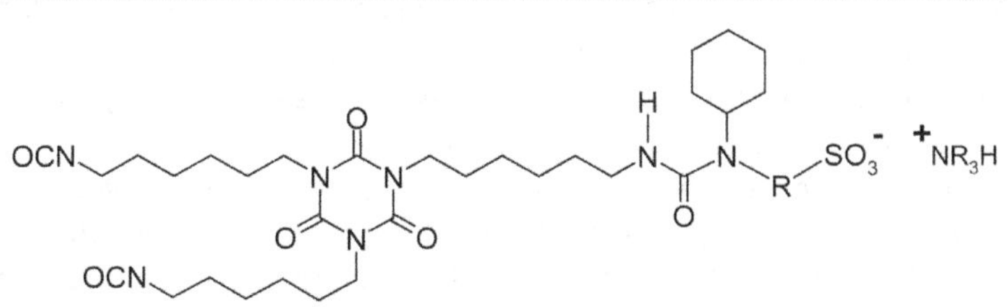

Figure 3.95: Ideal structure of an aminosulphonic acid modified HDI trimer [298, 299]

Reactive diluents

- Reactive diluents for radiation curing coatings
  The well-known and well-dilutable/emulsifiable reactive diluents such as ethoxylated trimethylolpropane triacrylate can be applied in aqueous, UV curing wood coatings. The degree of ethoxylation defines the compatibility with water.
- Reactive diluents for 2C polyurethane coatings
  Water-soluble polyols a reaction partner for polyisocyanates are applied in aqueous 2C polyurethane coatings in order to increase the solids content (non-volatile content), to increase the degree of crosslinking as well as to lower the application viscosity.

Other recipe ingredients

The pigments applied for water-borne coatings, fillers, matting agents as well as additives are sufficiently described in Chapter 3.2. The photoinitiators which are applied for aqueous coatings are described in Chapter 3.1.6.

3.1.7.2 Water-borne coating systems

Various water-borne coating systems are applied for the industrial wood coating. When applying water-borne coatings, the emission of organic solvents will be reduced significantly. Due to their inflammability, water-borne coatings can be applied as exchange products for solvent-borne coating systems in order to reduce the premium of the fire risk insurance in the firms. The following contribution presents the main types of water-borne coatings that are applied depending on the required quality and parameters of the coating process. The fundamental information about the film formation of water-borne coatings are discussed in detail in Chapter 3.2.

1C water-borne coatings

The single-component water-borne coatings are divided into self-crosslinking (SC) systems and systems which are not self-crosslinking (NSC). For both systems, the film formation or drying, respectively, is performed by evaporating the volatile components of the water-borne coatings (water, organic solvents). This also is referred to as physically drying water-borne coatings.

1C water-borne coatings (not self-crosslinking, NSC)

Usually, the not self-crosslinking water-borne coatings are built up on the basis of acrylic dispersions. Combinations with polyurethane dispersions are possible. Polyacrylates which are neutralized with sodium hydroxide also are applied. Water-borne coatings which are not self-crosslinking were the first coatings for the industrial coating of wood and wood-borne materials. As the water-borne coatings only dry physically and are not cross-linked chemically, the resistances of the resulting coating films are not resistant to PVC plasticizers, not solvent resistant and swollen with prolonged exposure to water or alcohol. Pigmented water-borne coatings are applied for the coating of hardboard panels for example (see Table 3.48). The applied neutralised polyacrylates (such as styrene containing polyacrylates, vinyl acetate containing polyacrylates) ensure a good roller transfer behaviour of the coating formulations in the coating process. In order to meet the chemical resistances, the water coating layer is painted with UV curing clear coatings (approx. 100 % non-volatile content) in the rolling process. Such coating surfaces then meet the resistance for furniture surface according to DIN 68861 part 1B or the requirements IKEA IOS-MAT 0066 according to R4 and R7, respectively. Generally, one-component water-borne coatings are used for the not strongly

Table 3.48: Pigmented single-component water-borne coating (primer) for the coating of hardboard panels for rolling

Component	**Function**	**Proportional weight**
Water soluble polyacrylate[1] (50 % in water)	Resin	18.10
Water	Solvent; adjustment of the viscosity	32.00
Defoamer[2]	Foam destroying effect	0.10
Amine (90 % in water) 2-amino-2-methyl-1-propanol	Neutralisation agent for the resin	0.20
Ammonia solution (25 % in water)	Neutralisation for the resin; Adjustment of the pH value to approx. 9	2.50
Wetting and dispersing additive[3]	Pigment wetting and stabilitzation of inorganic pigments	0.10
Butyl glycol	Solvent	3.80
Titanium dioxide, rutile type[4] (pre-treated with aluminium components)	White pigment, colouring	21.20
Chalk[5]	Filler	9.00
China Clay ASP 170[6]	Filler, anti-settling agent	13.00
		100.00

Viscosity (23 °C, 6 mm flow cup): approx. 40 to 60 s

1) Craymul 2229 (Cray Valley)
2) "Nopco" 8034 (Nopco Paper Technology)
3) "Borchi" Gen ND (Borchers)
4) "Kronos" 2300 (Kronos)
5) "Omyacarb" 2 BE (Omya)
6) Aluminiumsilikat ASP-170

stressed furniture components or primers. For higher demands on the resistances, one-component water-borne coatings are sealed with crosslinking coating systems.

1C water-borne coatings (self-crosslinking, SC)

The chemical resistances of the coatings can be approved significantly due to the application of self-crosslinking, one-component water-borne coatings. Acrylic dispersions (primary dispersions) and/or secondary dispersions are applied which may react with the carboxyl groups of the resins by chemical cross-linking (for example with dihydrazides) or by complexation of metal ions (Zn^{2+} amine complexes) with the carboxylic groups of the film forming resins (see Chapter 3.2). Depending on the selection of resins, the water-borne coatings almost are in compliances with the chemical requirements of DIN 68861 part 1B and the stress groups R7, R4 and R2 (IOS-MAT-0066) of the company IKEA for furniture surfaces. The water-borne coatings generally are applied as primers, finishing layer coatings are a good economic and technical alternative to the solvent-borne cellulose nitrate coatings. For the filming of the resins, the formulations contain 4 to 8 % organic solvents such as butyl glycol. Aqueous clear coatings are formulated with solids contents (non-volatile content) of approx. 30 to 40 %. Usually, conventional solvent-containing CN clear coatings or 2C PU coatings have a non-volatile portion of 20 to 25 %. When switching from solvent-borne to water-borne coating-systems, the wet application amount can be reduced by approximately 10 to 20 % due to the enhanced non-volatile fraction of water-borne fractions. The industrial application requires fast drying and stackable clear coatings which are stacked on top of each other or applied after approx. 30 to 45 minutes of forced drying (34 to 40 °C) followed by cooling to approx. 23 to 25 °C. Usually, the normal single-component water-borne coatings require a drying time of at least 12 to 16 hours at standard climate (23 °C/50 % relative air humidity) until they are stack-resistant and mechanically robust. When applying the self-crosslinking water-borne coatings as primers, these can be painted with conventional, solvent-borne CN coatings or 2C PU coatings. Depending on the ageing of the primed surfaces, a coating sanding has to be made in advance in order to guarantee a perfect adhesion.

UV curing water-borne coatings (1C systems)

Since 1976, the coatings industry deals with aqueous UV curing coatings. The most important reasons are as follows:
- Reduction of the emission of organic solvents
- Renunciation of the application of low molecular monomers due to their partly skin irritating and sensitising potential in conventional UV spraying coatings
- More efficient production processes compared to other one-component water-borne coatings
- Good chemical and mechanical resistance of the coating surface
- More simple matting in comparison to 100 % UV coating systems

At the end of the 1980ies, the manufacturers of profiled wood as well as the chair industry are the pioneers in the implementation of these coating systems.

Depending on the applied resin, a distinction is made between physically drying or non-physically drying UV curable water-borne coatings. At the beginning, conventional UV resins were transformed into aqueous emulsions by means of external emulsifiers or UV resins with hydrophilic structure under the influence of high shear forces. Drawbacks soon were recognized such as no freeze/thaw stability, low stability of the emulsions (change of the particle size distribution), great sensitivity to changes in pH as well as low UV reactivity. These types of resins do not dry physically prior to the UV curing. Thus, there is a great danger of inclusions of dirt and dust in the drying tunnel.

A major drawback of these resins is that no or only an insufficient UV curing is possible in the shadow zones of the UV lamps when applying three-dimensional UV coated substrates. The coating layer remains sticky. Further developments resulted in mixtures consisting of physically drying acrylic dispersions with UV curable oligomers. Due to various drawbacks such as instability of the mixtures, the systems quickly disappeared from the market. The third generation of the UV curable water-borne coatings based on UV curable polyurethane dispersions finally has achieved a breakthrough in the market in the last 10 to15 years. Subsequently to the physical drying, these water-borne coatings reveal tack-free films with good adhesive strengths. In addition, these water-borne coatings mostly are low viscous, free from co-solvents and reactive thinners, need low amounts of photoinitiators, are highly flexible and at the same time very hard, easier mattable and show good adhesion strength on various wood materials or plastics. Table 3.49 illustrates a reference formulation for a single-component, UV curable clear coating. The UV curable clear coatings applied in the market generally are characterized by the following points:
- No pot life
- Processing with all known spray processes (for example airmix, airless, bucket gun)
- The efflux time frequently amounts about 50 to 120 seconds in the 4 mm flow cup
- Drying at a temperature of 40 to 45 °C in approx. 6 to 12 min; subsequent curing by means of an UV lamp (mercury high pressure lamps with 80 W/cm) at a feed rate of approx. 3 to 5 m/min

Table 3.49: Single-component, UV-curing clear coat for the spray application [300]

Component	Function	Proportional weight
Acrylated polyurethane dispersion[1)] (non-volatile fraction approx. 39 %)	UV resin	70.00
Fumed silica[2)]	Matting agent	1.50
Dispersion of a modified polyethylene wax[3)] (non-volatile fraction approx. 55 %)	Improvement of the surface properties such as smoothness	2.00
Polysiloxane/hydrophobic filler in polyglycol[4)]	Defoamer	0.80
Benzophenone/1-hydroxy-cyclohexyl-phenyl-Ketone 5) (1:1 mixture)	Photoinitiator	0.70
Polyether modified polydimethylsiloxane[6)]	Subsurface wetting	0.20
Polyurethane-based associative thickener[7)]	Adjustment of the viscosity and rheology	0.60
Water	Solvent	24.20
		100.00

Non-volatile fraction (DIN EN ISO 3251): approx. 31 % viscosity (at 23 °C, 4 mm bucket cup): approx. 35 up t o 45 s

gloss level (60 ° measurement angle) at twice 100 g/m² wet application amount: approx. 10 min UV drying (50 °C, 1 to 5 m/s air velocity): approx. 8 to 10 min UV curing (1 UV mercury high pressure lamp, 80 W/cm): 5 m/min

1) "Bayhydrol" UV 2282 (Covestro)
2) "Acematt" TS 100 (Degussa)
3) "Aquamatt" 270 (Byk)
4) "Byk" 028 (Byk)
5) "Irgacure" 500 (Ciba)
6) "Baysilone" Lackadditiv 3739
7) "Borchi Gel" LW 44 (Borchers)

- Gloss levels from dull matt to glossy can be achieved more regular than with no physically drying, aqueous UV-coatings
- Chemical resistance according to DIN 68861 part 1 B is achieved
- Soft-PVC-solid
- Applications are furnitures for kitchens, bathrooms, living rooms and bedrooms, stair steps, wall coverings and ceiling coverings, doors as well as chairs

The physically drying, aqueous UV coatings are applied both as clear coatings or pigmented primers and/or topcoats. In the formulation of highly pigmented coatings, care has to be taken that the pigment concentration is not too high, since otherwise it may result in curing problems. This is expressed by a poor adhesion to the substrate, for example. UV coatings mainly are applied in the furniture industry as well as in the door manufacturing industry.

2C polyurethane water-borne coatings

Towards the end of the 1980ies, it was possible for the first time to process 2C polyurethane coatings from the aqueous phase [301]. This would not have been thought possible for a long time, since water reacts with the applied polyisocyanate compounds resulting in the formation of N-substituted polyurea structures and separation of carbon dioxide gas (CO_2). CO_2 results in foam formation in the coating film. By the implementation of fine-particle, water-borne, carboxyl- and hydroxyl-containing resins (polyacrylate) with emulsifying properties ensures the incorporation of the low viscous polyisocyanates and significantly reduces the side reactions of the polyisocyanates. Researchers of Bayer MaterialScience found out that a thin protecting polyurea membrane is formed around the polyisocyanate droplets when incorporating polyisocyanates which are specially modified for water-borne coatings. This polyurea membrane prevents a penetration of water molecules into the membrane. The hydroxyl content of the applied resins normally is between 2 to 4 % based on 100 % of non-volatile fraction, and the acid number is 20 to 30. The particle sizes are between 100 and 200 nm. The non-polar, low viscous and inert the polyisocyanate, the weaker the reaction with water [302]. Very hydrophilic polyisocyanates are easier to incorporate into the aqueous phase in any case, but tend to a stronger reaction with water. Numerous analytical studies have confirmed that the carboxyl groups of the resins hardly react with the applied polyisocyanates under elimination of CO_2 [303].

In order to obtain a dense network of urethane from the reaction of the resins with polyisocyanate despite side reactions, a polyisocyanate to resins ratio of 1.1:1 to 1.8:1 is applied.

The handling of aqueous 2C polyurethane coatings is different from the handling of solvent-borne coatings. In comparison to the solvent-borne 2C PU coatings, aqueous 2C polyurethane coatings have a clearly limited processing time (pot life) from 2 to 4 hours. The pot life cannot be tracked by means of a measurement of the viscosity in comparison to solvent-borne systems [302]. The polyisocyanate component has to be emulsified very finely into the aqueous dispersion of the resins shortly prior to processing so that the two components can react effectively with each other. Depending on the formulation and type of polyisocyanate, the viscosity increases. Only after 10 to 20 minutes, the viscosity decreases and stabilizes on a product-specific level. Reverse effects of viscosity also are known. The introductory process significantly is influenced by the following parameters:
- Interfacial tension (base coating, hardener)
- Energy input at the mixing of the components
- Viscosity and hydrophilicity of the hardener solution (polyisocyanate)
- Viscosity variations between components (base coating and hardener)

The application of hydrophilised polyisocyanates leads to a simple incorporation, but also to a significant deterioration of the resistances against alcohol and colouring reagents such as coffee, mustard and red wine. The high requirements of the kitchen industry and furniture industry require the application of hydrophobic polyisocyanates. A high mixing energy is necessary for an optimal mixing of the components. In practice, the components are mixed by stirring (compressed air stirrer) or with application of 2C mixing plants such as jet disperser [305]. A hand mixing often results in a bad filming and resistance of the coating film.

2C dosing systems and mixing plants [304]

Dosing systems and mixing plants shall feature various procedural unit operations. The exact dosage and mixing of the materials in a pre-set mixing ratio are the essential unit operation. In principal, the dosing is volumetric. The mixing takes place e.g. with a static mixer in the mixing block of the plant. The systems available on the market can be divided into two classes:
- Mechanically operating dosing systems
- Electronically controlled dosing systems

Mechanically operating dosing systems

In such cases, the pumping unit adopts the dosage. Dosing pistons or precision gear pumps are applied as dosing elements which work according to the displacement principle. In order to increase the width of the adjustable mixing ratio, the mixing ratios specified by mechanical clutch can be varied by the piston stroke in the case of the dosing piston pump or by the infinitely variable gear in the case of gear pumps.

Electronically controlled dosing systems

All these dosing systems all have in common is that the dispensing process is controlled by a computer. The computer can take over the function of the mechanical transmission and control the pumps for example in analogy to the mechanically dosing precision gear pumps. The surfaces coated with aqueous 2C polyurethane coatings (clear coatings and pigmented systems) have the same mechanical technological performance characteristics such as solvent-borne 2C PU coatings. Therefore, these are applied more successfully to substitute solvent-borne systems in the office furniture industry and kitchen furniture industry. The drying times for pigmented 2C PU coatings at standard climate (23 °C, 50 % relative humidity) amount approx. 12 to 16 hours. Only thereafter, the surfaces can be stressed or stacked, respectively. The time up to the de-stacking of the components can be reduced to approx. 6 to 8 hours by means of forced drying (approx. 40 to 45 °C). Under standard climate, the chemical curing (polyaddition) is completed only after 5 to 7 days. Only after this, the coating surfaces achieve their full chemical resistance according to DIN 68861 part 1B.

Aqueous, UV curing, 2C polyurethane coatings

In order to reduce the relatively long drying and curing times of pigmented aqueous 2C polyurethane coatings in the industrial coating of kitchens, the UV curing mechanism is combined with the polyaddition mechanisms of the 2C PU coatings in the so-called Dual-Cure systems (see Chapter 3.1.6). The combination of both curing mechanisms facilitates a formulation and safe curing of highly pigmented coatings. The combination of both curing mechanisms facilitates a formulation and safe curing of highly pigmented coatings. Particularly yellow and red colours cause problems for the pure Mono-Cure curing with UV light. UV curing and hydroxyl group-containing resins are applied in dual-cure formulations. The well-known isocyanates in the blend ratio base coating/hardener 100:5 or 100:10 are applied as a hardener. in the industrial processing, the coating systems are processed by means of 2C dosing plants/mixing plants. The total drying/curing time is about 6 to 20 minutes at forced drying (approx. 20 to 50 °C) depending on the component geometry (flat or profiled components). Once the components are UV curing and shortly cooled, the surfaces can be turned around or stacked on top of one another.

3.1.7.3 Processing of water-borne coatings

A rethinking is mandatory among the appliers with respect to the processing of water-borne coatings. The ideal conditions of processing are 20 to 23 °C / 40 to 65 % of relative humidity. Also, the temperature of the coatings as well as wood-based materials should be between 20 to 23 °C. If the temperature of the coating as well as the substrate temperature are less than 20 °C, this can lead to the non-optimal filming of the water-borne coatings. In addition, what should be kept in mind is that the wood equilibrium moisture content should be kept between 8 and 12 %. A workpiece to be dried can extremely accelerate the drying of water-borne coatings so that the filming cannot occur professionally.

In addition, the following topics should be considered:

Fluctuations in air humidity may strongly impact the rheological behaviour of water-borne coatings during the application process. Thus, air humidities of more than 70 % severely can delay the drying of the coating during the application. This may cause coating runs on vertical surfaces. For this reason, a relative air humidity of 45 to 65 % is advisable.

Due to the high surface tension of water-borne coatings, the spraying with water-borne coatings is more difficult than with solvent-borne coatings. But mostly one has to work with a higher spray pressure. Since water strongly swells wood, the wood has to be primed finely and dried quickly in order to reduce the raising of the wood fibres (approx. 60 to 80 g/m^2). An old painter wisdom says: "A doubling of the applied quantity at standard climate (23 °C/50 % relative air humidity) a fourfold increase in drying time". In water-borne coatings, it is advisable in principle better to coating twice a thin layer than to coating once a too thick layer. In many cases, the heating of the substrate prior to the coating process impacts positively the swelling of the wood. thus, the interaction between the wood fibre and water almost can be excluded. In order to reduce the swelling capability of the wood, in comparison to the solvent-borne coatings, the wood sanding prior to the application of a water coating often is carried with finer abrasive paper. Depending on the type of wood, the abrasive paper has a 180 to 220 grit. Depending on the type of material, the degree of swelling is pronounced variously. The coatings should be selected carefully especially for MDF panel which directly should be coated with water-borne coatings. MDF panels that contain a small amount of glue tend to a stronger straightening of wood fibres. Here, MDF panels with polyurethane glues offer advantages in comparison to MDF panels with urea glues. The thermos-smoothing has been proven in the reduction of the water swelling in the application of profiled MDF panels (see Chapter 3.1.9).

The alkalinity of the water-borne coatings may cause dis-colourations of tannin rich woods (such as oak, ash, limba, mahogany, teak, pine and spruce). Due to the ingredients, coniferous woods (pine) may show a yellow-greenish discolouration and disturbances in the adhesive strength in the region of knotholes. Here, it is necessary to perform a trial painting in advance. The stain formation or the discolouring, respectively, in the region of knotholes of pine trees is a natural process and can never be excluded completely despite special insulating primers. The age of the pine wood at the time of the coating has a tremendous impact on the bleeding of the wooden ingredients. The effect is promoted by a temperature load in the drying process or storage process of the components. For this reason, the parameters of the drying process should amount not more than 40 °C in the coating process of pine wood.

Water-borne coatings have a lower 'grain accentuation' than solvent-borne coatings. This is an advantage for lighter coloured woods such as maple due to the levelling of colour differences, in general. In dark woods such as cherry tree or mahogany where a good grain accentuation is required, the lower 'grain accentuation' has to be balanced with a wood stain or with a coloured lacquer.

The most water-borne coatings tend to a physical drying on the plant components such as application equipment and containers for dosing coatings. The physical dryings of the coating are resolvable difficultly and may lead to contaminations, if these physical dryings fall into the liquid coating. As a possible consequence, the coating surface features a spotty appearance, and the sieves and nozzles of the spray unit are blocked.

As it is well-known, water-borne coatings should be processed only with stainless steel equipment. Only tested metals such as stainless steel (V2A or V4A) should be applied in order to avoid chemical reactions of water-borne coatings with different metals such as non-ferrous metals. The same applies to supply lines, suction pipes and drying plants.

Water-borne coatings react sensitively to contaminated substrate surfaces. This is reflected by wetting problems or by the formation of craters, for example. Basically, it should be ensured that one should not work with silicone-based hand creams, aftershaves, and silicone-based lubricants in the coating rooms. Substrate surfaces such as melamine films should be cleaned and sanded well in advance.

The application of a compressed air stirrer is recommended in order to incorporate additives and thickeners during the production. This is very important in order to incorporate additives into the water-borne coatings very well. A non-homogeneous incorporation may lead to the formation of craters or other film failures.

Prior to the application of water-borne coating systems, a plant that previously was operated with solvent-borne coating systems, has to be purified very well. Incrustations of coatings have to be removed completely. All pumps and hoses have to be flushed thoroughly with a specific dilution, or have to be replaced. Afterwards, it has to be purified with a so-called relinking agent (e.g. based on alcohol) and water. When applying an automatic spraying gun, the belt cleaning has to be converted on water-borne coatings. Only once this is complete, one can work with water-borne coatings. It should be noted that normal solvent-containing diluents are not suitable for the dilution of water-borne coating systems. Improper applications of such dilutions enable precipitations of the coating components as well as device errors such as clogged filters and nozzles. After the application of water-borne paints, the working devices have to be purified thoroughly with tap water. Incrusted coating residues have to be removed with special cleaners.

The once opened packages of water-borne coatings have to be closed after application. Dried residuals of paint do not get into the packages. When appropriately stored in originally closed packages, the storage capability amounts between 3 and 6 months depending on the quality of the water-borne coating. For prolonged storage, the material should be subjected to a preliminary test prior to the application. Water-borne coatings should be stored frost-free and permanently not below a temperature of 5 °C.

Fundamentally it is known, that the drying of water-borne coatings is more difficult than the drying of solvent-containing coating systems since water has a high heat of vaporization, a high boiling temperature as well as a high surface temperature. The drying process in water-borne coatings highly depends on a variety of different factors (see Chapter 7). The drying behaviour of the water-borne coating highly depends on the wet application amount. At forced air drying, the water-borne coating first should evaporate in a vapour zone. If this does not happen, then there is the risk that the surface of the water-borne coating physically dries first, and the still existing residual amounts of water and any organic solvents cannot evaporate more evenly. Subsequently to the ventilation phase in the flash-off zone, the water-borne coating can be dried forcedly by means of circulating air dryers/jet dryers or perhaps combinations with coldness and sorption dryers. There also is the possibility of the application of short-wave and medium-wave infrared emitters. Another possibility of accelerated drying is the application of an adequate air flow without temperature input. A particularly low moisture level in the incoming air is not essential. The physical drying can be realised for example in the form of blowguns or ventilated rack trolleys. Here, the coating surface is blown with cold air. One nozzle consumes approximately 350 litre/minute air at a pressure of about 4 to 4.5 bar. However, too high flow rates of the air may lead to cracks in the coating surface. Here too, it is necessary to comply with a process window. In order to achieve good drying results, the parameters in the following chapter should be:

- Supply air temperature 20 to 50 °C
- Relative air humidity of the supply air < 70 %
- Flow rate at the coating surface approx. 0.5 to 1.2 m/s

Table 3.50: Procedure example for the hardboard panel coating of rear wall of furnitures

No.	Machine	Product	Drying/curing	Wet application amount	VOC content
1.	Grinding machine for the wooden sanding				
2.	Dedusting				
3.	Single rolling machine	Aqueous roller primer (white pigmented)		30 g/m²	2 g/m²
4.	Jet dryer		20 s at 120 °C circulation air temperature		
5.	Grinding machine for the coating sanding				
6.	Double roller	Aqueous roller primer (pigmented)		20 g/m²	1 g/m²
7.	Jet dryer		20 s at 120 °C circulation air temperature		
8.	Double roller	Aqueous roller primer (pigmented)		20 g/m²	1 g/m²
9.	Jet dryer		20 s at 120 °C circulation air temperature		
10.	Double roller	Aqueous roller primer (white (pigmented)		20 g/m²	1 g/m²
11.	Jet dryer		40 s at 120 °C circulation air temperature		
12.	Single roller (hard rubber roller)	UV clear coating (solvent containing)		8 g/m²	1 g/m²
13.	UV dryer with 3 mercury high-pressure lamps (each 80 W/cm)				
			Wet application amount	98 g/m²	6 g/m²

Depending on the temperature, humidity and wet application amount such as 1C water-borne clear coating are:

- Dust-dry after approx. 15 to 30 min,
- Grindable ('smoothable') after approx. 45 to 60 min and stackable once upon another after 8 to 12 hours
- The drying values specified are based on the following basic parameters:
- Applied quantity of 80 to 90 g/m² per operation (base coat and topcoat)
- Temperature 23 °C
- 50 % relative air humidity

Examples for industrial coating processes

The following Figures and Tables illustrate possible system concepts for eco-friendly coating with UV water-borne coatings. For the user, the geometry of the wood substrates, the belt coverage density, the number of coated wood substrates per day in square meter, the number of colour shades, the surface effects or surface qualities, respectively, to be achieved as well as the maximal costs per square meter for the coating determine the plant concept to be applied.

Coating of hardboard panels

The roller coating of hardboard panels or chipboard panels with aqueous base materials and subsequent overcoating with UV curing topcoats is operated for more than 25 years. The coated panels are applied for furniture blocks, rear panels of the kitchen and furniture, drawer bottom panels, table tennis tables as well as doors. Table 3.50 illustrates a typical process for the coating of furniture back panels. Pigmented, single-component water-borne coatings (NSC) are applied in several steps and dried with circulating air dryers/jet dryers. Often, a pre-heating of panels with infrared emitters will be carried out to minimize the swelling of wood fibre and to shorten the drying process. At the end of the coating process, an UV curable clear coating with approx. 10 to 20 g/m^2 is applied in the rolling process in order to meet the required chemical properties. Normally, DIN 68861 Part 1B is met. Those lines have a total length of approx. 200 to 2500 m and operate at line speed of 20 to 120 m/min.

Coating of kitchen furniture

Figure 3.96 illustrates an industrial plant concept for the efficient coating of coloured kitchen fronts with dual cure curing (2C PU/UV) water-borne coatings. As a general rule, MDF wooden materials that are coated with a white melamine foil as well as with a PVC foil are applied as substrate materials. The feed rate of the plant amounts 4 to 6 m/min. Prior to the coating, the coated substrate material is grinded and cleaned in order to achieve an optimal adhesion of the coating. The coating is applicated by spraying guns or spraying robots in batch mode. The coating processing should be made by means of a 2C dosing and mixing system. The

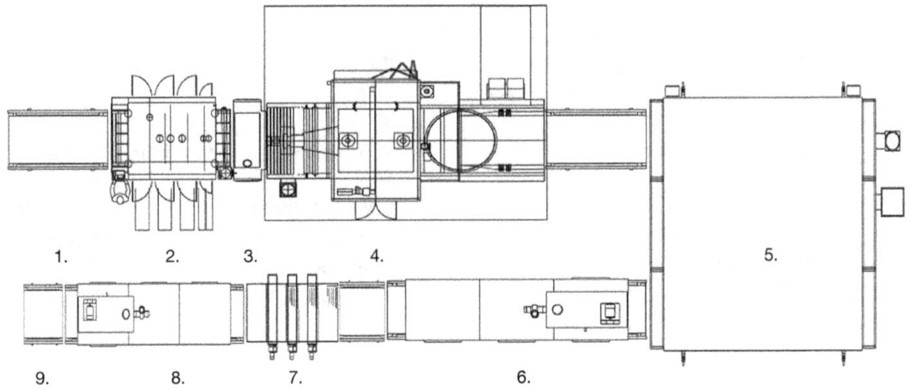

1. Loading belt
2. Grinding machine
3. Dedusting
4. Automatic spraying gun
5. Rack dryer 25–45 °C
6. Jet dryer 25–30 °C
7. UV dryers with 3 UV lamps (Ga/Hg/Hg) each 60–120 W/cm
8. Cooling compartment
9. Take off

Figure 3.96: Plant example of the application of pigmented 2C PU UV curing water-borne coatings in the kitchen furniture industry

Table 3.51: Procedure example for the electrostatic coating of chairs by means of UV curing water-borne

No.	Procedural process	Product	Remarks
1.	Manual hanging up of the chairs		Line feed rate approx. 4 m/min
2.	Overhead conveyor		
3.	Lighting systems		
4.	Omega loop	1C UV multilayer water-borne, transparent (wet application amount approx. 90–100 g/m²)	High speed rotational disk with recovery of the coating
5.	Manual spray booth	1C UV multilayer water-borne, transparent	Repair defects
6.	Air circulation dryer		40 up to 60 °C air circulation temperature; Air velocity approx. ca. 3–5 m/s; Drying period approx. 30–45 min
7.	UV lamp unit 80–120 W/cm power		Curing of the primer; the UV lamps are arranged in an omega loop
8.	Coating sanding		Manual or mechanical coating grinding for example in a rotating drum filled with glass beads; approx. 20–30 seconds per chair, for the application in the coating grinding, the chairs have to be taken off from the overhead conveyor
9.	Manual hanging up of the chairs		Overhead conveyor
10.	Omega loop	1C UV multilayer water-borne coating, transparent (wet application amount approx. 90–100 g/m²)	High speed rotational disc with recovery of the coating
11.	Manual spray booth	1C UV multilayer water-borne coating, transparent	Repair of the bare spots
12.	Air circulation dryer		40 up to 60 °C air circulation temperature; air velocity approximately approx. 3–5 m/s; drying period approx. 45 up to 60 min
13.	UV lamp unit 80–120 W/cm power		Curing of the topcoat; the UV lamps are arranged in an omega loop
14.	Stacking		By hand on the corresponding pallets

wet application is 120 to 130 g/m². The drying is performed in a rack dryer with three different drying zones. In the first zone, the components are pre-dried for 10 to 15 minutes at a temperature of approx. 35 to 45 °C, while the air velocity is increased to > 2 m/s. The residence time is approx. 10 to 15 minutes. In the third zone, the temperature amounts approx. 30 to 35 °C, while the air velocity is > 2 m/s.

Here, the drying time is approx. 10 minutes. Subsequently a jet dryer with approx. 30 °C warm air is connected downstream for an optimal drying of the profiled kitchen fronts. Thus, 'picture frame effects' due to enclosed water in the profile area can be avoided. Finally, the surfaces in an UV dryer are UV cured with one gallium doped mercury emitter and two normal mercury high-pressure lamps and subsequently cooled in a nozzle dryer (25 °C) prior to the withdrawal. Now, the surfaces can be further processed or stacked one on top of the other. The total processing time amounts between 30 and 45 minutes depending on the geometry of the kitchen formats.

Coating of chairs

The industrial coating of chairs have special demands on the coating material and the coating technology. The chairs are coated in separate components or in an assembled system. These frequently are coated in a hanging or standing state in several steps. Due to the three-dimensional geometry of the wooden chairs, electrostatic application processes with high speed rotational discs (Omega loops) and/or airmix spraying are applied. Electrostatic spray discs are coating vaporizers which rotate around a vertical axis and thereby are moved continuously up and down by means of a lifting appliance. Thereby, the workpieces are driven around on an omega-shaped circular orbit around the rotating, up and down swinging disk. The rotational speeds vary between 20,000 and 30,000 U/min. depending on the rotation disc and coating system. Applying three centres of rotation inside the omega loop, the chairs are rotated by a quarter rotation. Thus, every point of the chair can be coated evenly according to chair model. Often, a subsequent injection is performed manually subsequently to the high speed rotational coating in order to coat possible defects. The robot coating also is applied. Here, the chair is rotated four times by 90 °, so that the spraying robot easily can achieve all areas of the chair. The electrostatic processes require a constant humidification of 55 to 65 % and a moisture content of the raw wood of 10 to 12 %. The electrostatic appliances with recovery of the coating exhibit an application efficiency of approx. 80 to 90 %. 1C UV curing clear coatings and/or 1C water-borne coatings (SC) are applied for the coating of chairs depending on the claim. Table 3.51 presents an example of the coating of chairs with UV curing water-borne coatings. The line speed is approx. 3 to 5 m/min depending on the procedure.

Coating of stair steps

Fast-drying 2C PU coatings yet are applied for abrasion resistant and robust coatings of stair steps in a spraying process. The processing solids of this coating systems typically amounts 25 to 40 %. The processability (pot life) subsequently to the addition of the hardener usually amounts at least 8 hours. The stair steps horizontally are coated twice. The edges totally are coated four times. In this procedure, the processing time of a stair step from the raw grinding to the finished coated stair step amounts approx. 12 hours. The drawback of the 2C PU coatings is that the overspray cannot be recycled well on the spray application. Depending on the geometry and filigree nature of the components, the overspray amounts approx. 40 to 50 %. These coating losses have to be disposed expensively as a coating sludge.

Especially the 1C UV water-borne coatings permanently have grown in significance as an eco-friendly and low-solvent alternative for the solvent-dilutable 2C PU spraying coating. In comparison to conventional 2C PU spraying coatings, a great advantage of the UV water-borne coatings is that the wooden substrates are stackable and packable subsequently to the evaporation of the water and subsequently to the UV curing. Figure 3.97 and Table 3.52 illustrate a procedure with 1C UV curing water-borne coatings. This procedure utilizes a 1C UV water-borne coating with a non-volatile content of 38 % as well as a solvent content of

Table 3.52: Coating of stair steps with UV curing water-borne coatings

No.	Process	Product	Wet application amount [g/m²]
1.	Rough grinding	150/180er abrasive paper	
2.	Base coat (surface + edge)	1C UV water-borne multi-layer coating (silky-matt)	100
3.	Drying	Air circulation drying 3–45 °C approx. 1–20 minutes	
4.	UV curing	2 mercury high pressure lamps 80 W/cm	
5.	Coating sanding	Brushing machine covered by abrasive paper strips	
6.	Topcoat (surface + edge)	1C UV water-borne multilayer coating (silky-matt)	100
7.	Drying	Air circulation drying 3–45 °C approx. 20 minutes	
8.	UV curing	2 Mercury high-pressure lamps 80 W/cm	
9.	Destacking		
	Sum	Total wet application amount [g/m²]	200

> 2 %. This application is accomplished with a surface spraying machine with a belt conveyor and an automatic doctoring of the overspray (Figure 3.97). Nearly 20 to 25 % of the thus recovered overspray is added to the UV water-borne coating which is used exclusively as a primer. The coating efficiency is increased to approx. 90 % due to the recovery [307]. Thus, the

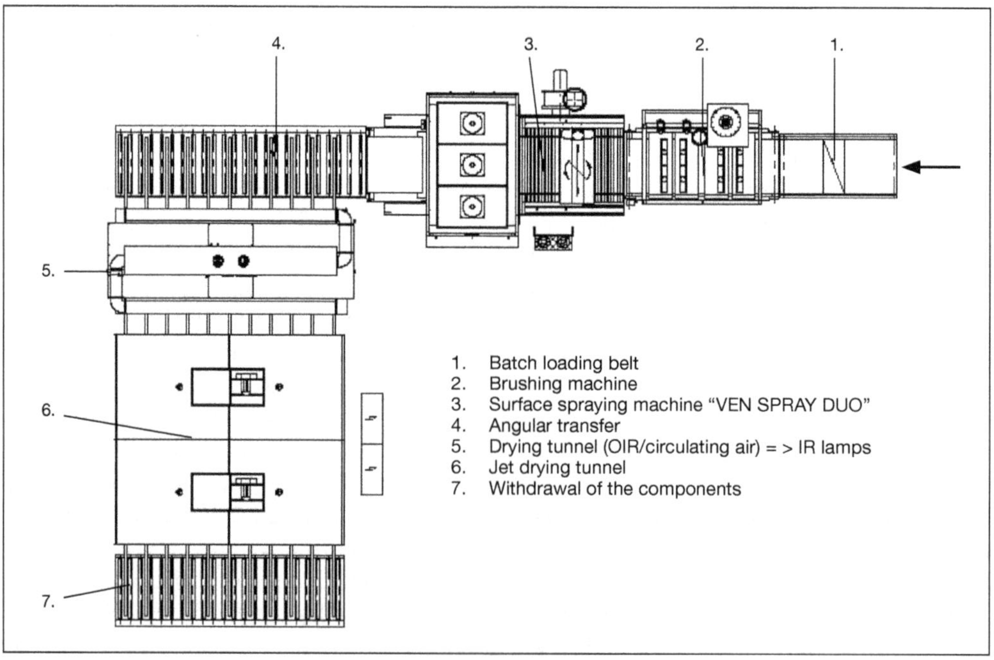

Figure 3.97: Plant example for the coating of stair steps with UV curing water-borne coatings [306]

disposal costs for the coating sludge can be saved. In comparison to the 2C PU coatings, the chemical curing is performed in fractions of a second. This reduces the processing time for the coating up to the final coating of a stair step several times. The abrasion resistance and the visual appearance of the stair steps coated with UV water-borne coatings are comparable with 2C PU coating surfaces. In comparison to cellulose nitrate coatings or conventional 2C PU coatings, the often criticized 'bad grain accentuation' of water-borne UV coating systems could be improved significantly in recent years by means of new raw materials in the UV water-borne coating formulations. In comparison to the solvent-containing 2C PU coatings, the benefits of the UV curable clear coatings for stair steps are the following [307]:

- Significantly lower amounts of coating (approx. 35 %)
- Reduction of the total amount of the solvent by > 90 %
- Amount of the coating waste is reduced by approx. 80 %
- No deterioration of the coating technical properties
- Processing times significantly are reduced due to the UV technology

3.1.8 Oils, waxes and natural resins

In recent years, the demand for natural surfaces [308–311] based on renewable raw materials has evolved rapidly due to a consumer awareness as well as by providing competitive products and procedures. Due to the increasing environmental awareness, furniture elements, interior elements and parquet floorings with oil products, wax products and natural resins based on natural raw materials are becoming popular increasingly. The main motivation of the coating industry to develop and to use coatings based on renewable raw materials ("RRM") is still the expectation that in future these products become more economical by constantly rising oil prices. The finite nature of the crude oil and the high level of dependence on the OPEC countries associated with this are reasons to find alternative resources for the production of coating materials. Furthermore, aesthetic factors as well as health factors of the coating materials used play an important role. The final consumers prefer coating products with a sustainable environmental effect and based on renewable raw materials. It is often referred to a Bio-Boom (LOHAS movement → "Lifestyle of Health and Sustainability"). From the treatment of wood with oxidatively crosslinking alkyd resins and oil cookings the consumers expect a better indoor climate and lower residual emissions (TVOCs, fission products) in the living area under residential-hygienic aspects. In addition, the treated wood should feature the nature of raw wood ('see, feel and smell wood'). Meanwhile oiled surfaces gained high popularity. In the meantime, approximately 40 % of the wood floors are being oiled. An increased demand for high-solids or 100 % oil containing products can be observed also in the furniture sector. In the do-it-yourself (DIY) segment a trend towards oil cookings and alkyd resins containing high solids are registered in order to meet the legal requirements (Decopaint Directive) and the technical requirements. In parallel, solvent-rich alkyd resins are recognisable (cf. Figure 3.2). Figure 3.2 (EMEA) clearly illustrates the market development or the redistribution of the technologies, respectively. Here, a significant increase in the application of alkyd resins / oil cookings (high solids), aqueous emulsions of alkyd resins, alkyd resins (aqueous) modified with acrylates as well as polyurethane dispersions modified with fatty acids were predicted. Simultaneously, the solvent-borne alkyd resins (high solvent content) as well as the pure polyacrylate dispersions (primary dispersions) tend to decrease. Today, there exists a variety of natural products for the treatment of wood. The products offered in the market differ considerably in their composition as well as in their application and do not always

Table 3.53: Solvent concepts of the manufacturers in coating systems based on natural resins[311]

Raw materials base	Alipathic hydrocarbons	Isoalipathic hydrocarbons	Terpenes	Solvent-free
Solvent	Test fuels (turpentine substitute), surgical spirit, test fuels free of aromatic substances	Isoaliphates (such as isopar)	Ethereal plant-based oils, balsam turpentine oils, citrus peel oils	No solvent or water
Origin of the solvent	Oil both of animal and plant origin	Oil both of animal and plant origin	Plant-based balsams, fruit peels (renewable raw materials) plant-based origin	Spring water, natural origin
Manu-facturing procedure	Distillation	Distillation in many reaction steps	Distillation	Water purification
Dis-advantages	Residues of suspected carcinogens, skin-degreasing, long-term risks non-excludable, possible negative impacts on central nervous system (in the absence of protective measures in the coating → respirator mask), often only low solubility	Low on odour (thus missing warning function), negative impacts on the central nervous system in the absence of protective measures in the coating, skin-degreasing, poor dissolving capacity for natural resins, this normally requires the addition of terpene	Skin-degreasing, possibility of allergic reactions, negative impacts on the central nervous system in the absence of protective measures in the coating	Inter-mediate sanding unavoidable
Advantages	Produced by simple distillation, therefore cheap; warning effectiveness by a strong smell	Freed of detrimental aromatics such as benzene (fraction < 1 %)	Warning effectiveness by a strong smell	No harmful solvents

meet the definition of natural coatings according to DIN 55945. According to DIN 55945, natural coatings are coating materials which consist of components resulting from the nature. These materials subsequently are not modified chemically nor modified in their natural structure and do not contain artificially manufactured components or additives. This definition says nothing about the harmlessness of natural coatings since natural coatings may contain substances hazardous to health [312]. In practice, unmodified forms of natural products hardly have been applied as resins for coatings. In order to produce a consistent quality, the coating materials have been converted by chemical or physical processes.

Even today, organic solvents are added to natural resins, oils, wax-oil mixtures and waxes to some extent in order to improve the processing properties. In addition to synthetic solvents, turpentine oils of natural origin are applied which are classified as harmful to health according to the Ordinance on Hazardous Substances. Table 3.53 exemplary illustrates different concepts [311].

In recent years, a variety of substitute products to the classic oil preparations and wax preparations (solvent content mostly > 50 %) is produced which are water-borne or characterized by a highly reduced amount of solvents. For some products one may not waive to solvents. Sometimes, solvents are necessary in order to transfer some products in a processable state and also to keep in this state.

3.1.8.1 Oxidatively curing oils and waxes

During the process of film forming, the coatings based on alkyd resins and/or oil cookings undergo a cross-linking reaction induced by atmospheric oxygen. Metal ions catalyse the oxidative cross-linking in order to increase the polymerization rate. This is precisely what leads to controllable drying times at ambient temperature. Among the drier metals commercially available, cobalt salts (primary drier) feature the highest and most universal drying activity. Thus, cobalt salts are the standard drier in the market worldwide. Due to the additional siccativation, the systems which are applied in the market are reactive and release a lot of heat energy during the process of curing [314]. Therefore, oil-containing cotton clothes should be dried carefully after application in order to prevent spontaneous combustion. Furthermore, these systems may not be processed with cellulose nitrate containing coating materials since coatings based on cellulose nitrate are highly combustible. The classification and labelling of such additives under REACH or CLP, respectively, is a vital important process for the industry. The cobalt consortium has announced the classification final points for cobalt driers; this classification has been reported in the CLP classification and labelling inventory. The industrially relevant cobalt driers (cobalt octoate, cobalt naphthenate and cobalt neodecanoate) are registered for the final points carcinogenicity, mutagenicity and reproductive toxicity (cmr) with 'missing data' and 'no classification'. Also in the issue of TRGS 905 from May 2014, organic cobalt siccatives such as cobalt octoate and cobalt naphthenate still are excluded [493]. Conventional dryers as cobalt salts or manganese cobalts are to be substituted by physiologically harmless iron salts (novel complexes). Iron-containing complex compounds are known which are able to accelerate oxidative crosslinking of unsaturated oils. In doing so, effects were observed which are not achieved with conventional desiccants (faster drying). The novel complexed iron salts have been tested in selected commercial products. It has been established that different drying effects up to complete failure may occur depending on the product composition. These findings extremely inhibit the general distribution of these environmentally friendly siccative variant, since it is not known exactly in what circumstances or conditions reproducible results can be achieved (nature and composition of the unsaturated fatty acids, ambient conditions, temperature, humidity of the air, oxygen supply, influence of the wood substrate/ingredients). Currently, this is an extreme disadvantage in comparison to the very widespread cobalt siccatives which provide reliable cross-linking outputs in a reasonable time largely independent of the criteria mentioned above.

Oils penetrate deeply into the wood and fill the pores. The surfaces are mechanically and chemically resistant. However, the surfaces do not achieve the excellent properties of 2C PU lacquers or UV lacquers for example. Oil-finished surfaces are extremely sensitive to dirt, since the surfaces feature open pores completely. Due to the open-porousness, water can penetrate into the wood substrate, and the wood swells up. Under the impact of light, the 'oil-finished' surfaces also tend to discolour, especially if oils with an enhanced amount of unsaturated fatty acids are applied. Also, products of decomposition arise contributing to residual emissions (see Chapter 9.2).

Especially carnauba wax or beeswax are applied as waxes. Beeswax is a building material for honeycombs being excreted from the abdominal glands of the bees. For the treatment of wood, it is applied in wax trains, furniture polishes as well as for furniture waxes and floor waxes. Carnauba wax is produced in large quantities from the leaves of the 6 to 12 m high

Table 3.54: Recommendations for the application of oxidatively curing oils and waxes [311, 313]

Evaluation class	Fields of Application	Example	Proposed Treatment	Effect
Very low	After-treatment, care of wooden furnitures in the residential area		Wax once or twice	Not accentuated (nature type)
Low	Initial treatment, not applied wood components at household furniture	Wardrobe interior	Wax twice; or oil wax once	Nature type
Less	Initial treatment, surfaces exposed to normal stress	Cupboards in the residential area	Blocking primer (shellac) once and wax once; or oil once and wax once	Nature type or light accentuated
Medium	Initial treatment, surfaces stressed with abrasion	Table plates, sideboards, household furniture fronts	Oil wax once, wax once; or oil wax twice; or each once oil and wax	Lightly up to moderately accentuated
Strong	Initial treatment, surfaces stressed with strong abrasion and cleaning agents	Floors, stairs, seating furniture, kitchen front	Oil wax twice; or hard oil twice and wax once	Accentuating
Very high	Initial treatment, surfaces stressed with strong abrasion, cleaning agents and water vapour	Bathroom furniture and kitchen furniture, floors and stairs	Hard oil twice up to three times; (in addition, possibly wax once) oil wax twice up to three times	Strongly accentuating

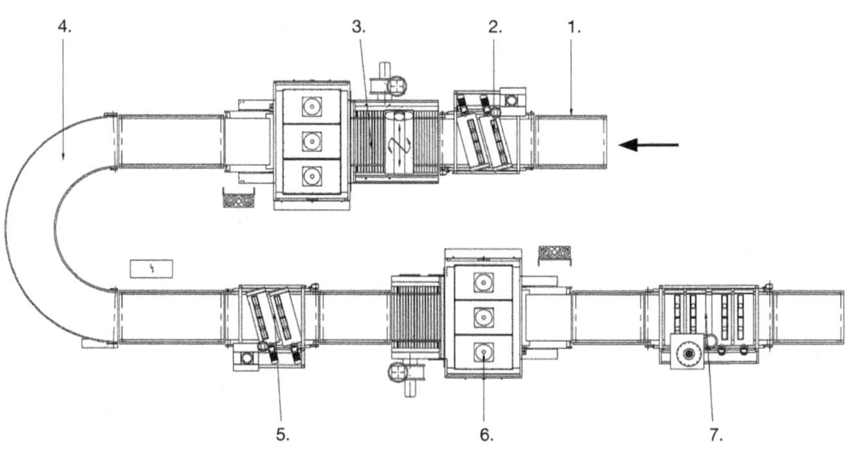

1. Conveyor belt
2. Brushing machine VBS
3. Surface spraying machine VEN SPRAY DUO
4. Belt curve
5. Brushing machine VBS
6. Surface spraying machine "VEN SPRAY DUO"
7. Brushing machine VBS

Figure 3.98: Industrial line for the application of oil/wax combinations *Source: Company Venjakob*

Brazilian Carnauba palm tree. Among other things, this tough and water-repellent product is applied in polishes, furniture and floor waxes. Carnauba wax is harder than beeswax.

Waxes form a very thin layer on the surface resulting in an increased resistance to abrasion and dirt repellent and water repellent effects. The surfaces have a nice haptics and mostly are low resistant to chemicals. A combination of oils and waxes unifies the advantages of both materials and provide a stable, breathable surface. Thus, these combinations are suitable for the industrial coating of furniture. Table 3.54 illustrates recommendations for the application of oils and waxes depending on the classification. The classifications range from very low (only for the care treatment of wood furniture in the private living area) to very high stress with water-repellent effect (for bathroom furniture and kitchen furniture, floors and stairs).

Often, solid wood is coated by means of a wide variety of application procedures such as manual application, spraying and rolling (see Figure 3.98) [308 3017] The total applied quantity is approximately 10 g/m² for a high solids final coated wax surface. As a primer, commonly oil-containing products based on linseed oil for example are applied which provide a stabilization of the wooden structure in addition to the priming properties. An oil/wax finish is achieved with approximately 30 g/m² oil-based product and approximately 10 g/m² wax-based product. The non-optimal chemical resistance and water resistance of oils significantly is improved by the hydrophobisation with waxes.

3.1.8.2 Natural resins

In the scope of this book, shellac as a natural resin briefly is described here exemplary.

Shellac

Shellac is a metabolic product of the female of the scale insect Tachardia lacca [315] and is soluble in alcohols as well as in aqueous alkaline solutions. Chemically, shellac is a low molecular weight polyester (molecular weight Mn of about 1,000 g/mole) having a relatively high acid number. The main ingredients are shellolic acid and aleuritic acid. The films dry pure physically and thus also can be processed with cellulose nitrate coatings in the same application equipment [316]. The surfaces are not highly resistant to water and alcohol. In combination with a primer based on carnauba wax, these natural resins can absolutely be applied for industrial coating of furniture surfaces. In order to achieve even higher resistances, shellac is linked chemically with polyisocyanates as a hardener component. Shellac is well polishable and grindable. Shellac undergoes a certain thermosplasticity by friction. Shellac also is characterized by its good sliding properties.

3.1.9 Powder coatings

3.1.9.1 Introduction and history

The powder coating is an example for the establishment of a nearly emission-free and low-waste coating process on the market. Although there was a drop in prices of the powder coatings in the 1990ies, in recent years the property profile of powder-coated surfaces could be improved enormously by means of intense scientific output [318–327] as well as development work. The powder coating of wood-based materials as well as the development of the technology of powder clear coatings for the automotive industry just are two examples of how the powder coating could open-up new markets and applications. In terms of economy and ecology, a great step forward should be expected from the powder coatings. In the industrial metallic coating, powder coatings often are superior to many liquid coatings with respect to their level of performance. So, powder coatings could conquer a market share of 15 to 20 %. The main advantages of the powder coating technology can be summed up easily:

- Virtually no solvent emissions (< 0.2 %, VOC)
- Almost no waste (no coagulation of the overspray)
- A material application > 95 % is possible by means of the recovery of the overspray
- High thicknesses of the layer without drips
- Very good functional and optical properties of the coating
- Single-layer application is sufficient often
- Good automatable process

The comparison of the different wood coating technologies for interior application (see Chapter 3.1) makes clear that water-borne coating systems as well as UV liquid coating systems have an advantageous position with regard to innovation with respect to the powder coating. Thus, in Austria the powder coating of MDF substrates was for the first time established in 1994 at *"Hali" Büromöbel*[328]. It was a 2-layer system consisting of an electrically conductive liquid paint primer and a low temperature thermosetting powder. Already in the year 1996, *"Hali"* re-introduced the powder coating since the two-step process wet coating/powder coating was too costly. Thus, powder coatings for wood-based materials remained a niche product. The barriers mainly were in the hitherto necessary high temperatures of > 140 °C for the melting process and cross-linking process of the conventional powder coatings. The insufficient electrical conductivity of wood and wood-based materials, the blistering at the coating surface due to escaping of water vapour as well as the insufficient decorative properties of the powder coatings were more hindrances for the commercial application of the powder coating technology in the wood processing industry until the late of 1990ies. But the scientific output and development efforts of the manufacturers of raw materials and coatings as well as of plant manufacturers and applicator manufacturers were able to improve the conditions for the application of this new technology. The innovation spirit of the furniture manufacturers and coaters still was and is necessary to promote the powder coating technology in this segment. The fact, that of course only the close cooperation of all participations evolved in the process may lead to the success, particularly also is applied to the implementation of a new technology such as the powder coating of wood-based materials.

In the meantime, however, already more than 70 plants[329] (including pilot plants) exist, where the range of coated products is manifold. This includes elements for radio and TV cabinets (RTM furniture), office furniture, especially table tops and desk pedestals, sales facilities, counter terminals, bathroom furniture, kitchen furniture and children's room furniture

Testing standards for furniture surfaces			Low-temperature powder coating	Powder coating standard	Liquid coating			DIN 68930 Kitchen furniture Requirements and testing
					2C PU coating	UV coating	Water-borne coating	
Layer thickness			110 µm	110 µm	100 µm	70 µm	120 µm	90 – 100 µm
DIN 68861 modified DIN EN 12720	Part 1	chemical resistance	B	B	B	B – C	C	≤ 1C
	Part 2	abrasion resistance	B	C	A	C	C	≤ 2E
	Part4	scratch resistance	B – C	D	E	B – D	C	≤ 4E
	Part 7	dry heat	B	C	A – C	B – D	D	≤ 7C
	Part 8	humid heat	A	B	A – B	B – D	C – D	≤ 8B
DIN EN ISO 2409		cross-cut adhesion test 2 mm cutting distance	GT 0	GT 0	≤ GT 1	≤ GT 1	≤ GT 1	≤ GT 1

Figure 3.99: Product performance comparison of different coating technologies compared to powder coatings[476]

as well as garden furniture. While the first applications still were subject to thermo-reactive powder coatings, a milestone was set with the first coating system for phono-furniture with UV powder coatings at Stilexo in Wales (1999) [330–333].

The development of the powder coating of wood-based materials was accelerated by the ever-increasing production figures of the MDF materials. Although economic difficulties in the furniture industry and construction industry faced this high growth in the recent years, the powder painting technology could conquer new market segments. This was enabled by the new requirements for the design specially for the office furniture. The three-dimensional design of wooden surfaces such as profiled narrow surfaces, indentations in the plate as well as breakthroughs in plates require not only a MDF being able for deep cutting but also an appropriate coating process. The powder coating with its advantages can be a cost-effective alternative which also may guarantee the quality which is wished by the customer. Although the first powder coating system for MDF furniture in Germany is installed since mid-2002 [334–337], many users in Germany are sceptical about this technology. Some plants are in planning throughout in Europe and Germany. By then, the tense economic situation at the beginning of the Millennium prevented investment costs on the part of the furniture industry. Furthermore, still some technical difficulties for the general application during the coating process need to be overcome. In recent years, the development of ultra-low-temperature powder coatings (3 to 5 minutes at a stoving temperature of about 120 °C to 130 °C) allows first of all a process-reliable coating of thermally sensitive wood substrates [475]. These wood substrates facilitate a fast cross-linking process at simultaneously lower and non-abrasive stoving temperatures. Unlike traditional coatings which can be applied for example on MDF only by means of complex multi-layer processes, a single-layered application is possible with newer powder coatings [475]. The single-layered material meets the function of a primer and at the same time the function of a high-quality topcoat. This application exhibits similar good adhesion properties and film flexibilities as a primer as well as the high abrasion resistance and chemical resistance of a topcoat. Thus, the MEK tests as well as pot band tests are passed easily [475]. Only due to the application of single-layered systems, the coating costs and processing costs significantly could be reduced in comparison to two-layered powder coatings. The significant reduction in costs and the fulfilment of the mechanical and chemical properties of coating surfaces, has enabled this technology of application.

3.1.9.2 Powder coating materials

According to DIN 55945, powder coatings are defined as coating powders which result in a coating after applying and fusing on the support. It is distinguished between duromer (thermoset) and thermoplastic powder coatings. Only the coating powders are of interest for the coating of wood and wood-based materials. In addition to the chemical composition, the grain size distribution is an important key figure of the powder coatings. The crushing of the powder coating chips in the manufacturing process produces a characteristic grain spectrum. This grain spectrum affects the flow behaviour of the powder coatings and determines indirectly the gradation in addition to the minimum layer thickness of the cured powder coating film.

Thanks to a supply of thermal energy (stoving powder coatings) or electromagnetic radiation (UV powder coatings), thermosets form a close meshed network with a high chemical resistance. The applied binders and hardeners essentially determine the properties of the powder coatings and their fields of application. The classic powder coatings are largely based on polymers such as epoxy resins, polyester resins, epoxy-polyester mixtures (hybrides), polyurethane resins and acrylate resins.

Although powder coatings have a lot of advantages in terms of ecology and economy compared to the conventional wet coating systems, the development of the powder coatings was not strongly promoted enough in the late 1980ies and early 1990ies. About 20 years ago, a forced development for water-soluble and aqueous coating systems was initiated. This development was required by a legal framework initiated by the automotive industry. The raw material industry and the coating industry as well as manufacturers of plants and equipments strongly were committed to the technology of water-borne coatings. A targeted research and development has been performed in this sector, so that the technology of water-borne coat-

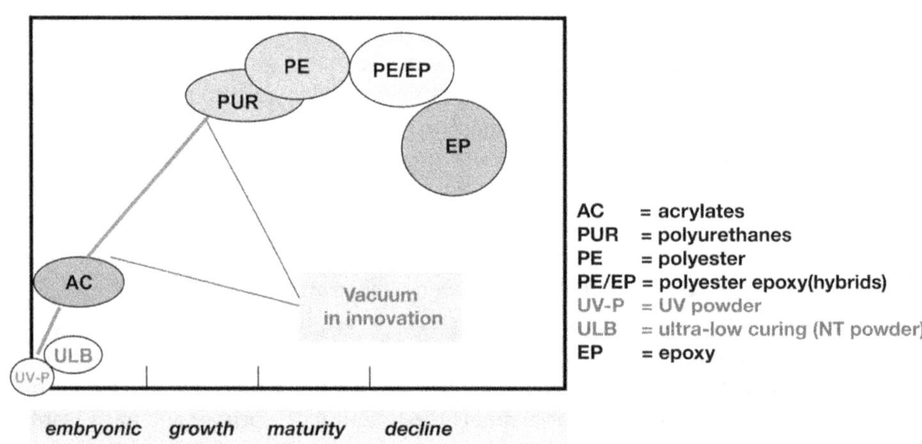

Figure 3.100: Actual lifecycle of powder coatings (as measured by future prospects/quality of performance characteristics) [331]

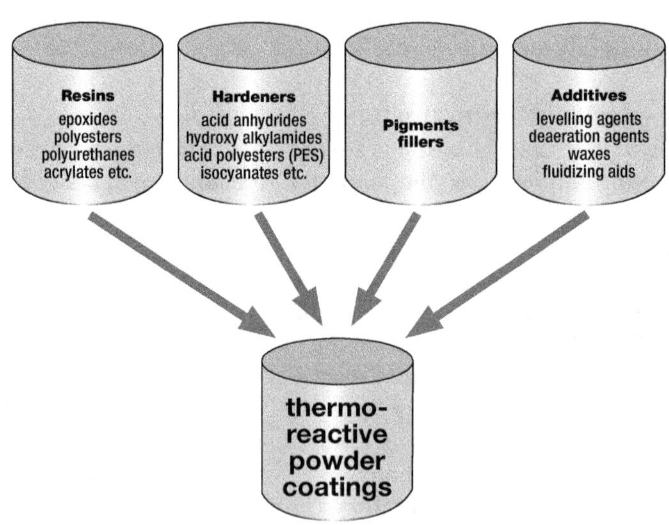

Figure 3.101: Composition of thermo-reactive powder coatings

ings had a significant development lead towards the technology of powder coatings. At that time, an empirical, slowly progressing development prevailed in powder coatings, both concerning to the coating as well as concerning to the technical side. The property profile for powder coatings does not correspond to the expectations of further chapters of the industry such as the automotive industry and many other industries. Powder coatings were not yet applied in areas such as coating of wood, wood-based materials and plastics. The life cycle curve of the powder coating systems makes this clear (see Figure 3.100). An innovation gap had to be overcome between the mature classic powder coating systems and new systems such as LT (low temperature) or ULT (ultra-low temperature) and UV powder coatings.

The prerequisite for the coating of heat-sensitive supports such as MDF panels was the development of powder coating materials which have to meet the following requirements:

- The melting temperature of the powder coating has to be as low as possible (below of 100 °C) in order to ensure a sufficient and homogeneous film formation
- Prevention of outgassing effects from the subsurface by excessive supply of heat
- Shortening of the time of the melting and networking process from minutes to seconds
- The cured powder coating must meet the requirements of the furniture industry in terms of decorative properties and chemical/mechanical properties
- Separation of the gradation and curing by analogy with liquid coating systems

Thermal-reactive powder coatings

The film-forming components of thermally curing powder coatings (stove powder coatings) generally consist of resins, hardeners, fillers, pigments and additives (see Figure 3.101). Predominantly, natural minerals such as barytes, feldspar, and chalk are used as fillers. As liquid coatings, also powder coatings generally are formulated in all shades of colour. However, the release window ΔE for shades of colour is wider than the release window ΔE for liquid coatings. These additives include gradation agents, deaerator agents, matting agents, waxes for influencing the surface hardness and structure additives, catalysts, flow enhancers and charge control tool e.g. at tribological powder coatings. For example, aluminium oxides and pyrogenic silicas are used as flow enhancers. Flow enhancers prevent an adhesion and clogging of the powder coatings, so that these remain pourable during an extended storage under the pressure of its own weight. These are important for a smooth workability of the powder coatings and increase the fluidisibility of powders. These can be transported and applied more easily.

Due to the enhanced cross-linking temperature, polyurethane systems leave out for their application on thermo-sensitive subsoil. Among the thermosetting powder coatings, only the following systems of resins actually are suitable for the coating of wood-based materials:

- Epoxy resin 110 to 140 °C 10 to 30 min (convection)
- Epoxy resin/polyester resin (hybrid) 130 to 150 °C 3 to 5 min (convection)
- Acrylate resin 130 to 140 °C 30 to 40 min (convection)

Acrylate systems on the basis of glycidyl methacrylates (GMA) have not yet been developed to such an extent that these can be applied for wood-based materials. Due to their excellent technological properties, acrylate systems mainly are applied as weather-resistant clear coatings in the automobile industry. Above all, the high price of these resins is a disadvantage. The high prices additionally make their application in the processing of wood-based materials more difficult. Furthermore, attention needs to be paid that a common processing of conventional powder coatings with powder coating systems on the basis of acrylates

Table 3.55: Reference formulation of a polyester/epoxy hybrid system curing at low temperatures (70/30)
Source: Zimmermann, DSM Coatings Resins

Component	Formulation 1	Formulation 2	Formulation 3
	white	high filled	brown
Polyester resin	460.0	395.0	414.0
Epoxy resin	197.0	169.0	177.0
Titanium dioxide	328.0	282.0	295.0
Barium sulphate		141.0	
Flow additive	10.0	8.6	10.3
Deaerator additive	5.0	4.4	5.1
Yellow pigment 1			7.8
Black pigment			11.9
Red pigment			29.7
Yellow pigment 2			49.1
Coatings test methods			
Gloss level 60°	97	97	63
Gloss level 20°	89	75	63
Pendulum hardness according to König	206	195	182
Reversed impact, 1 day [in/lb]	160	160	160
Erichsen cupping	> 9	> 9	> 9
Gel time at 160 °C [s]	100	102	103
Powder stability (10 is best)			
Start	9	9	9
Powder stability (4 weeks, 40 °C)	8	8	8

results in surface defects as well as in craters. Furthermore, the storage stability of acrylate powder is drastically reduced. Storage temperatures of more than 50 °C rapidly result in an agglomeration of the powder particles. Possibly, the storage as well as the transport require a cooling of the powder.

The classical resins for LT powder coatings are epoxides. Phenolic hardeners can be applied for the formulation of powder coatings fulfilling the stoving conditions for the application on wood-based materials. Generally, the surface of powder coatings based on epoxy resin are very smooth and glossy. The main disadvantage of epoxy resin powders lies in its tendency to yellowing during over-burning as well as in its vulnerability to chalking. The exposure to UV light as it is available in the sunlight spectrum leads to chalking and yellowing of the surfaces. Thus, their application for decorative purposes is limited to interior areas. The requirements of the furniture industry are not fulfilled with these systems.

Apart from the epoxy resins, saturated polyester resins are the most important resins for the thermosetting powder coatings. Solid, linear or lightly-branched polyester resins with a glass transition temperature of more than 50 °C are used. In combination with epoxy resins (hybrid systems), saturated polyester resins establish an optimal mixture consisting of LT powder coating and resin for the application on wood-based materials. Carboxy functional polyester resins (acid polyester, PES) with relatively low medium molar masses (Mn approx. 2,000 to 2,800) and an acid value between 50 and 90 mg KOH/g are applied. These resins are combined with epoxy resins based on bisphenol A. The molecular weight of the epoxy resin preferably is between 2,000 and 4,000 with an epoxy equivalent between 400 and 800.

The glass transition temperature (T_g) of powder coating resins should be above a value of 50 °C, so that the extruded powder coating can be grinded perfectly and does not sinter during storage. On the other hand, a too high glass transition temperature accounts for an enhanced melt viscosity and thus a more difficult processing in the extruder as well as a poor performance when melting.

Normally, the optimal mixing ratio of polyester resins to epoxy resins amounts 70:30. These systems are suitable to meet the requirements for horizontal surfaces, but not the requirements for kitchen worktops. Especially in the edge area as well as in the peripheral zone, the danger of the crack formation by dehydration is very large. Special emphasis has to be placed on a uniform density profile of the MDF plates in order to avoid dry cracks after heat treatment in the curing process. Table 3.55 presents reference formulations and technological data of such a LT hybrid powder [338, 339]. The range of baking times of these systems are between 3 minutes at a temperature of 150 °C and 5 minutes at a temperature of 130 °C.

The fact that powder coatings are free of solvents makes the process of film forming of the thermosetting systems more complex in comparison to liquid coatings. Thus, this involves higher requirements on the technology of processing. The film formation of the thermosetting powder coatings starts with the fusing of the powder to a flowable molten mass which results in a coating film due to cross-linking reactions in the case of a further increase of the temperature. This process is coupled with an increase in viscosity over several orders of magnitude. The molten mass passes through a viscosity minimum (see Figure 3.102 and 3.103). Generally, this viscosity minimum should be as low as possible, and the time for the solidification of the molten mass (gel time) should be long enough in order to achieve a sufficient wetting of the substrate. Only in this way a good adhesiveness to the coating is possible. Furthermore, the air as well as gases and vapours being present in the powder layer may escape from the subsurface, and the melting process of the powder should proceed well. The Chapter 3.1.9.5 describes the curing process of the powder coatings more exactly.

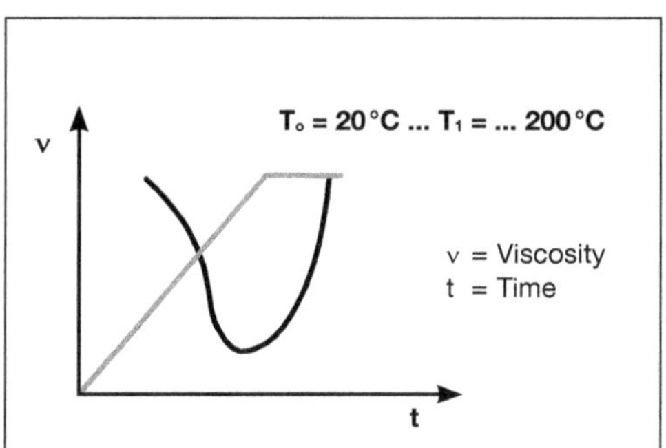

Figure 3.102: Schematic flow curve (black curve) and temperature cure (grey curve) of a thermosetting powder coatings [331]

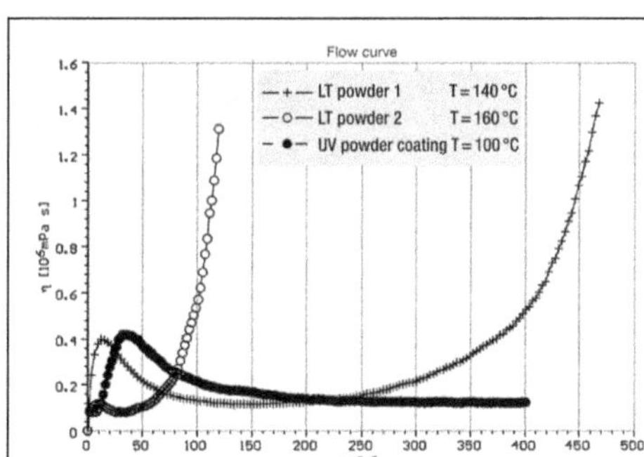

Figure 3.103: Investigation of the flow behaviour of low temperature powder coatings/UV curing powder coatings [232]

UV powder coatings

A major break-through came with the application of UV curing powder coatings [340–348] which were used since the end of the 1990ies. Depending on the composition and curing parameters, the UV curing powder coatings meet the high requirements of the furniture industry and kitchen industry. The UV curing powder coatings enable a decoupling of the melting process from the curing. Since the curing of the powder coating melt by means of UV radiation still requires seconds (approx. 20 s), the heat exposure on the wooden substrate can be reduced. Figure 3.103 shows very well that UV curing powder coatings are more suitable for heat sensitive materials such as wood-based materials already in the pre-heating phase in comparison to thermo-reactive powder coatings, since the required temperatures are up to 30 K (Kelvin) lower in comparison to LT coatings. As long as there no UV light impacts on the melt, a change in viscosity hardly takes place. This allows a control of the deaeratoring and the flow process.

New trend-setting perspectives for the application of powder coatings in the wood processing industry were created with the development of UV powder coatings. The acceleration of the process of film formation is an important economic factor. The reduction of the performance time in comparison to LT powder coatings amounts nearly 10 to 20 %. Now, the time-determining factor is no longer the curing, but the melting of the powder.

Similar to the thermosetting powder coatings, the film forming components of the UV curing powder coatings are resins, fillers, pigments and additives. Similar to the liquid UV systems (see Chapter 3.1.6 and Figure 3.104), photoinitiator still complete the coating.

Meanwhile, different resins are available for UV powder coatings. These resins contain acrylated, methacrylated polyester, polyacrylates, fumaric acid based polyesters alone or in combination with solid vinyl ethers. Acrylated methacrylated epoxides and unsaturated polyurethanes are described in the patent literature [349–356]. The applied powder coatings usually

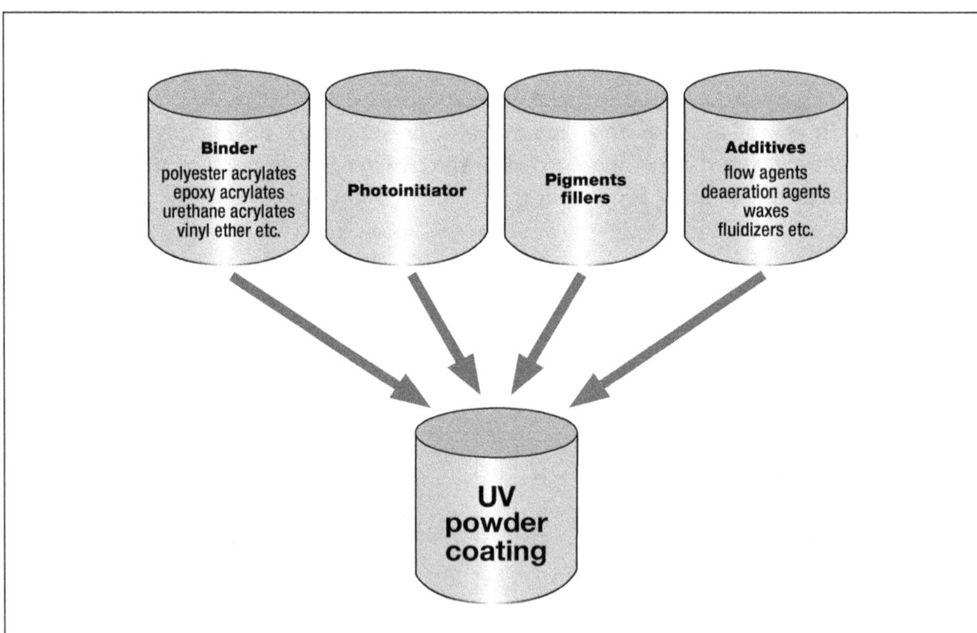

Figure 3.104: Composition of UV curing powder coatings

Table 3.56: Reference formulations of UV curable powder coatings

Components	**White – mat**	**Clear coat**
(Meth)acrylated epoxy/polyester resin	66.3	
(Meth)acrylated polyester resin		81.7
(Meth)acrylated polyurethane resin		9
Titanium dioxide	25	
Photoinitiator (BAPO)	1	
Photoinitiator (AHK)	1	2
Matting agent/texture additive	3	
Flow agent	1	6.8
Deaerator agent	0.5	0.3
High disperse silica (pourability)	0.2	0.2

are fused at temperatures of approx. 130 °C. As well as in the LT powder coatings, a storage stability at temperatures of minimally 35 to 40 °C should be guaranteed in the case of UV powder coatings. The melt viscosity as well as the storage stability depend on the glass transition temperature (T_g) of the coating formulation. Nowadays mostly crystalline and amorphous resins are mixed [357] in order to guarantee a low melt viscosity and a good storage stability. Thus, the amorphous resin shall guarantee the storage stability, and the crystalline resin shall ensure a low melt viscosity. However, some of the crystalline resin dissolves in the amorphous resin and has a plasticising effect so that the resins have to be matched exactly. Many of the resins used today feature glass transition temperatures of 45to 60 °C.

The ideal photoinitiator for pigmented UV powder coatings [358–360] must have an enhanced activity. It must not impact plasticising on the coating (for a good storage stability) und must not generate any volatile decomposition products. In practice, a combination of bis(2,4,6-trimethylbenzoyl)-phenyl-phosphine oxide (BAPO) for the curing and 1-[4-(2-Hydroxyethoxy)-phenyl]-2-hydroxy-2-methyl-1-propane-1-one (-hydroxy ketone derivative, AHK) for the surface curing has been established. The concentration should be totally < 3 % and naturally depends on the pigmentation. The ratio of the photoinitiators may vary between 1:1 and 1: 4 (AHK:BAPO) whereby an increase up to a certain quantity of BAPO (bisacyl phosphinoxide) may result in an improved depth of cure. But an increase in the amount of BAPO also leads to a yellowing of the coating. This so-called effect of photo-bleaching is explained in the Chapter 3.1.6. Since most of the yellow pigments absorb in the same wavelength range, actually any good curing, yellow UV powder coatings are available.

It is also important to know that the addition of photoinitiators based on α-hydroxy ketone (AHK) may lower the glass transition temperature (T_g) of the powder coatings by approx. 2 Kelvin per % of the used photoinitiator. Usually 1.5 to 2 % photoinitiators are applied for clear coatings.

If UV powder coatings for matt surfaces are formulated, it must be remembered that the matting agents being suitable for thermo-reactive powder coatings here have no effect. The UV powder coatings behave similarly to 100 % UV liquid systems (Chapter 3.1.6). Due to the short curing times the matting additives have not enough time to reach the surface. Meanwhile, there exist some possibilities to generate mat and textured surfaces with UV powder coatings such as:

Table 3.57: Property profile of different coating systems for the application to furniture surfaces [322]

Furniture specific performance characteristics	Test standards	Requirements for work fields in kitchen furniture according to DIN 68930
Chemical resistance	DIN 68861 Part 1	1C
Abrasion resistance	DIN 68861 Part 2	2E
Scratch resistance	DIN 68861 Part 4	4E
Behaviour on dry heat	DIN 68861 Part 7	7C (100 °C)
Behaviour on humid heat	DIN 68861 Part 8	8B
Impact resistance (5 mm ball)	EN 438 Part 2	>20N (>10N according to the German Association of Quality Furniture
Resistance to change of climate	Ihd factory standard	0–1
Light fastness (behind window glass)	DIN 53231 Process 2	Degree of light fastness 5–6 for pigmented systems

- Special matting agents for UV powder coatings (such as micronized waxes, spherical particles)
- Dry mixture of differently pigmented powders
- Powder with special particle size distribution
- Application of short-wave UV radiation (excimer lamps)

Nowadays, UV powders for different gloss levels are available commercially. Table 3.56 exemplary describes reference formulations of both, a white mat clear coating and a glossy clear coating. These coatings can be molten with IR radiation (32 kW/m^2) whereby surface temperatures of approx. 130 °C can be achieved. The curing process proceeds with an UV-dosage of 1,100 to 1,500 mJ/cm^2. The currently available UV coatings fulfil the quality requirements of the kitchen furniture industry (see Table 3.57) and are a good alternative to melamine surfaces.

3.1.9.3 Fundamental requirements for wood and wood-based materials

The above mentioned technical difficulties especially consist of the control of an electrostatic coating process [318] such as the powder coating on a very badly conductive subsurface such as wood. The electrostatic powder coating requires a material with a specific surface resistivity of $\sigma \leq 10^{10}$ Ω or a specific volume resistivity of $\rho \leq 10^{8}$ Ωm, respectively. Wood and wood-based materials meet these requirements only at a moisture content of more than 7 to 8 %. If the moisture content lower than 7 to 8 %, then the conductivity has to be increased by means of suitable pre-treatment processes. Enhanced temperatures are a further problem for the curing of powder coatings. Actually, these temperatures still are above 100 °C. The resinous coniferous woods are eliminated a priori for the powder coating. Already at substrate temperatures above 50 °C, the wood components are expelled from the surface. Furthermore, the balancing humidity amounts 9 to 10 % at standard atmosphere (23 °C/50 % r.F.). The dan-

Wet coatings						**UV powder coatings**		**LT powder**
CN solvent system	UPE solvent system	UV 100 % system	2C PU solvent system	2C PU water-borne system	Acrylate water-borne system	One-layer structural powder	Primer and powder coating	Epoxy system
1B–1C	1B–1C	1B–1C	1B	1B	1B–1C	1B	1B	1B–1C
2E	2D	2C	2C–2E	2B	2D–2E	2B–2C	2B	2B–2C
4E	4E–4C	4B–4D	4B–4E	4E	4E	4B–4C	4B–4C	4C–4E
7B–7D	7B–7D	7B–7D	7A–7C	7A–7C	7C–7D	7C	7B–7C	7C–7D
8B–8C	8B–8C	8A–8C	8A–8B	8A–8B	8B–8C	8B–8C	8A	8B–8C
3–30N	12–14N	4–31N	15–30N	15–30N	3–45N	>20	>20	No data
0–3	0–3	0–2	0–1	0–1	0–1	0–4	0–1	0–3
5–6	5–6	5–6	5–7	5–7	5–6	6–8	6–8	6–7

ger of coating defects due to withdrawal of water vapour is too great for a technical application. Especially the diffuse-porous deciduous trees with small pore diameters behave more favourable. Thus, on a pilot-plant scale very good results can be achieved for example on the face of a longitudinal cut of a beech or an alder. But, cross-sectional areas (end-wood) as well as anomalies in the growth as well as secondary mutations in the wood still are problematic zones. According to the current level of technology, the steamed beech shows the best range of properties for the coating with powder coatings. this is relatively poor in wood components and has no large pores. In a laboratory, beech wood could be powder coated at a melting temperature of > 100 °C without blistering [327]. It also is reported on auspicious experiments regarding to powder coating of deciduous woods at a laboratory scale [345]. An industrial application of the powder coating in the solid wood is not to be expected in the foreseeable time.

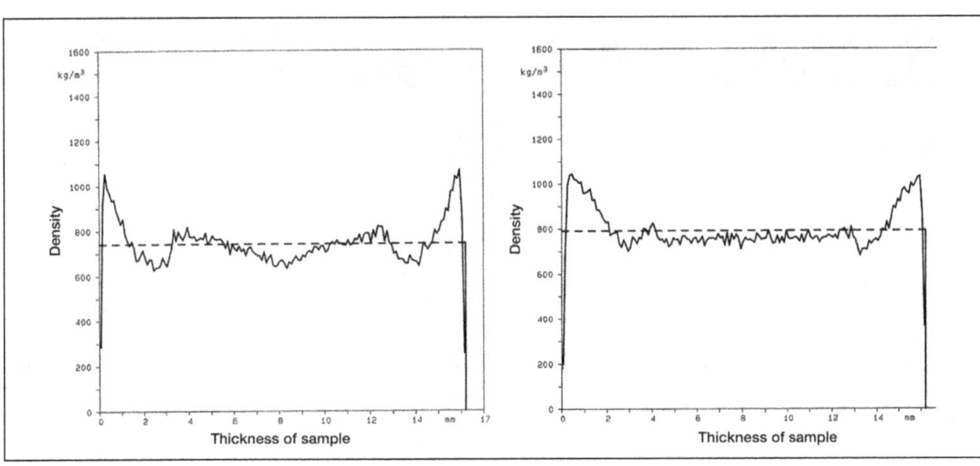

Figure 3.105: Bulk density profiles of MDF, manufactured with different rates of compression [321]

Wood-based materials, especially medium density and high density fibreboard considerably are more favourable of the powder coating process. The advantage of MDF is that it is a nearly area isotropic material. Furthermore, a favourable balancing humidity of approx. 6 % for the powder coating can be easily adjusted. This balancing humidity is somewhat lower than the above mentioned balancing humidity of 7 to 8 % being used for an electrostatic powder application. However, at a lower balancing humidity the danger of a crack formation during the curing process of the powder coatings is lower.

But also, the MDF plate is not entirely unproblematic since the MDF plate consists to 85 to 95 % of wood fibres with hydrophilic properties. The influence of the air humidity on the electric properties in the application of powder coatings as well as on the degassing behaviour during the process of film formation are correspondingly large. Furthermore, there is a danger of crack formation in use due to the swelling of wood. Whether the MDF substrate is suitable for the powder coating, is determined by the process of manufacturing to a vast extent along with the correct pre-treatment.

The main processes in the manufacturing of MDF process are the following:

- Roundwood is processed to wood chips
- The wood chips generally are boiled at temperatures above 150 °C
- Grinding in the refiner
- Blow line gluing
- Drying of the glued fibres
- Pressing of the board

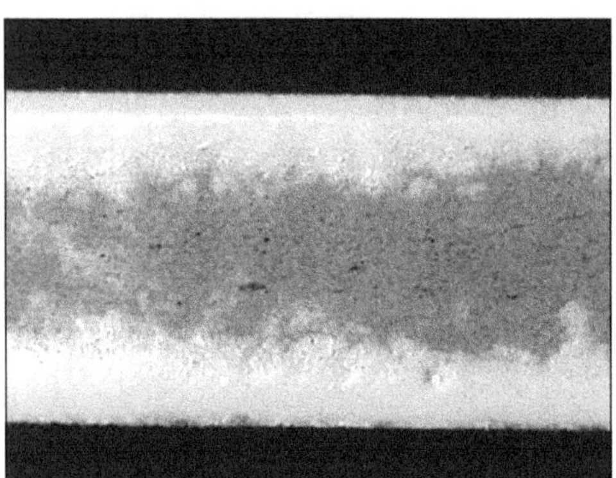

Figure 3.106 (left above):
Powder film on a narrow edge of a MDF with a too low density in the depth [321]

Figure 3.107 (right above):
REM picture of beech fibres [321]

Figure 3.108 (right below):
REM picture of spruce fibres [321]

The feature of MDF depends on press parameters such as rate of compression, warming up of a fleece, and rate of evaporation. From this, in addition to the desirable properties, also undesirable may result for the MDF. Figure 3.105 exemplary presents two grades of MDF which distinguish due to the different bulk density profiles caused by different rates of compression.

For example, for the direct coating, the initially decreasing and subsequently increasing bulk density in the left bulk density profile is less suitable than the more homogeneous profile since the pronounced minimum in the bulk density profile in the left diagram may favour cracks in the coating. Figure 3.106 presents a powder coated narrow edge of a MDF plate having a too low density in the depth. Here, the untreated narrow edge cannot be painted completely with a powder coating film since the coating penetrates up to 300 µm into the MDF plate.

Depending on the type and manufacturer, the MDF substrates not only have a different bulk density distribution. The following items have a significant influence on the quality of the MDF substrates:

- Bulk density profile (gradation in density vertical to the plane of the MDF plate)
- Adhesive system (PMDI resin, UF resin, MUF resin or PF resin)
- Fibrous structure
- Used hydrophobing agent

A good wetting behaviour of a substrate is a prerequisite for a faultless, optically attractive surface. Not least, the morphology of the wood fibre is responsible for this (see Figures 3.107 and 3.110). In this example, the fibre length of the spruce wood amounts approx. 3 mm while the fibre length of beech wood amounts approx. 1 mm here. The gluing of the wood fibres in the blow-line only is performed selectively, so that the longer spruce fibres may dissolve themselves from the composite due to an electrostatic application of the powder coating and straighten up. This results in unintended surface defects.

Thus, a fine fibre quality is an important feature of quality of a powder coatable MDF plate. Recent findings [361] show that the wood species is not crucial but above all the geometries of the fibres. Actually, a single-layer coating with a sufficient quality only can be realized by application of a MDF adapted to a powder coating.

At present, a lot of manufacturers of MDF boards work intensively on the production of powder coatable types of MDF boards which are characterized by the following specifications:

- Fine fibre quality
- Mean bulk density of at least 780 kg/m^3
- Preferably regular bulk density profile
- Electrostatic conductivity at a low moisture content, for example by application of modified resins or by using additives
- 24 h swelling in thickness (EN 317) < 6 %
- Transverse tensile strength (DIN EN 317) > 1,0 N/m^2 (Note: This also is valid for wet coatings)

It can be summarized that the finer the fibres were ground in the production of MDF, the better were the surface quality, medial layer and the homogeneity of the plate. For the powder coating of MDF, currently (as of August 2014) the Egger MBP-L is recommended to the favourites of MDF[475]. Meanwhile, on the market MDF qualities being very well suited for a powder coating are available. However, these MDF boards are subject to certain manufacturing tolerances which can be and must further restricted in the future.

3.1.9.4 Pretreatment of wood-based materials

The pretreatment of the MDF has a decisive impact on the coating result and thus on the achievable quality of the refined work piece. The surface has to be as smooth as possible in order to ensure a minimal consumption of coating material as well as to ensure an enhanced manufacturing reliability. The powder coating cannot compensate for a coarser roughness; they favour the formation of a so-called orange peeling. Due to the electrostatic powder application, special requirements are placed for the grinding process. Any long wood fibres have to remain on the surface since the wood fibres may straighten up under the impact of an electrical field and subsequently stand out of the powder coating film.

The Figures 3.109 and 3.110 describe the impact of the mechanical wood pulp on the coating result. The more fine-grained the applied abrasive paper, the smoother surfaces can be achieved naturally, whereby a grinding with the granulation P1200 on wood will not be performed in practice. As an example, MDF panels can be grinded with a surface grinding of

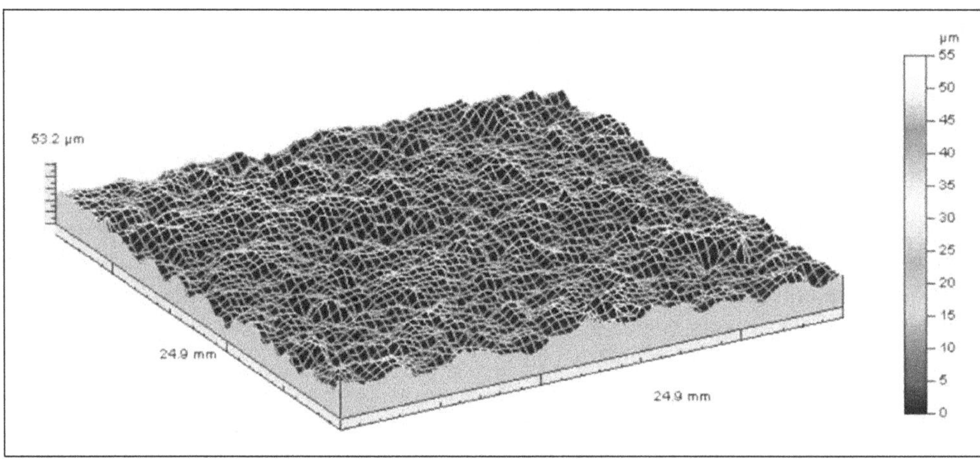

Figure 3.109: Powder coated MDF, grinding with the granulation P180 [321]

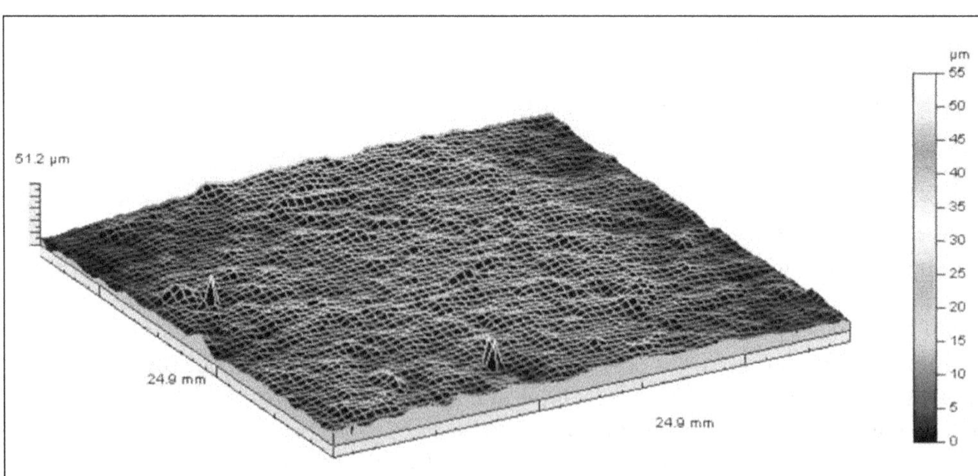

Figure 3.110: Powder coated MDF, grinding with the granulation P1200 [321]

at least 220 granulation. The edges have to be at least broken (radius > 1 mm). By grinding and subsequent cleansing with compressed air, the surface is smoothed finely and evenly as well as rid of impurities, minor scratches, dust, grease etc.

An optically high-value single-layer powder coating only can be achieved on a perfectly smooth underground. Limits have been set to the MDF due to its morphology. The thermo smoothing particularly is recommended for deeply milled MDF elements [362–364]. It is particularly suitable since it has not only smooth the milled profiles and narrow edges, but also compresses the surface additionally. The principle of this process is based on a pressing of wood fibres exposed during the milling process into the subsurface as well as on a permanent attachment on the subsurface. This is done by applying a heatable smoothing tool which plastifies the wood fibres by exposure to heat and pressure. CNC-controlled surface milling machines are suitable for interior profiles. Outer profiles can be smoothed at an enhanced feed rate using format processing machines (Figure 3.111).

The **fixed smoothing** was developed for the application at CNC machining centres, while the **roller smoothing** was developed for the processing of narrow edges in through feed machines.

Meanwhile, research projects report on feed rates of 40 m/min during the **roller smoothing**. Actually, further tests under industrial conditions are performed. The smoothing is a dust-free procedure. Therefore, the smoothing does not require suction removals. Blades smoothly milled surfaces with completely removed dust and chips are a prerequisite for the smoothing since otherwise dust and chips are pressed into the surface under formation of irregularities in the surface. The smoothing densifies the surface in such a way that a change of liquid coatings or powder coatings can be prevented. This results in a reduction of the required coating materials and the necessary steps of work as well as to an improvement of the surface quality. With regard to the powder coating this means that one may dispense with the primer coat. Thus, a single-layer powder coating is possible. Furthermore, the intermediate sanding can be saved since the straightening/swelling of the wood fibres during the liquid coating is lowered. The comparison of the different methods for the pretreatment of surfaces (Figures 3.112 to 3.115) shows the effectiveness of the thermo smoothing.

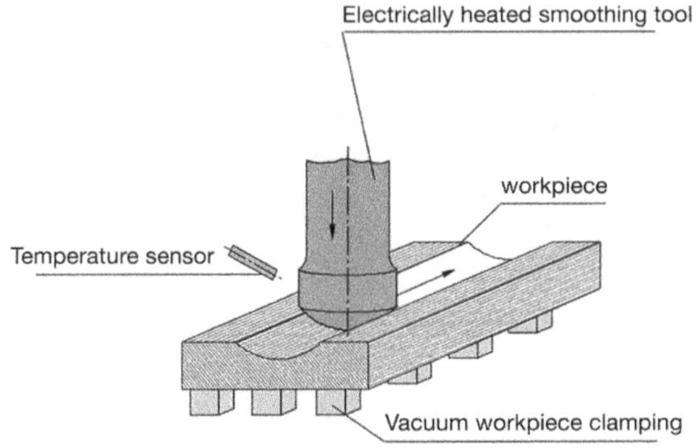

Figure 3.111: Schematic diagram of thermo smoothing [362]

Meanwhile, quite a few furniture manufacturers apply this procedure for the pretreatment of edges of kitchen fronts which however are coated with liquid coating.

Besides a smooth surface, naturally the electrical conductivity of the MDF also is an important factor. This only can be ensured by a sufficiently high balancing humidity of the MDF plate. The humidity of the air in the manufacturing facilities of the furniture industry is so low that the MDF materials are relatively dry. Thus, the electrical conductivity often is too low with regard to an electrostatic coating. There is often a comparatively short residence time at the workpiece loading station since the near surface layer being relevant for the electrical behaviour dries out quickly.

A lower moisture content would be desirable for the melting and cross-linking of UV curable powder coatings in order to prevent bubble formation by means of escaping water vapour during film formation. Generally, MDF plates with a moisture content of 6 to 7 % are applied. The resin also has an impact on the electric conductivity of the surface of wood-based materials. Thus, phenolic resin or MDF plates attached to PMDI have a higher conductivity than melamine-formaldehyde plates or MDF plates attached to melamine-urea. In order to guarantee a sufficiently high conductivity for a safe process of powder coating, the conductivity of the wood-based panels has to be enhanced. For this purpose, there are different approaches:
- Temporary increase of the conductivity by moistening of the MDF
- Application of an electrically conductive primer
- Dielectrical heating in the high-frequency field or microwave field
- Pre-heating of the MDF with IR radiation or by convection
- Addition of a conductivity additive at the production of MDF

Disturbances due to dehydration of periphal zones or other surface zones can be avoided by moistening with water. A conductivity-enhancing additive may support this measure. The investigations still are ongoing. Especially, the swelling of the fibres has to be avoided. If an electrically conductable primer is used to increase the conductability, then normally an intermediate grinding is necessary prior to the application of the topcoat. For cost reasons, this method is not applied in the powder coating of wood-based materials.

Another possibility to enhance the electrical surface conductivity for a sufficient long period of time is the warming of a coating carrier by means of high-energy electromagnetic waves (high frequency field or microwave oven). Here the moisture content of the MDF

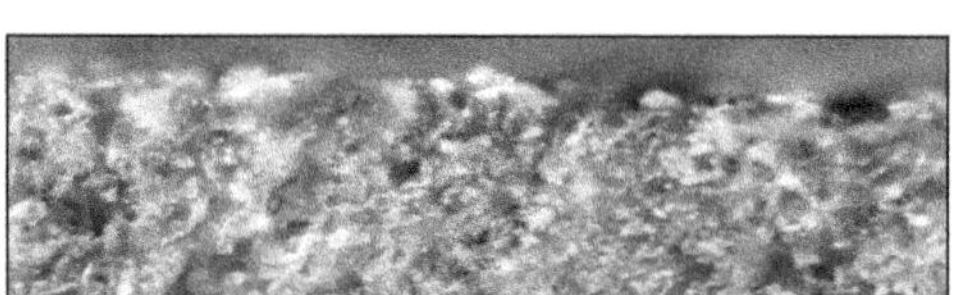

Figure 3.112: Milled surface MDF[363]

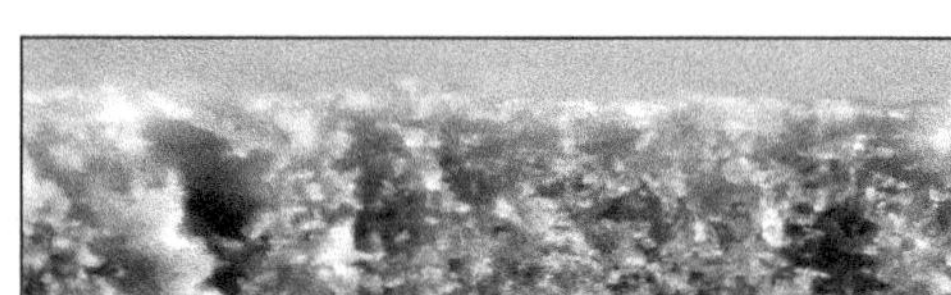

Figure 3.113: Polished surface MDF[363]

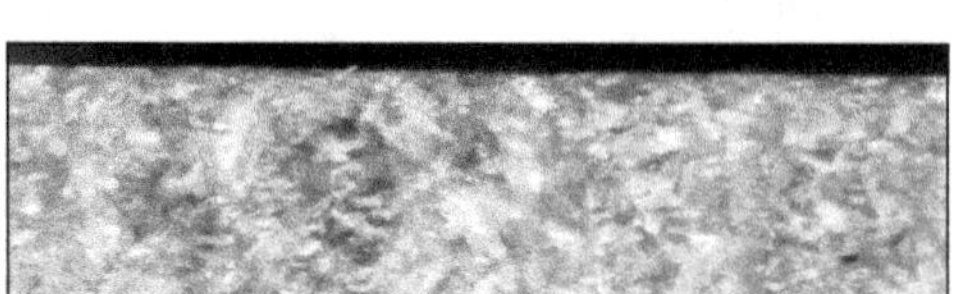

Figure 3.114: MDF smoothed (midth)[363]

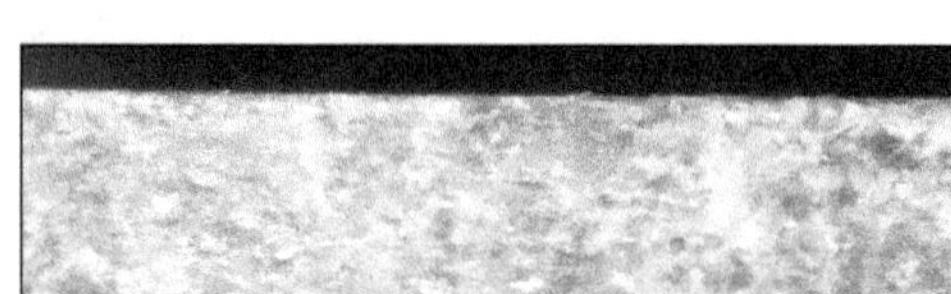

Figure 3.115: MDF smoothed (edge)[363]

hardly is varied. With this method, the high-frequency alternating field directly affects the water molecules in the MDF. Water is a dipolar molecule, and thus it is excited to vibrate. Thus, the greater the product of dielectric constant ε and loss angle tan δ, the greater the heat being formed. The temperature gradient from the inside to the outside supports the migration of the water to the surface. Over there, the water is concentrated in the cooler area near to the surface and increases its conductivity up to a value which is sufficient for an electrostatic application process. However, an industrial application is not possible so far.

A preheating with medium wave IR lamps [365, 366] is favoured in order to increase the conductivity of the MDF in the industrial application at surface temperatures not larger than 90 °C. In the process, water evaporates in the top layer while another part of the water migrates from the cool medial layer to the surface. Within the subsequent zone, the surface cools down, and these liquid condensates in the near surface MDF layer resulting in a sufficiently enhanced electrical conductivity.

However, a lot of investigations indicate that this type of increase in conductivity still does not facilitate satisfactory results with regard to the finish quality. Cracks, blisters as well as pinholes can be the outcome (see Figures 3.116 and 3.117). Experiences [367] show that the surface structure of the MDF plates can be modified strongly by means of the preheating since the fibres at the surface swell. The resulting differences in the electrical conductivity result in a more pronounced manifestation of the fibre structure at the surface.

Meanwhile, a new generation of MDF plates with qualities [361] are available at the market which are suitable for a powder coating even without preheating. With those substrates, a conductivity-enhanced additive is admitted to the process of manufacturing. Thus, an additional pretreatment step is avoided. This seems to be an economic alternative with a great potential in the future.

3.1.9.5 Application and curing of powder coatings

The application of powder coatings [368] is performed by an electrostatic charging of the powder particles in special powder spraying devices with a subsequent separation at the grounded work-piece. The electrostatic powder application requires a strong electric field between the charged powder cloud and the material surface to be coated. Two different processes are applied for charging the particles:

Figure 3.116: Outgassing/Blisters and small pinholes [361]

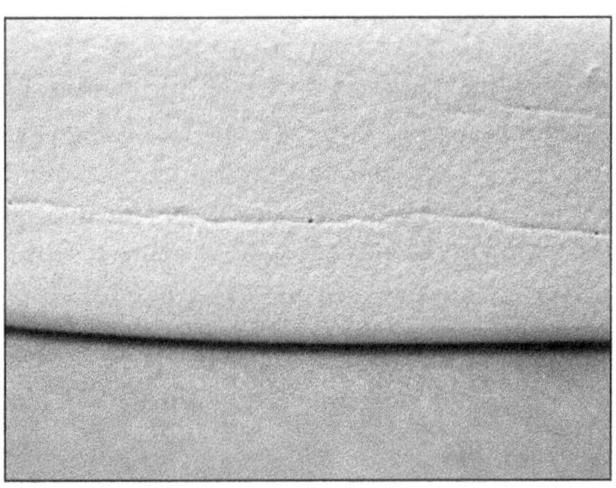

Figure 3.117: Edgecracking and craters

Picture: Stephani [361]

- Tribostatic charging by means of friction of the powders in the charging channel of the powder gun
- Corona charging by means of accumulation of free ions in the field of corona discharge

Within the process of tribostatic charging, powder particles are grinded with high velocity along the surface of the loading pipe of the tribo coating pistol so that the powder particles are charged. Teflon or glass were used as friction partners for charging the usual powders. Thus, the powder is charged positively. It is advantageous, that the tribo coating pistol produces only a small Faraday effect. Thus, niches and corners are accessible relatively simple by means of this technology, since only the orientation of the air flow influences the direction of the powder. The different resins are chargeable to different extents so that several powder additives are required in order to make the powder capable for tribostatic charging. A further disadvantage is, that metallic powder coatings and effect powder coatings cannot highlight their effect optimally.

In the area of corona technology, the charging of the powder coating particles exclusively is performed by means of ion bombardment. A high-voltage cascade is used to generate voltages of 40,000 to 100,000 V which are supplied to an electrode at which gaseous ions are produced which adsorb at the powder particles. The strong electric field supplies an effective charging and ensures that the charged powder particles directly fly to the grounded workpiece. However, the electric field lines are prone to transport the powder particles directly to the nearest surface and edges. These electric field lines do not act in cavities (Faraday cage).

In avoidance of orange peel effects due to back ionisation from the material surface, actually ion-reduced (ion-deficient) corona pistols are applied. Corona pistols shall generate a powder cloud without free air ions. This is achieved for example by means of a grounded supplementary electrode in an insulation spacing behind the corona electrode. Due to their good control characteristics also independently of the powder chemistry as well as due to a higher area performance towards the tribo coating pistols, actually ion-deficient corona pistols mainly are applied for the powder coating of wood-based materials.

Two different technologies as well as their combinations are available for the melting and curing of thermoreactive powder coatings:
- Convection heat
- IR radiation
- Combination of IR radiation and convection drying

Using convection heat, it can be seen that for wood-based materials in spite of longer residence times (such as 140 °C/6 min) a low temperature is more favourable than a higher temperature (such as 140 °C/6 min) with regard to an avoidance of blisters and craters in the film. However, if powder coatings shall be melted and cured rapidly and more efficiently without damaging the wood or wood-based material, infrared heating-technology is more suitable than the conventional method using hot air. Investigations [318] have shown that wood-based materials are more heavily charged when stoving in a circulating-air oven in comparison to a comparable stoving period of time by means of IR radiation. The gelation of the powder by means of the IR radiation is considerably faster than the gelation in a circulating-air oven and thus reduces the temperature load of the substrate especially in low-melting UV powder coatings as well as in highly reactive LT powder coatings. Powder coatings absorb IR radiation very well, and the powered mass is warming up faster in comparison to conventional heating methods such as circulating-air ovens.

Principially, the short wave IR radiation as well as the medium wave IR radiation (up to 2.8 µm in clear coatings and up to the medium wave sector 3.5 µm in pigmented systems) are particularly suitable for the melting of powder coatings [318]. However, the short wave IR radiation is less suitable for the profiled parts since large temperature gradients may occur between the base area and the milled profiles of the MDF plate during the application of the IR radiation. These differences in temperature increase the tensions in the coating film which possibly may result in coating defects. Thus, medium wave and long wave electrical or gas catalytical lamps are more suitable for profiled MDF plates. In order to have a larger process reliability, IR lamps also can be combined with convection heat ("Triab", DuPont). This provides an enhanced number of degrees of freedom for the control of the melting process.

A typical coating process with thermoreactive powder coatings [331] contain three processing steps:

1. Preheating of the surface of the MDF plates with medium wave IR lamps from all sides up to a surface temperature of maximally 90 °C.
2. Coating by means of corona spray pistols (60 to 80 kV) or tribo coating pistols within a time period of 1 to 2 minutes after the first step.
3. Melting with IR lamps during a time period of 30 to 60 sec at a surface temperature of 130 to 150 °C with subsequent maintaining the object temperature for cross-linking the powder coating by means of convection heat for a time period of 3 to 5 minutes.

Until now, the preheating time for the warming up of the substrate from ambient temperature to a temperature of 140 °C is not considered and naturally depends on the substrate and on the plant design.

Unfortunately, the spectrum of characteristics of single-layered LT powder coatings does not satisfy the requirements of the high-quality furniture surfaces. However, UV powder coatings are more suitable for this purpose. Exactly like the thermally curing systems, the

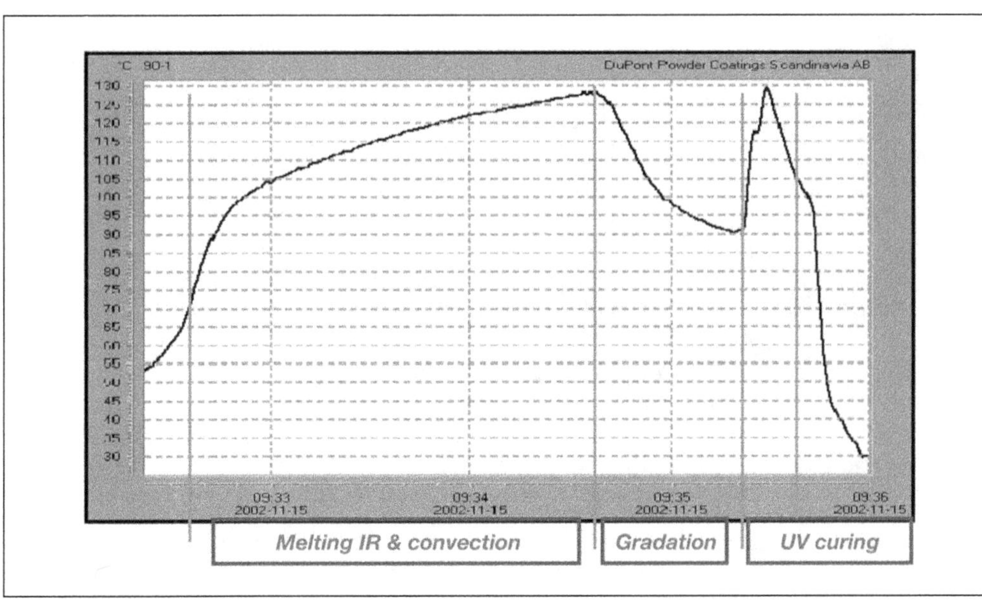

Figure 3.118: Typical temperature profile (coating temperature) of a UV powder coating process [331]

application of UV powder coatings largely is performed by means of ion-deficient corona pistols. Compared with the stoving technology, a major advantage of the UV technology is the decoupling of the melting process and cross-linking process. Due to the absence of a thermally activated cross-linking, subsequently to the fusing one may wait until an optimal melting process performs. Even melting times of 1 to 3 minutes at a temperature of approx. 110 °C are sufficient. In comparison with thermoreactive powder coatings, the short-term and moreover the relatively low temperature effects reduce the danger of blistering enormously. Subsequently to the pre-heating, the powder coating is cured by means of UV light within a period of time between 30 and 60 seconds. For short periods of time only temperatures of maximally 130 °C at the plate surface are measured during the total melting process and curing process. Figure 3.118 illustrates a typical temperature curve of the coating process. The selection of the suitable UV lamp is done similar to liquid UV coatings and is described in the Chapter 7.4.

Figure 3.119: Cracks in a powder-coated MDF plate

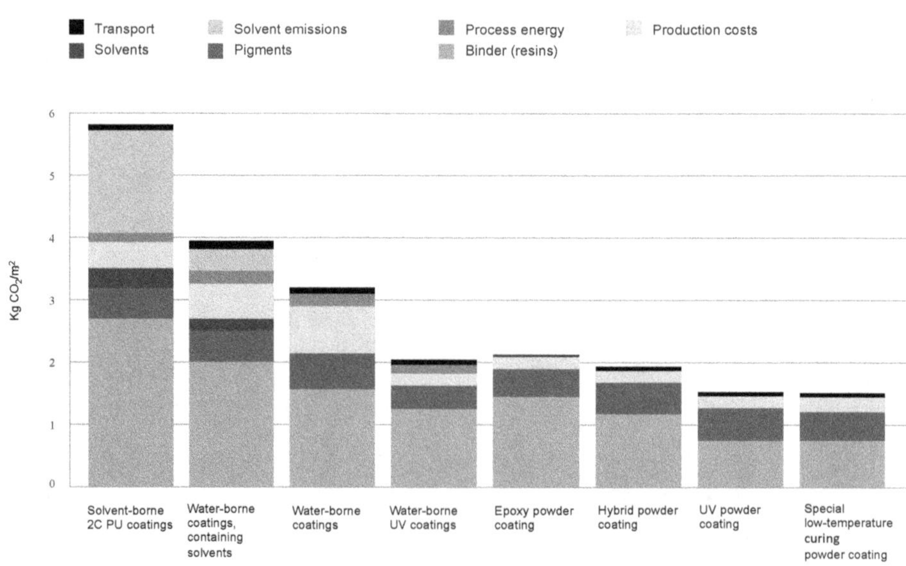

Figure 3.120: CO_2 footprint of various coating systems – 1 m² coated MDF (150 μm) [476]

A typical process is described as follows:
- preheating zone (approx. 60 °C surface temperature of the MDF plate)
- application zone (80–110 µm layer thickness)
- melting zone (IR and convection dryer, approx. 120 °C)
- UV dryer (Hg lamp, for pigmented systems gallium doped)
- conveying velocity approx. 1,5 m/min.

UV powder coatings can be used in order to obtain surface qualities which are comparable to melamine surfaces.

3.1.9.6 Prospects

Despite the outstanding spectrum of characteristics of the coating with UV powder coating [369–376], this technology has not made progress. Firstly, compared to a thermoreactive powder, the much higher price of an UV powder impedes its application in the serial production in Germany. On the other hand, still not all technical problems are resolved in order to guarantee a single-layered UV powder coating with a consistent quality in the line cost-effectively.

Thus, a coating technology of 3D furniture fronts also requires a 3D UV technology. Critical sectors are not only narrow edges but also deep profiles in which the powder coating may penetrate deeper due to the enhanced porosity of the wood-based material and possibly crosslink no longer. In particular, the demanding requirements of the kitchen furniture industry on a maximal colour deviation of $\Delta E \leq 0.2$ to 0.3 (LAB system) are no longer met. As in the wet coating sector, not yet all colour shades can be cured optimally by means of the UV technology. Contrary to the UV wet coating technology, the powder coatings have to be kept in a melting phase by means of heat supply. Establishing tensions under non-optimal temperature control may result in edge losses and pittings. For the avoidance of edge losses, especially the UV radiation has to be performed in a very regular, completely continuously progressing process. If the MDF plates and temperature control are not always consistent, then coating defects which are shown in the Chapter 3.1.9.4 arise. Due to the technological difficulties in the UV powder technology, thermoreactive powder coatings are favoured on the actual plats for the powder coating on wood. Here again, even in the edge sector optimisations still are required which however are solved for the most part. Here, cracks are typical defects (see Figure 3.119) as *Prieto* et. al [377] have found in coating experiments. These are single-layered powder coatings.

This crack formation in the narrow edge results nearly inevitably when not pre-dried MDF plates convectively are heated for melting and curing of thermo-reactive powder coatings. These cracks partly are spanned by the powder coating film. Components with deep profiles near by the peripheral zone are critical, since here the cracks begin at both ends and meet in the middle. However, when properly selected the MDF and when designing the plant exactly, it is possible to coat furniture surfaces in good quality also with UV powder coatings, as examples from France and Australia will show [378, 379]. However, UV powder coatings compete with conventional UV coatings and aqueous UV coatings.

Already today, single-layered or two-layered coatings with LT powders or ULT powders are possible in good quality in the case of MDF substrates being selected at an optimal process control. The powder coating of wood-based materials is a modern and complicated process which depends on many factors and requires motivated co-workers in order to practices this technology. A fine tuning still is necessary, in order to further push this technology and in order to be competitive technologically and economically just in the sector of qualitatively high-quality furniture surfaces. The goal must be a single-layer powder coating.

An important milestone for the further establishment of powder coatings for the coating of wood is the commitment of the world's largest furniture store 'IKEA' to apply powder coatings extensively [479]. A large Ikea suppliers in Slovakia (company Ekoltech) applies low-temperature powder coatings for IKEA. However, other improvements such as smoother surfaces or the reduction of the 'orange peel' are desired by users and consumers. In the case of LT powder coatings, recent studies such as from DSM 'Carbon Footprint Study' as well as publications by Ramseier Woodcoat and IGP powder substantiate a significantly better eco-efficiency and Life Cycle Analysis as in the past adopted [476–478].

3.1.10 Coating of decor finish foils

In general, decor finish foils are saturated synthetic resin, decoratively printed or unprinted paper webs which are coated with so-called finish coatings or foil topcoats for the protection of the paper surface against mechanical and chemical stresses. Subsequently, in the furniture industry the decor finish foils are pressed onto chipboards or MDF plates, respectively.

According to the quality standard "decor finish foil" of the IHD from the year 2010, the following definition was used: Decor finish foils are special, printed or coloured papers with a coated surface. The quality of the future decor finish foil is determined by the carrier material, pressure and especially by the coating [494, 495].

The coating of the printed paper is known for more than 75 years. The patent specification by the company MASA from the year 1928 reports for the first time on the "Procedure for the coating of printed paper" [380]. The decor finish foils have started their triumph in the furniture industry just approximately 35 years ago. With the development of the decor finish foil, the furniture industry received the opportunity always to apply consistent and equivalent surfaces with wood finishes or fantasy decors in large amounts [381]. When regarding the globally consumed square meters (approximately 15 billion m^2), the decor finish foils are the most important coating materials of the furniture industry [382]. In the decor finish market 2015, decor finish foils have a market share of approximately 10 % (see Figure 3.121) [383]. Europe is the most important consumer of decor finish foils (impregnant and pre-impregnant) which is followed by Asians who mainly use low weight papers (LBWP) (see Figure 3.121). In practice, there are often synonyms such as decoratively printed paper, furniture foils etc. for the term of the coated decor finish foils (finished effect). Decorative printed papers also are used for the production of melamine foils or melamine films, edges, short pressure laminates and high-pressure laminates for other areas of application which are not described further here. Figure 3.122 illustrates the areas of application and applications of decor finish foils in Europe in the year 2015. The conditions on the market have changed due to Ikea's strategic decision a few years ago to focus more and more on decor finish foils and to build up wood-based panels even over its subsidiaries [481]. Ikea demonstrates that laminating also can be worthwhile for a wood-based product manufacturer if it is integrated towards the finished furniture or even to the sale. The once frowned cheap product decor finish foil based on pre-impregnates is still not yet sophisticated technologically and yet a high-tech product already today. It offers the possibility to the decor printers to present itself as a full-service provider in the matter of surface as well as to gain more independence from the wooden material customers. Today, everything is possible from the simple and inexpensive mass product to rather haptic elaborated surface equipped with all sorts of features confusingly similar to a real veneer. Modern laminating systems have made the decorative film perhaps the most suitable for the industry for modern production of standard furniture. On the contrary, there is much evidence that the melamine surface will lose market shares to the foil [482]. Even the today's possibilities of the equipment provide many advantages for the decorative film on the basis of pre-impregnated decorative paper with regard to the achievable haptics. The optically and haptic greater proximity to the real veneer means a considerable advantage over the melamine surface for the decorative film. The newly installed production systems for acid-curing decorative finishes are high-performance lacquering machines with a production speed of up to 450 m/min and with an operational width of 2.75 m. With the joint-free transition of the decorative foil from the surface and the edge, the door to almost unlimited design freedom is open for the furniture manufacturers. All available paper thicknesses, carrier materials and surface types (structure, high gloss etc.) are suitable. For example, a flawless surface finish with profiled chipboard edges or an enormous shock resistance of the edges is striking which additionally accommodates the growing demands on the quality of the furniture [483].

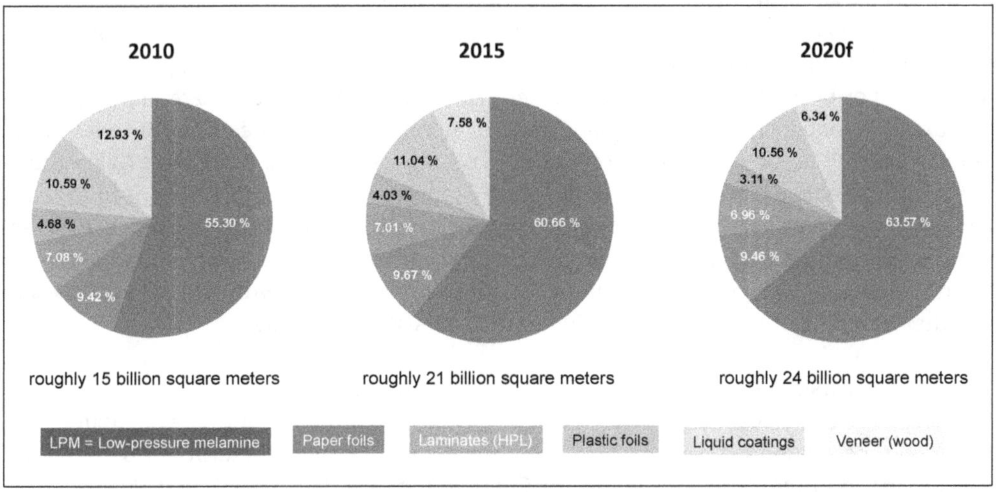

Figure 3.121: Global decorative surfaces market – paper based 2010 to 2020f [480]

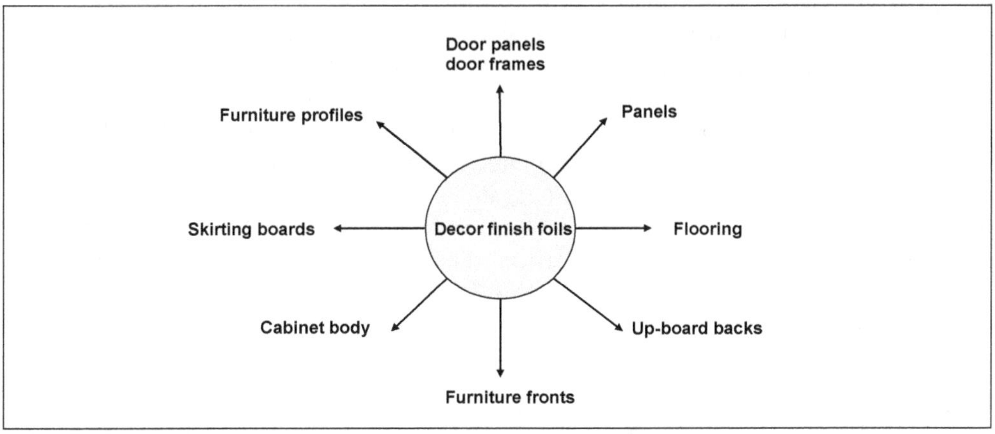

Figure 3.122: Application areas of decor finish foils in Europe 2015 [481]

3.1.10.1 Historical development of decor finish foils

At the beginning and in the mid of the 1960ies, furnitures such as beds, wardrobes, tables and doors mainly are manufactured individually. The progressive rationalisation of manufacturing processes and materials led to a strengthened involvement of chipboard panels. However, the former quality of the chipboard panels were characterized by a relatively coarse chip structure. The coating of such chipboards required a multiple coating in order to cover the chip structure and irregularities. For saving coating steps, monochromatic prime paper foils were developed and implemented whereby these prime paper foils were grouted on chipboards [384].

Initially, it were amino resin impregnated papers with a surface weight of approximately 120 g/m^2 and a cellulose nitrate coating. Subsequently to the grouting on chipboard panels, these served as a basis for coating and were polished to the first time and subsequently coated with coating systems.

Sometimes later, also printed paper films with wood decor design were established as an alternative to veneer surfaces in order to avoid extensive work processes on the veneer preparation. Thus, all forms of wood design and fantasy design could be reproduced typographically. Some years later, an improved kind of impregnated papers – the finished foil – were applied for a further simplification of the production processes. These are impregnated monochromatic or printed paper films which were equipped with a coating layer already at the manufacturer of paper films subsequently to the impregnation [384]. The decor finish foils thus obtained as well as edging films for a continuous coating of edges are supplied as rolled goods to the furniture industry as well as to the panel industry. The processors only need to tailor and grout the decor finish foils and not to coat these foils. At the beginning of the 1970ies, the implementation of decor finish foils drastically reduced the prices per square metre in the production process in comparison to the furnitures veneered with real wood. This was just the beginning of the finished coatings which persists until today.

3.1.10.2 Base papers for the decor finish foils

The actually oldest known paper was found in the People's Republic of China and dates back from the time around 60 before Christ. The process for the manufacturing of paper firstly was described in the People's Republic of China in the year 105 after Christ. At that time, tree bark, hemp and rags were applied. Today, base papers for decor finish foils are high-tech products. These consist of selected fibres, fillers, pigments and certain chemical additives (such as wet strength agents and others). In addition, this paper also features different sorts of cavities such as pores, capillaries and channels. These cavities may account for up to 50 % of the volume of the decorative printed paper and are responsible for the important properties such as the absorption of resin and penetration. Short fibres and long fibres were used for the manufacturing of decorative printed papers. The amount of long fibres improves the tensile strength as well as the tear propagation strength [385]. Thus, the splitting strength of the decorative printed papers significantly is increased for example by addition of straw pulp fibres [386]. Depending on the area of application, a distinction is made between different base papers and basic weight[40] (see Table 3.58). The so-called Japan papers are distinguished from the above-mentioned papers by the raw materials base as well as by their de-

40 Weight per unit area of papers: The weight per unit area of a paper is defined as the weight in g/m^2 under controlled conditions. The total mass is the sum of the mass of the fibrous materials, fillers, auxiliary materials and water.

Table 3.58: Overview of the applied base papers

Paper	Level of impregnation (proportion by total weight)	Weight of paper [g/m²]
Impregnate: standard	30–40 %	60–140
edging paper	30–40 %	140–250
Pre-impregnate	5–35 %	40–80
Japan paper	<6 %	≤40

gree of resin coating). The manufacturing takes place in Asia by means of local fibres. The paper often is reinforced by means of amino resins which are added to the paper during the manufacturing process. The Japan papers are not impregnated additionally. As a low-priced alternative, the Japan papers compete with the pre-impregnates.

Impregnation of decorative printed paper

The term impregnation of decorative printed papers refers to a filling of cavities of a capillary-porous material. The impregnation ensures a mechanical reinforcement and thus an enhancement of the tensile strength of the paper. Thus, apart from the impregnation process, the choice of the wood paper, the material processing as well as the machine parameters when manufacturing paper mainly impact the quality and properties of the binding force between the fibres, by the composition of the impregnated resins and by the impregnation process. The splitting strength describes the properties of the paper strength in z direction [387].

Depending on the type of the impregnation procedure, a distinction is made between decoration foils with a fully impregnated core (offline process) and decoration foils based on pre-impregnates (online process) at which a finish coating can be performed subsequently. Thus, decor finish foils with a finished surface can be produced.

Impregnates

When manufacturing impregnates, first of all a raw paper is produced in a multistage process at the paper manufacturer. Thereafter, the raw papers are sold to printing companies and/or producers of decor finish foils. Here, the raw papers are printed in a decor as required in a gravure printing process in a second step. Subsequently, in a third process step the printed paper is impregnated with special resin compounds at an impregnation line. In particular, the paper webs are pulled through impregnation containers filled with resin solutions, and are applied with a double doctor blade. Subsequently, a rest distance – also designated as a breath distance – succeeds so that the resins of the solution of the impregnating resins may penetrate into the paper web. The drying and the curing normally is performed in a floating drying system at a temperature of approximately 140 to 180 °C whereby the floating drying system allows a contactless process control by means of hot air nozzles. Compared with pre-impregnates, the impregnate contains longer fibres and thus requires high levels of impregnation in the process of impregnation in order to produce decorative papers resistant to splitting. Among the pre-impregnates, the splitting strength of the decorative papers mainly are achieved by a lower and finer length of fibres. Compared with the pre-impregnates, the impregnates have an enhanced air permeability, a lower surface compaction, an enhanced total strength of the paper, are relatively brittle and smoothed on one side. The fundamental difference between impregnates and pre-impregnates consists of the point in time at which the impregnation is performed.

Pre-impregnates

At the end of the 1970ies, diverse paper manufacturers tried to realize the impregnation in one working operation in the paper manufacturing. With this, the foundation for the generation of pre-impregnates were laid [388]. Pre-impregnates are decorative papers which are impregnated prior to the decorative printing and finish coating. Only at the beginning of the 1980ies, pre-impregnates came on the market. Pre-impregnates serve primarily serve as cost-effective alternatives to fully impregnated finish foils [389]. The pre-impregnates consist of fully wood-free pulp. In the early years, the optical quality and the haptics of the pre-impregnates were not comparable to those of the fully impregnated finish foils.

In the last 20 years, the basic material as well as the printing quality improved significantly. Compared with classic finish foils, the price advantage further strengthens the demand for pre-impregnates in the furniture industry. Already in the paper machine, decoration foils based on pre-impregnates are impregnated immediately after the sheet formation and drying with a resin mixture depending on an area related mass of the base paper. Within the process of drying and smoothing often within the paper machine, the impregnation resins have to react with each other (chemical cross-linking) in the paper in order to get paper which is more resistant to splitting. This guarantees that there is no separation of fibres within the decor finish foils in the subsequent processing steps. Further requirements for the pre-impregnates for the further processing are an enhanced covering power, surface quality, good printability in gravure rotary printing, good coatability, good flatness and adhesion with chipboard panels and MDF boards [390]. The applied, thermally cross-linkable, aqueous mixtures of impregnation resins are resin solutions based on urea resins, melamine resins or phenolic resins [391]. The more use of these resins results in relatively hard and brittle products. Thus, additionally to the above mentioned impregnation resins, thermally cross-linkable polymer dispersions based on acrylic acid, acrylic acid ester, styrene, butadiene, vinyl acetate or acrylonitril are applied. Among the pre-impregnates, the resin absorption amounts between 5 and 35 weight percent. By contrast with aqueous polymer dispersions, the classical impregnation reins (urea resins as well as melamine resins) intrude deeper inter the raw paper. The polymer dispersions accumulate more and more at the upper and bottom side of the decorative printed paper and ensure a good smoothability as well as surface tightness of the paper surface. A smooth and compacted surface is crucial for an excellent printability and coatability [392]. The patent literature shows that the penetration depth of polymer dispersions in the group of impregnation resins significantly can be influenced by different additives such as water-borne polymers based on polyvinyl alcohols, gelatine, starches, cellulose derivatives, alginates and their mixtures as well as maleic acid co-polymerisates consisting of (meth-)acrylamide and (meth-)acrylic acid [393–396].

Different requirements are posed for the pre-impregnates [383, 397]:

- Visual aspects
- Compliance with the requirement E1 (emission of formaldehyde)
- Operating characteristics on printing machines and coating machines
- Printability (smooth decorative side, wettability by means of the printing ink)
- Coatability (suitable surface tension, regular impregnation, defined suction capability, porosity)
- Bondability (such as coatability, compatibility with different adhesive systems, roughness of the back side, suction capacity, wettability for the water-borne glues, flatness)
- Suitability or further processing (scotch tape resistance, depending on the degree of impregnation, splitting strength)

The pre-impregnats continue to grow steadily since an improved as well as increasingly realistic reproduction of wood designs and wood structures on papers is possible since some years [398]. A further aspect is the improved printability with respect to the printing quality and printing speed. Also, low surface weights with a sufficient opacity as well as pre-impregnates with a better flexibility are producible. Pre-impregnate films are very flexible in the application of modern adhesive joints [399]. Furthermore, the pre-impregnates meet the required concentrations of formaldehyde according to the DIBt-regulation 100 without problems. In the future, there will be more providers of pre-impregnates [400]. The quality could be further enhanced. The trend towards 'thinner decor layer' will undoubtedly and inexorably strengthened. A growing together of pre-impregnate and LBWP (low weight paper) is foreseeable. In the future, even more decorative paper printer will exist which also perform the coating of decor finish foils.

3.1.10.3 Coating materials for finishing of decor finish foils

Different coating materials have been established worldwide for the coating of decor finish foils (see Figure 3.123). The applied coating systems are designated as film topcoats or finish lacquers. Depending on the quality requirements for the cured coating film, it may be distinguished between resistant and high resistant decor finish foils. The aqueous, acid-curing coating systems dominate the markets in Europe. In contrast, in North America solvent containing, acid-curing as well as UV curing and electron beam cured coating systems are applied in the ratio of approximately 1:1. In Asia, solvent-borne 2C polyurethane coatings predominate. The total amount of coating materials worldwide for the coating of decor finish foils is estimated to approximately 50,000 tons. Approximately 60 % of these are aqueous, acid-curing coating systems. The remaining amount mainly is distributed among UV curing as well as solvent borne 2C polyurethane coatings. In recent years, the electron beam-curing lacquer systems have continued to gain in significance and quantity since these have enhanced chemical and mechanical-technological resistance.

The properties of the coating decisively determine the quality of the furniture surfaces in the production of decor finish foils. In the following sections the applied coating technologies are introduced:

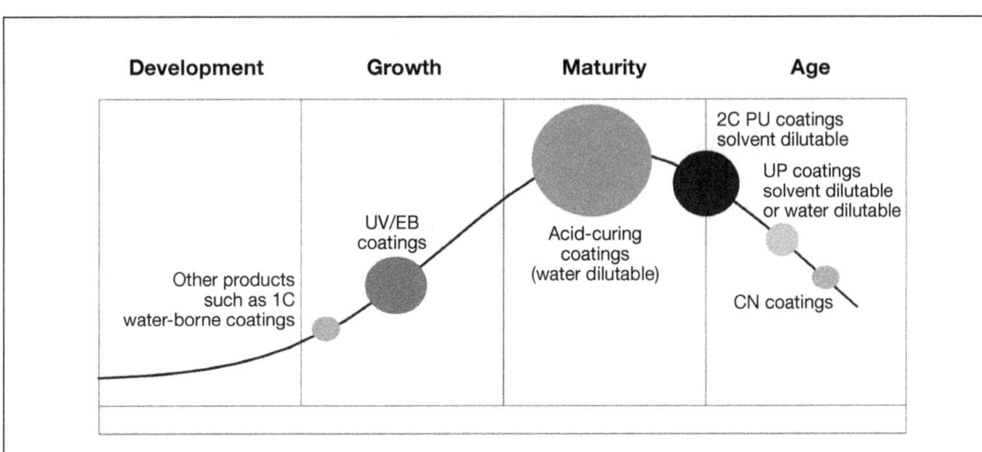

Figure 3.123: Life curve of the coating technologies for the coating of paper decor foils

Cellulose nitrate coatings (CN coatings)

The first film finish coatings were made from cellulose nitrate. Due to the enhanced amount of solvent of the formulations as well as due to the insufficient chemical and mechanical properties, today the film finish coatings are no longer of importance (see Chapter 3.1.2). The low temperature resistance is a disadvantage of cellulose nitrate coatings. This low temperature resistance does not permit any processing on short-cycle presses at temperatures above a value of 80 °C. In Asia, the CN coatings today still are used for the priming of paper films.

UP coatings

Today, the unsaturated polyester coatings have no significance for the coating of decor finish foils. In the former German Democratic Republic, the UP coatings mainly were used for the coating of decor finish foils. Depending on the composition, the UP coatings tend to a volume shrinkage during the chemical curing (radical polymerisation). This volume shrinkage generally results in a curling tendency on the paper medium. This curling tendency is very obstructive for the further processing of decor finish foils. Moreover, the coating films have a tendency to yellowing. In addition to very low amounts of solvent borne products, also aqueous UP coatings are used for special applications. The basic raw materials for the formulation of UP coatings are listed in Chapter 3.1.5.

Acid-curing coatings

Today, the aqueous acid-curing coatings dominate the European market for the coating of decoration foils. At the beginning of the 1970ies, the first aqueous decor finish coatings were developed and established into the market. These are 2C systems based on urea resins and/or melamine resins and an acid hardener, normally p-toluene sulfonic acid (see Chapter 3.1.3). Water-borne, hydroxy-functionalized polyester, polyether, water-borne acrylate dispersions and others were used as a reaction partner for the amino resins. Figure 3.124 il-

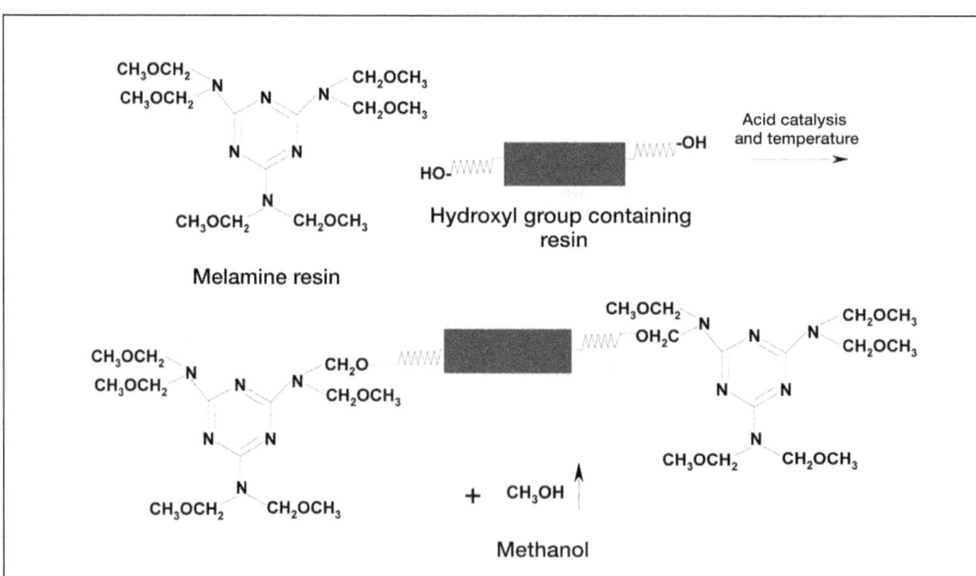

Figure 3.124: Schematic illustration of the polycondensation at water-borne, acid-curing topcoats

Table 3.59: Water-dilutable, acid-curing topcoat for decor finish foils [402]

Component	Function	Parts by weight [%]
Water-borne, hydroxyl group containing acrylate dispersion[1)] based on acrylic acid esters and methacrylic acid esters as well as styrene	Resin, reactant for amino resins	71.23
Triethanolamine solution 1:1 in completely demineralized water	For adjusting the pH value of the acrylate dispersion to approx. 8.2	1.77
Melamine resin[2)] with methanol partly etherified (> 69 % delivery solid content)	Resin; reactant for the acrylate dispersion	21.35
Butyl glycol	Solvent; reduction of the viscosity	4.75
Polyurethane thickener[3)]	Setting of the viscosity	0.55
Tenside[4)] (2,4,7,9-tetramethyl-5-decin-4,7-diol), 50 % in butyl glycol	Improvement of the substrate wetting with defoaming properties	0.35
Sum		100.00

Solid content (percentage of non-volatile components of the coating): approx. 47 %
viscosity (4 mm cup at 20 °C): approx. 20 s
Hardener component: p-toluene sulfonic acid 50 %
water 50 %
Sum 100 %
Mixing ratio/processing instructions: Base coat 100 parts by weight
Hardener 5 parts by weight
Viscosity with hardener (4 mm cup at 20 °C): approx. 20 s pot life (at 20 °C): < 1 day, process immediately
Wet application amount: approx. 10 g/m² drying parameter/curing parameter: 30 s at 150 °C
1) "Luhydran" S 937 T (BASF SE); OH value (calculated on 100 % solid content): approx. 100 mg/KOH/g; particle diameter (mean value): approx. 0.2 µm
2) "Luwipal" 063 (BASF SE)
3) "Surfynol" 104, 50 % in butyl glycol (Air Products Chemicals)
4) "Collacral" PU 75, 25 % (BASF SE)

lustrates schematically the polycondensation reaction at aqueous, acid-curing topcoats with elimination of methanol. Also, formaldehyde scavengers such as urea or casein were applied often, in order to reduce [401] the emission of formaldehyde below the legally prescribed limit of ≤ 3,5 mg formaldehyde/h*m² for laminated chipboards, determined according to the gas analysis method EN 717-2.

Modern topcoats significantly fall below the value of 3.5 mg formaldehyde/h*m² (DIBt[41] Guideline 100). The Tables 3.59 and 3.60 illustrate reference formulations for water-borne topcoats and shall elucidate the principal composition of topcoats [402, 403]. In the practice, the coatings are formulated with solids between 60 and 90 %. The processor blends the coatings with the hardener component and tap water regarding to the corresponding application viscosity. The production rates can be enhanced to a value of 180 to 190 m/min by an optimisation of coating formulations as well as by an application of high-performance dryers and pre-impregnates as a substrate. The aqueous acid-curing coatings have gained acceptance worldwide due to their long pot life, enhanced reactivity and high quality of the coated surfaces for decor finish foils. In North Amerika, solvent-borne, acid-curing coatings still are applied which gradually are exchanged by acid-curing coatings or UV/EB coatings.

41 DIBt (Deutsches Institut für Bautechnik) Directive 100: Directive for the classification and monitoring of wood based panels according to the emission of formaldehyde, version June 1994

Table 3.60: Water-borne, curing topcoat for decor finish foils [403]

Component	Function	Parts by weight [%]
Polyester resin[1)] approx. 90 % in water (containing hydroxyl groups and carboxyl groups)	Resin; reaction partner for amino resins	27.80
Melamine resin[2)] (> 98 % delivery solid content), hexamethoxymethyl melamine	Resin; reaction partner for the polyester resin	50.00
Isopropanol	Solvent; reduction of the viscosity	5.00
Butyl glycol	Solvent; reduction of the viscosity	4.00
Water (distilled)	Solvent; reduction of the viscosity	12.70
Silicone additive[3)] 10 % in butyl glycol	Removal of surface defects and improvement of the wetting to the underground	0.50
Sum		100.00

Solid content (percentage of non-volatile components of the coating): approx. 75 %
proportion polyester/melamine resin: approx. 1:2
viscosity (4 mm cup at 20 °C): approx. 33 s
viscosity after 8 weeks (4 mm cup at 20 °C, with an addition of 10 % hardener): approx. 51 s
hardener component: p-toluene sulfonic acid 40 %
water 60 %
Sum 100 %
blend ratio/processing information: base coat 100 proportional weights
hardener 10 proportional weights
viscosity with hardener (4 mm cup at 20 °C): approx. 24 s pot life (at 20 °C): approx. 15 days
wet application amount: approx. 10 to 20 g/m² drying parameter/curing parameter: 20 to 30 s at 140 °C
1) "Worléepol" V 450, 90 % in water (Worlée Chemie) acid value (DIN EN ISO 3682): maximal 15; amount of hydroxyl groups (OH amount): approx. 8,5 %
2) "Cymel" 303 (Cytec)
3) "Worlée-Add" 330 (Worlée Chemie)

2C polyurethane coatings

The chemistry and composition of the solvent borne 2C PU coatings are well known by the classic coating of furnitures (see Chapter 3.1.4). The 2C polyurethane coatings mainly are used in Asian countries for the coating of decor finish foils. Primarily, so-called low weight papers such as Japan foils are coated with 2C PU coatings. The coatings are characterized by an excellent penetrating capacity on the decorative printed paper, very good chemical resistances as well as adhesive strength properties. Depending on the coating formulation and hardener component, non-yellowing, elastic and hard coating films can be manufactured. The finishing solid content as well as the hardener component and process diluent often is between 30 and 50 %. The coating plants only can be operated with feed rates of 50 to 100 m/min due to the slow drying and curing (polyaddition) of the 2C PU coatings. Subsequently, the coated paper rolls (coils) are stored in separate drying rooms for at least 48 to 72 hours at temperatures between 40 and 50 °C in order to accelerate the chemical curing (polyaddition). Only thereafter, the paper rolls are delivered to the customers.

Water-borne coatings

1C water-borne coatings are applied, for example, for the back coating of papers. These 1C water-borne coatings can be formulated in such a way that these coatings can be applied for the development of a flexibility as well as for the impregnation of paper. Furthermore, the

1C water-borne coatings help to reduce the suction behaviour of the papers. The advantage of this is that a less amount of decor finish coating or good surface effects, respectively, are achieved in the subsequent coating of the front side of the paper. This is called a good top-coat gloss. Often self-crosslinking acrylate dispersions and/or polyurethane dispersions (see Chapter 3.2) are used. The chemical resistances significantly are lower in comparison to the thermally curing, acid-curing water-borne coatings.

UV/EB coatings

The thermal curing of aqueous acid-curing coatings reaches its limits due to the increasing cost pressure in the coating of decor finish foils. High gloss levels (60 to 90 units in the 60° measurement angle) only are realisable in the 2-layer structure or 3-layer structure. The existing coating plants may achieve a maximal feed rate of 250 m/min. The application of radiation curing finish coatings thus may open up new perspectives. The formulation of UV curing or EB curing decor finish coatings uses the same raw materials as the classical UV-curing wood-based coatings (see Chapter 3.1.6).

The photoinitiator often is delivered separately in order to adjust the amount added to the respective conditions of the plant. UV curing finish coatings which are deposited with 6 to 10 g/m^2 and are implemented in an oxygen atmosphere using high feed rates of 100 to 350 m/min often present an odour nuisance. This only can be achieved when the UV curing process is performed under exclusion of atmospheric oxygen. In these conditions, cleavage products of photoinitiators hardly are formed. Own extraction experiments have shown that any photoinitiators or their cleavage products as well as not consumed monomers such as tripropylene glycol diacrylate are detectable. As a rule, UV topcoats which are cured under atmospheric oxygen, have a lower scratch resistance than UV finish coatings which were cured with exclusion of atmospheric oxygen. It is for this reason that only those UV coatings are applied for the coating of decor finish foils which are cured under exclusion of oxygen or under a reduced oxygen content. Investigations on UV curing systems illustrate that the greatest increase in reactivity is in the range between 2.000 and 100 ppm of residual content of oxygen [404]. In the same way, electron-beam curing systems

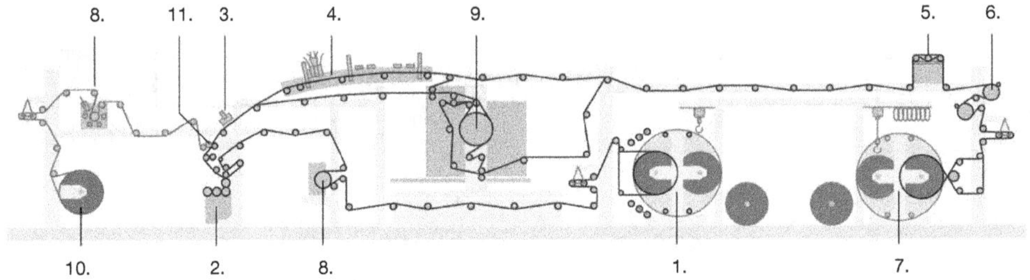

1. Automatic unwinder
2. Coating unit
3. IR lamp unit (short-wave)
4. UV curing unit
5. Edge guiding unit
6. Cooling unit
7. Automatic winder
8. Corona station
9. Electron beam curing unit
10. Automatic unwinder
11. Wet lamination unit

Figure 3.125: Basic concept for a UV curing and/or electron beam curing line *Source: Polytype, Switzerland*

(EB coatings) which only can be cured under an inert atmosphere (see Chapter 7.5) are applied for the coating of paper. Figure 3.125 illustrates the basic concept of an UV curing or electron-beam curing plant. In comparison to the acid-curing topcoats, the application of UV curing or EB curing coatings being cured under exclusion of atmospheric oxygen has the following advantages:

- No emission of formaldehyde or other residual emissions
- Very high application solids (non-volatile amount) are possible (approximately 100 %)
- Formulations contain a very low amount of organic solvents
- High production rates are possible (> 200 m/min)
- Compact coating plants due to the elimination of dryer zones
- High gloss levels are more easily achieved
- More scratch-proofed coating surfaces
- better resistance to chemicals

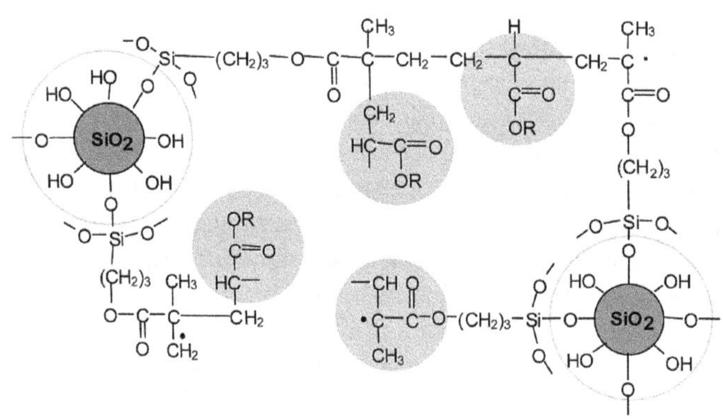

Figure 3.126: Schematic structure of an UV resin which is modified with inorganic nanoparticles [405-407]

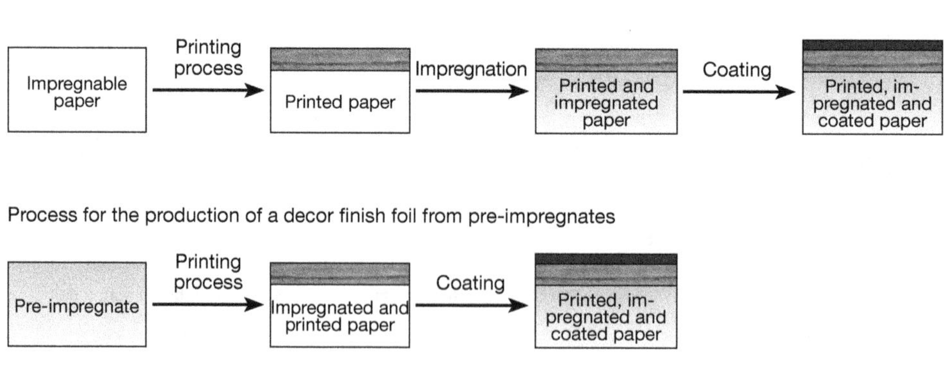

Figure 3.127: Important process for the coating of decor finish foils

The initial problems in achieving very matt gloss levels of < 5 (60° measurement angle) could be eliminated for example by application of excimer lamps ('physical matting', see Chapter 3.1.6) or by application of optimised coating formulations. The good scratch resistance of radiation curing finish coatings can be enhanced significantly by incorporation of hard nano-scale as well as micro-scale particles. A special process makes it possible to organophilize inorganic nanoparticles so that these nanoparticles can be chemically bounded into resins with up to 50 % of proportional weight [405-407] (see Figure 3.126). The coating formulations distinguish oneself by relatively low viscosities and a low abrasiveness.

3.1.10.4 Process for the production of decor finish foils

Figure 3.126 illustrates the most important manufacturing processes of decor finish foils. The first of these (multistage procedure) is used to imprint an impregnable paper for the first time and subsequently to impregnate on a separate line (impregnation technology). Afterwards, the coating with a foil topcoat is performed. The multi-stage procedure becomes less important for the rational production of decor finish foils since the printing process, the impregnation process as well as the coating process are performed separately from each other and at relatively low production rates [382]. The second process uses an already impregnated paper (pre-impregnate). Within the production process, this pre-impregnate can be imprinted and immediately topcoated with a finish coating in the same or in a separate plant. The imprinting and coating in a plant and in one work step is referred to as the online production procedure. This saves enormous processing costs and process times since the step of impregnation is omitted for the producer of decor finish foils. As a general case, the paper manufacturer may perform the impregnation more economical than the producer of decor finish foils. This facilitates significantly higher rates of production in comparison to other procedures. The coating application mostly is carried out with roller assemblies equipped with a putty or with anilox roller to roller (paper coils). In the case of thermally curing finish coatings, the feed rate for the coating lines amounts between 10 to 90 m/min for edges and 100 to 450 m/min for pre-impregnated papers (pre-impregnate).

3.1.10.5 Requirements for decor finish foils

In 2010, leading manufacturers of decor finish foils joined forces with the Institut für Holztechnologie Dresden GmbH in order to create a uniform guideline for testing and declaring the quality of their products with regard to surface and environmental parameters. Uniform, reproducible and differentiated test methods were to be derived by means of test methodological investigations and verification in ring tests. The aim of the investigations was to ensure a uniform declaration of the product properties of the decorative film (regardless of the later application requirements) of all participating companies (see Table3.61 [484]).

3.1.10.6 Embossing of decor finish foils

The term embossing of decor finish foils refers to an imitation of the natural pore structure of the wood [408]. This can be done with of a purely mechanical or physical-chemical route.

Embossing by mechanical means

The effect of the wood pores can be imitated by using structured press plates or press belts with wooden structure [409]. Further developments led to an embossing on a stamping station using embossing calendars. In order to avoid mechanical stresses specifically in case of thicker papers, it is advantageous to humidify the decorative printed papers prior to

Table 3.61: Specification for declaring the properties of decorative foils [484]

Properties/test methods		Unit/category	Specification of the property/result (example)
1	**Thickness** acc. to IHD-W-454 (06/2009)	µm	70
2	**Grammage** acc. to IHD-W-455 (06/2009)	g/m^2	80
3	**Degree of gloss** acc. to IHD-W-456 (06/2009)	At a measuring geometry of 60°	34 (60°)
4	**Opacity** acc. to IHD-W-458 (06/2009)	Grey-scale degree	3.5
5	**Surface tension** acc. to IHD-W-459 (06/2009)	mN/m	44
6	**Chemical resistance** acc. to:		
	DIN 68861 T1/ IHD-W-460 (06/2009)	Stress group 1C	Fulfilled
	IKEA Standard IOS-MAT 0066/ IHD-W-460 (06/2009)	stress group R2/R4	fulfilled
	FIRA-Standard 6250 (horizontal stress)/IHD-W-460 (06/2009)	FIRA-RA-6250 (horizontal)	fulfilled
7	**Abrasion resistance** acc. to IHD-W-461 (06/2009)	Class	3
8	**Scratch resistance** acc. to IHD-W-462 test method 1 (06/2009)	N	0.3
9	**Tesa adhesive strength** acc. to IHD-W-463 (06/2009)	Assessment level (time of exposure)	1–5/(16 h–21 d)
10	Adhesive strength acc. to IHD-W-464 (06/2009)	Cross-cut parameter	1–5
11	**Light fastness** acc. to EN 15187	Level	> 6; = 6 or < 6
12	**Formaldehyde emission** on the basis of EN 717-2	Reference value (≤ 1.3 mg HCHO/h m^2)	Is observed
13	**Migration behaviour** acc. to EN 71-3	Limit values for heavy metals in mg/kg	Are observed

embossing [408]. A considerable disadvantage of the classical mechanically produced pores is that the printed and embossed 'pores' not always lie on the top of each other. New processes such as optical systems (camera systems) enable an embossing of pores where pores also exist at natural wood [410]. In this context one of 'synchronous pores'.

Chemical/physical embossing

In the mid-1970ies, a new generation of water-borne finish coatings (acid-curing coatings) was developed in order to imitate the pore structure of natural wood on decor foils. These finish foils were referred to as pore coatings or as finish coatings with synchronous pore effect. It is also referred to 'chemical pore' or 'real pore' which is produced to differences in the surface tension of the applied coating materials. Usually, printing inks containing special silicon oils are applied. Subsequently to the application of the printing ink (wooden structure), the corresponding finish topcoat is applied. Due to the different surface tensions of the printing ink and finish topcoat, the topcoat is not wetted in the domain of printing inks [411–413]. In Japan, the significant procedures or patents, respectively, are developed for the coating of thin film papers (Japan paper). This procedure supplies outstanding pore structures in comparison to those pore structures which were generated by means of purely mechanical procedures. Compared with the classic mechanical embossing, the advantage of this procedure is in the compliance between the pressure (imitation of the wooden pores) and the pore effect (synchronous pore[42]). The generation of pore structures on a physical-chemical way also is applied successfully in the field of printing and coating of chipboards and board-of-frame plates, since the achievable pore structure is similar to the wood structure in comparison to a mechanically generated pore structure (see Chapter 3.1.6.9).

Calendering process

Even if the calendering process may no longer be applied in this manner, the calendering process shall be explained briefly due its simplicity as well as due to the field of application of solids containing and environmentally friendly coating systems, and thus to give inspirations. The essential feature is the chemical curing of the pre-dried coating film with a heated calender[43] using the contact procedure. This is achieved by application of highly concentrated, aqueous, acid-curing coatings (clear coatings) similar to the currently used acid-curing foil coats. The 'solid content' amounted approximately 70 to 100 % [414]. The high solid content containing coating systems repeatedly were applied by means of rollers with a total amount of approximately 10 to 30 g/m². A drying/pre-condensation process in a jet dryer is performed subsequently to the roller application. The drying parameters with a jet dryer amounted approximately 30 s at a temperature of 100 °C while the drying parameters with an infrared drying unit (short-wave IR lamp) amount approximately 5 s. The final coating occurred in a double calender or triple calender (calender temperature approximately 190 °C) under pressure and without an intermediate sanding. When cured, the surface is smoothed simultaneously, whereas the wood fibres were embedded into the coating so that an intermediate sanding was not necessary. A high-quality, durable and very scratch-resistant surface was achieved with several consecutively connected coating processes with intermidiate drying. The surfaces are stackable subsequently to the exit from the calender. The main feature is the simplification of the line, significant reduction of the coating costs as well as the application of environmentally friendly water-borne coatings. The coating lines were operated with feed rates of approximately 15 to 20 m/min. The coating of panels, veneer edges, door frames as well as furnitures were the fields of application. This procedure was launched on the market by the company Hymmen using the designation 'Thermo-Kontakt-Verfahren TKV 1982/1982' [415].

42 A synchronous pore is characterized by the fact that the three-dimensional pores arise where these are found in natural wood on where the printing ink is applied, respectively.

43 A calender is referred to as a temperature controllable rolling mill consisting of three or more alternatively arranged and polished steel rollers whereby the substrates pass the narrow gap between these steel rollers at high pressures and temperatures

3.2 Coating of wood and wood-based materials for outdoor applications

Wood for outdoor applications is very important to humans for many centuries. The measures for a sufficient constructive protection initially have to be considered (see Chapter 2.4.2.2) in order to protect wood from environmental influences. Additionally, weather protection with coatings (physical wood protection) and/or chemical wood protection with coating materials may be required [416, 417]. The physical wood protection shall avoid the absorption of storm water in the wood since on the one side the wood moisture quickly compensates for the ambient air humidity. On the other hand, ultraviolet rays may destroy the wood. Apart from the technical challenges, the applied coating also should be decorative. However, the surface treatment cannot compensate errors of an inadequate construction protection or errors resulting from a chemical wood protection. The key adversaries of the wood are the moisture (dew, fog, rain, snow, ice, condensation water), fungi (stain fungus, mould fungus), insects, UV radiation, change of temperature (day and night, thunderstorms, seasons) as well as mechanical stress (hail, wind). However, one should be aware that a durable outer coating cannot consist unharmed without a regular care of the coating even when using extremely resistant woods [496].
Today, a current wood protection means:

- Constructional-constructive measures and selection of suitable types of wood as well as of a suitable quality of wood
- Targeted utilisation of chemical ingredients
- Selection of suitable, protective and decorative coatings
- Regular care and maintenance

Wood is hygroscopic and adapts its moisture balance to the ambient climate with changes of its volume. The wood products have an average wood moisture of 12 to 21 % in the outdoor area. The wood moisture should amount approx. 13 % (DIN 68360) prior to the coating of wood in the outdoor area.

For example, the manufacturing methods for wood components changed with the time. Industrially pre-fabricated coated components as well as modern building constructions consisting of wood with an improved thermal insulation are now state of the art.

The European VOC Directive 1999/13/EU or 2004/42/EU (see Chapter 9), respectively, are the trigger for low-emission coatings and processes. Additionally, an increasingly significant durability of the coating and thus the wood construction are required. These properties usually are not realised by resins such as alkyd resins. This is, why today aqueous low-emission coatings are applied for industrially manufactured wood components such as wood windows. Primarily water-borne acrylate co-polymerisate dispersions (referred to as acrylic dispersions), water-borne alkyd resins, partially polyurethane dispersions or acrylated polyurethane dispersions are applied as resins. Mixtures or combinations of these resins also are applied. A variety of new specifications and test standards inter alia for the selection of wood, storage and testing of coating materials were elaborated in order to offer a durability with external weathering as long as possible to the coated wood components. With the European Standard DIN EN 927 being established officially in Germany since March 2002, an internationally recognized standard for the categorization, testing and classification is available for the European coating industry and their end users. DIN

EN 927-2 specifies the performance criteria/minimum requirements for coating systems on wood in outdoor areas. The performance criteria are defined with regard to the three application categories „without dimensional accuracy", „limited dimensional accuracy" and „dimensional accuracy" (DIN EN 927-1) and specified on the basis of the results of several compulsory tests – verification of outdoor weathering according to the regulation DIN EN 927-3 as well as the verification of the hydraulic permeability according to the regulation 927-5. Additional tests are provided that may be used by manufacturers or for specific purposes in order to obtain additional information on performance issues on a standardised basis. It is a combination of different test methods. In addition, the outdoor weathering and the artificial weathering, the absorption of liquid water, the wet grip resistance as well as the water vapour permeability are tested. In combination to a known coating standard, only the combination of different test methods (confer VdL-RL 14) enables an objective basic statement on durability.

Apart from the wood substrate as well as apart from the total coating system (three-layer process/four-layer process), in the formulation of coating materials and in the application technology (painting, flowing, spraying) the resins have a decisive impact on the durability of the components. For many years, solvent-borne and water-borne coating systems are applied for the physical wood protection in the outdoor area. Fundamentally, one may differentiate between clear, glazing and opaquely pigmented coatings. These coatings are adjusted to the industrial applications and application methods or to wood substrates and coordinated with one another. These coatings can be classified as follows:

Solvent-borne systems

Clear and translucent systems:
- Solvent-borne systems based on alkyd resins
- Solvent-borne systems based on natural resins

Opaquely pigmented systems:
- Solvent-borne systems based on alkyd resins
- Solvent-borne systems based on natural resins

Water-borne systems

Clear and translucent systems:
- Water-borne systems based on acrylic dispersions and/or water-borne alkyd resin emulsions (hybrid systems), occasionally addition of polyurethane dispersions (for example improved hail resistance)
- Water-borne systems based on alkyd resin

Opaquely pigmented systems:
- Water-borne systems based on acrylic dispersions and/or water-borne alkyd resin emulsions (hybrid systems), occasionally addition of polyurethane dispersions
- Water-borne systems of alkyd resin emulsions
- Water-borne acrylate- or vinylacetate dibutyl maleinate dispersions for weather protection colours (painter utilization)

3.2.1 Components of solvent-borne coatings

Worldwide, solvent-borne coating systems still play a role for the industrial application. But solvent-borne coating systems are substituted more and more by water-borne systems. In

Europe, solvent-borne systems increasingly are on the retreat. Figure 3.128 illustrates the general composition of solvent-borne coating formulations for the outdoor application.

3.2.1.1 Resins for solvent-borne coatings

Mainly alkyd resins based on unsaturated fatty acids are applied for solvent-borne impregnations, primers, intermediate coatings and final coatings. Apart from the alkyd resins, also modified fatty oils[42] or modified alkyd resins (such as styrenised or urethanised alkyd resins) are applied partially. The polyester content involves the physical drying and the weather resistance (gloss retention, freedom from yellowing and similar), while the oil content involves the smoothness of the coating films (internal plastification), wettability of pigments and in particular, the ability to an oxidative cross-linking. Normally, the paint chemical classification of alkyd resins refers to the fatty acid content converted to the oil content or triglyceride content as well as on the type of oils or fatty acids, respectively. The appointment according to the oil content (triglyceride content) in the recipe is made in accordance with the following nomenclature:

- Less than 45 % oil (in the German-speaking countries: lower than 40 %): 'short oil alkyd', lean alkyd resin
- 45 to 60 % oil (in the German-speaking countries: 60 %): 'medium oil alkyd', medium strength alkyd resin
- More than 60 to 70 % oil: 'long oil alkyd resin'
- More than 70 to 85 % oil: 'very long oil' or super-fatted alkyd resins (The information on the 'oil length' normally is more common.)

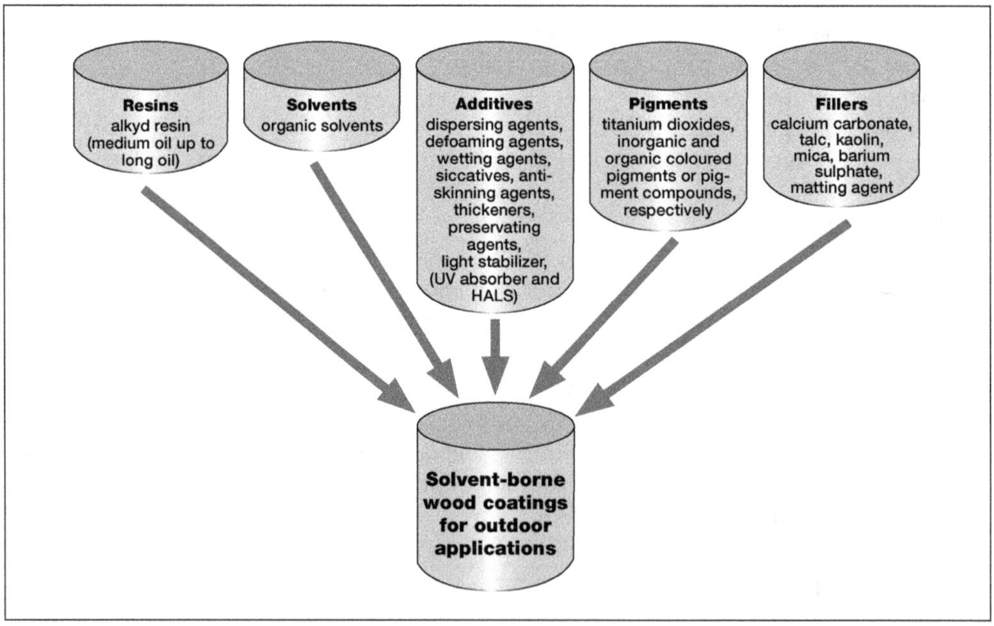

Figure 3.128: Overall composition of solvent-borne coating formulations for outdoor applications

42 Fatty oils are triglycerides of fatty acids, that means esters of glycerine. The fatty acids are aliphatic compounds containing double bonds in different numbers and position and mainly consists of C_{18} acids.

With regard to the type of oil or the type of built-in fatty acids, respectively, a categorization into 'drying' and 'non-drying' is performed. Fish oil alkyds, linseed oil alkyds, soybean oil alkyds as well as tall oil alkyds are important alkyd resins for outdoor applications.

Apart from the raw material components, the application technical properties of the alkyd resins mainly are influenced by the manufacturing process (obtained molar mass and molar mass distribution). Low-molecular, polar components improve for example the dispersibility, but simultaneously deteriorate the gloss retention [418] and extend the process of oxidative curing. The reason behind this is that the hydrophilic components interact with the humidity. It can be justified by the fact that the hydrophilic components interact more intensively with the moisture.

The high molecular and non-polar compounds ensure a good weather resistance and gloss retention of the coating. The non-polar compounds are sites of enhanced probability for other non-polar components. Thus, silicon oils for example may enrich at these sites in an irregular distribution. In extreme cases, this may result in silicone craters.
Selected advantages of alkyd resins:
- Oxidative curing in the presence of siccatives for example at ambient temperature (20 °C/ 50 % relative humidity of the air) single-component systems
- Good pigment wetting
- Usually good flow characteristics and resulting from this a good brushing behaviour of the coatings for example
- Good price performance ratio

Disadvantages of alkyd resins:
- Low chemical resistance (especially against alkali)
- Depending on the type and content of the fatty acid a yellowing (especially a dark yellowing in the dark) occurs by and by
- Relatively rapid loss of gloss (chalking)
- Relatively slow drying (especially for a high oil content or fatty acid content, respectively)

High solid alkyd resins

Due to the high solvent content of the conventionally oxidatively cured alkyd resins, efforts always were available to provide high-solid types with a solid content of > 85 to 100 % on the market. Due to the low molecular weight as well as due to an oil length of up to 80 %, the resins have a lower viscosity in comparison to conventional alkyd resins. Thus, also solvent-free formulations also are possible since these formulations have a processible viscosity also in its 100 % form. However, due to their lower and closer molecular weight these resins usually have longer drying times and curing times. These resins also tend to a stronger yellowing due to the high oil content. Such resins mainly are applied in the sector of painting. These resins play a minor role for industrial applications on wood.

3.2.1.2 Organic solvents

According to the standard DIN 55945, a solvent is a liquid consisting of one or more components which may solve the resin or in which chemical reactions can be performed. In practice, solvents also are referred to as diluting agent or diluent. The applied organic solvents have the function to make the coating material processable and to promote the filming and the progress of the coating film. The following solvents may be applied for medium oil and long oil alkyd resins:

- Aromatic hydrocarbons (xylene)
- Aliphatic hydrocarbons (petrol, white spirit, crystal oil)
- Alcohols such as n-butanol to reduce the viscosity

The compatibility with aliphatic hydrocarbons increases with increasing oil length, while the compatibility with aromatic hydrocarbons decreases with decreasing oil length. In 'natural products', occasionally terpene hydrocarbons (turpentine oil) are applied which may result in allergic reactions of the employees.

3.2.1.3 Pigments, fillers and matting agents

In principle, all inorganic and organic pigments being suitable for outdoor applications can be applied. The role and the properties of the pigments are outlined briefly in Chapter 3.1. Mainly micronized iron oxide pigments are used for the formulation of transparent glazes which simultaneously are protected against UV light. All known fillers such as calcium carbonate, barium sulphate, kaolin, talc, mica, aluminium silicate etc. are applicable. Synthetic pyrogenic and precipitated matting agents often were used as matting agents.

3.2.1.4 Additives

Siccatives

Siccatives are metallic salts of natural organic acids (metal soaps) of certain metals such as cobalt, manganese, calcium, zirconium and others. Erroneously, the siccatives also are referred to as dry matter. A differentiation is made between primary dryer and auxiliary dryer. Primary dryers such as cobalt salts or manganese salts accelerate the surface curing while zirconium salts accelerate the full curing. For example, calcium salts shall bring about synergistic effects in combination with other metal salts and are an essential feature of formulations based an alkyd resins. The siccatives significantly accelerate the decomposition of peroxides and thus initiate the chemical cross-linking of the unsaturated fatty acids [419–421]. Among the commercially available metal containing dryers, cobalt salts primary dryers) exhibit the highest and most universal drying activity and thus are the standard dryer in the market worldwide [485, 486]. The classification and labelling of such additives under REACH or CLP is a crucial process for this sector. For cobalt dryers, the Cobalt Consortium has announced the final points for this classification. This classification has been notified to the CLP classification and labelling inventory. Industry-relevant cobalt dryers (cobalt octoate, cobalt naphthenate and cobalt neo-decanoate) are reported with „data missing“ and „no classification“ for the final points carcinogenicity, mutagenicity and reproductive toxicity (CMR). Organic cobalt siccatives such as cobalt octoate and cobalt naphthenate still are excluded also in the TRGS 905 issue of May 2014 [493]. Hitherto, the available research results are not robust enough. Thus, the consortium needs to collect additional data. At present, only for cobalt octoate the environmental classification will aggravate on R50/53 (previously R51/53). According to the current state, it is to be expected, that cobalt salts being used in the coating industry are classified as a CMR substance (carcinogen, mutagen, reproduction-technical). A CMR classification im-

$$\left.\begin{array}{l} Me^{2+} + R\text{-}OOH \; Me^{3+} + R\text{-}O\cdot + OH^- \\ Me^{2+} + R\text{-}OOH \; Me^{2+} + R\text{-}OO\cdot + H^+ \end{array}\right\} H_2O$$

Figure 3.129: Acceleration of the decomposition of peroxides with siccatives [419]

mediately would result in restrictions of the sale of coating agents for buildings and other coatings for do-it-yourself applications as well as sheet-fed printing inks. In preparation for the enormous effect, the European Council of the Paint, Printing Ink and Artist's Colours Industry (CEPE) as well as the VdL (Verband der Deutschen Lackindustrie) have informed their members in order to use the time to find an equivalent substitution for cobalt dryers or to begin the changeover to other binder systems. Conventional dryers such as cobalt salts and manganese salts shall be substituted by sanitary harmless iron salts (novel complexes). At the same time, it is to be examined whether the utilisation of the new siccatives may dispense with the use of anti-skinning agents. Iron-containing complex compounds are known which may accelerate the oxidative cross-linking of unsaturated oils. In doing so, effects were observed which are not achieved with conventional siccatives (faster through-drying). The novel complexed iron salts have so far been tested on selected commercial products. It was found that different drying effects up to a complete failure may occur depending on the product composition. These findings extremely inhibit the general dissemination of this environmentally friendly variant of siccatives since it is not known under what circumstances or requirements reproducible results can be achieved (type and composition of the unsaturated fatty acids, temperature, atmospheric humidity, oxygen supply, impact of the wood substrate/ingredients). At present, this is an extreme disadvantage with respect to the very widespread cobalt siccatives which provide reliable network outputs within a reasonable time largely independently from the criteria mentioned above. The manufacturer of the newly complexed iron salts (OMG Borchers) confirms that currently there are no systematic investigations between the composition of the fatty acids in oil boiling systems or alkyd resins and the mode of action of the new iron salts. Moreover, the reaction mechanism is unclear. It also could be observed that in comparison to cobalt salts the film hardness already is achieved after 24 to 48 hours and cannot be developed. The question why oxidatively cross-linking products exhibit this behaviour in combination with the new iron complex is still unclear [487].

Anti-skinning agent

So-called anti-skinning agents are added in order to prevent a chemical reaction during the storage of the coatings in the container. These are reducing agents which reduce the oxygen which is diffused in und thus prevent the oxidative curing. Depending on the selection of the anti-skinning agent (and ketoximes), some substances have the capability of complexing metal salts (siccatives). These agents also are referred to as antioxidant. Along with the metal ions of the siccatives, the anti-skinning agents form complexes so that these agents may prevent redox processes. After application of the coatings, the anti-skinning agents vaporate into the environment. Oximes or phenolic substances are applied as anti-skinning agents. Here, substituted phenols such as di-tert.-butyl-cresol operate as a radical scavenger, while ketoximes act as a complexing agent. In this context, it must be borne in mind that the most frequently used ketoxime (methyl ethyl ketoxim) should no longer be applied due to toxicological aspects. In recent years, alternative anti-skinning compositions often consisting of dialkyl hydroxyl amins and ß-diketones were developed and applied [488].

Other Additives

The usual surface-active and dispersing agents, defoamers and levelling agents based on silicon oil, substances influencing the rheology such as hydrogenated castor oil, bentonite and so forth. Furthermore, a wide range of different types of waxes were applied for the surface smoothness as well as for the reduction of the gloss level. UV absorbers such as benztriazoles, derivatives of triazine as well as sterically hindered amines (HALS – Hindered Amine Light Stabilisers) are added to the formulations in order to obtain an optimal UV protection of the wood and an optimal UV protection of the resin. Since the mid-eighties of the 20th century, there is a trend to light-coloured or colourless clear coat which usually cannot be formulated with transparent iron oxide pigments due to the colouration (see Chapter 8) [422]. Recent investigations exhibit that in addition to the selection of the binder a correct combination of light protection additives is crucial for a good durability of colourless (pigment-free) coatings during weathering. The application of HALS compounds in the primer contributes to a significant increase of the colour stability of coated wood surfaces in the outdoor application [489]. The previous shortcomings of the colourless clear coating systems such as a rapid yellowing of the wood substrate, embrittlements of the coating film, crack formation, adhesion problems and exfoliation can be avoided with an optimal formulation and a sufficient dry film thickness (at least 30 µm) [489].

3.2.2 Components of water-borne coating formulations

In Europe, water-borne wood coatings are increasingly used for the industrial coating, since these have significant procedure advantages compared with solvent-borne systems such as a faster drying, improved block resistance as well as an enhanced weather resistance. Furthermore, environmental friendliness and a good recycling capability also are important factors in the coating process. Figure 3.130 illustrates the general composition of water-borne formulations for outdoor applications.

3.2.2.1 Resins for water-borne coating formulations

Water-borne alkyd resins

Among all other resins, especially in the recent years water-borne alkyd resins have grown in significance. These alkyd resins were applied as sole resin or often also as combination partner with acrylic dispersions, for example.
Fundamentally, the following classification may occur:

- Externally emulgated alkyd resins
- Water-soluble and emulsifying alkyd resins
- Hybrid systems

External emulsified alkyd resins

Thereby, the components of the alkyd resins are emulsified in water when hot by admixing of special ionic and non-ionic emulsifiers without addition of organic auxiliary solvents or neutralizing agents. Figure 3.131 illustrates schematically the difference between external emulsified and self-emulsifying water-borne alkyd resins.

Thanks to the production process, the external emulsified alkyd resins feature a rough average particle size. Due to the rougher particle size, the gloss level of the dried coating film is often lower in comparison to self-emulsifying alkyd resins.

As distinguished from self-emulsifying alkyd resins, external emulsified alkyd resins result in a higher solid content of the resin (> 50 % solid content). This was one of the most important reasons for the development of external emulsified alkyd emulsions. In practice, the external emulsified alkyd resins often are combined with acrylic dispersions in order to enable improved drying properties and high coating solid. New developments have focused on an incorporation of solvent-free alkyd resins (70 to 80 % solid content) in acrylic dispersions for achieving a high content of coating solid [423]. In practice, these systems are referred to as hybrid systems.

Water-soluble alkyd resins

Medium oil to long oil alkyd resins with incorporated carboxyl groups (acid value > 40) are referred to as real dissolved resins. These alkyd resins contain a relatively high proportion of organic amines or ammonia as a neutralizing agent as well as a certain amount of co-solvents (organic solvents). These alkyd resins also are referred to as molecular dispersed or ion dispersed systems (real solutions), respectively, with a particle size under 1 nm. The relatively high susceptibility to saponification is problematic with the classical resin solutions.

Self-emulsifying alkyd resins

These represent resins with a reduced acid value (often < 30) and incorporated hydrophilic groups. These groups may exhibit a steric or ionic mechanism of action. It is also frequently the case, that both groups are incorporated (see Figure 3.132).

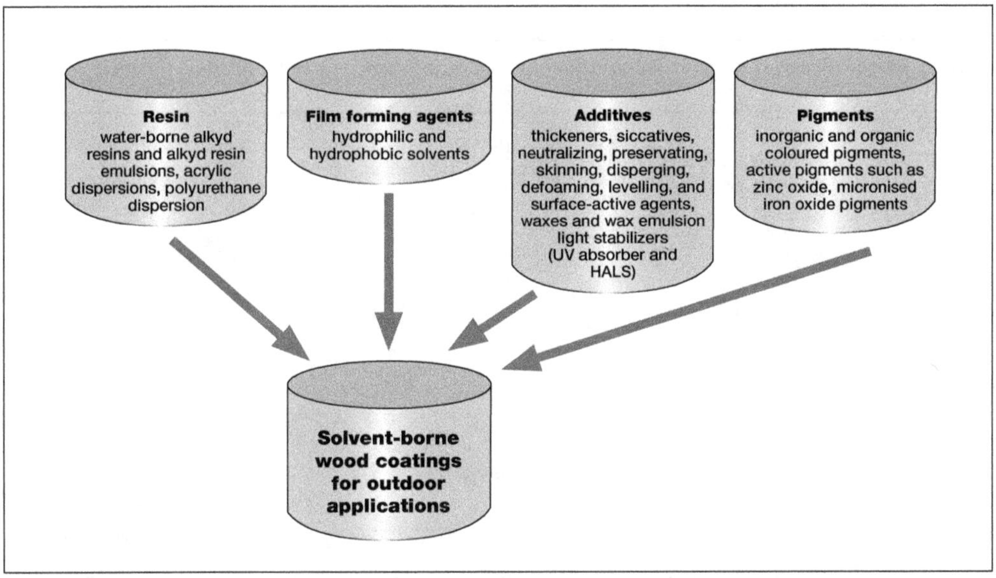

Figure 3.130: General composition of water-borne formulations for outdoor applications

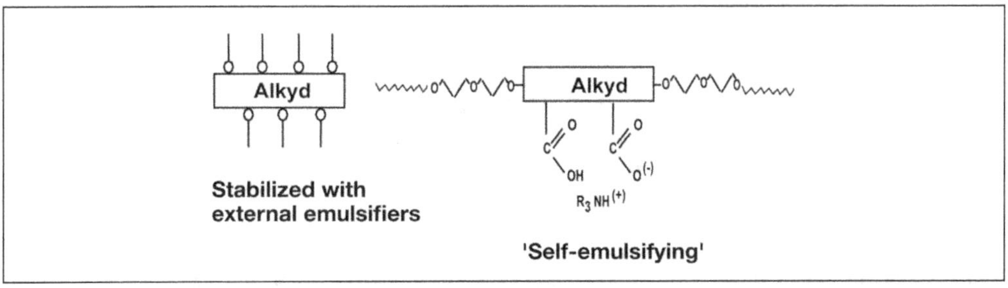

Figure 3.131: Difference between external emulsifiable and self-emulsifying water-borne alkyd resins

Thus, an emulsification without addition of external emulsifiers is possible (self-emulsifying) and are also referred to as hydrogels. These are colloidally dispersed systems with a particle size between 1 and 100 nm. Carboxyl group containing compounds are applied for the ionic stabilisation whereby these compounds are neutralized prior to the addition of water and subsequent emulsification with amines or ammonia. The content of neutralizing agents, however, is significantly lower compared to alkyd resin solutions. Usually, any addition of co-solvents is necessary. These systems are characterized by a very good wetting of the wood substrate as well as a good penetration into the wood matrix. It has to be taken into account that only a minimum of such hydrophilic groups is incorporated in order not to influence the water sensitivity of the resulting resins too negatively. It should be noted that the hydro-

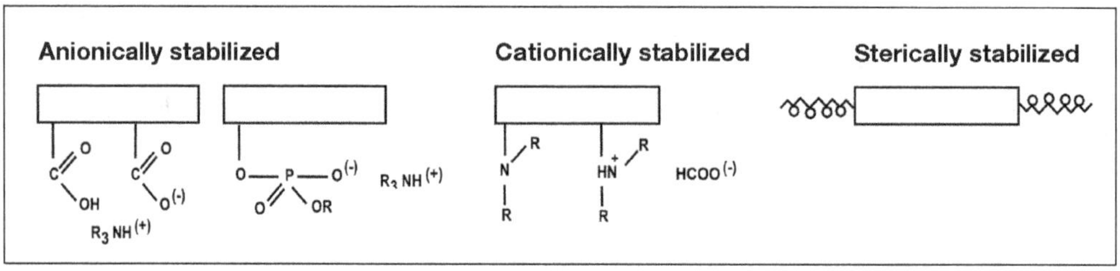

Figure 3.132: Different systems of stabilisation for self-emulsifying water-borne alkyd resins

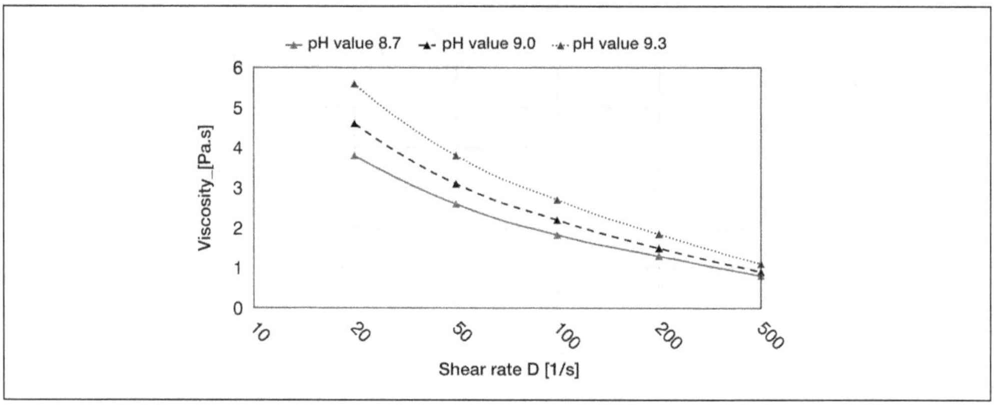

Figure 3.133: Impact of the pH value on the viscosity characteristics of self-emulsifying alkyd resins[424]

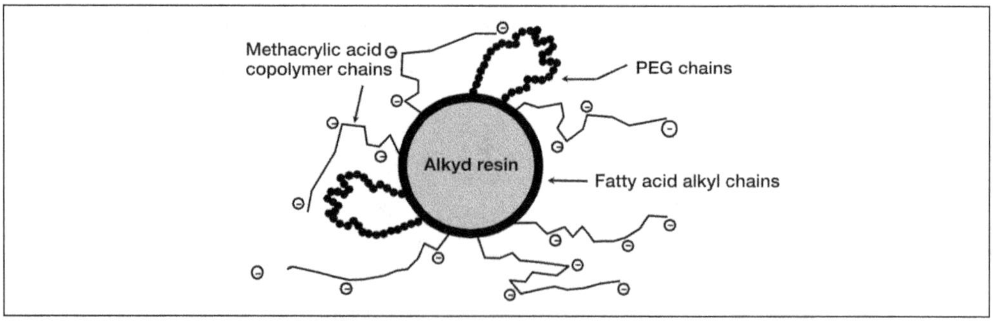

Figure 3.134: Schematic illustration of the structure of alkyd resin emulsions modified with a methacrylic acid containing co-polymerisate[424]

philic groups support the pigment dispersion. That means, the water-borne alkyd emulsions are suited excellently for grounding or dispersing of pigments, respectively.

Cationic stabilised systems are possible, too. These systems are applied for special water-borne sealer against wood components (such as tannins) (see Chapter 2.4.1). The self-emulsifying alkyd resins have another rheological behaviour in comparison to solvent-borne alkyd resins. Thus, the self-emulsifying alkyd resins have more likely a viscous structure in comparison to the Newtonian flow behaviour of the solvent-borne alkyd resins. The pH value has a large influence on the viscosity characteristics (see Figure 3.133) [424].

Thus, an increasing pH value results in an enhanced number of carboxyl ions and, as a consequence, to smaller particles. This results in more transparent emulsions with an increase in viscosity. In this context, the increase in viscosity can be explained by means of a swelling of the hydrophilic shell [425].

However, the pH value does not influence the fundamental course of the flow curve – the flow curve is simply moved in parallel. Therefore, it is very important, to control the pH value of coating systems based on self-emulsifying alkyd resins. Beside its influence on the viscosity, the pH value also has a large impact on the storage stability as well as wetting properties of the resin or the coatings made therefrom, respectively. Thus, the pH value should lie in the alkaline region (approx. 8.5) slightly. In addition, this is important in the case that these resin emulsions are combined with acrylic dispersions, for example. Apart from the usual components, the formulation of coating systems especially requires the addition of a suitable siccative as well as an anti-skinning agent.

Hybrid systems

Hybrid systems may consist of physical mixtures of water-borne alkyd resin emulsions and acrylic dispersions. The properties of both resins can be combined ideally with a corresponding mixture ratio between the alkyd resin emulsion and acrylic dispersion. The modification of alkyd resin emulsions in the manufacturing process with acrylate components such as methacrylic acid containing co-polymerisates is another possibility. Normally, manufacturers may manage finished combinations easier since the manufacturer of the resin may adjust the compatibility and the storage stability appropriately. The modification accelerates the physical drying while maintaining a good wetting of the wood substrate. Furthermore, the liquid coating materials produced therefrom feature a good freezing stability. Furthermore, the tendency to yellowing is reduced. The coatings are characterized by an excellent re-coatability.

Water-borne acrylic dispersion

Acrylic dispersions for outdoor applications commercially are available since the mid-sixties of the 20th century. Today, the water-borne acrylic dispersions are the most important group of resins worldwide for the water-borne wood coatings. Usually, acrylic dispersions are anionic dispersions which are manufactured with regard to the process of emulsion polymerisation. The characteristics of the acrylic dispersion can be varied over a wide range by varying the acrylic monomers, the stabilisation system as well as the particle morphology. Previously, single-phase acrylic dispersions – also referred to as single-stage polymerisates – were applied for the industrial outdoor coating. Recent developments of the last years have led to an implementation of multi-stage polymerisates (core-shell dispersions) with improved property profile [426, 427]. Thus, core-shell-dispersions have gained great significance for industrial outdoor applications of wood today. Single-stage polymerisates are applied here to an always lesser extent, even more frequently in combinations (such as water-borne combinations).

Table 3.62: Comparison of the properties of different resins in water-borne white coatings [424]

Type of resin	Gloss 20°/60°	Haze	Levelling	Tack-free time [h]	Drying behaviour after 24 h drying at ambient temperature laboratory assessment	Degree of dryness after 24 h drying at room temperature TG according to DIN 53150
Acrylated alkyd emulsion	70/85	14	14	2	24	6
Hybrid system (alkyd/acrylate dispersion)	60/85	20	20	0.5	20	4
Solvent-based long oil alkyd	80/90	12	12	5	30	3

1) Levels of assessment: 10 = very good 50 = bad

Table 3.63: Particle size and optical appearance of dispersions

Particle diameter [μm]	Optical appearance
> 1 μm	Milky white dispersion
1 mm–0.1 μm	Blue-white up to brown-white dispersion
0.1–0.05 μm	Grey-white, semi-transparent
< 0.05 μm	transparent

Important parameters of influence for the properties of the acrylic dispersion

The manufacturing process and the composition have a big impact on numerous parameters such as glass transition temperature, degree of cross-linking, molecular weight, particle size, particle size distribution as well as particle morphology which impact the property profile of the dispersion or polymer film. This will be explained in more detail, hereafter.

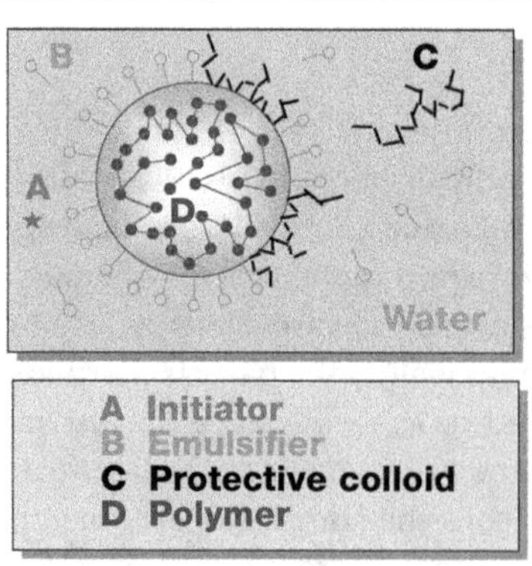

Structural parameter

Polymer property
- Chemical composition
- Glass transition temperature (T_g)
- Cross-linking
- Molecular weight
- Functional groups

Colloid properties
- Emulsifier, type, amount
- Electrical charge of the latex cover
- Electrical charge in the serum
- Protective colloid

Particle properties
- Particle size (50–200 nm)
- Particle size distribution
- Particle morphology

Figure 3.135: Structure parameters of a dispersed particle

Particle size and particle size distribution

Often, there are no homogeneously large particles. Instead, one finds a more or less wide particle size distribution which corresponds to the normal Gaussian distribution. If there are particles of only one defined size, it is referred to as a monodisperse or as a monomodal dispersion. However, if the dispersion contains particles with two different particle sizes, such a system is referred to as a bimodal dispersion. In contrast, multimodal dispersions are characterized by a very broad particle size distribution. Depending on the manufacturing process, the polymer particles of acrylic dispersions may have particles sizes between 50 and approx. 500 nm. Depending on the particle size, the corresponding dispersions appear as an almost transparent to milky white liquids. Due to the optical appearance, the polymer dispersions can be classified as listed in Table 3.63.

The particle size and the particle size distribution of polymer dispersions are influenced by the stabilisation system. Coarse-grained dispersions often are stabilised by protective colloids while fine-particle protective colloids are stabilised by emulsifiers. In comparison to coarse-grained dispersions, fine-particle dispersions generally are something more sensitive with respect to the stability due to the larger particle surface. Different principles are applied for the stabilisation. Dispersions can be stabilised either sterically (mostly by protective colloids or non-ionic emulsifiers) or electrostatically (mostly by anionic or cationic emulsifiers) with corresponding anionic monomers.

Industrially manufactured dispersions mostly show a combination of the three mentioned types of stabilisation. Figure 3.135 illustrates this schematically.

Instead of the earlier commonly used alkyl phenol ethoxylates (APEO), mostly fatty alcohol ethoxylates now are used as non-ionic emulsifiers due to toxicological and environmental considerations. In wood coatings, the particle size and the average particle size distribution influence important properties such as transparency of the liquid dispersion or the resulting coating films, the penetration capability into the substrate and, therefore, the adhesive strength (see Figure 3.136). Furthermore, fine-grained acrylate dispersions with the same glass transition temperature as coarse-grained acrylate dispersions film better and result in a lower level of whitening of the coating after water exposure. Especially due to the penetration capability similar to water-borne alkyd resins, usually very fine dispersions are preferred for wood coatings and thus specifically for primers.

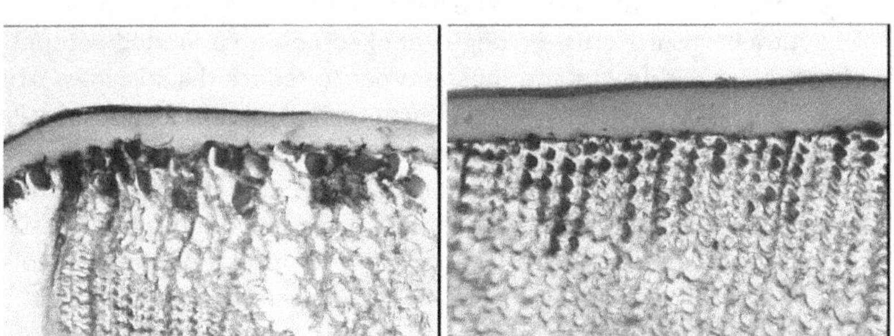

Figure 3.136: Impact of the polymer architecture of acrylate dispersions on the penetration depth in pine wood (light microscope) [428]

Properties of the polymer of acrylate dispersions

The different types of the applied monomers determine the polymers properties. Here, a distinction is made between hard monomers, soft monomers as well as stabilising or cross-linking monomers, respectively.

The glass transition temperature (T_g) of the polymer or the minimum film forming temperature (MFT) of the dispersion, respectively, can be adjusted with a selection of the monomers as well as with the monomer ratio. The glass transition temperature (T_g) is a measure for the softness of the polymer film. The lower the T_g, the softer the polymer film and the lower the MFT.

In case of acrylates, acrylic acid esters yield more flexible polymers in comparison to methacrylic acid esters. As will be discussed later, one wants to achieve highly flexible and simultaneously non-sticky (block resistant) films for wood coatings. Flexible films are quite sticky due to the low T_g. In extreme cases, this results to a total adhesion of the coating surface in the coating process when stacking or pressure loading.

Furthermore, the MFT shall not become too high due to an increase of the rigid monomer content, since an enhanced concentration of organic solvents would be required once more in order to secure a good filming of a coating. This has a negative impact on the VOC content as well as on the blocking of the components. The right balance between T_g, MFT and elasticity has to be found always. It will be frequently experienced that the optimal properties in view of elasticity such as elongation at break and block resistance cannot be achieved by the variation of the monomer ratio or a selection of the monomers, respectively. This can be solved by means of two manufacturing possibilities:

- By application of cross-linking agents
- By variation of the manufacturing process and by producing of multi-phase polymer particles, that means by variation of the particle morphology (core-shell dispersion).

Cross-linking of polymer dispersions in the manufacturing process and coating process

Cross-linking reactions can be induced during the emulsion polymerisation or foremost during the drying of coating films. Usually, di-functional monomers were polymerised in order to crosslink polymer particles during the manufacturing process. Thus, a homogeneous network arises within the polymer particle. This process also is referred to as an intra-particular cross-linking. The resulting network can be described solely by means of the cross-linking density [429]. The intra-particular cross-linking is applied only to a limited account in an enhancement of the molecular weight and thus in order to reduce the stickiness of the corresponding polymer films. This is due to the fact, that a strong intra-particular cross-linking impedes or completely inhibits the filming of the corresponding particle, respectively.

Due to a systematic incorporation of functional groups (such as carboxyl groups) acrylate dispersions may crosslink during the filming process by adding multivalent metal ions (Zn^{2+}, Zr^{2+} oder Al^{3+}). Often, this type of cross-linking (complex formation) is reversible depending on the pH value. The cross-linking has a strong impact on the mechanical behaviour of the coating films [430]. The usually applied Zn^{2+} and Zr^{2+} containing salts are amine complexes which only are stable in a solution with a sufficient ammonia concentration. The metal salts may form complex compounds with the carboxyl groups due to the evaporation of ammonia during the drying process. Thus, cross-linked dispersions (ionic mechanism) preferentially are applied as a sealer on wood against discolourations and bleeding of wood components. In most cases, a common disadvantage is the insufficient resistance of the coating films against alkaline compounds.

Chemical cross-linking by means of condensation reactions during the film formation

The technical literature holds numerous reports regarding cross-linking reactions which lead to an improvement of the mechanical properties of polymer films due to a formation of chemical bondings. In this context, it is relevant to refer to the summaries by *Bufkin* and *Grawe* [431] as well as *Ooka* and *Ozawa* [432].

During the process of film formation, also condensation reactions can be applied for the cross-linking of polymer chains. This is referred to as self-crosslinking acrylate dispersions. Difunctional or multifunctional cross-linking agents of low molecular weight in the form of water-borne solutions are added to the acrylate dispersion subsequently to the polymerisation. A well-known example is the reaction of dihydrazides with carbonyl groups situated along the polymer chain (see Figure 3.137). With alteration of the pH value, functional groups such as N-methyl groups being incorporated in the polymer chain may crosslink with each other by elimination of formaldehyde and water or by a simple condensation with themselves.

The cross-linking systems described above especially are suitable for the enhancement of the chemical resistance of the coating films. Thus, among other things the wet adhesion and the chemical resistance are influenced positively by means of a suitable mechanism of cross-linking [433]. A good wet adhesion is an important prerequisite for a durable exterior wood coating. Cross-linkers of this sort also are referred to as wet adhesion promoters. these are relatively expensive specialty monomers which are copolymerised. Apart from the compounds mentioned above, mostly monomers containing amino groups, acetoacetate groups, cyanoacetate groups, urea groups, thiourea groups or cyclic urea groups based on (meth) acrylate, maleate, allyl ether or vinyl ether are applied [434–435]. These monomers effect a solid

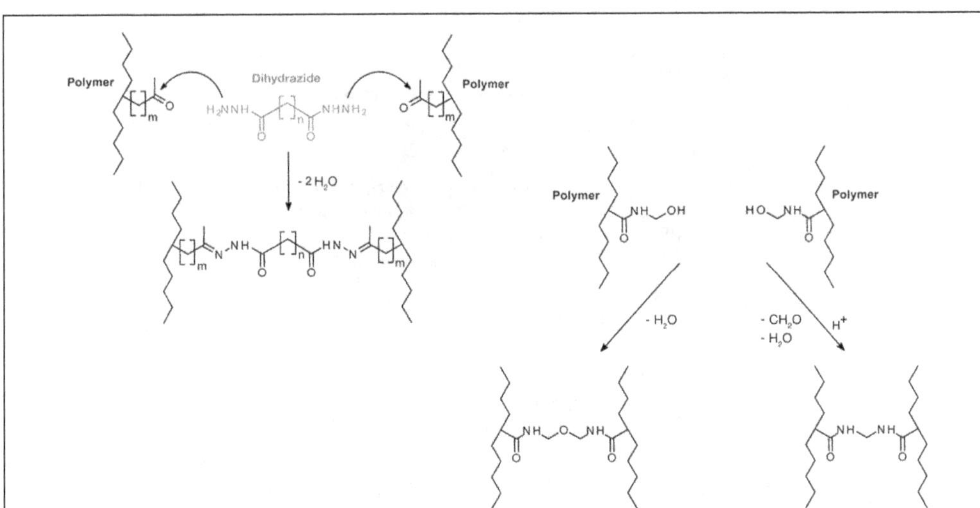

Figure 3.137: Cross-linking of polymer chains with condensation reactions

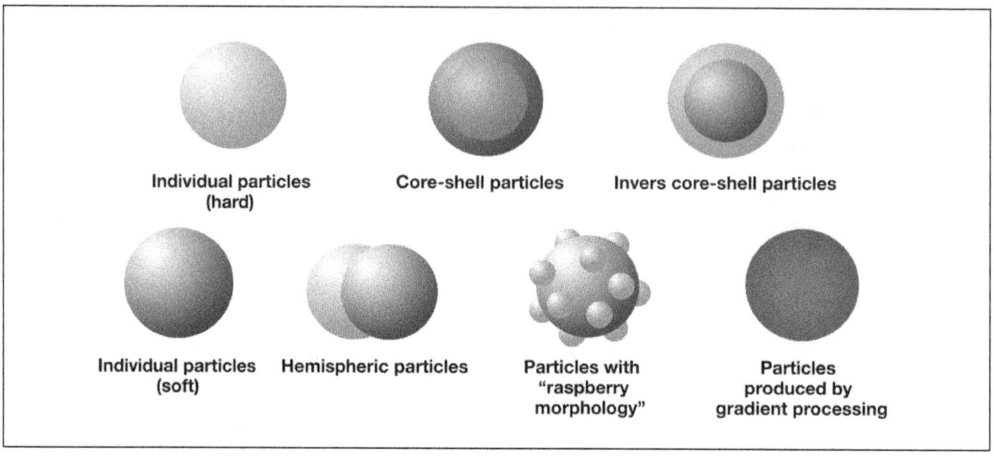

Figure 3.138: Particle morphology of single and multi-stage emulsion polymers

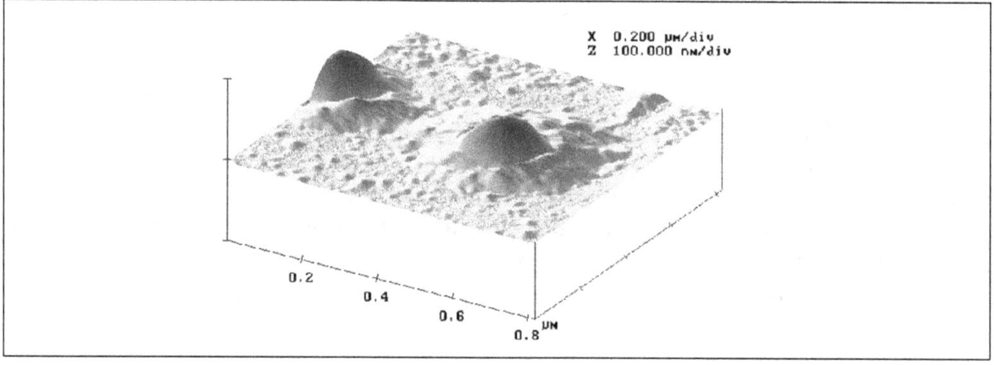

Figure 3.139: AFM micrograph of a multi-stage particle of an acrylate dispersion

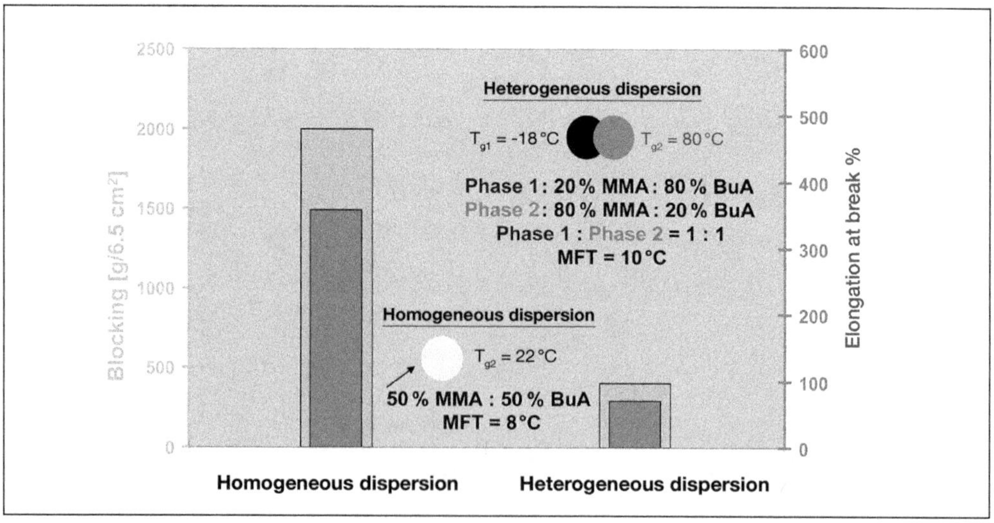

Figure 3.140: Blocking and elongation at break of a homogeneous and heterogeneous dispersion with the same total composition (50 % by weight MMA and 50 % by weight BuA)

anchorage of the polyacrylate dispersion coating for example by means of the formation of hydrogen bonds and with acid base interactions with the support. This also applies to the adhesive strength of ancient coatings based on alkyd resins.

Particle morphology

Apart from the single-stage procedure as a classical process, there is the possibility to produce dispersions in a two-stage procedure [436]. This procedure enables the production of polymer particles having a multi-phase structure. Usually, these polymer particles consist of a soft and hard polymer phase. The soft polymer phase is responsible for the film formation process, while the hard polymer phase is responsible for the block resistance or non-stickiness, respectively, of the resulting polymer film. The properties of these heterogeneous dispersions shall be described using heterogeneous model dispersions as an example. These are produced by a successive dispensing of both, separately produced monomer mixtures with a subsequent polymerisation. Dispersions with a multi-phase structure are maintained by this procedure. These also are referred to as heterogeneous dispersions or core-shell dispersions. Some of the possible structures are featured schematically in Figure 3.138 [437].

The upper structures of the figure are idealised structures. Under the practical conditions of the manufacturing process, mostly complicated structures such as a strawberry structure or hemispheric structures exist in commercially available dispersions.

The different structures (also referred to as particle morphologies) can be produced during the process of polymerisation by means of a variation of the monomer composition (such as hydrophilic/hydrophobic monomers), stabilisation system or manufacturing procedure (such as gradient mode of operation). On the one hand, both polymer phases have not to be compatible with each other too good, as otherwise the phase transition soft/hard or vice versa is no longer efficient, and as the filming characteristics as well as the block resistance can be influenced negatively. On the other hand, it is important that both polymer phases are compatible during the drying phase insofar as interferences among each other resulting in a loss of gloss or in an insufficient film formation are avoided. The structure of the particles can be captured and presented by the so-called atomic force microscopy (AFM) [438].

Figure 3.139 illustrates an AFM exposure of a heterogeneous dispersion. It is easily recognisable, that the soft outer phase is brought on the hard inner phase. Both phases are connected with each another.

In comparison to the dispersions prepared by means of conventional procedures (also referred to as homogeneous dispersions), the heterogeneous dispersions offer many advantages. The following comparison shall illustrate this. In doing so, the heterogeneous model dispersions are compared with homogeneous dispersions [439]. Both types of dispersion had, more or less, the same minimum film formation temperature of approx. 8 °C and the same total monomer composition of 50 % methyl methacrylate (MMA) and 50 % butyl acrylate (BuA). Figure 3.140 illustrates the values of the dispersions in a non-pigmented glaze formulation.

The heterogeneous dispersion has a significantly better block resistance (lower value = lower separating force of the blocked samples = good block resistance) and König pendulum hardness in comparison to the homogeneous dispersion at the same minimum film formation temperature (MFT). The elasticity or elongation at break of the free coating film with the heterogeneous dispersion is lower than those of coating films with a homogeneous dispersion. The elasticity still is sufficient so that any crack formations on wood arise during exposure to weathering. The heterogeneous dispersion should have a sufficient hailstone resistance for the industrial coating of windows. This is necessary so that any damages at windows may occur due to appearing thunderstorms. The elasticity of the polymers or coating films, respectively, under shock-like stress is decisive for this property. However, the hailstone resistance thus correlates with the elongation at break of the films. A poor hailstone resistance results in a formation of hairline cracks in the final coating on the exterior wood component such as wood windows if hailstones impinge upon the impact site. Damages due to moisture as well as subsequently fungal infestation and spallings may occur at this hairline cracks. This test is simulated in a lab by means of the falling ball method in which a steel ball drops onto a wood substrate from a height of 1 m. A gypsum plasterboard also can be applied as a test substrate. A dispersion or a coating, respectively, is classified as suitable (preferably free of cracks) or unsuitable (many cracks) by means of the number of cracks round the formed indentation.

This property can be adjusted appropriately by means of an optimisation of the manufacturing process of the dispersion, for example. One option to do this is the gradient mode of operation. This mode of operation may generate particle morphologies with a continuous phase transition (see Figure 3.138). This morphology may have advantages in the hailstone resistance or flexibility, respectively, since one may imagine that there were no predetermined sites of fracture from the soft phase to the hard phase due to the continuous phase transition.

A high block resistance of the coatings with adequate flexibility and preferably low-solvent formulation is a fundamental prerequisite for industrially coated wood windows or furniture components, for example.

Only modern, tailor-made resin systems correspond to these high requirements especially for the industrial application for exterior timbers. Here the water-borne acrylate dispersions (acrylates without or with a low amount of styrene) have become extremely important due to their properties and possibilities of variation. Also, the combination possibilities with water-borne alkyd resins and polyurethane dispersions allow the formulator to develop tailor-made finished products. In comparison with solvent-borne alkyd resins and water-borne alkyd resins, acrylate dispersions feature a clearly faster drying behaviour as well as a clearly better early block resistance.

Whitening and water uptake

The whitening is initiated by the penetration of water and the solvation of water-soluble fractions and hydrophilic groups in the polymer film. The whitening not only depends on the initi-

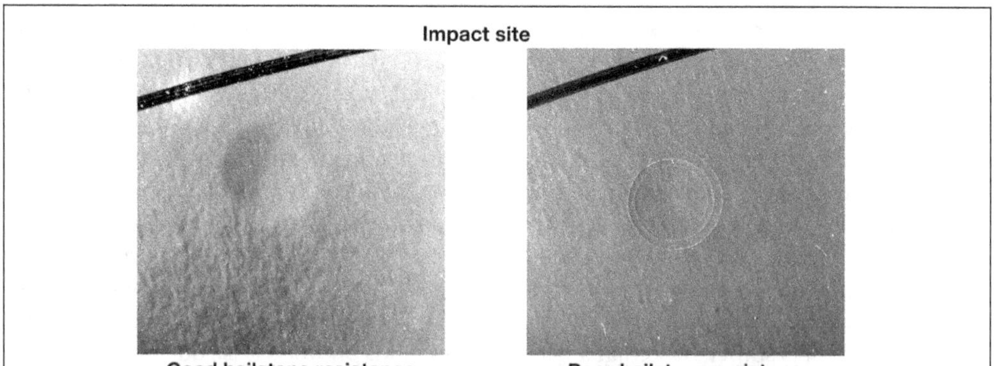

Figure 3.141: Hail test of water-borne wood coatings based on acrylate dispersions

ators used for the polymerisation, on the emulsifiers or protective colloids, on the used monomers, but also on the particle size of the dispersion and quality of the film formation process. Protective colloid stabilised dispersions usually exhibit an increased whitening.

The water uptake is defined as the increase in weight of a dispersion film when in contact with water. It has to be considered that whitening and water uptake are not directly connected with one another. Dispersion films feature an enhanced elasticity during water uptake but also a reduction of the mechanical resistance. This is caused by a reduction of the adhesive strengths of the polymer depending on the amount of the absorbed water. The amount of water uptake is mainly determined by two factors:

- The composition of the copolymer with regard to its hydrophilicity. Of particular importance in this respect are functional monomers such as co-polymerisable acrylic acid.
- Water soluble salts which can be implemented between the latex particles during film formation. Upon immersion in water, the salts create an osmotic pressure which results in an enhanced swelling of the dispersion film.
- Overall, it is apparent that the water uptake increases with increasing hydrophilicity of the polymer films [440].

An optimal film formation is important, since otherwise the whitening or the water uptake, respectively, are impacted unfavourably. The selection of the film-forming agents here also play an important role. Acrylate dispersions for wood coatings and especially for final coatings shall feature a minimum of whitening as well as a low water uptake. These is ensured through modern types of dispersion.

Water-borne polyurethane dispersions

In recent years, water-borne polyurethane dispersions [441, 442] have increasingly gained in importance for wood coatings. One of the reasons of this are their special properties. The good mechanical resistance is a pronounced fundamental property of the polyurethane dispersion. Depending on the composition, the films produced are rigid and block resistant. The relatively high price of the aliphatic polyurethane dispersion is disadvantageous. These polyurethane dispersions are widely used especially in parquet coatings as well as in furniture coatings. These also are incorporated in formulations as combination partners for the acrylate dispersion. As a sole resin, polyurethane dispersions usually play a minor role. Rather acrylate dispersions modified with isocyanates (acrylated polyurethane as a mixture or incorporated in a polymer) are applied.

The synthesis of the first water-borne polyurethane dispersions (PUD) has been successful in the 1960ies. Polyurethane dispersions are commercially available since the beginning of the 1970ies. In contrast to the water-borne acrylate dispersions, PUDs usually are water-borne secondary dispersions. Subsequently to the production, the polymer is emulsified or dispersed, respectively, in water with auxiliary materials.

Raw materials used in the production of polyurethane dispersions

The production starts from diisocyanates and diols or polyols, respectively.

Diisocyanates

Aromatic and aliophatic diisocyanates are applied. For example, methylenediphenyl diisocyanate (MDI), toluene diisocyanate (TDI) or isophorone diisocyanate (IPDI) are applied as aromatic diisocyanates. Hexamethylene diisocyanate (HDI) is applied as an aliphatic diisocyanate.

Polyols

Different types are applied as polyols. In doing so, low molecular diols such as ethylene glycol or 1,4-butanediol result in relatively hard and brittle, mostly high viscous prepolymers with an enhanced amount of urethane groups. However, most frequently higher molecular diols or polyols are applied. Depending on the basis, a distinction is made between polyether diols, polyester diols or polycarbonate diols. Products based on polycarbonate polyols have a good hydrolysis resistance, resistance to light, weather resistance as well as heat resistance, but these products are relatively expensive. Depending on the structure of the polyester component, polyurethane dispersions based on polyester polyols are more sensitive to hydrolysis, but lower priced.

Emulsification

In order to enable an emulsification as well as a preferably small impact on the resistance to water, internal emulsifiers (acid group containing compounds) are used. Dimethylol propionic acid which is incorporated according to the following scheme is applied very often:

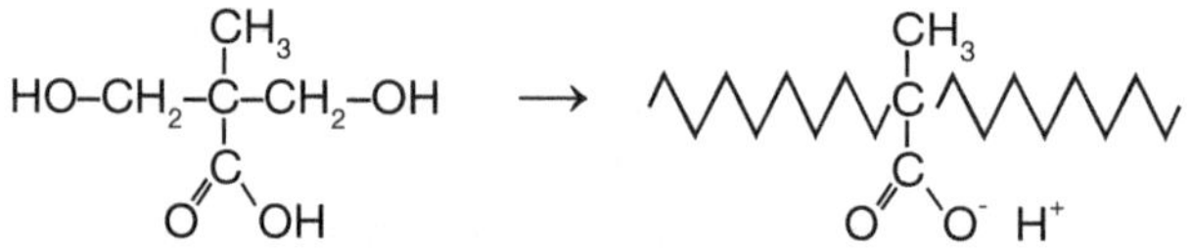

Dimethylol propionic acid

The acid groups thus incorporated into the prepolymer then are neutralised with amines for example. Mostly tertiary amines such as triethanolamine (TEA) dimethylethanolamine (DMEA) are applied.

With the aid of the mentioned components, now a prepolymer can be synthesized which is emulsifiable or dispersable in water, respectively.

Depending on the composition and depending on the molecular weight of the prepolymer, a solvent still may be necessary for an easier handling and for an optimal emulsification. For this purpose, depending on the manufacturing process, different solvents such as acetone, methyl ethyl ketone or N-methyl pyrrolidone (NMP) are used. Due to the toxic properties, increasingly alternatives such as N-ethyl-2-pyrrolidone (NEP) are applied instead of NMP. It will turn out whether the substitution favoured by some producers will prevail. One must also await whether NEP can also can be harmful to health from a later point of view.

Chain extension

A chain extension has to be performed in order to increase the molecular weight since usually a relatively low molecular and NCO terminated prepolymer (an oligomer with NCO end groups) results from the polyaddition. The addition reaction of isocyanates and amines is applied for this. This can be bifunctional amines such as hydrazine or ethylenediamine, for example. These amines may react with each other according to the following scheme:

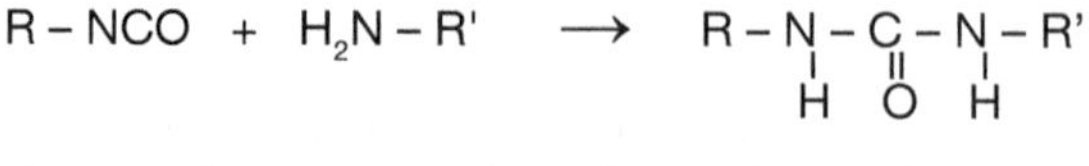

The NCO terminated oligomers are joined together to form a urea group. An extension of the chain also is possible by a reaction with water. The chain extension with diamines may take place in the prepolymers both after dispersion in water as well as prior to the addition of water. It depends on the production process which method is chosen.

Manufacturing process

Two selected procedures are given in order to explain the production of the polyurethane dispersions schematically. In Figure 3.142, the two most common methods are compared with one another.

Advantages of the prepolymer ionomer method
- High space-time yield
- Very fine particles depending on the amount of NMP
- Particle size ≤ 30 nm
- Cheaper method
- No distillate

Advantages of the acetone method
- Low solvent dispersions
- Larger variation range of the particle sizes
- Complete molecular development prior to the dispersion

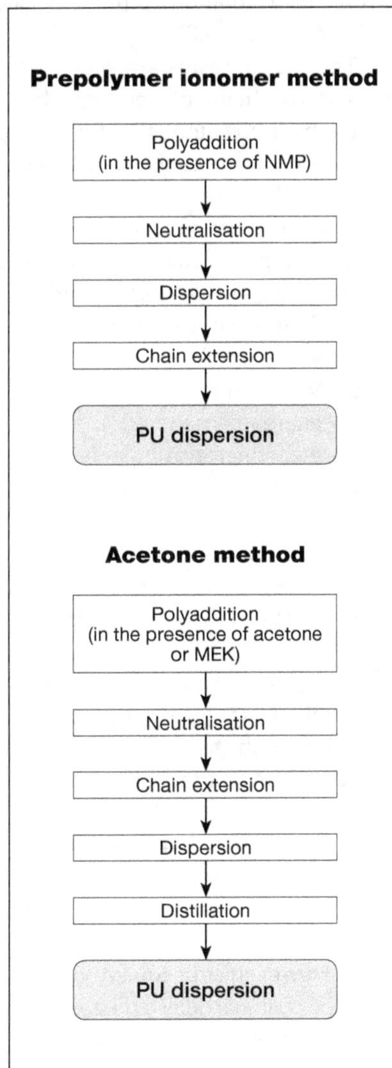

Figure 3.142: Manufacturing procedure of polyurethane dispersions

Properties

The outstanding properties of the polyurethane dispersions are:
- High elasticity
- Low-temperature impact strength
- Abrasion resistance
- Barrier effect towards softening agents
- Resistance to chemicals

As expected, the light resistance of the polymer films is not optimal when using aromatic reaction components, the polymer films tend to yellowing. The yellowing does not occur any more when aliphatic compounds are used. However, the films resulting from aliphatic compounds feature a fairly poor chemical resistance as well as mechanical resistance. Some aspects of achieving the described properties will brief be summarised. The enhanced chemical resistance is an important property which can be derived from the structure of the polymers. The simplified featured resonance structure (or tautomerism) shows that the formation of hydrogen bonds is possible.

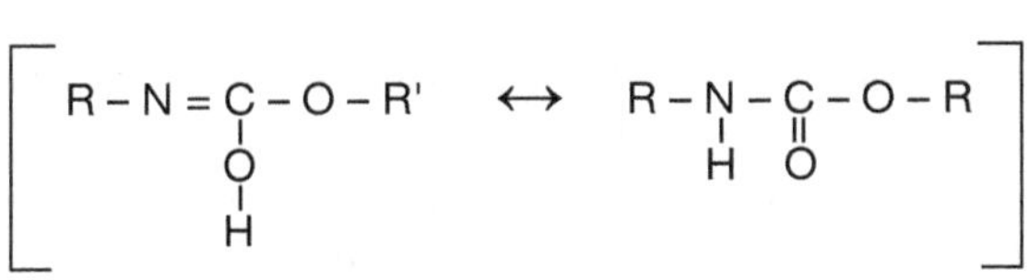

These result in an enhanced chemical resistance with simultaneously good adhesive strength as well as a high level of hardness. Thus, the self-crosslinking also in combination with acrylate dispersions or the further development of NMP free PU dispersions and the incorporation of PU amounts in acylate dispersions are a further challenge. Also, products based on renewable raw materials such as oils are further developed.

UV-curable resins

In the light of the experiences of the furniture sector, approx. 20 years ago one began to develop and test UV-curable water-borne resins [443] for an industrial coating of windows. Thus, the resins were modified and optimised in their composition so that these resins feature a sufficient flexibility and thus a weather resistance. Even the coating plants and in particular the UV irradiation technology became attuned appropriately to the systems (see Chapter 3.1.6 and 3.2.3.2).

The typical resins are special UV cross-linking polyurethane acrylate dispersions. Similar to polyurethane dispersions, the polymer is emulsified in water as a secondary emulsion. Mostly, hydroxyketones and bisacyl phosphinoxides (BAPO) are applied as photoinitiators. The developments concerning this currently provide promising results. In particular, in terms of the weather resistance of the coating systems resulting from this, there are good results so far which promise a good long-term stability [443].

The advantages of the UV curable resins for outdoor applications would be for example:

- Fast curing and drying – rapidly proceeding process
- Fast final block resistance
- Low water uptake
- Low VOC content
- Low energy demand
- Low wet adherence
- Flexible, but still scratch-proofed

However, there are also disadvantages:

- High investment costs and raw material costs
- Curing at problem areas (3D geometries) are difficult

Currently, these overall higher costs just are difficult for many window manufacturers. The first products are established as a complete system at the market, and practical experiences are gained with these products. Therefore, it remains to be seen, how this technology could be used in the future.

3.2.2.2 Film formation and film-forming agents (coalescents)

For acrylate dispersions or generally polymer dispersions, the process of film formation is quite different from solved or colloidally solved resins. Polymer dispersions consist of particles which slowly penetrate into each other during the evaporation of water and thus create a film. However, subsequently to the film formation light structures of the originally available dispersion particles still are apparent.

Thus, every polymer dispersion requires a certain temperature during the process of drying and film formation in order to develop an optimal film. This temperature also is referred to as minimum film forming temperature MFT. The most recipes based on acrylate dispersions or generally polymer dispersions require a small addition of solvents which ensure a good film formation at the corresponding application temperatures. These solvents also are referred to as film-forming agents or coalescents. Since the film formation process for polymer dispersions, also referred to as coalescence, is essential for achieving the desired final property, this process and the relationships connected with this shall be elucidated more precisely.

Glass transition temperature T_g

The glass transition temperature (also referred to as glass temperature) and the MFT are closely linked. A high glass temperature of the polymer results in a high MFT of the polymer

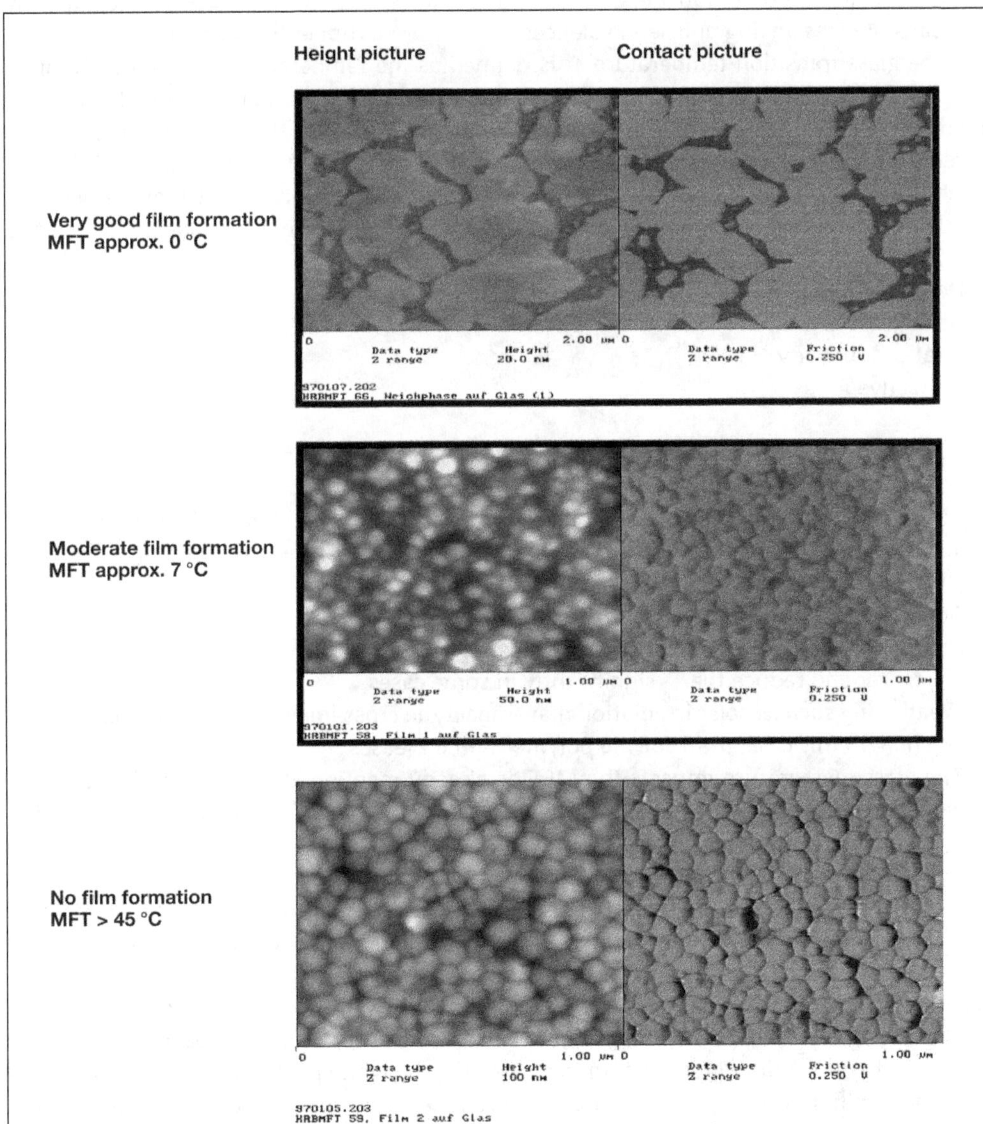

Figure 3.143: Electron micrograph of the film formation of dispersion particles

dispersion derived from this. While low molecular, solid substances exhibit strong melting points when heat is supplied (the molecules move as a whole in the melt), macromolecules feature an unsharp awakening. Initially, the mobility of the flexible side chains of the polymer increase with increasing temperature. Similar to the melting process, these additional movements of the site chains require space which is created by these movements under formations of a free volume [444]. Thus, the specific volume becomes larger, and the body expands. Therefore, one may distinguish between polymers in a:

- Glassy state where the macromolecules are frozen and where the chain segments are fixed
- Glass transition range or softening range where the movement of the polymer segments is enhanced
- Viscoelastic state or a rubber elastic range where the macromolecules are held together by means of cross-linking and semi-valences and feature a rubber-like behaviour.

The glass transition temperature T_g is defined as the temperature at which the molten polymer again merges into the hard glass state during cooling. In fact, it is merely not a defined temperature, but a transition region of about 10 to 20 °C. In this respect, this also is referred to as the freezing temperature region. Within this range, many properties of the polymers vary discontinuously so that the measurement of these properties may be used for the determination of the T_g value. Depending on the temperature, the following properties can be tracked:

- Density of the polymers
- Specific heat
- Heat conductivity
- Refractive index

The thermomechanical analysis (TMA) as well as differential scanning calorimetry (DSC) are important measurement methods.

In general, the higher the glass transition temperature of the polymer, the harder the consistency at ambient temperature. Other parameters relating to the amount of T_g have to be considered for polymers which are manufactured from dispersions:

- Added solvents (solvent retention) still present in the film
- Humidity in the film. Depending on the structure of the resin, water molecules act as a plasticizer and reduce the T_g significantly in some cases
- Weathering such as solar irradiation may impact the cross-linking reaction in the polymer reaction. In this case, the T_g of the polymer film increases
- The glass temperature impacts the MFT as already mentioned above. A high glass temperature of the polymer also results in an enhanced MFT value of the polymer dispersion

Minimum film forming temperature

The determination of the minimum film forming temperature (MFT) and the white point (WP) is performed in accordance with DIN 53787. Thereafter, the minimum film forming temperature is defined as the lowest temperature, at which the dispersion forms a continuous film without showing film cracks subsequently to the drying process. The white point is the highest temperature at which the dispersion does not form a film, but leaves a milky layer subsequently to the drying process. In former times, the white point has been interpreted as the minimum film forming temperature. However, numerous investigations have shown that the polymer film still has fine cracks above the white point. Thus, the white point always is some degrees Celsius below the minimum film forming temperature.

The minimum film forming temperature of a dispersion can be influenced by different substances:
- Solvent or film-forming agent (results in a more or less pronounced lowering of the MFT depending on the physical characteristics)
- Plasticizers (lower the MFT)
- Hydrophilic polymers such as protective colloids (result also in a lowering of the MFT because of water retention)

The Figure 3.144 schematically illustrates the process of film formation.

Film-forming agent (coalescent)

A good film-forming agent should fulfil the following properties [58]:
- The film-forming agent should be effective for the given polymer. That means, a maximal decrease of the MFT should result with a minimum addition
- The film-forming agent has to be participated easily into the system and absorbed appropriately by the polymer particles without reduction of the stability of the system (for example formation of specks during storage)
- The residence time of the film-forming agent in the polymer has to be sufficiently long in order to enable a homogeneous film formation. A rapid evaporation is desirable in solvents after coalescence so that the film remains not too long soft and thus not block resistant, susceptible to soiling or sensitive against mechanical damage
- The film-forming agent must not tend to hydrolyse in a weakly alkaline medium
- It should be low on odour, not or little toxic as well as environmentally compatible

When formulating and selecting the film-forming agent it should be considered that when using only water-miscible film-forming agents there is the risk that on very absorptive wood supports the film-forming agents penetrate with water very rapidly into the underground.

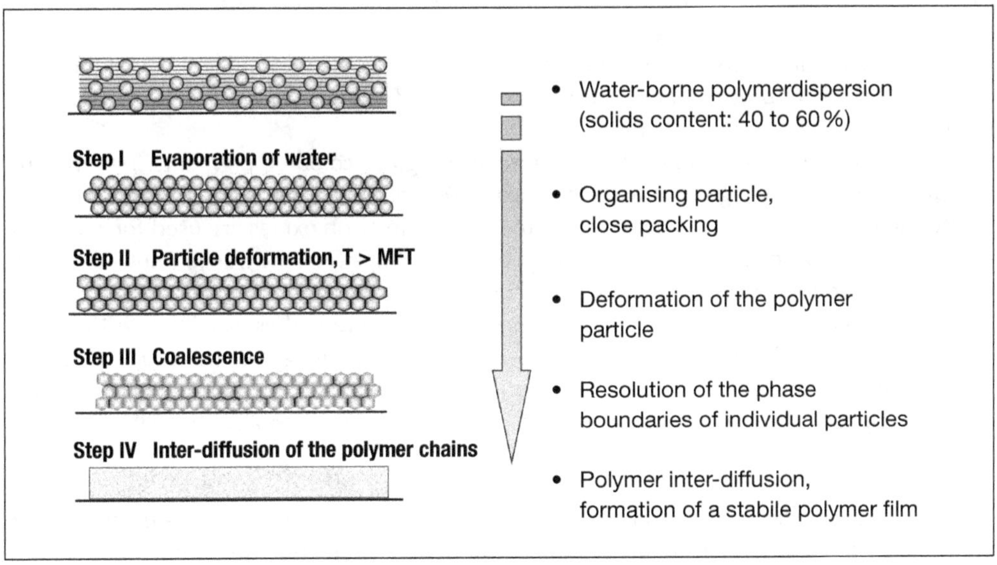

Figure 3.144: Schematic illustration of the process of film formation of polymer dispersion

Thus, the film formation can be disturbed which then manifests for example in fogging, cracking et cetera. Combinations of different film-forming agents (such as hydrophilic with hydrophobic) frequently are used in order to adjust balanced drying properties and film formation properties (see Figure 3.145).

The hydrophobic solvents preferably diffuse into the latex particles while the hydophilic solvents rather affect the surface of the dispersion particles. Especially in coatings, glazes and low pigmented systems, the solvents frequently are premixed or mixed with water in order to avoid a solvent shock and formation of specks resulting from this. Some common solvents are white spirit, methoxy propanol, texanol, butyl glycol, butyl diglycol, butyl diglycol diacetate and dipropylene glycol butylether (DPnB). Dimethylphthalate, diisobutylphthalate, benzoates or adipates, dimethyl succinate, dimethyl glutarate preferably are applied as a plasticizers.

It should be noted that the viscosity rises due to a swelling of the polymer particles if the concentration of the film-forming agents increases. Hydrophobic film-forming agents (such as white spirit) may have an anti-foaming effect and thus support the effectiveness of anti-foaming agents. Depending on the application, the added amount of solvents always is in the order of 2 to 10 %. Today, it partly can be dispensed with the addition, since the core-shell dispersions for example may feature sufficiently low minimum film forming temperatures.

3.2.2.3 Fillers, pigments and matting agents [445]

Pigments

Depending on the application, inorganic white pigments as well as inorganic and organic coloured pigments are applied. Transparent, fine-particle pigment preparations (mostly iron oxide pigments) are applied for the glazes whereby these pigment preparations are mixed with a low amount of normal coloured pigment preparations in order to adjust the desired colour shades.

Today, almost exclusively titanium dioxides in the crystal modification rutile are applied as a white pigment, since titanium dioxide has the highest light scattering effect of all white pigments and features a very good weather resistance.

As coloured pigments, the usual light-stable organic pigments widely are used in form of aqueous pigment preparations. There are a variety of manufacturers who also offer VOC-reduced and APEA free systems today.

In particular, iron oxide pigments are used as inorganic coloured pigments. Transparent, fine-particle iron oxide pigments with an approx. 10-fold lower particle size (mean particle diameter of approx. 0.05 μm) in comparison to the covering iron oxides are used for the transparent pigmented wood stains. In addition to the colouring effect, the transparent iron oxides mainly are used to achieve a good UV protection in the wood-preserving coating (see Figure 3.146). Usually, aqueous pigment preparations are used here and especially with the transparent iron oxides. At least, 0.5 and usually 1–5 percent by weight of transparent iron oxide pigment (with regard to the total formulation) are necessary if any additional UV absorbers were applied. Thus, an UV absorption of < 400 nm is given (see also item UV absorber).

Pigments

Partially, fine-particle titanium dioxides, silicates or active zinc oxides also are used as colloidal particles or nanoparticles in order to improve the UV protection predominantly. This especially is performed in unpigmented, clear coatings. However, this usually does not result in completely clear films. In recent years, effect pigments such as aluminium metal effect pigments or also interference pigments and pearlescent pigments are added for wood coatings with special effects for design trends.

Functional pigments

Functional pigments primarily are not used due to their optical properties, but because of other properties. Thus, functional pigments often are used in water-borne sealers for example in order to improve the barrier effect against wood components. In doing so, for example zinc oxide or special phosphate containing mixed phase components from anti-corrosive pigments are applied. Calcium-/magnesium phosphosilicate would be some examples of this.

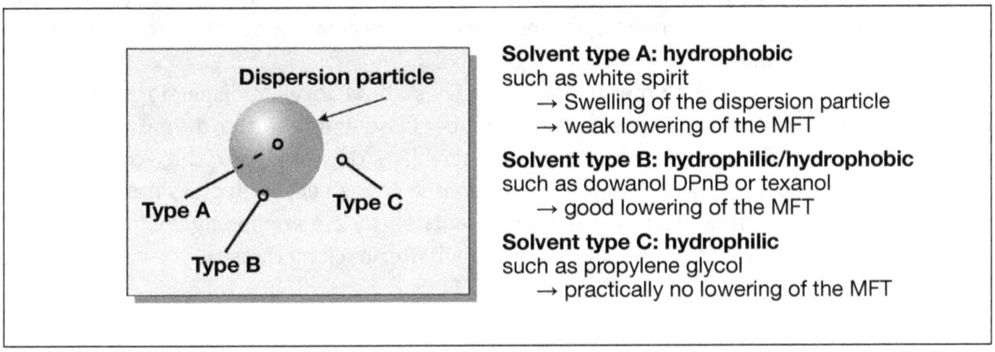

Figure 3.145: Interaction between dispersed particles and different types of solvent

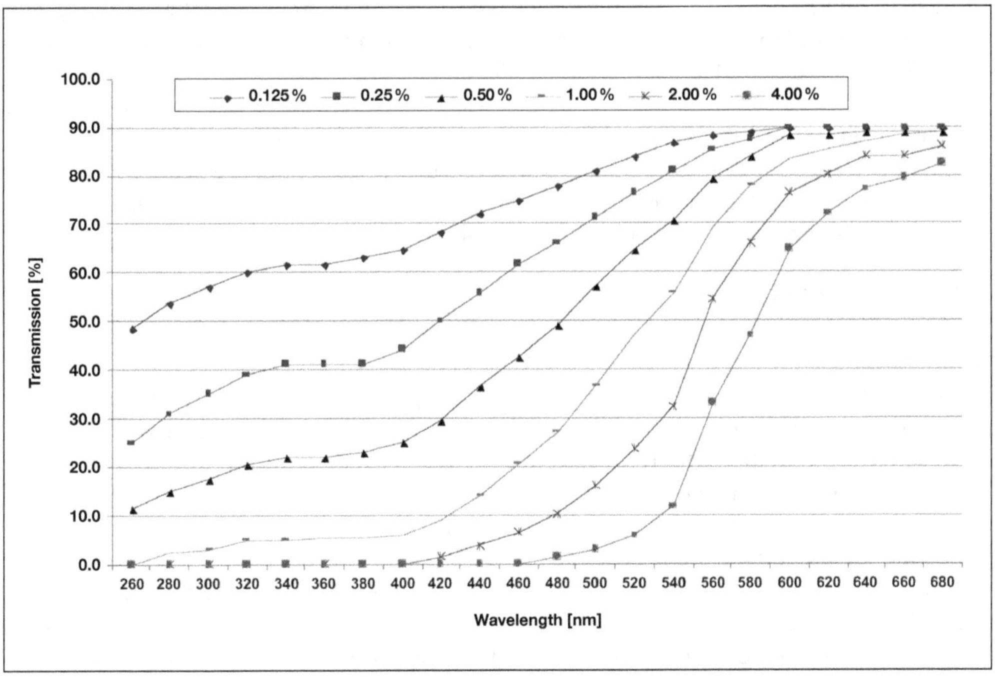

Figure 3.146: Transmission of transparent iron oxide red at different concentrations percent by weight related to the total recipe [446]

Fillers

Fillers have a direct impact on the gloss, weather resistance, abrasion resistance, matting, rheology, sedimentation, barrier effects (for example against moisture, gases as well as partly wood components) and cracking. The fillers also contribute significantly to the qualitative and economic formulation of coating materials. Fillers mainly are produced from naturally occurring minerals occurring to the substance class of carbonates, silicates, sulphates and oxides. The most important representatives of this substance class are the naturally occurring calcite, chalk, dolomite, kaolin, quartz, talc, mica, diatomaceous earth and baryte. In addition, there are also synthetically produced fillers such as precipitated chalk, precipitated baryte or precipitated silicate. The application of fibre fillers may be useful in special cases. In the selection of the filler, the most important criteria for the optimal distribution of the pigments are the particle size, the particle size distribution and the particle shape of the filler. In addition, the surface, the need for resins, hardness, weather resistance, purity, colour shade and price also play an important role.

Furthermore, there are polymer fillers. These are styrene/acrylate dispersions with a cavity which initially is filled with water and irreversibly filled with air after drying. This causes an additional refraction of light. However, polymer fillers do not significantly contribute to the wet opacity. These also are not film formation due to the very high glass transition temperature at ambient temperature. These components partly are applied against wood ingredients also in sealers due to their contribution to a better packing density.

Matting agents

The term matting agent generally refers to fillers which impact the surface of a coating in such a way that the gloss level thereof decreases (matting). The matting agent often cause a targeted roughness of the coating surface which results in a diffuse light scattering and makes the surface look dull. The type and amount of the matting agent determine the level of matting. With pigmented coatings, this effect also can be controlled with the filler composition. In this case, for example fine-particle silica (usually with special surface treatment matched to water-borne applications), waxes (as powder or emulsions) or polymer powders based on polyurea for example, are applied in analogy to solvent-borne coatings and especially the transparent coatings. Matting agents as well as silicas may negatively influence the whitening of transparent coatings based on water-borne resins.

3.2.2.4 Additives

Dispersants

The most commonly applied wetting agents and dispersants are salts of polyacrylic acids, non-ionic fatty alcohol ethoxylates and polyphosphates. Another disadvantage is an excess of dispersants effecting the water resistance of the coating agent. It also results in an increased foaming. The foam formation generally is a problem in the production and processing of water-borne coatings. Emulsifiers in the dispersion or the applied additives as well as wetting agents and dispersants can be considered as an originator for the foam. In addition, cellulose derivates such as ethyl hydroxyethyl cellulose or methylcellulose with pronounced hydrophilic and hydrophobic groups in the molecule are more powerful foaming agents as the less hydrophobic Hydroxy ethyl celluloses. In particular in the event of layer-forming wood coatings, a preferably foam free film should exist after application for an optimum protection and weather resistance of the coating film.

Defoaming agent

Defoaming agents are added to the coating components in order to minimize the foaming tendency. These often are complex mixtures consisting of actual active substances (such as mineral oils, silicone derivatives, waxes, fatty acid), partly water as a continuous phase as well as further additives. Defoaming agents act by penetrating into the foam lamella, destabilising the lamella and causing them to burst. The defoaming effect of such mixtures often is increased by addition of hydrophobic, highly dispersed solids (such as pyrogenic silica). Surface defects may occur in the transparent systems during application in the case of overdosage or in the case of a wrong selection of a defoaming agent. Especially, with regard to the thick-layer systems being sprayable with the Airless process, often several defoaming agents or the air vents as a combination defoaming agent being developed some years ago.

Rheologically active additives

Thickeners are applied for the controlling the viscosity in all shearing zones and the associated characteristics such as draining behaviour, stability, course, processability, coating resistance, storage stability and settling behaviour. Furthermore, thickeners impact the water retention, open time, water resistance as well as gloss level. The rheologically active additives can be partitioned into three main groups:

- Thickeners based on cellulose, polysaccharides or alginates and xanthan
- Synthetic thickeners
- Inorganic thickeners

Synthetic thickeners mostly are applied in water-borne wood coatings based on dispersions. Mainly acrylic thickeners and/or polyurethane thickeners are used. depending on the modification and efficiency, acrylic thickeners are classified in ASE (alkali swellable emulsions) and HASE (hydrophobically Alkali swellable emulsions) or HEURASE (hydrophobic modified ethylene oxide urethane modified acrylic swellable emulsions), respectively. Polyurethane thickeners also are referred to as HEUR type (hydrophobically modified ethylene oxide urethane).

Viscosity profiles similar to those of cellulose ether compounds can be achieved by polyacrylate thickeners of ASE type. Polyacrylate thickeners thicken the aqueous phase and are used to adjust the viscosity at low shear rates (low shear viscosity). Polyacrylate thickeners often generate a yield point. Thus, polyacrylate thickeners prevent the coating material from draining-off at application in thick layers. The rheology is obtained with associative acrylic thickeners (type HASE or HEURASE) in case of low pigmented coating systems. Herein, a physical interaction is established between polymer particles and the thickener molecules. Usually, acrylic thickeners are involved in trade as a 30 % acid dispersion. With the addition of basic compounds, the particles initially swell, then to solve completely in an alkaline medium and to develop the thickening effect.

Polyurethane thickeners generally are within the class of associative thickeners. In contrast to acrylic thickeners, polyurethane thickeners act over a broad range of pH value. The thickening effect highly depends on the system and is affected by other formulation components such as solvents, resins and solid content.

In wood coatings, cellulose ethers of methyl cellulose and hydroxyl ethyl cellulose are applied only in specially pigmented formulations. The former is somewhat more stable to enzymatic degradation. The cellulose ethers remain soluble in the paint film. Thus, the proportion of cellulose ethers in wood coatings should be kept as low as possible. The thickening effect is based on the binding of the water molecules by means of hydrogen bonds. That means, the aqueous phase is thickened. The thickeners based on alginates, polysaccharides or xanthan are highly effective, also produce a high yield produced and often are applied as a co-thickener.

Inorganic thickeners are sheet silicates on the type of montmorillonite such as bentonite (natural or synthetic, aluminium silicate) or hectorite (magnesium silicate). Their thickening effect is based on the spatial orientation of the platelets. Compared to cellulose ethers, the inorganic thickeners only slightly affect the water resistance and primarily are used as anti-settling agents as well as for controlling the low shear viscosity. The sheet silicates as well as the acrylic thickeners and polyurethane thickeners have any water retention capability. A slight thixotropisation of the paints can be achieved by an addition of titanium chelate complexes. However, thus one relies on dispersions which are stabilised with cellulose ether. Polymer films resulting from cellulose ether-stabilised dispersions feature a stronger whitening or hydrophilicity, respectively. Dispersions manufactured from these polymer films have a slightly higher low shear viscosity even without addition of thickeners.

Conservatives

The application of biocides in coatings essentially takes place for two reasons: The first aspect relates to the preservation of the liquid paint and especially of the water-borne coating systems in order to guarantee a greater storage capability. This applies to the so-called in-can preservation of the coating material. The second intended purpose of microbicides is the dry film preservation in order to protect the cured coating against microbial attack by algae or fungi. Both in the liquid and in the dried paint film, water-borne coating components based on dispersions are subject to a permanent attack by microorganisms such as bacteria, fungi or algae, respectively. Thereby, mainly protective colloids, thickeners, emulsifiers and defoamers serve as a breeding ground. Thus, these products need a protection against an attack by microorganisms. It should not be conserved, if the pH value of the coating material is in excess of 11.

In doing so, the storage conservation as well as the in-can preservation require a broad range of efficacy against bacteria and fungi. Commercially, there are a variety of products on different chemical bases. Mainly various derivatives of isothiazoline are applied in water-borne coatings.

In recent years, the protection against microbial infestation has become problematic due to the reduction of the proportion of volatile compounds such as residual monomers and solvents in the paints as well as due to the establishment of more mild conservatives. The effect against fungi and algae is paramount for the conservation. A broad effective algaecide and fungicide often are added to the dried paint film. Today, products commercially are available which have a significantly lower tendency against leaching during exposure to weathering and also are less problematic with regard to the labelling. Both are achieved for example by 'encapsulation' of the active ingredients.

Wood preservatives [447]

Since wood is susceptible to fungal infection on exposure to moisture in the outdoor area, appropriate fungicidal compounds are added to the impregnating agents in order to achieve a best possible protective effect of the total paint structure. These primers would then usually achieve a protective effect against blue stain and rot with regard to the standard DIN 68800 part 3 or as a certified wood preservative with regard to RAL (see Chapter 2.4). In doing so, combinations consisting of propiconazole and iodopropynyl butyl carbamate mostly are applied.

Thus, the active compounds shall have the following characteristics:
- High efficacy and broad spectrum of effectivity against wood-discolouring and wood-destroying fungi
- Algicidal effect
- Lowest possible toxicity
- Biodegradability
- Worldwide registration as an additive for wood preservatives

Insecticides are added only in special cases. Insecticides generally are not applied at customary wood preservative primers for timber structures such as windows or wooden panellings.

Light stabilisers and UV protection

The lignin obtained in the wood as well as the celluloses are attacked by the UV components of the sunlight and destroyed (depolymerisation). In doing so, light woods are coloured dark and degraded. the thus occurring cracks and greyings are particularly problematic due to the leaching of water-soluble lignin degradation products.

Thus, lignin has an absorption maximum in the UV-B range at approx. 280 nm. The celluloses, hemicelluloses as well as the wood ingredients absorb at wavelengths of under approx. 200 nm, but also feature an additional absorption in the UV-A range up to approx. 400 nm. Since not only UV light but also visible light with a wavelength of up to approximately 540 nm initiate photooxidation processes on the wood surface, it is also advantageous to keep short-wavelength portions of the visible light away from the wood surface in order to prevent chemical reactions. For clear coats, UV absorbers and transparent pigments which both are combined with free-radical scavengers (HALS) are available. The thinner the layer of the wood stain, the higher the proportion of the UV-absorbing components in the glaze. This particularly applies to colourless glazes. Alkyd resins have a certain absorption effect here. Polyacrylates are completely transparent to UV light. UV absorbers (e.g. benzotriazole), partly combined with sterically hindered amines – so-called HALS compounds (HALS = hindered amine light stabilisers such as tetraalkyl substituted piperidine) – are added in order to provide a sufficient UV protection to the very bright or non-pigmented lacquers based on polyacrylate dispersions [448]. In some cases, also UV absorbing nanoparticles consisting of titanium dioxide or zinc dioxide are added.

Thereby, the HALS compounds primarily shall delay the radical photo-oxidative degradation of the lignin. The amount of UV absorber should be selected such that approximately 95 % of the UV radiation is absorbed. In non-pigmented or bright glazes, respectively, typically 1 to 5 wt. % based on the total formulation are added. This is a good compromise between cost and effect.

Other additives

Water-borne coatings based on water-borne alkyd resins contain siccatives and anti-skinning agents usually matched with water-borne systems. Thereby, the chemical basis corresponds to the siccatives and anti-skinning agents as described for solvent-borne alkyd resins (see chapter 3.2.1.4). Topcoats usually contain additions of wax emulsions (also with fine particles or nanoparticles) or special silicones in order to hydrophobise or to improve the block resistance or scratch resistance, respectively, or in order to achieve special surface effects (beading effects or lotus effects). The topcoats still contain specific wetting agents usually in order to improve the wetting of water-borne coatings on wood substrates.

3.2.3 Tasks and functions of wood coatings for outdoor applications

An adequate physical protection of wood can be achieved by translucently pigmented, UV-absorbing or opaque coating systems [449]. Furthermore, today a modern "organised" wood protection involves the application of constructive structural measures, the selection of suitable wood species and wood quality, the targeted utilisation of chemical agents only where it is really necessary, as well as the regular care and maintenance of the overall coating system [489]. Colourless coatings only are useful, if these properly are equipped with UV protection agents and radical scavenger. If the coating is damaged, grey to black spots at the damage points. The wood coatings essentially have the following tasks and function:
- Decorative effect
- Protection against UV light
- Humidity protection
- Protection against blue-stain and mould infestation
- Reduction of the swelling and shrinking movements

3.2.3.1 Coating systems for dimensionally stable and unstable components

In Europe, the requirements of the relevant standards and guidelines to grant a shelf life as long as possible [450–452] vary by region and the corresponding market conditions. In Southern Europe for example, a stronger focus often is put on the hardness or block resistance, respectively, as well as on the appearance of the coating in wooden windows for example. This may have negative properties on the weather resistance or durability, respectively. Another problem is that the wood qualities increasingly deteriorate (for example shorea bracteolata or pine with a high proportion of sapwood), or new cost-effective tropical woods are established on the market whereby these tropical woods rather are unsuitable for the construction of windows. Thus, it is a challenge for many manufacturers of coating materials and window manufacturers to develop a durable coating system or to achieve a durable coating system.

Wood species and wood protection

The selection of wood species [449] for the window production, for example, is described in the fact sheets HO 02 and HO 06 [453, 458] (wood species list) or in the corresponding table in the BFS leaflet no. 18 [461] as a guidance. Knot free profiles should be used. this is realised by laminated profiles or profiles which are produced with finger-jointed assemblies in length. One should pay attention to a proof of suitability. The variety of wood species provides a material for almost any field of application. Apart from the mechanical properties, the content of wood ingredients and consequently the resistance classes are decisive. The basic standards hereto are described in Chapter 2.4.2. If timbers whose resistance for the corresponding intended application are insufficient are selected, these timbers have to be coated with a prophylactic chemical wood protection according to DIN 68800-3 (Wood Preservation for building construction – Part 3: Preventive chemical wood protection). However, this only is prescribed for structural components according to the Building Act. These wood preservatives must have a general building inspection approval of the Deutsches Institut für Bautechnik (DIBt). The standard DIN EN 460 (durability of wood and wood-based products) describes which wood protection is required. A wood protection against blue stain (B) and wood-

Table 3.64: Levels of application and examples for dimensionally stable, restricted dimensionally stable and non-dimensionally stable components

Levels of application	Allowed dimensional stability	Typical examples of the levels of application
Non-dimensionally stable	Non-restricted	Overlapping wooden panelling, fences, garden sheds
Restricted dimensionally stable	Approved to a limited extent	Wooden panelling with tongue and groove, wooden buildings, garden furniture
Dimensionally stable	Approved to a very limited extent	Wood components incl. windows and doors

destroying fungi (P) is required according to the standard DIN 68800-3 for non-structural components such as new windows and exterior doors, if the timber applied do not correspond to the durability class 1 or 2, respectively, and if the standard DIN 68800 is agreed between the contractor and the client. This is approved in the VOB for carpentry works. Thereby, examinations for blue staining are performed according the standard DIN EN 152.1 (Testing of Wood Preservatives; Part 1: Application in coating method), while examinations for wood-destroying fungi (rot fungi) are performed according to DIN EN 113 (Testing of Wood Preservatives). In the industrial wood coating, a penetration depth of a few microns up to a few millimetres is obtained by impregnation depending on the type of the wood. The suitability of these wood preservatives can be detected by using a mark of conformity of the RAL Quality Mark Association Wood Preservatives. By way of derogation from Germany, in Austria mainly the protection against blue-staining is required.

Thereby, the wood preservative has to be compatible with the subsequent coating system (see manufacturer specifications). In the event of anti-blueing agents on pine tree sapwood, the penetration depth (blue stain free zone) has to be greater than 1.5 mm (DIN EN 152). This is particularly important in order to have a sufficient security against blue stain with emerging higher moisture contents of the wood. Thus, an increased infestation by blue stain may result at wood moisture contents of more than 20 % (as in many other types of sapwood) as well as at an insufficient protection against blue staining.

Components

The selection of the coating and coating system depends on the selection of the wood for the components and the intended application or use of the component, respectively. Thus, the selection of wood is an important basis for the functionality and durability of the intended system-compatible coating system. It practically forms the basis for a long life of the whole construction. In this regard, a distinction is made between different 'classes of component' depending on the function.

Components whose usability does not depend on the compliance with tight shape tolerances are referred to as non-conformance components. These include fences, overlapping wooden panels, freestanding pillars and pergolas. The experts disagree whether such components are to be painted with diffusion permeable or diffusion impermeable coatings if they are even coated.

Components which require an enhanced moisture protection in comparison to non-dimensionally stable components, are designated as dimensionally stable in order to remain fit for use. These are wooden panellings (tongue and groove), for example.

Table 3.65: Standard DIN EN 927-1: Classification according to stress groups

Construction	**Climatic conditions**		
	Moderate	Strict	Extreme
Protected	Weak	Weak	Medium
Partly protected	Weak	Medium	Strong
Not protected	Medium	Strong	Strong

Table 3.66: Standard DIN EN 927-2: Threshold values for evaluation criteria – outdoor weathering and water uptake.

	Dimensionally stable	**Limited dimensionally stable**	**Dimensionally instable**
Blistering	0.3	0.7	1
Crack formation	0.7	1.7	3
Exfoliation	0.3	0.7	1.3
Adhesive strenght	0.7	1.7	2.7
Maximal total value	6	12	18
Maximal difference, so that the test is valid	2	3	4
Value of water uptake according to EN 927	≤175 g/m²	≤250 g/m²	No threshold value

However, a clear distinction between these two types of components is not always possible, in practice. High quality fences consisting of heavy tropical timber and correspondingly coated in comparison to cost-effective fences made of spruce wood with dipcoat-priming are an example. In the first case, the component can be described as dimensionally stable, while in the second case the component is not dimensionally stable.

Components which tolerate only small dimensional changes in order to preserve their usability are referred to as dimensionally stable components. The standard DIN EN 942 (wood in carpentry works – general requirements) as well as the DIN EN 133047 and the leaflet HO 02 of the Association of Window and Facade Manufacturers are valid for dimensionally stable components such as windows. Dimensionally stable types of wood are described here. The dimensionally stable components require a special moisture protection, for example.

Requirements for coatings

The specific requirements on wood coatings for dimensionally stable components [454–457, 459, 460] such as wooden windows or wooden doors are described in the guidelines of the Institut für Fenstertechnik (also referred to as Rosenheimer guidelines) as well as in the guidelines of the Association Window Fascade (VFF) – for example leaflets HO 01 and HO 03 – or the Initiative *'Pro Holzfenster'*, which significantly was elaborated by the Wilhelm Klauditz Institute (WKI Fraunhofer Institute, Braunschweig). The standard EN 927-1 up to 8 as well as the standard ÖNORM C-2350 coating materials for coatings on dimensionally stable outdoor components consisting of wood – minimum requirements and verifications) describe the key components of the catalogue of requirements for wood coatings in the outdoor area.

Additionally, recommendations for the factory final treatment (coating) of wooden windows and wooden front doors were elaborated by the working group Surface of the IPH (Initiative Pro Holzfenster) in March, 2003 [469]. In addition, a RAL quality mark was established for wooden windows in order to improve and protect the quality of the wooden window. The BFS leaflet No. 18 'Coatings on dimensionally stable external construction components consisting of wood, especially windows and exterior doors' contains application instructions with checklists for painter applications, especially for reconstruction.

Climate stresses

The durability as well as the protective effect of the overall wood coating depend on the intensity of the stress caused by different climatic zones. In Central Europe, it is believed that the weather stresses on the north side of buildings are lower than the weather stresses on the southwest side, the so-called weather side. Thereby, the construction of the object as well as the possible protective effect resulting from this also play a role. These aspects also are considered in the standard DIN EN 927 Part 1, and the selection of the wood coatings are divided in three stress groups as it is illustrated in Table 3.65. Simplified, one has defined three climatic conditions in the standard DIN EN 927 Part 1. Thereby, the structural conditions also are considered.

Humidity protection and layer thickness

One of the main tasks of the coating system is the adequate protection of the structure against the penetration of liquid and vapour water. This shall ensure the dimensional tolerance of the dimensionally stable woods. A moisture penetration of more than 20 % as well as the related risk of fungal infestation and cracking have to be avoided. For this purpose, there are a variety of investigations. Liquid water usually is kept from. On the one hand, the fundamental property of the resin plays an important role, while the coating formulation determines the water uptake as well as the diffusion behaviour of water vapour on the other hand. The testing procedures as well as the details concerning performance indicators are described in the Standard DIN EN 927 Part 4 and 5. Table 3.66 presents the currently valid limiting values.

In doing so, the water uptake usually is expressed in g/m^2. In the case of the diffusion of water vapour, the permeability of water vapour can be expressed in kg/m^2 after 14 days (WD 14) analogous to the standard DIN 927-4A or as the diffusion resistance factor μ or p and – calculated from this – diffusion equivalent layer thickness Sd in m (Sd value = μ value x layer thickness in m) (see also DIN EN ISO 12572:2017-05). Apart from the composition of the coating material, the dry film thickness greatly affects the moisture behaviour. Sufficient layer thicknesses for a good moisture protection are important especially at the edges. A curvature of the edges (radius of curvature $r > 2$ mm) as well as lowest possible air inclusions (lower dry film thickness of air bubbles) have to be considered.

The water vapour permeability on the inside of wooden windows shall not be larger than the water vapour permeability on the outside in order to avoid a penetration of moisture and to avoid blistering on the outside. In principle, solvent-borne systems behave less problematic than water-borne systems. The initially good protective effect of alkyd systems against moisture is reduced in the course of weathering with degradation phenomena followed by cracking and spallings. Water-borne systems based on polyacrylate dispersions are permanently elastic. The initially pronounced hydrophilicity is reduced in the course of weathering due to leaching of the hydrophilic ingredients (emulsifiers, dispersing agents etc.).

Table 3.67: Minimum dry layer thickness of coatings for windows and front doors

	Opaque	**Translucent**
Intermediately coated windows before installation at final coating at the object	Not less than 50 µm	Not less than 50 µm
Blind frame, construction connection range	Not less than 50 µm	Not less than 50 µm
Hidden surfaces such as glass rebate	Not less than 30 µm	Not less than 30 µm
Blades, blind frame	Not less than 100 µm	Not less than 80 µm

Table 3.68: Categorization of the coatings according to plenty (layer thickness) analogous to EN 927

Minimal	**Medium layer thickness below 5 µm**
Low	Medium layer thickness between 5 µm und 20 µm
Medium	Medium layer thickness between 20 µm und 60 µm
High	Medium layer thickness above 60 µm

The humidity also affects the adhesion strength. Thus, a wet adhesion of the coating material is required for a good weather resistance and adequate protection against moisture. The risk of blistering and spalling thus are minimized. Here, too, there are a number of investigations [452, 454, 463, 464]. This disadvantage of water-borne systems was almost compensated by modifying the polymers, for example by incorporation of wet adhesion promotors (such as silanes) in polyacrylic dispersions as well as by a relatively low particle size. The today's water-borne systems offer a better long-term protection against moisture in comparison to traditional alkyd systems.

The total layer thickness is important for the functionality of the coating structure and, in particular, the moisture protection. Here, the non-dimensionally stable or limited dimensionally stable components require a lower layer thickness in comparison to the dimensionally stable components. For the latter, usually a 3 to 4-layered coating system is required. A two-layer coating system is sufficient for non-dimensionally stable components and limited dimensionally stable components. In this context, dry film thicknesses of coatings have been defined for the factory treatment of wooden windows and wooden front doors are defined [447].

However, the layers should not be significantly thicker than 150 microns and, thus, penetrated moisture also may diffuse out again. It has to be ensured that the moisture content is not permanently above 20 % (risk of blue-stain fungal infestation with subsequent problems).

Many ring trials have shown that the water uptake through the coating as well as the wet adhesion are crucial criteria. This applies to the currently used water-borne coating systems. For this, various coating systems have been tested in the project AIR (1994 to 1998) with European research institutes. The focus was on low-solvent, that means water-borne coating systems. Systems based on polyacrylate dispersions and alkyd-polyacrylic hybrids achieved the best results. Thus, the usability of the water-borne systems was confirmed. In the standard DIN EN-927, the coatings are classified in classes according to the fullness (layer thickness) as illustrated in Table 3.68.

It therefore follows that only coatings with a high abundance, i.e. a high film thickness, meet the requirements for an industrial coating of windows according to the Rosenheim specifications for wooden windows. However, in the new VFF leaflets these layer thicknesses are enshrined only for industrial applications since these layer thicknesses for wooden windows, for example, only can be achieved with a disproportionately high effort for application in coating, or the coating systems are not practical for this purpose due to the different rheological properties. Generally, it should be noted that the type of wood usually has a greater impact on the moisture behaviour of a component in comparison to the component.

Colour shade of the behaviour

At the latest during the tendering of a building, the colour shade of the surface coating is set. During the exposure to the sun, the selection of a suitable wood for wooden windows, for example, is associated with the surface heating induced by the colour shade of the coating material. Both have to be coordinated in order to ensure a long lifetime of the component. Very dark coatings may attain surface temperatures of approx. 80 °C. The wood as well as the total construction including the adhesive joints thus are subject to a higher strain. Due to the poorer UV protective effect, bright glaze colours usually have a lower durability and require shorter maintenance intervals (also see 'ift leaflet'[454]).

Wood processing, wood working and preparation (frame connections)

The processing of the wood prior to the coating is a prerequisite for a durable and long functional coating system. The quality and testing regulations RAL-RG 424/1 have been proven for wooden windows. Thus, the profiles of the wooden windows are processed with slicing (hydro slicing). It should be noted that smaller wood fibres are not removed completely by grinding. This only is possible after an impregnation and subsequent intermediate sanding. Restored wood damages (such as screw anchors) have to comply with the standard DIN EN 942 (wood in carpentry works – general requirements). All frame connections must have an adhesive bonding corresponding to the stress group D 3 with regard to DIN EN 204 (classification of thermoplastic wood adhesives for non-structural applications). The frame connections also have to be sealed (see also the 'ift guideline' Bonding of wooden windows – Part 2: Bonding of frame corner connections). The proof of the temperature resistance according to the standard DIN EN 14257 (WATT '91) is connected with this.

Colourless wood impregnations

As an outdoor application, wood impregnations are applied as a first surface treatment and adapted to the type of wood. The standard DIN 68800 part 3 requires an impregnation against blue stain and fungal attack for all types of wood which are not classified in the resistance classes 1 and 2 with respect to the standard DIN 68364 (characteristic values of wood species – bulk density, modulus of elasticity and consistencies). The efficacy has to be demonstrated and documented with respect to the standards DIN EN 152 and DIN EN 113 as well as by awarding the quality mark RAL GZ-830. This is true for wooden windows directly exposed to weather. Wood-aluminium windows do not require this protection; here, a pure protection against blue-staining is sufficient. These wood impregnations are non-pigmented low viscous and have a low non-volatile content (between 3 and 10 percent by weight). Non-resistant types of wood contain blue stain inhibitory as well as rot fungus inhibitory additives. These shall penetrate into the wood as much as possible, so that the active substances such as fungicides or also the hydrophobising agents may penetrate good into the wood matrix and solidify the wood matrix. In addition to the wood-protective agents, these materials may contain additives of special algicides and fungicides in special formulations for the so-called film protection.

Thus, already here the wood fibres are bonded well in order to smooth the wood surface even optimally in the subsequent grinding process after application of the primer or intermediate coating, respectively, if necessary. The wood fibres are applied prior to the application of colourless or transparent as well as opaquely pigmented coating system. However, it must be ensured that the surface of the wood does not become too hydrophobic after drying of the primer – for example by application of hydrophobic additives – in order to avoid adhesion problems with their unpleasant consequences (such as detachments) during the application of subsequently applied water-borne coating materials. Usually, the impregnations are applied with the immersion method, flow coating procedure or also by the spray process. It should be considered that the minimum drying times specified by the paint manufacturer have to be fulfilled after application of the primer. A distinction is made between solvent-borne and water-borne impregnations.

Solvent-borne impregnations

Solvent-borne primers have a very good penetration capability. The anti-fugal or fungicidal active ingredients also are dissolved in the primers very well and thus are distributed in the wood

Components	Parts by weight [%]
Alkyd resins, long or medium oil 100 % 1)	60.0
Combination of dryer (cobalt, zircon, calcium)	Approx. 5
Butyl diglycol	100
Wood preservatives (biocides) 2)	Approx. 4–20
De-aromatised hydrocarbons 180–220	815–831
	Approx. 1000

1) For example "Worleekyd" L 7904 (producer: Worlée)
2) Specifications as active ingredient concentration. The added amounts depend on the type of biocide or biocides, respectively, as well as on the application quantity of the impregnation (depending on the requirements for example according to the standard EN 113 or EN 152). The active ingredient has to be well predissolved or premixed prior to its addition to butyl diglycol, for example. Usually, the addition of a viscosity stabilising, oxime containing additive is essential.

Figure 3.147: Schematic formulation of solvent-borne, non-pigmented impregnation

Components	Parts by weight [%]
Acrylate dispersion, fine-particle (approx. 50 nm), approx. 34 % 1)	295
Wetting agent AMP 90	2
Substrate agent/water (1:9)	5
Defoaming agent	
Water	624–640
Preservative agent	2
Wood preservative (biocide) 2)	Approx. 4–20
Butyl diglycol	50
	Approx. 1000

The impregnating properties even can be improved by an addition of water-borne alkyd resins (producer: Cytec).

1) for example Mowilith LDM 7667 (producer: Celanese Emulsions GmbH)
2) Specifications as active ingredient concentration. The added amounts depend on the type of biocide or biocides, respectively, as well as on the application quantity of the impregnation (depending on the requirements for example according to the standard EN 113 or EN 152). The active ingredient has to be well predissolved or premixed prior to its addition to butyl diglycol, for example.

Figure 3.148: Schematic formulation of water-borne impregnations

matrix subsequently to the application. These active ingredients must have the above mentioned. As a resin, these active ingredients usually contain medium oil or long oil alkyd resins and partly also modified linseed oil. The substances which are mentioned in Chapter 3.2.2.4 mainly are used as fungicidal active ingredients. Figure 3.147 illustrates a schematic formulation.

Water-borne impregnations

Water-borne impregnations feature a slightly poorer penetration capability than solvent-borne impregnations. But today, especially with the special resins the penetration capability is sufficient so that the requirements of the corresponding approval tests are met. The active ingredients thus are dissolved in water, or may be added in an emulsified form. It also is possible to emulsify the active ingredients into the solvent. The active ingredients are analogous to those organic compounds which are mentioned in the chapter 3.2.2.4. Fine-particle, not too hard acrylate dispersions which are also referred to as hydrosols (mean particle diameter approx. 50 nm) as well as water-borne alkyd resins dissolved in water or in emulsions are used as resins. Combinations of hydrosols and alkyd emulsions (hybrids) also are customary. Thus, the alkyd resins wet the wood substrate somewhat better than the hydrosols and support the penetration behaviour of the active ingredients or the impregnation, respectively. Figure 3.148 illustrates an example recipe.

Translucent or opaque pigmented primers with or without fungicidal properties

Apart from colouring, the primer shall build a tight contact to the wood fibres in order to form an optimally smoothed wood surface in the subsequent grinding process. In special cases, the primers even may contain fungicidal additives against blue stain whereas these fungicidal additives are tested as a wood preservative according to the standard DIN EN 152: 2012-02. Often, biocides also are applied as a film protection. Primers are formulated with a somewhat increased solid content. The non-volatile amount of non-pigmented or translucent primers mostly is between 10 and 18 %, while the non-volatile amount of opaque pigmented primers is between 25 and 35 %.

Usually, primers also contain pigments. For this purpose, on the one hand, the primer shall equalise the absorbance of the wood for the subsequently applied layers, while the primer shall give a certain colour shade to the wood on the other hand. This applies particularly to transparent primers. The colour shade of the primer is adjusted to the later applied coating system. The levelling effect prevents an uneven appearance of the coating effect of the coated component especially in the event of translucent colour shades. Thus, the primer already shall provide a certain UV protection for glazings. The primers shall ensure a good network of adhesion strength of the entire coating system to the wood in order to minimise the risk of peeling. The primer should dry as soon as possible, and it should be grindable as fast as possible. If necessary, a grinding after application of the interim coating hereinafter is possible in order to prevent a looping through. Ideally, subsequently applied coatings should not provide a straightening of the wood fibres any more. However, in the primer the applied film-forming agent should not have a too high hardness in order to prevent a cracking during the weathering.

In general, the primers also are applied by using a dipping process or a flow coating process. In doing so, a distinction is made between solvent-borne and water-borne primers. It should be noted that also the aqueous primers now have a great significance. Thereby, depending on the region in Europe, or depending on the philosophy of the producer, the primers even are formulated with water-borne alkyd resins or only based on polyacrylate dispersions. Thereby, the water-borne alkyd resins may improve the wetting on certain types of wood, adhesion as well as the support of the barrier effect against wood ingredients (such as resin, waxes, fats or tannic acid).

Solvent-borne primers

These primers are based on alkyd resins. Depending on the application, transparent or opaque structure, the primers are transparent or opaque, usually white pigmented. In glazed structures, the primer already is coloured for the subsequent glazed structure (see Figure 3.149).

Water-borne primers

Water-borne primers contain polyacrylate dispersions or styrene acrylate dispersions or water-borne alkyd resins, in some cases as a combination. Polyacrylate dispersions, water-borne alkyd resins or combinations from both resins are applied in transparent primers.

Polyacrylate dispersions, styrene acrylate dispersions and also water-borne alkyd resins are applied in opaque, white pigmented primers. Since a barrier effect against wood ingredients is necessary, the example recipe is discussed separately in the next chapter.

Pigmented primers with barrier function against wood ingredients (insulating primer)

These are special, mostly white pigmented primers which are required for a water-borne, light opaquely pigmented coating composition on woods rich in tannin or rich in wood ingredients (for example oak, merbau, red cedar, larch). Primers also are applied on oak or highly resinous, conifers such as pines whereby especially the knotholes are critical. Without the application of blocking primers, more or less severe discolorations in water-borne topcoats occur in such resinous or tannic woods. The reason for this mainly is the fact that water-soluble, chromophore compounds such as tannin pass through the coating film after some time. Particularly in the presence of high humidity, the migration of tannins is accelerated. This leads to a discolouration of the coating surface especially in tannin-rich woods such as oak, merbau and others [490]. Its main component is penta-metadigalloyl glycoside which is a derivate of the tannic acid (3,4,5 trihydroxybenzoic acid). Colourless solutions of tannic acid and

Components	Parts by weight [%]
Alkyd resin long oil 100 % 1)	180
or alkyd resin medium and long oil 80 % in "Shellsol" RD 2)	225
Combination dryer (cobalt, zircon, calcium)	Approx. 15
Anti-skinning agent	2
"Dowanol" DPM	10
De-aromatised hydrocarbons 180–220	793–748
Film protection optional	3
	Approx. 1000

1) for example "Worleekyd" T 7313 (producer Worlée)
2) for example "Vialkyd" AS 754 / 80 SD 60 (producer Cytec)

If necessary, wood preservatives (biocides) even are added to the primer depending on the desired profile of requirements.

Figure 3.149: Schematic formulation for solvent-borne, transparent primers

especially its alkali metal salts discolour to brown on contact with oxygen of the air. In the case of alkyd emulsions, the oxidative curing can be delayed because of a deactivation of the siccatives by the complexing properties of the tannins.

To date, there exist no satisfactory possibility of formulation of water-borne topcoats in order to waive a blocking primer. These discolourations may occur within a short-time frame after the coating or after a certain exposure time on weather. These discolourations often are in stripe form or punctiform, that is, not necessarily consistent. This is caused by water-soluble tannins or resin acids which particularly may leave in the alkaline area preferably. The sealer shall fix these components permanently and thus avoid the bleeding as well as the perfusion in the water-borne topcoat.

In the past, exclusively solvent-borne primers based on polymerised resins or alkyd resins. Due to the environmental problems, today the solvent-borne primers are increasingly less in use. Due to their hydrophobic effect and properties (resin solution), the solvent-borne primers practically have trapped and blocked the ingredients. This also applies for solvent-borne topcoat constructions based on alkyd resins for example, which have an adequate barrier effect also without a blocking primer. In the past, various attempts have been made to solve the problem of passing through in water-borne wood coatings. One approach was the use of anionic polymer dispersions in combination with selected compounds of zinc, aluminium or titanium [490]. Moreover, certain functionalities such as urethane groups or alkyl urea groups are apparently capable of interacting with tannin. This resulted in improved barrier properties of the basic binders [490].

Special water-borne polymer dispersions such as polyacrylate dispersions and styrene acrylate dispersions (such as anionic core-shell polymers) partly in combination with other resins such as water-borne alkyd resins are applied for water-borne sealers which increasingly are used today. In doing so, the polymer dispersions shall feature a good water resistance. Furthermore, the polymer dispersions have a special networking technology which complex or crosslink the wood ingredients, respectively. In addition to the selection of a suitable resin, the amount of the resin (pigment volume concentration PVC not to high), the selection of the pigments (surface treatment of the titanium dioxide types, if necessary functional pigments), filler composition (most compact package, partly platelet-shaped fillers), the amount and type of dispersants and thickeners as well as the type of solvent are import in the recipe

Components	Parts by weight [%]
Acrylate dispersion, fine-particle (50 mm), approx. 34 % [1)]	294
Wetting agent AMP 90	2
Substrat wetting agent/water (1:9)	5
Defoaming agent	2
Preservative agent	2
Water	692
Film protection (optional)	3
	1000

1) For example "Mowilith" LDM 7667 (Producer Celanese Emulsions) The primer also may contain a conventional acrylate dispersion (solids content for example 50 %), for example "Mowilith" LDM 7714, instead of an extremely fine dispersion.

Furthermore, also wood preservatives (biocides) or transparent iron oxide preparations can be included for the transparent coating structure.

Figure 3.150: Schematic formulation for water-borne transparent primers

development. Generally, one has to pay attention to a 'hydrophobic' formulation as possible. Also, the solid content of the primer should not to be too low. Mostly, the values are about 50 to 60 %. This exemplary recipe contains a cross-linking dispersion, so that one usually may dispense with the addition of functional pigments such as zinc oxide. If zinc oxide is applied in primer formulations, it should be considered that this pigment has to be pre-stabilised or neutralised very well in order to prevent compatibility problems with polymer dispersions such as thickening during the storage. The insulating primer even may contain special algaecides and fungicides as a film protection in order to impart a sufficient film protection also to the subsequent coating structure, and thus to minimise the risk of an unsightly surface change during weathering due to a possible growth of algae or fungus growth.

Intermediate coatings

In industrial applications, particularly in the window construction, usually so-called intermediate coatings are applied for qualitatively high-value coating constructions which provide a long-term guarantee of the total window construction including the coating construction of up to 10 years depending on the type of wood and application. The intermediate coatings are formulated somewhat richer in solids (transparent pigmented usually between 20 and 30 %, opaque pigmented between 30 and 40 %) in comparison to primers and already often contain the topcoat resin. The aim of this is the completing of those natural defects in the wood which were not fully covered by the primer in order to ensure an additional security for the adhesion or water resistance of the overall construction, respectively.
The intermediate coating mainly fulfils three functions:

- Colourless or very low pigmented intermediate coating is chosen for glazes. Thus, the looping through of the primer (main colouring) is prevented for slight grinding.
- The grinding exposure is reduced significantly.
- A good pore filling takes place for porous woods. In coniferous woods, a sufficient layer of the paint film also is formed in end-grain areas, V-groove, corner joints as well as in sprout connections.
- Also, topcoats applied in a thin layer mostly are applied as intermediate coatings. Intermediate coatings also may be non-pigmented if these are formulated separately.

Components	Parts by weight [%]
Acrylate dispersion or styrene acrylate dispersion, networking approx. 48 % 1)	650
Water	72
Wetting agent/dispersing agent	3
Titanium dioxide 2)	125
Thickener (PU thickener)	2
Anti-foaming agents	3
Calcium carbonate 2 µm	125
Preservative agent	2
Solvent	18
	1000

1) for example "Mowilith" LDM 6151 (producer Celanese Emulsions GmbH)
2) for example "Kronos" 2059 (producer: Kronos)
The formulation even may contain a film protection and zinc oxide depending on the applied type of dispersion

Figure 3.151: Schematic formulation for water-borne pigmented primers with blocking function against wood ingredients

Clear, translucent pigmented or opaque pigmented topcoats (wood glazes)

The topcoats practically are the weather protection shields for the wood. Thus, the topcoats also are referred to as weather protection coatings. In addition, the coating has to prevent a too rapid uptake of rainwater by the wood for example. Furthermore, the coating shall prevent that the wood humidity rapidly is adjusted to the ambient humidity, and that the destructive UV light achieves the wood. Depending on the location of the wood component, the topcoats are exposed to a more or less severe stress caused by the weather. A distinction is made between clear coats, transparent laser systems and opaque pigmented coating systems. The weather protection colours as a specialty are intended for painting applications, and thus will not be considered here. In addition to the technical requirements, the weather protective coatings also have a decorative function. The following requirements profile is valid for the topcoats:

- Good weather resistance
- Sufficient elasticity (even during hailstorm) and permanent elasticity (no embrittlement)
- Good adhesive strength and wet grip (even on ancient coatings, for example aged alkyd resin)
- Good water resistance (low swelling capability of the water, low whitening in the case of transparent systems)
- Sufficient water vapour permeability
- Good block resistance (for dimensionally stable components)
- Environmental friendly (compliance with the relevant guidelines)
- Permanent UV protection by preventing the lignin degradation
- Protection against destructive fungi (adjustment of the wood moisture on values below 20 %)
- Compatibility with sealants and sealing profiles
- Resistant to cleaning agents (alcoholic) and plaster (alkali-resistant)
- Uniform layer structure without micro-foam, if possible
- Universal colouring
- Easily renovated
- Favourable processing properties (for example low flow behaviour, foam-free application[463])
- Compliance with the accepted standards and regulations for the end-use application (for example wooden windows)

Clear or translucent topcoats (wood coatings)

It refers to non-pigmented, transparent or semitransparent wood coatings containing transparent, usually very fine pigments. Thereby, the glaze shall emphasise the grain as well as the appearance of the wood. The glaze also shall have a good protective function from the effect of the weather. Thus, the amounts of pigments are chosen such that the structure of the wood still is recognised on the one hand while on the other hand the UV protection (also supported by light stabilisers) still is sufficient. In practice, this may cause problems, if one selects still partly brighter, very weakly implemented or colourless wood stains due to optical aspects. If the glaze has an insufficient UV absorption, the underlying wood is attacked and destroyed.

Fine-particle, micronised iron oxide pigment usually is applied for transparent pigmented wood stains. Clear glazes require an addition of UV absorbers (see also Chapter 3.2.2.3 and 3.2.2.4). Wood stains can be divided in thin film coatings and thick film coatings depending in the amount of solids (Table 3.69). Here, the new BFS leaflet no. 18 distinguishes between impregnating solutions (dry film thickness $\leq$ 5 microns) and film formation glazes (dry film thickness $\geq$ 5 microns).

Table 3.69: Schematic formulation of water-borne wood coatings

Components	Thin film coatings proportions of active compounds in parts per weight [%]	Thick film coatings proportions of active compounds in parts per weight [%]
Binding agent solid	10–20	25–45
Solvent	0–10	0–10
Pigments (transparent)	0–3	0–3
UV absorber	0–3	0–3
Additives[1)]	0.5–3	0.5–5
Water	60–80	35–60
Solid content	< 30	30–60

1) Thickener, levelling agent and substrate wetting agent, if necessary film preservative, matting agent, defoaming agent, wax emulsions

Thin layer coatings

Thin layer coatings have a low content of solids and a low viscosity. Usually, dry film thicknesses of about 25 microns are applied. Thin film coatings are not suitable for dimensionally stable components as a final coat, since these coatings provide only an inadequate protection against moisture due to the small layer thickness. The film layer coatings penetrate deeper into the wood than thick film coatings. Thin film coatings are applied on fences, for example. The thin layer coatings may contain wood preservatives depending on the application.

Solvent-borne thin layer coatings

Alkyd resins which often also contain wood preservatives are applied as resins. The alkyd resins are less suitable for industrial applications.

Water-borne thin layer coatings

Preferably fine-particle polyacrylate dispersions or also water-borne alkyd resins or combination of both, respectively, are applied as resins. The application of polyacrylate dispersions is referred to as impregnation coating. Also, impregnation coatings are not applied in the industrial area.

Thick layer coatings

Thick layer coatings have a high proportion of resins. Thick film coatings are adjusted to a higher viscosity in order to achieve wet film thicknesses of 300 microns at least in industrial application processes (such as airless spraying). Thick layer coatings mainly are applied on dimensionally stable components such as doors and windows. For this, dry film thicknesses of at least 80 microns for coatings and 100 microns for opaque systems are necessary (Rosenheimer Guidelines, HO leaflets). This only is possible with a thick-film spray application.

Solvent-borne thick layer coatings

Solvent-borne thick layer coatings have only a minor role in the industrial sector. Thus, details of the recipe are not described here.

Water-borne thick layer coatings

Water-borne thick layer coatings based on polyacrylate dispersions feature a good moisture protection. Due to the enhanced viscosity, the water-borne thick film coatings penetrate less into the wood and thus may feature adhesion problems when directly applied to the wood. For this reason, a primer always should be applied prior to the application of the thick film coating. In order to ensure a sufficient adhesion, dispersions are used in thick layer coatings. The polymer scaffold of these dispersions contains components which improve the adhesion in order to guarantee a sufficient adhesion (wet adhesion promotor) [464]. This particularly is important in the case of refurbishment coats so that a good adhesive strength on old coatings of alkyd resins is given, for example.

A good block resistance of the coating is important for the industrial application. For this reason, usually polyacrylate dispersions with a multi-phase polymer structure (core-shell dispersions) are applied. Combinations of such dispersions with water-borne polyurethane dispersions (partially polymerised into the polymer) or with water-borne alkyd resins also are used for this purpose. Occasionally, also dispersions with a single-phase polymer structure (homogeneous dispersions) or appropriately combined (for example as in renovation paints) are applied. In order that the thick film coatings also feature a good durability, in comparison to thin film coatings, the thick film coatings must have a pronounced elasticity in order to give in to the occurring stresses and strains during the exposure of the wood to the weather. This particularly is important for a sufficient flexibility to hail. The application of too hard or too inflexible polyacrylate dispersions may result in a cracking especially at corners and edges or cracks in hailstorms. Heterogeneous dispersions require a balanced compromise between elasticity and block resistance. Modern polyacrylate dispersions have thus properties.

The minimum film forming temperature of the applied heterogeneous dispersions is as low as possible (usually below 10 °C) in order to have a smallest possible addition of film-forming agents since the film-forming agents may impact the early block resistance negatively. For example, dipropylene glycol n-butyl ether, butyl glycol and small amounts of texanol are applied as film-forming agents.

For example, propylene glycol can be added in order to adjust the sprayability and improvement of the spray pattern. Waxes, mostly in the form of wax emulsions, may be incorporated in order to improve the hydrophobicity, mechanical resistance and block resistance. However, disturbances in the reworking or for reducing the intermediate adhesiveness may result (for example by excessive flowing of the wax to the surface) under unfavourable circumstances or at a too high amount of waxes. Thickeners should be applied only in moderation in order to improve the rheology (adjustment of the flow behaviour, adjustment of the yield point in the airless spraying process) and to improve the edge coverage, since otherwise the thickeners make the total system too hydrophilic. This may result in an increased whitening of the transparent coating, for example.

Transparent iron oxides are applied for the pigmentation, since these iron oxides offer the best protection against UV radiation. The pigment content should amount approx. 1.5 to 2 % referring to the coating. This corresponds to approx. 6 % of a commercial pigment preparation. From this, a pigment concentration of 2 g/m^2 results at a coating application of 100 g/m^2 per layer. The measurement of the UV absorption in such a coating results in an absorbance of 60 % at a depth of 15 microns, in an absorbance of 70 % at a depth of 25 microns and in an absorbance of 100 % at a depth of 50 microns. The addition of an UV absorber is recommended for lower contents of iron oxides or bright colours. Here, a content of

1 to 3 parts by weight is required. Nanoscale pigments such as transparent titanium dioxide partly are applied here in order to achieve a further improved UV absorption (Figure 3.152).

Opaque pigmented topcoats

This term describes coatings which are opaquely pigmented white or coloured tones. A distinction is made between solvent-borne and water-borne topcoats. Here too, the solvent-borne coatings become less important for the industrial application. Thus, solvent-borne coatings are discussed only briefly.

Solvent-borne opaque pigmented topcoats

The solvent-borne and opaque pigmented topcoats mainly are formulated using alkyd resins. Mostly of them are high solid types. These topcoats have an increasingly marginal significance for industrial applications and should not be discussed in further detail here.

Water-borne opaque pigmented topcoats

Similar to the thick layer coatings, the coatings contain products similar to resins, since here the requirements in particular according to the elasticity and block resistance are similar. However, here one may have a broader range of selection due to the pigment content or filler content, respectively, since for example the block resistance can be adjusted more easily by using the pigmentation. For this reason, in addition to acrylate dispersions with multi-phase polymer structure (such as core-shell dispersions), also acrylate dispersions with a single-

Components	Parts by weight [%]
Component 1	
Core Shell acrylate dispersion approx. 50 % [1)]	740.0
Ammonia, 25 %	2.0
Component 2:	
Water	93.5
Matting agent (fumed silica)	15.0
Defoamer/deaerator	4.0
Wetting agent/dispersing agent	3.0
"Dowanol" DPnB	20.0
Methoxy butanol	20.0
Thickener 1: alginate or xanthan gum	0.5
Thickener 2: associative PU or acrylate thickener	5.0
Preservative agent	2.0
Component 3:	
Transparent iron oxide preparation [2)]	40.0
Water	15.0
Wax emulsion (30 %)	40.0
	1000.0

Component 2 is separately dispersed (dissolver) and then stirred into the component 1. The component 3 is admixed little by little by stirring.

1) For example "Mowilith" LDM 7416 (Manufacturer: Celanese Emulsion GmbH)
2) For example "Luconyl" Orange 2416 (Manufacturer: BASF SE)

A film protection agent may be contained optionally.

Figure 3.152: Schematic formulation of a water-borne wood coating for wooden windows, airless sprayable

phase polymer structure are applied which also contain wet adhesion promoters or cross-linking agents, respectively. Also, combinations of both types or combinations of such dispersions with water-borne polyurethane dispersions (or polyurethane containing copolymers) are possible. The advantages of the acrylate dispersions are:
- Good weather resistance
- Low tendency of yellowing
- High water resistance
- High film hardness with sufficient elasticity whereby a sufficient resistance to mechanical damages is given
- Low surface adhesion

The formulations also may contain water-borne alkyd resins as a combination partner. Thus, a slightly better ability to renovate is given on the one hand, while on the other hand a somewhat stronger yellowing may occur.

The coatings provide a high UV protection due to the opaque pigmentation. The pigment volume concentration is in the range between 10 and 25 %, while the content of the non-volatile components is between 40 and 50 %. Only a little amount of "free" water is available for the coating dispersion due to the high percentage of dispersion. The minimum film forming temperature or the glass transition temperature, respectively, of the applied conventional acrylate dispersions are around 20 °C. In the so-called core-shell dispersions, the minimum film forming temperature also can be 0 °C. An application of film-forming agents is required in every case. Texanol, dipropylene glycol n-butyl ether, butyl diglycol as well as butyl diglykol acetate proved as the most active compounds. All compounds remain often long in the film. This can become problematic if an earliest reduction of the surface adhesiveness is desired. In this case, highly volatile and water-miscible solvents such as methoxy butanol or butyl glycol are applied. However, the exclusive use of water-miscible solvents for application on porous surfaces may create problems as these penetrate into the substrate together with water and are no longer available for the film formation. In dispersion coatings (2 to 10 %), the solvents are not only used as film-forming agents, but also as a rheology modifier and for regulating the open time during spraying. A disadvantage is their block tendency in the first drying phase. The eco-label *'Blauer Engel'* can be received by means of small amounts of film-forming agents.

If matting is required, the usual matting agents or appropriate fillers can be added. Subsequently to their production, the coatings can be adjusted to a pH value between 8 and 9 in order to obtain a sufficient storage stability as well as to obtain an optimal gloss in the case of glossy coatings. Ammonia or 2-amino-2-methyl propanol are the most suitable neutralising agent since these agents result in a high gloss and in a good water resistance. Other neutralising agents such as sodium hydroxide or potassium hydroxide behave unfavourably. Waxes, mostly in the form of wax emulsions, can be incorporated in order to improve the hydrophobicity, mechanical resistance and block resistance. However, this can result in disturbances in the over-coatability and intermediate adhesion under unfavourable conditions or at too high contents of waxes.

For the coating of wooden windows, the formulation is modified as required by application of fillers or matting agents and according to the drying time (reduction of the amount of propylene glycol). These coatings also may contain so-called functional pigments in order to improve the barrier effect of the entire system structure.

Special coatings for joint protection, joint filler, end-grain protection

The parapet joint of V-groove is the critical area of a wooden window. The main reason for this is a gluing which mostly is not performed all-over the surface as it is required by the regulations since resin spots have to be avoided. During spray application of intermediate coatings or topcoats, the V groove only is coated insufficiently due to the rebound effect of the coating material. The manufacturers of window coatings have developed products for the critical points at wooden windows in order to have an additional protection against any penetrating moisture (infiltration of moisture and water behind the critical points) at sub-optimal implementation of the window construction to the possible defects. These products were developed for water-borne coating structures since these mainly are applicable, as described.

An elastic joint protection based on acrylate dispersions or acrylate PU dispersions exists for open parapet joints or V-groove or exposed end-grain areas, respectively. This material is a type of high viscous, elastic and colourless thick film coating which is injected from a cartridge, for example, to the joint sealants in the joint. This joint protection has a hydrophobic effect and usually is non-pigmented. Furthermore, these materials also can be applied to protect any possibly exposed end-grain. These materials also are referred to as end-grain protection. Also, materials with a lower viscosity is offered as an end-grain protective agent for applications by brush. Subsequently to the drying process, these protective materials should be coated as soon as possible with the respective coating system. So-called joint fillers mostly based on acrylate dispersions exist as a second special product.

These fillers are applied for an elastic sealing of crack joints, cutting kerfs and mitre joints, for example. There are partially different formulations: with more rapid drying (easily filled, thus less transparent) on the one hand, and with somewhat slower drying (unfilled, thus very good transparency) on the other hand. Depending on the type of wood and topcoat (transparent or pigmented), the specific material then is selected. It should be noted, that these joint fillers are not suitable for connection joints as these joint fillers are less elastic.

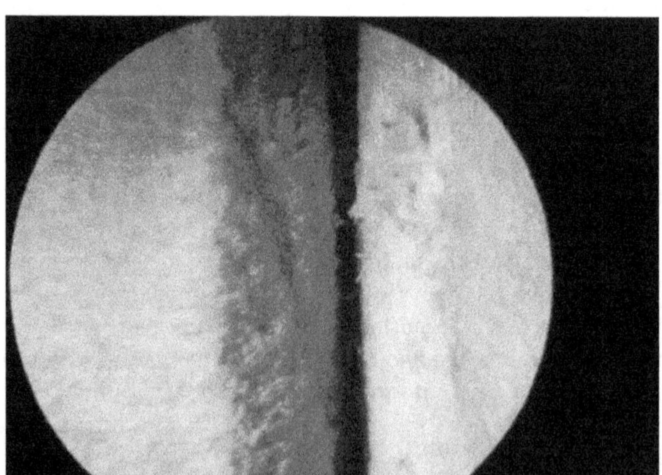

Figure 3.153: Insufficient layer formation above a V-groove

3.2.3.2 Examples of application for the industrial coating of windows and front doors

In the factory coating of wooden windows, for example, one has to distinguish between pre-coated windows with a final coating on the object and the factory finished coated wooden windows. The first procedure almost exclusively is prevailed in The Netherlands. In the second procedure, one distinguishes between a conventional manufacturing of wooden windows or a coating of a finally produced window framework, respectively, and the production as well as a complete coating of individual parts of windows which are assembled subsequently to the coating process. The manufacturing and coating of the entire window framework is the commonly applied method. The method of single-item production is applied only in special cases partly due to the technology which is not fully developed yet. In any case, a rational coating process mainly depends on the conveyor technology used. Here, the application of overhead monorail systems or circular conveyors in craft-based industries even is quite reasonable. In the case of an industrial coating of windows, the plant manufacturers often recommend a so-called Power & Free conveyor system even at relatively low flow rates. Despite higher investment costs, Power & Free conveyor systems belong to the standard equipment of transport in the sector of coating of windows due to their advantages. Thus, the main advantages are the flexibility or the ease of implementation of a distribution process for variable processes (for example the starting-up of different flow coating zones or two passes through spray booths with double application of the topcoats), respectively.

A distinction is made between the coating systems depending on the recommendation of the manufacturers. The four-layer structure (impregnation and three coatings) features a much smaller water uptake even after 40 months of outdoor exposure at pinewood as a two-layer structure (primer and topcoat) or three-layer structure (impregnation, primer and topcoat). Thus, the two-layer structure primarily is suitable only for wood-aluminium windows.

Flow coating line combined with a manual spraying station and recycling system [465]

The coating concept presented in the Figure 3.154 was implemented at a manufacturer of windows which predominantly is focused on an object-related individual production of windows. It consists of two separately operating systems for flowing and spraying and is designed for maximally 90 frames per layer. As indicated in the Figure 3.155, the flow coating line equipped with a flow coating zone already is prepared for the retrofitting of a second flow coating zone. After flowing, the work pieces pass a dry buffer, where they are dried at indoor climate. Optionally, a second or third flow coating cycle can be performed. For grinding, the dried parts are removed from the traverses and subsequently applied to the traverses of the spray coating line. This includes two manual spraying stations with a dry precipitation as well as a circulating-air dryer. For paint recovery, two cooling walls were installed whereby these cooling walls can be moved on rails in front of the spray wall. Optionally there is the possibility to perform the coating process also without recovery if the coating process has not been considered profitable due to an insufficient lot size.

Flow coating combined with a single cabin solution and "Prolac" system [465]

Figure 3.155 illustrates a plant concept which is installed in a company which has evolved from its handicrafts-based roots. This plant concept consists of a flow coating line as well as a coating line. The flow coating line consists of three throughput flow zones for white and translucent primers as well as for colourless impregnations or colourless primer fillers. Furthermore, the flow coating line consists of a dryer and a power & free conveyor system.

The window coating plant has a capacity of 150 finished windows à 2 frames per layer. It should be noted that always it is flowed twice before the top coating process: For

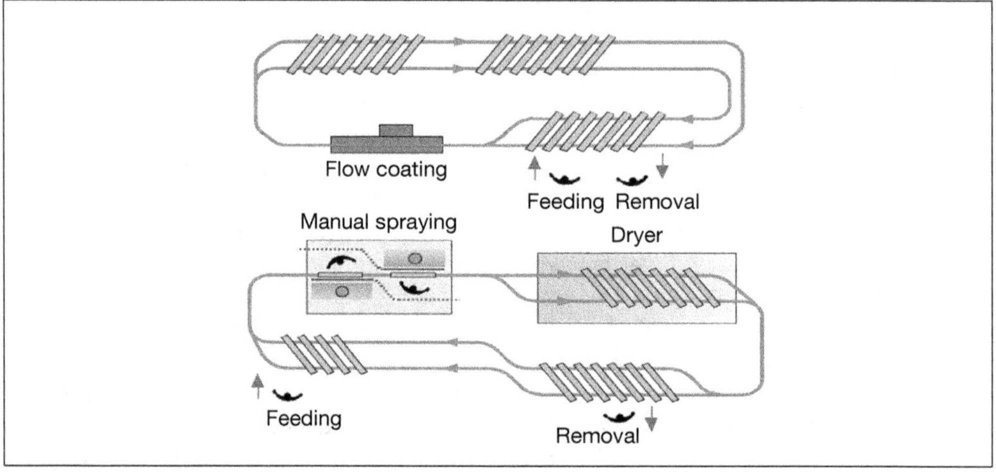

Figure 3.154: Industrial coating of window in a flow coating line

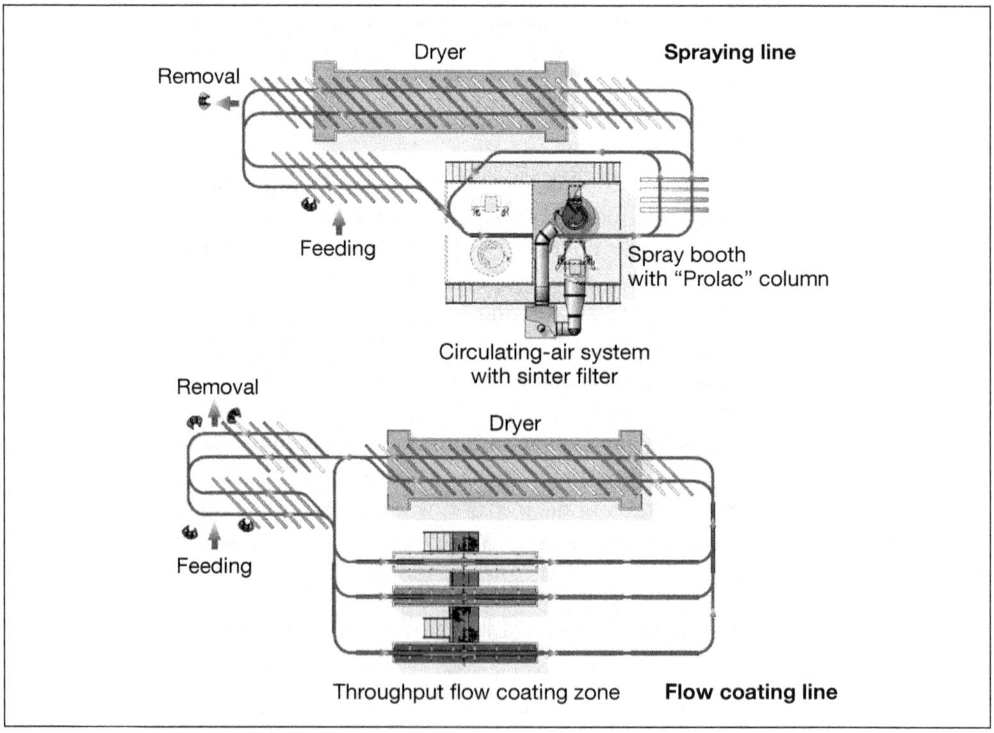

Figure 3.155: Industrial coating of windows by flow coating and spraying[465]

soft woods, there exists a flow coating one each in impregnation and primer. For meranti, primer and filler are flowed successively prior to the application of the topcoat. The plant operator expects a better wood protection and lower the grinding expense due to the multiple flowing.

Combination of flow coating and spaying systems [465]

The following practical example presents a further concept with overspray recycling. The investment made here is even more extensive since the operator emphasizes the flexibility in order to implement special customer requirements rapidly. The approx. 300 m long Power & Free system which is equipped with a comfortable control connects the flowing section with the spraying section whereby the flow coating area is equipped with two optionally retractable throughput flowing zones, evaporation line and air circulation dryer. Two manual spray dispensers which are intended for customer parts, automatic cabins equipped with "Prolac" columns and an opportunity for a manual pre-spraying as well as an evaporation line and a dryer belong to the spray section. The route of the workpieces is programmed within this task. Thus, one may determine which of the both flow coating sections has to be applied. Furthermore, it has to be determined whether one has to transport to the grinding at the hanger or to the intermediate inspection with a separate grinding. The front and back side top coating optionally is performed in the manual spray section or automatic spray section. It is also possible to program a new run for the second topcoat application.

Electrostatic industrial coating of wooden windows [467]

Industrial coating of wooden windows by combination of a flow coating method with an electrostatic spraying process (White primer – non-resistant wood meranti). Here, an example from The Netherlands will be discussed. In the Netherlands, the windows usually are delivered only with an industrial pre-coating (primed). Subsequently to the installation, the painter performs the final coating. The window frames previously are impregnated separately with a wood preservative. For this described coating process, two white pigmented primers are applied in two individual operations. Thus, the first primer contains an equipment against blue stain according to the standard EN 152.1 while the second primer functions as a barrier against wood ingredients. The impregnated window frame thus is mounted into a Power Free conveyor system. It is transported into a flow-coat plant. In a first step, it is flowed by using 32 nozzles and a pressure of 1.5 bar each. Thus, a complete coating is ensured, including the grooves and seams. In the evaporation zone, the workpieces are brought into an inclined position by means of a unilateral lowering of the traverses so that excess coating material may run-off. Then, the run-off material is collected for recovery. In the evaporation zone there is a relative humidity of about 85 % in order to achieve a consistent run-off and levelling of the primer or a sufficient wetting of the wood pores. The throughput takes approx. 18 minutes. Then, the drying is performed in a convection oven at a temperature of 35 °C as well at an air velocity of approx. 0.5 m/s. Subsequently, it is grinded, and the components are conveyed to the spray station. Two identical painting lines with an evaporation zone and exhaust dryer are operated due to the required throughput of 300 components daily. The dry film thickness of the second primer is about 60 microns. A fully automatic coating line is applied. Thus, an automatic product recognition as well as a process control of the electrostatic plant takes place. In order to increase the electrical conductivity, the window elements are sprayed with water joining the spray booth. The system is controlled by means of an infrared contour detection. Respectively six electrostatic airless spray guns then spray with an optimal wrapping-around of the coating. Spray processes supported by the airless avoid the generation of the Faraday effect. Both sides of the window components are coated in succession. The plant is designed by three colour supplies and the option of detaching the high voltage in such a way that frequent colour changes are possible. Subsequently to the coating process, the workpieces pass through the evaporation zone and then are dried in a convection drier. Then, most of the primed windows are mounted in the building shells subsequently to the final assembly. After completion of the structural work, the primed windows are provided with a topcoat. Figure 3.156 features this spraying process in a simplified manner.

Windows coating white pigmented for large series production; continuous coating process [466]

This plant is designed for coating for large series production and primarily is used for the application of white pigmented intermediate coatings and topcoats. The plant is designed for a continuous coating process (Figure 3.158).

The correspondingly impregnated and primed window frames then are put into the plant and subsequently coated in a central coating booth each alternatively by the flow method (intermediate coat) or spraying method (topcoat). The advantages of this process are:

- The colour application takes place at the vertically hanging and respectively rotating workpiece
- The individual part obtains an all over coating also of all end-grain zones
- The central coating booth enables an assembly throughput
- Direct workpiece coating by means of an automatic control of the flow coat nozzles and spray nozzles
- Material cycle with low application amount and a careful load; there a minimal effort in colour change
- Recycling of all oversprays for example by means of the "Coolac" method
- Central control of the humidor climate (incl. evaporation zone)
- Forced drying by adsorption

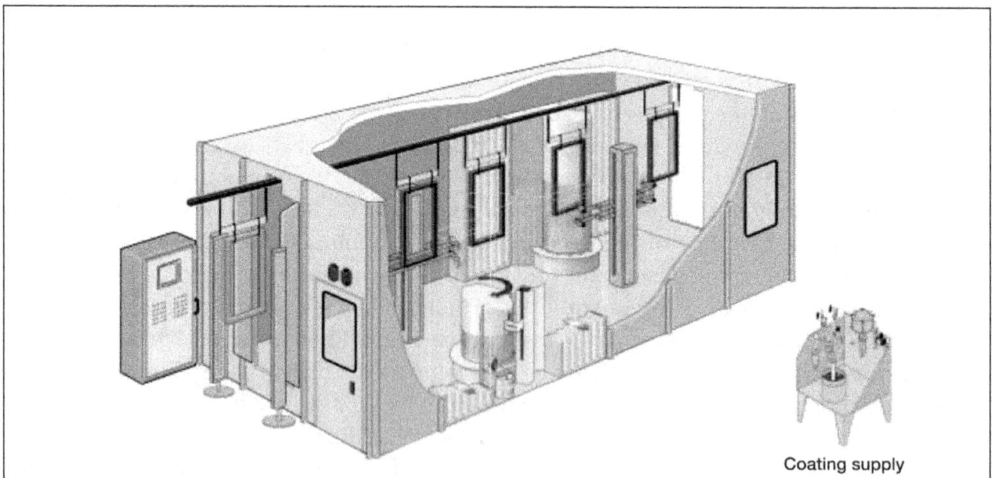

Figure 3.156: Industrial window coating with electrostatic spray method [467]

Windows coating for small series [471]

This plant is designed for the optional implementation of experiment series and contract manufacturing of small series (see Figure 3.158). The manual spraying station for applying the topcoat is provided with an electrostatic spraying system. The overspray still is relatively high due to the different forms of wooden windows. A "Teflon"-coated cooling wall has been installed in order to keep the loss of coating as low as possible. Then, the arising overspray is scraped off manually. If a higher throughput is desired, one may convert to the already above-mentiond "Prolac" column. The non-collected overspray is removed from the exhaust

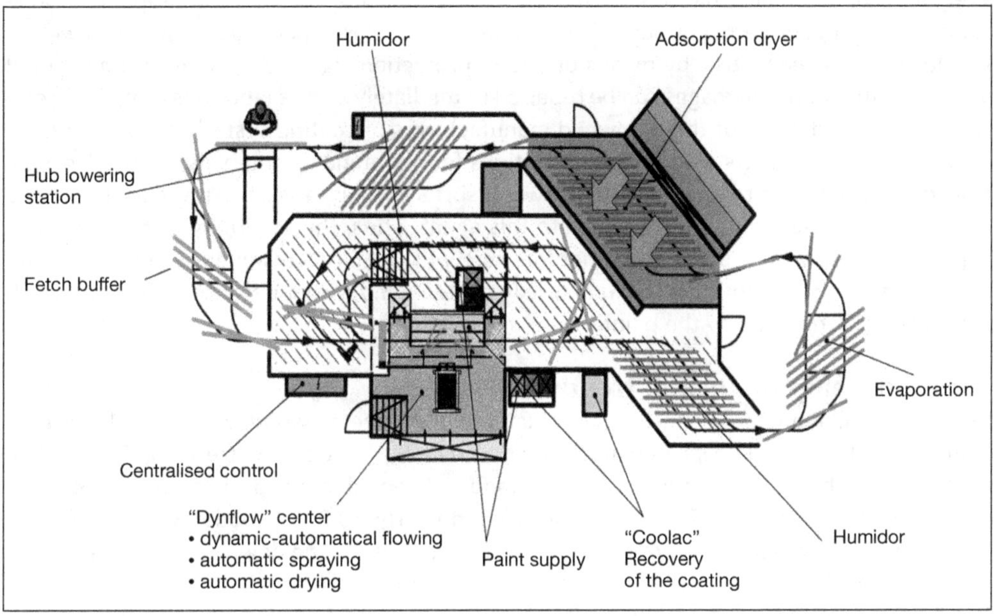

Figure 3.157: Industrial coating of windows for large number of items *Source: Plant of Oberflächencenter*

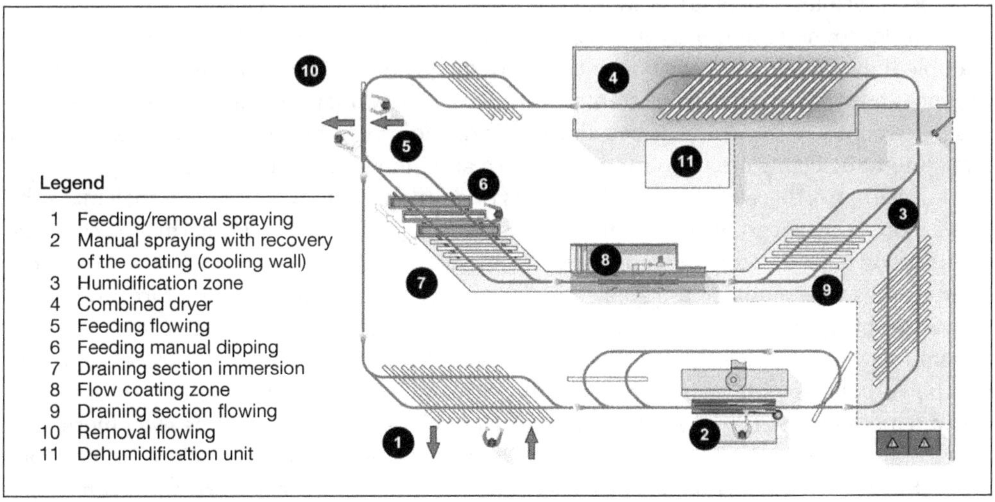

Figure 3.158: Variable window painting line with manual spraying station and flow coat system [471]

air by means of an effective ventilation system (injection of conditioned fresh air) with dry deposition. The drying process takes place in a convection dryer which optionally may be operated with dehumidified air.

3.2.3.3 Maintenance and Care

Despite the numerous regulations, today the quality of the wood often leaves something to be desired. The focus is on windows consisting of pine wood with a high proportion of sapwood and meranti quality is better suited for cigar boxes as for windows. Whether connected to industrial plants or top coated by brush on the object, the durability of the coating requires a regular care and maintenance [470]. This will ensure that small damages are repaired immediately before major damages occur. The painter should close open parapet joints as well as V-groove during renovation by means of a joint protection signet. A service contract would be ideal so that minor damages can be repaired immediately before laborious remedial measures are required. Most of the renowned manufacturers of coating systems for windows offer such service contracts with recommendations for maintenance intervals. A so-called 'care milk' has proved for a regular care of the wood surface (twice a year). After cleaning of the window, the frame surface is rubbed with this agent. Thus, the durability of the surface is improved. These special products do not interfere with a later refreshment paint. These care products are suitable not only for industrially coated windows but also for windows which are coated or renovated by the painter.

Future developments and trends

Nearly 126 million window units (frame material: PVC, metal, wood, wood/metal) were required in 2014 in the European window market (EU 28, Norway, Switzerland, Turkey, Russia, Ukraine). The market share of wooden windows (wood and wood/metal) has steadily decreased in recent years; in 2013, it still amounted nearly 20.2 % (wooden windows frame) and nearly 6 % (wooden frame/metal frame). For some years, the downward trend seems to be stopped as the wooden windows is increasingly being chosen again by the consumers, in particular due to the long-term guarantees and the improved quality concept combined with the beautiful optics. A low gloss of the wooden surface is always in the trend in the coating of wooden facades, balconies as well as of wood panelling.

In order to increase the acceptance of wood by consumers for outdoor applications, for example the wooden windows or the corresponding coatings, respectively, are marketed in combination with a long-term guarantee as well as in combination with a windows passport or service contract, respectively (see above). More and more obsolete, manually operated or semi-automatic plants are retrofitted to modern and innovative coating lines.

Furthermore, there exist research projects or development projects which shall make the wood more resistant to water and fungi by means of a corresponding thermal or chemical modification such as acetylation or by the DMDHEU process. It remains to be seen to what extent woods thus modified are applied in the window construction in the future. Compared to the previously used solvent-borne coatings, modern, qualitatively high value low-solvent or solvent-free water-borne coating systems have been developed which can be processed by application processes with a higher application efficiency. Here, the flowing coating process will become more efficient [468]. Especially, the recovery of the overspray will be even more optimised in the future in order to have a circulation of the coating material as closed as possible.

Actually, the resin technology provides trends in optimising the performance with regard to the block resistance and higher elasticity. The reduction of the foaming of water-borne systems as well as a further improvement of the UV protection due to nanoscale products let expect interesting products in the future. It is important that the coating systems and the wooden windows are developed in partnership with the coating manufacturers and plant manufacturers. There are challenges for the industries. Since the year 2016, the first self-repairing coating systems for windows were offered on the market. These are microcapsules which are integrated in the coating system. When the surface of the coating is damaged by hailstorm for example, the capsules burst and the liquid contents leak out. This seals damages and prevents damages such as exfoliation, blistering or discharge of tannic acid. This is an interesting starting point to further increase the durability of wood coating systems [473]. For a couple of years, the "Hot Coating Technology" has been discussed in the coating of window profiles since a very high flexibility and resistances in the overall coating can be achieved by the use of special components of a polyurethane hot melt. This technology may open-up new possibilities in the sheathing of window profiles for outdoor applications [474].

3.3 References

[1] Market study: THE WESTERN EUROPEAN MARKET FOR INDUSTRIAL WOOD COATINGS, Information Research (IRL), 5th Edition, April 2004

[2] Asher, J.: Industrial Coatings – A time of change, Financial Times, Management Report, London, 1993, p. 27

[3] Brenni, S.: Wood Coatings must adapt new technologies, The Coatings Agenda 1994, p. 105

[4] unknown: Holzlacke in Westeuropa, Welt der Farben, September 1994, p.14–17

[5] unknown: Industrially applied coatings for wooden furniture, board, doors etc., excl. Joinery, Presentation DSM, 2006

[6] Paulus, W.: Status of UV/EB in Europe, Radtech Europe Conference in Barcelona 2005

[7] Prieto, J., Kiene, J.: Oberflächenbehandlung von Holz im Innenbereich, Lehrbuch der Lacke und Beschichtungen, Kittel Band 9, p. 319–351, S. Hirzl Verlag Stuttgart, Leipzig, 2004

[8] Sappel, M.: Holzbeizen – Ein Vergleich von historischen und modernen Beizen, Vorträge am Mittwoch, Informationsblatt 30, Freilichtmuseum Hessenpark, 04.05.2005

[9] Michaelsen, H., Buchholz, R.: Vom Färben des Holzes – Holzbeizen von der Antike bis in die Gegenwart, Michael Imhof Verlag GmbH & Co. KG, Petersberg, 2006

[10] unknown: Nitrocellulose als Bindemittel für Holzlacke bewährt, Holz-Zentralblatt, Sonderdruck, No. 122, 2002

[11] de Gaulmyn, G.: Wood coatings Industry in Europe – Is the decline a fatality?, Nitrocellulose in wood coatings Environmental and Safety Aspects, 21st Annual Conference of World Nitrocellulose Manufacturers – Hamburg 2002

[12] Hoppe, L.: Celluloseester, Chapter 1.4.2.4.2.2 in Kittel – Lehrbuch der Lacke und Beschichtungen, Vol 1, S. Hirzel Verlag Stuttgart Leipzig, 1998, p. 227–237

[13] Stoye, D., Freitag, W.: Cellulosederivate, Chapter 9.4.1 in Lackharze – Chemie, Eigenschaften und Anwendungen, Carl Hanser Verlag München Wien, 1996, p. 364–370

[14] Prieto, J., Kiene, J.: Holzlackieren, Chapter 9.2.5 in Kittel – Lehrbuch der Lacke und Beschichtungen, Band 9, S. Hirzel Verlag Stuttgart Leipzig, 2004, p. 317–346

[15] Müller, B., Poth, U.: Lacke auf Basis von Cellulosenitrat, Lackformulierung und Lackrezeptur, Vincentz Network, Hannover, 2003, p. 76–81

[16] Poersch-Panke, H.-G.: Walsroder Nitrocellulose für edle Oberflächen, Firmenschrift der Wolff Cellulosics GmbH & Co. KG

[17] Vollhardt, K.P.: Kohlenhydrate, Chapter 23 in Organische Chemie, VCH Verlagsgesellschaft mbH, Weinheim 1988, p. 1096–1097

[18] Karsten, Lückert, O.: Lackrohstofftabellen, 10. Edition, Curt R. Vincentz Verlag Hannover, 1996

[19] Atkins, P. W.: Physikalische Chemie, VCH Verlagsgesellschaft mbH, Weinheim, 1990, p. 187–188

[20] Olsen, J. H.: Säurehärtende Lacke – Möglichkeiten und Entwicklungen, Umweltfreundliche und emissionsarme Möbel, WKI-Bericht No. 31, 1995, p. 9–18

[21] Roth, G.: Einsatz von Reaktivlacken zur Beschichtung von Holzwerkstoffen, Umweltfreundliche und emissionsarme Möbel, WKI-Bericht No. 31, 1995, p. 19–30

[22] Anonym: Krank durch „Billy", Stern No. 45, 1992

[23] Lengsfeld, W.: Über die Strukturbutylierter Aminoplaste, Farbe und Lack, 75. Vol., No. 11, 1969, p. 1063–1065

[24] Ott, J.: Chapter 6.2. Amido- und Aminoharze, Stoye/Freitag – Lackharze, Carl Hanser Verlag München Wien, 1996, p. 104–126

[25] Kjellquist, K., Rassing, J., Wesslén, B.: Surface Reactive Acetal Functional Waterborne Microparticles. XIXth International Conference in Organic Coatings Science and Technology, 281–298, 1993

[26] Klug: Was wissen wir von säurehärtenden Lacke? 27. Essener Diskussionsabend, dafazet, 32. Vol. No. 1, 1978

[27] Bayer, O.: DRP 728 981 (1937)

[28] Krauß, W., Dünnwald, W.: Chapter 2.2 Polyaddukte, Vol. 2: Bindemittel für lösemittelhaltige und lösemittelfreie Systeme, Kittel – Lehrbuch der Beschichtungen, S. Hirzel Verlag Stuttgart, Leipzig, 1998, p. 193–265

[29] Müller, B., Poth, U.: Lackformulierung und Lackrezeptur, 2. Edition, Vincentz Network Hannover, 2005, p. 98–111

[30] Wieczorrek, W.: Polyisocanatharze, Lackharze – Chemie, Eigenschaften, Anwendungen, Hrsgb.: Stoye, D und Freitag, W., Carl Hanser-Verlag München Wien, 1996, p. 183–229

[31] Mondt, J.: Reaktionskinetische Untersuchungen von Zweikomponenten-Acrylat-Systemen am Beispiel Macrynal/Isocyanat. In: Kundeninformationen, Fa. Hoechst

[32] Prieto, J., Kiene, J.: Oberflächenbehandlung von Holz im Innenbereich, Lehrbuch derLacke und Beschichtungen Vol. 9, Hirzel Verlag Stuttgart, 2004, p. 325–350

[33] Kittel, H.: Lehrbuch der Lacke und Beschichtungen, Vol I/2, W.A. Colomb Verlag, 1980, p. 778–793

[34] Goldschmidt, B. Hantschke, E. Knappe, G. Vock, Glasurit-Handbuch Lacke und Farben, 11. Edition, Curt R. Vincentz Verlag, Hannover1984

[35] Prieto, J., Kiene, J.: Lehrbuch der Lacke und Beschichtungen, Kittel Vol. 9, p. 319–351, S. Hirzel Verlag Stuttgart, Leipzig, 2004

[36] Deutsches Lackinstitut: Klavier- und Geigenlacke treffen glanzvoll den guten Ton, Informationsdienst Deutsches Lackinstitut No. 61, Dezember 1999

[37] Prieto, J.: Moderne UV-härtbare Möbellacke, I-Lack 67, p. 584–588, 1999

[38] Müller, M., Ungesättigte Polyesterharze, Lackharze, Carl Hanser Verlag München–Wien, 1996

[39] Mlezivs, Novak, Kunststoffe 9, 1984

[40] Tranckner, H-J., Pohl, H. U., Untersuchungen zum Trocknungsmechanismus lufttrocknenderungesättigter Polyesterharze, Die angewandte Makromolekulare Chemie 108, p. 61–78 (No. 1751), 1982

[41] Riese, W. A.: Zur Formulierung ungesättigter Polyesterharzlacke, Farbe und Lack 1966, 72. Vol., No. 2, p. 140–148

[42] Gebhardt, W., Herrmann W., Hamann, K.: Die Verdunstung des Monostyrols aus Lacken auf Basis ungesättigter Polyester, Farbe und Lack 1958, 64. Vol., No. 6, p. 303–309

[43] Selbeck, H., Dhein, R., Rudolph, H., Reuter, K.: Enolisierende Beta-Ketoverbindungen als Zusatzbeschleunigerin ungesättigten Polyesterharzen, Farbe und Lack 1979, 85. Vol., No. 1, p. 14–17

[44] Demmler, K., Schlag, J.: Das Verhalten der Radikale in gehärteten Polyesterharzen, Farbe und Lack 1971, 77. Vol., No. 3, p. 224–231

[45] Demmler, K., Schlag, J.: Über die Kalthärtung ungesättigterPolyesterharze mit dem System Benzoylperoxid/tertiäres aromatisches Amin, Source unknown

[46] Demmler, K.: Derstrukturelle Aufbau von Polyesterlackfilmen und sein Einfluss auf die Eigenschaften, Farbe und Lack 1966, 72. Vol., No. 10, p. 971–981

[47] Technisches Informationen: GLOSSALL, Fa. Georg M. Langer& Co., Ritterhude, Bremen

[48] Novak, J., Mleziva, J., Eichler, J.: Einfluss von Inhibitoren auf die Lagerfähigkeit von ungesättigten Polyesterharzen, Die Angewandte Makromolekulare Chemie 128 (1984), No. 2023, p. 123–132

[49] Elias, H-G.: Makromoleküle, Vol. 1 Grundlagen, Hüthig & Wepf Verlag Basel – Heidelberg – New York, p. 528–538

[50] Alfrey, T., Price, C. C.: J. Polym. Sci., No. 2, 1947, p. 101–106

[51] Schoger, H.-A.: Synthese und Eigenschaften an aphiphilen Copolymeren mit periodischen Strukturen und deren Wechselwirkung an Grenzflächen, Dissertation, Universität Stuttgart, 2000, p. 19–24

[52] Firmeninformation der Fa. Menzerna über Industriepoliermittel, 2006

[53] Fries, K.-W.: Die Alternative zum Lack, Plastverabeiter, 52. Vol., 2001, No. 1, p. 24–25

[54] Kirsten, T.: Mehr PS für die Oberflächenbeschichtung, Pressebericht der Fa. Hennecke, 2001

[55] Ondratschek, D., Hruschka, R., Chatziapossidis, I.: Einsatz lacksparender, elektrostatischer Hochrotationsglocken beim Lackieren von Holz und Holzwerkstoffen, Forschungsbericht FZKA-BW-PLUS, Mai 2004, Universität Stuttgart

[56] Patent EP 0943637 A1, BASF AG, 1999

[57] Sauer, F.: Substitution of cobalt accelerators in unsaturated polyestersystems, Borchers GmbH, Monheim

[58] Steinert, A.: Der Ersatz von Kobaltsikkativen und Hydrochinonderivaten in Offsetdruckfarben, Borchers GmbH, Monheim

[59] Srna, C.: Problematik von Kobaltsikkativen, Farbe + Lack, 104. Vol., 8/1998, p. 10
[60] Flory, P., J.: Principles of Polymer Chemistry, Cornell University Press, Ithaca/London, 1973
[61] Barzynski, H., Penzlen, K., Volkert, O.: Photopolymerisation, Chemiker Zeitung, 96. Vol., 1972, No. 10
[62] McCloskey, Chester M., Bond, John: Photosensitizers for Polyester Vinyl Polymerisation, Industrial and Engineering Chemistry 1955, p. 2125–2129
[63] Charlesby, A., Wycherley, V., Greenwood, T. T., Proc. R. Soc. London, Ser. A 224 (1958), 54
[64] DAS 1.195.491, Glasurit-Werke, M. Winkelmann AG, 1963
[65] DAS 1.927.320, Bayer AG, 1970
[66] US 2.448.828, Du Pont, 1946
[67] DAS 1.297.269, Herbol-Werke, Herbig-Haarhaus AG, 1960
[68] DAS 1.571.081, Müanyagipari Kutató Intézet, 1965
[69] unknown: UV-Härtung für farbige Lackierungen, Moderne Holzverarbeitung, 1970, p. 86–87
[70] Giegold, H.: Neuzeitliche Holzoberflächenverfahren, Industrie Lackierbetrieb 1972, 40. Vol., No. 2, p. 66–71
[71] unknown: Polyesterlackhärtung mit Ultraviolett-Strahlungsquellen Industrie Lackierbetrieb 1968, 36. Vol., No. 11, p. 479–482
[72] Müller, M.: Persönliche Mitteilung. Fa. Bayer AG
[73] PPG Industries – Industrial Coatings – History of Liquid/UV, Website PPG, 23.05.2005
[74] Kistler, R.: Verfahrensbeispiel aus Glasurit-Sonderverfahren 14. Kundeninformation Glasurit-Werke M. Winkelmann AG, Hiltrup 1970
[75] Patheiger, M., Fuhr, K: UV-Härtung von Beschichtungsmaterialien auf Basis ungesättigter Polyester, Farbe und Lack 1975, 81. Vol., No. 3, p. 209–215
[76] Brushwell, W.: Lackhärtung durch Strahlung, Farbe und Lack 1976, 82. Vol., No. 12, S. 1127–1131
[77] Kaminski, E.: UV-härtende Holz- und Möbellacke als umweltfreundliche Lackieralternativen, Industrie Lackierbetrieb 1984, 52. Vol., No. 12, p. 482–484
[78] Kaminski, E., Krause, H.-G.: UV-härtende Lacke als umweltfreundliche Alternative für die Möbellackierung. I-Lack 1985, 53. Vol., No. 6, p. 202–204
[79] Chang, C.-H., Mar, A., Tiefenthaler, A.: Photoinitiators: Mechnisms and Applications, Handbook of Coatings Additives, Vol. 2, Leonard J. Calbo, Ed., 1992, Marcel Dekker, Inc.
[80] Brushwell, W.: Strahlenhärtbare Beschichtungen, Farbe und Lack 1985, 91. Vol., No. 9, p. 812–815
[81] US No. 524,953, PPG, 1966
[82] A 59996, Vianova Kunstharz AG, 1974
[83] DOS 2314814, Reichhold Chemie AG, Wien, 1973
[84] Hauser, P., Osterloh, R., Jacobi, M.: UV-Härtung pigmentierter Lacke. BASF Aktiengesellschaft, 6700 Ludwigshafen am Rhein, Release date unknown
[85] Kremer, W.: Deckend pigmentierte UV-härtbare Lacke nach dem Double-Cure-Prinzip, Farbe und Lack 1988, 94. Vol., No. 3, p. 205–208
[86] DE 2830927 A 1, BASF AG, 1980
[87] DE 2909994 A 1, BASF AG, 1980
[88] unknown: Die unendliche UV-Geschichte, Industrie Lackierbetrieb 1992, 60. Vol., No. 6, p. 228
[89] unknown: Möbellackierung in der großen Abschlussdiskussion, Industrie Lackierbetrieb 1989, 57. Vol., Nr.11, p. 409–410
[90] Kremer, W.: UV-härtende, deckend pigmentierte Lacke für Holz und Holzwerkstoffe. Firmenunterlage der Firma Bayer AG, 26.02.1988
[91] Meyer-Jungnick, M., Trageser, D. A., Wiesner, L.: Derindustrielle Einsatz von Großstrahlenquellen. Atomwirtschaft, No. 6, 1967
[92] Meyer-Jungnick, M., Trageser, D. A., Wiesner, L.: Electron Curing of Paint Films. London OCCA Discussion, Paint, Oil and Colour Journal, 1967
[93] Meyer-Jungnick, M., Trageser, D. A., Wiesner, L.: Electron-Beam Curing. Paint Technology, No. 3, 1968
[94] Tawn, A. R. H.: The Electron-Beam Curing of Coatings. OCCA Journal, September 1968
[95] Hoffmann, A. S., Smith, D. E.: Electron Radiations Curing of Monomer/Polyester Mixtures. Mod. Plastics, June 1966
[96] Deninger, W., Patheiger, M.; Lackhärtung mit Elektronen-und UV-Strahlen, Farbe und Lack 1968, 74. Vol., No.12, p. 1179–1184
[97] Lobert, M.: Elektronenstrahlhärtbare Systeme. Sonderdruck Holz-Kunststoff 1995,
[98] Garratt, P.: Strahlenhärtung, 1. Edition, Curt R. Vincentz Verlag, Hannover, 1996
[99] Häring, E.: Lackhärtung mit Elektronenstrahlen und ultraviolettem Licht, Emissionsarm lackieren, Curt. R. Vincentz Verlag, Hannover 1987
[100] Baumann, H., Timpe, J-H.: Photopolymere, Prinzipien und Anwendungen, 1. Edition, Deutscher Verlag für Grundstoffindustrie, Leipzig 1988
[101] Prieto, J.: Untersuchungen zur Mattierung von Beschichtungsstoffen, die mit UV-Strahlung gehärtet werden, Auszug Diplomarbeit, Universität GH-Paderborn 1994
[102] CEFIC, UV/EB Acrylates Sector Group.: Guide to the classification and labelling of acrylates, Version of 7. Dec. 1999

[103] Larbig, H.: Rohstoffe fürdie radikalische UV-Härtung, 4. Esslinger Oberflächen Seminar, September 2004
[104] Kühl, G.: UV-Härtung von Lacksystemen, Metalloberfläche, 3 (1988), 139–145
[105] Leppard, M.: Chemical advances in UV curable epoxy acrylate resins, Polymers Paint Colour Journal, date unknown
[106] Kemmerer, R.: A Novel Approch to Reducing the viscosity of UV/EB Curable Oligomers, Polymers Paint Colour Journal, 179, 1989
[107] DE 4040290 C2, Synthopol Chemie Dr. rer. pol. Koch GmbH & Co KG, 1996
[108] DE 3836370 C2, BASF Lacke + Farben AG, 1992
[109] Beck, E., Lokai, M., Keil, E., Nissler, H.: Studies into the emissions from radiation-cured coatings, Marketing Dispersion BASF AG, Ludwigshafen
[110] DE 3704098 A1, BASF AG, Ludwigshafen
[111] Fouassier, J.P. et al.: Radiation curing in Polymer Science and Technologie, Volume 1: Fundamentels and Methods, Elsevier science publishers, 1993, p. 232
[112] EP 0586849 A2, Bayer AG, 1994
[113] Kollaritsch, K.: Bindemittel für strahlenhärtbare Lacksysteme", Diagramm 3 (1989), S.1–12
[114] DE 3706355 A1, BASF AG, 1988
[115] Holman, R.: U.V. and E.B. Curing Formulation for Printing Inks, Coatings and Paints, Sita-Technology, London 1984
[116] Möller, F.: Houben Weyl, Vol. 11/1 (1957), p. 277–280
[117] Organikum, 16. Edition, 1986, p. 502–505
[118] EP 0903363 A1, BASF AG, 1998
[119] EP 0902040 A1, BASF AG, 1998
[120] EP 0902040 A1, BASF AG, 1998
[121] DE 4232013 A1, Bayer AG, 1994
[122] EP 0424745 A2, Bayer AG, 1990
[123] EP 0547462 B1, Bayer AG, 1992
[124] Stoye, D., Freitag, W.: Lackharze: Chemie, Eigenschaften und Anwendungen, Carl HanserVerlag München Wien, 1996, p. 215–216
[125] Beck, E.: Rohstoffe für die radikalische Strahlungshärtung, UV-Seminar, Fachhochschule Esslingen, 26/27.02.2002
[126] DE 2809715, 1977
[127] US 4128600, General Skinner, 1978
[128] DE 19800528 A1, Bayer AG, 1999
[129] EP 1103572, Bayer AG, 2001
[130] WO 0168777, BASF Coatings AG, 2001
[131] WO 00/39183, BASF AG, 2000
[132] Fouassier, J.P., Radiation Curing in Polymer Science and Technology. Vol. 1–3, 1993, p. 181
[133] Peters S., Overview of Dual-Cure and Hybrid-Cure Systems in Radiation Curing, Radiation Curing in Polymerscience and Technology, edited by Fouassier, J. P., Rabek, J. F., Volume III, 1993, p. 177
[134] Steinrücken, R.: Grundlegende Untersuchungen neuartiger Dual-Cure-Härtersysteme in konventionellen 2K-PUR Lacken, die dadurch sowohl polyaddierend, als auch UV-härtend vernetzbarsind, Diplomarbeit Universität-Gesamthochschule Paderborn, 2000
[135] Mackreth, P.: Dual-Cure, fast and colourful, Furniture Journal, November1999, p. 35
[136] Bruchmann, B., Beck, E.: Ein neues niedrig viskoses Isocyanatoacrylat – Perspektiven für Dual Cure Anwendungen, BASF AG, 08.04.2004
[137] Fischer, W., Weikard, J.: Dual Cure, the combination of superiorproperties, Bayer AG, Date unknown
[138] DE 4424101, BASF Coatings AG, 1994
[139] Prieto, J.: Strahlungshärtung, DFO-Seminar-Paderborn, 2003
[140] Schwalm, R., Binder, H., Funhoff, D., Lokai, M., Schrof, W., Weiguny, S.: Vinyl Ethers in UV-Curing: Copolymers with Acrylates and Unsaturated Polyesters, Vortrag RadTech Europa 1999, Berlin, p. 103–109
[141] Udding-Louwrier, S., Witte, F., Sjoerd de Jong, E.: Strahlenhärtende Pulverbeschichtungssysteme, DSM Resins BV, Zwolle, NL, Date unknown
[142] Bartlett, P., Nozaki, K.: J. Am. Chem. Soc. 68, 1495, 1946
[143] Olson, K., Butler, G.: Macromolecules, 17, 2486, 1984
[144] Elias, H.-G.: Makromoleküle-Grundlagen, Struktur, Synthese und Eigenschaften, Bd. 1, 5. Edition, Verlag Hüthig & Wepf, Basel 1990
[145] Wehman, E., Westerhof, H., Rietberg, J.: UV-Curable wood coatings; a comparison of a 100 % solid maleate vinyl ether system and a water borne polyurethane dispersion, DSM Resins BV, Zwolle, NL, date unknown
[146] Prieto, J.: Vinyletherhaltige Lacksysteme für die Möbelbeschichtung?, Vortrag nicht veröffentlicht, 1997
[147] DE 19903725 A1, BASF Coatings AG, 2000
[148] DE 19835849 A1, BASF Coatings AG, 2000
[149] DE 19835867 A1, BASF Coatings AG, 2000
[150] Meij, E.: Some practical aspects of Maleate/Vinylethersystems in Wood Coating application, Presentation RadTech Europa 1995, Maastricht, p. 389–393
[151] Draize, J. H.: Appraisal of the Safety of Chemicals in Foods, Drugs and Cosmetics. Dermal Toxicity, Association of Food and Drug Officals of the United States, Texas State Department of Health, Austin 1959
[152] Gelbke, W., Zeller, H.: Zur Hautreizung von Hexandioldiacrylat, Farbe + Lack 86 (1980), 876–879
[153] Fies, M., Endres, H., Klagge, R.: Novel Higher Molecular Weight HDDA Homologues for Rad-Cure Applications, Cognis Deutschland GmbH, Düsseldorf, Germany, Date unknown

[154] Beck, E., Keil, E., Lokai, M.: Strahlungshärtung und Wasser, Farbe + Lack 98 (1992), 165–170
[155] Kühl, G.: UV-Härtung von Lacksystemen, Metalloberfläche, 3 (1988), 139–145
[156] Baumann, H., Timpe, H-J.: Photopolymere: Prinzipien und Anwendungen, 1. Edition, DeutscherVerlag für Grundstoffindustrie, Leipzig 1988
[157] Kirchmayr, R., Berner, G., Hüsler, R., Rist, G.: Vergilbungsfreie Photoinitiatoren, farbe + lack 88 (1982), p. 910–916
[158] Grebe, W.: Die UV-Technologie in Druck-und Packstoffveredlung, Sonderdruck aus Papier+ Kunststoff-Verarbeiter, 1981
[159] Vrancken, A.: Mechanismus der Strahlenhärtung bzw. Strahlungstrocknung von Überzügen und Druckfarben, farbe + lack, 83 (1977), 171–179
[160] Kirchmayr, R., Berner, G., Rist, G.: Photoinitiatoren für die UV-Härtung von Lacken, farbe + lack, 86, 1980, p. 224–230
[161] Chang, C., Mar, A., Tiefenthaler, A., Wostratzky, D.: Photoinitiators: Mechanisms and Applications, Handbook of Coatings Additives, Volume 2, Marcel DekkerInc, 1992
[162] Jakubke, H-D., Jeschkeit, H.: Fachlexikon ABC Chemie, Vol. 2, 3. Edition, Verlag Harri Deutsch, Thun, Frankfurt/ Main 1987
[163] Bassi, G., Cadona, L.: New Developments on Benzophenones in the Photoinitiated Polymerisation, Vortrag der Fa. Fratelli Lamberti S.P.A. Research Laboratories, Date unknown
[164] Wiesing, R.: Untersuchungen zurVergilbung von strahlenhärtbaren weißpigmentierten Beschichtungsstoffen, Dipolmarbeit, Fachbereich Chemie und Chemietechnik derUniversität–Gesamthochschule– Paderborn, 1999
[165] Kremer, W.: UV-härtende, deckend pigmentierte Lacke für Holz und Holzwerkstoffe. Firmenunterlage der Firma Bayer AG, 26.02.1988
[166] Kolczak, U., Rist, G., Dietliker, K., Wirz, J.: Reaction Mechanism of Monoacyl- and Bisacylphosphine Oxide Photoinitiators Studied by 31P-, 13C- and 1H-CIDNP and ESR, Journal of the American Chemical Society, 1996, Volume118, No. 27
[167] Rutsch, W., Dietliker, K., Leppard, D., Köhler, M., Misev, L., Kolczak, U., Rist, G.: Recent Developments in Photoinitiators, Date unknown, Ciba-Geigy, Schweiz
[168] Rad Book, Version 3.1, Ciba Specialty Chemicals Inc., 1998
[169] DIN 55943, Farbmittel – Begriffe, Ausgabe: 2001-10
[170] Goldschmidt, A., Hantschke, B., Knappe, E., Vock, G.: Glasurit-Handbuch Lacke und Farben, 11. Edition, Curt R. Vincentz Verlag, Hannover1984
[171] Müller, M.: Technischer Stand derheutigen UV-Pigmentlacke, Industrie-Lackierbetrieb, 61, 1993, p. 47–53
[172] Fouassier, J.-P., Rabek, J.-F.: Photoinitiating Systems, In: Radiation Curing in Polymer Science and Technology. Vol. II, Elsevier Science Publishers Ltd., 1993
[173] Ferch, H.: Mattierungsmittel für die Lackindustrie, Firmenschrift No. 21 der Degussa AG
[174] Christian, H.-D.: Aktueller Stand zur Mattierung von UV-Lacken, Degussa AG, Date unknown
[175] Garratt, P.: Die Mattierung von UV-härtenden Lacken auf Basis ungesättigterPolyesterharze, Farbe + Lack 96, No. 8 (1990), p. 587–591
[176] Götz, R.: Fragen derMattierung säurehärtender Lacksysteme, Deutsche Farben-Zeitschrift No. 11 (1968)
[177] Kraus, A.: Mattlack, Handbuch derNitrocelluloselacke, Teil II
[178] Feig, P.: ZurFrage derFormulierung von Mattlacken". Farbe und Lack No. 9 (1968)
[179] Simpson, L. A.: Factors Controlling Gloss of Paint Films, Progr. Organic Coatings, 1978, 6, S.1–30
[180] Eisenmenger, E., Ferch, H., Seibold, K.: Zur Wirkungsweise synthetischer Kieselsäuren als Mattierungsmittel in Klarlacken, 13th FATIPEC Congress, 1978, p. 193–205 and Ferch, H.: Mattierungsmittel für die Lackindustrie, Firmenschrift No. 21 der Degussa AG
[181] Field, R. J.: Matt Paint Film: Real-Time Observation of Film Shrinkage and Associated Roughening of the Surface During Drying, PRA/Vincentz Verlag, 3rd Nürnberg Congress, New Technologies forCoatings and Inks, 13.–14.03.1995
[182] Ferch, H.: Mattierungsmittel für die Lackindustrie, Firmenschrift No. 21, Degussa AG
[183] Kittel, H.: Lehrbuch der Lacke und Beschichtungen, Vol. III, Verlag Colomb, 1976
[184] Schneider, H., Kent, D. J.: Parameters affecting the matting of UV-Curing Acrylate-Based Systems, Grace GmbH, Worms, Germany, Date unknown
[185] Aldcroft, D., Earl, G. J.: Advances in the Matting of UV-Curable Systems, Crosfield Limited, Date unknown
[186] Garratt, P. G.: The Flatting of Radiation Curable Paints, Fatipec Kongreß Venedig 1986
[187] Höndgesberg, H.: Mattierungsmittel und Effektgeberfür die Lacke und Farben auf Polymerbasis, Coating 10/89, Martingswerk GmbH, D-5010 Bergheim/Erft, p. 378–382
[188] Gallouedec, F., Valot, E., Bernard, M.: The use of ultra fine microporous polyamide powders as reinforcement and matting agent for UV-Cure systems, Waterborne, High-Solids and Powder Coatings Symposium, 22.–24.02.1995, New Orleans, LA, USA
[189] Kent, D. J., Schneider, H.: Parameters Affecting the Matting of UV-Curing Acrylate-Based Systems, 18th FATIPEC-Congress, Venice, 1986, p. 353–384

[190] Kent, D. J.: Matting of UV-Curing Coating Systems, Polymers Paint Colour Journal, 1990
[191] Kent, D. J.: Matting of UV-curing coating systems, Polymers Paint Colour Journal, 09.05.1990, Vol. 180, S. 305–307
[192] Garratt, P. G.: Ultraviolett Radiation Dual-Cure Processes for the Production of Low-Gloss Films, RADCURE EUROPE 1987, Conference Proceedings, 04.–07.05.1987, München, Germany
[193] Askienazy, A., Zwanenburg, R.: Adhesion Optimisation Through Oligomer/Polymer Selection, Radtech Europe, 02.–06.05.1993
[194] McAuliffe, G.: Optimising the Adhesion of UV-Coatings, Paint Polym. Colour J., 1995
[195] Häring, E., Ahlers, K., Holtmann, G.: Strahlenhärtbare Hybrid-Systeme, Industrie-Lackierbetrieb, 48 Vol., 1980, p. 449–451
[196] Mehnert, R., Roth, A.: Physikalische Mattierung von Acrylatbeschichtungen durch Bestrahlung mit 172 bzw. 222 nm Photonen, Institut für Oberflächenmodifizierung e.V. (IOM) und Heraeus Noblelight GmbH, Date unknown
[197] Freid, J. R.: Gloss-Reduction mechanisms of Radiation Cure Coatings, Radiation Curing, Februar 1982, p. 19–25
[198] Hahn, E. A.: Dual Cure – Gloss Controll of Radiation Polymerized Coatings, Radiation Curing, 14, Mai 1974, p. 13–17
[199] DE 2244327, PPG Industries, 22.03.1973
[200] Skinner, D.: Building a process forlow gloss UV-Curable Coatings, Radtech Europe, Maastrich 1995, p. 549–556
[201] Kiene, J., Prieto, J.: Kittel – Lehrbuch der Lacke und Beschichtungen, Vol. 9, zweite Edition 2004, S. Hirzel Verlag Stuttgart, p. 317–350
[202] Nargiello, M.: Flatting Agent Performance Critical To New Formulations, Modern Paint and Coatings, February 1990
[203] Feig, P.: Qualitative und quantitative Erfassung der Mattierungswirkung synthetischer Kieselsäuren und deren Einfluß auf die Beständigkeit von Oberflächenbeschichtungen, Fatipec Congress 1970
[204] URL:http://www.iom-leipzig.de/html/deu/Schwerpunkte/exci.cfm
[205] Excimer UV-Strahler, Unterlage der Firma Heraeus Noblelight GmbH, Hanau, Deutschland
[206] Schubert, R., et al.: Vortrag über den Einsatz von Excimer UV-Strahlern, 9. UCB Symposium, Garmisch-Partenkirchen, 1998
[207] K-aktuell: Informationen aus dem Kunststoff-Cluster, Ausgabe 2/2005, p. 2–3
[208] Biannual Report 2004/2005, Leibniz-Institut für Oberflächenmodifizierung e.V. (IOM), p. 31
[209] Fried, M., J.: Stabilisierung von Acrylmonomeren, Farbe + Lack, 100. Vol., 8/1994, p. 604–609
[210] Kurland, J. J.: Quantitative Aspects of Synergistic Inhibition of Oxygen and p-Methoxyphenol in Acrylic Acid Polymerization, J. Polym. Sci., Polym. Chem. Ed. 18, 1980, p. 1139–1145
[211] Cutie S. et al.: The Effects of MeHQ on the Polymerization of Acrylic Acid in the Preparation of Superabsorbent Gels, J. Appl. Polym. Sci. 64, 1997, p. 577–589
[212] Levy, L. B.: InhibitorOxygen Interactions in Acrylic Acid Stabilization, Plant/Operations Progress 6 (4), 1987, p. 188–189
[213] Becker, H.: Polymersiationsinhibierung von (Meth-)Acrylaten – Stabilisatorund Sauerstoffverbrauch-, Fachbereich Chemie der technischen Universität Darmstadt zur Erlangung des akademischen Grades eines DoktorIngenieurs, 2003, p. 49–50
[214] Henrici-Olivé, L., Olivé, S.: Polymerisation: Katalyse-Kinetik-Mechanismen, Verlag Chemie 1969
[215] Bevington, J. C.: Radical Polymerization, Academic Press Inc., New York 1961
[216] Boguslavslavskaya, L. S., Chim. Prom. 43, 1967, p. 177–183, in Russian language
[217] Henrici-Olivé, G., Olivé, S.: Makromol. Chem. 24, 1957, p. 64–75
[218] Henrici, G., Schulz, G. V.: Reaktionskinetik der Polymerisationshemmung durch molekularen Sauerstoff, Die Makromolekulare Chemie 18/19, 1956, p. 437–454
[219] Studer, K., Decker, C., Beck, E., Schwalm, R.: Overcoming oxygen inhibition in UV-curing of acrylate coatings by carbondioxide inerting, Part II, Progress in Organic Coatings 48, 2003, p. 101–111
[220] Personal nessage Dr. Erich Beck, BASF AG, Ludwighafen, Germany
[221] Winkler, E.: Dünne Schichten mit großer-Wirkung, Effektstoff-Forschung – Von der chemischen Strukturzum physikalischen Effekt, Broschüre der BASF AG, 1997, p. 32–41
[222] EP 1323525A3, Heinrich J. Kesseböhmer KG, 2003
[223] Studer, K., Decker, C., Beck, E., Schwalm, R.: Overcoming oxygen inhibition in UV-curing of acrylate coatings by carbondioxide inerting, Part I, Progress in Organic Coatings 48, 2003, p. 92–100
[224] EP 0540884, Herberts & CO GmbH, 12.05.1993
[225] EP 0559500 B1, Osaka Sanso Kogyo Limited, 08.03.1993
[226] DE 2425217 A1, Union Carbide Corp. US, 23.01.1975
[227] Walia, J.: Nitrogen Inerting With Low Levels of Photoinitiator, Fusion UV Systems, Ltd., Date unknown
[228] Misev, L.: Nitrogen Inerting Benefits Thin UV Coating Cure, Ciba Specialty Chemicals Inc., Date unknown

[229] Aehlig, K.: Umweltfreundliches Lackieren von Holz und Holzwerkstoffen durch UV-Härtung unterInertgas zurErzielung emissionsarmer, qualitativ hochwertiger Beschichtungen, BMWi-Projekt, Date unknown
[230] Starzmann, O.: Entwicklungsmöglichkeiten der UV-Inert-Technologie, I-Lack 3/99, 67. Vol., p. 141–144
[231] Lobert, M.: Elektronenstrahlhärtbare Systeme, Sonderdruck aus HK 12/95
[232] „Larolux" – UV-Lackhärtung in Kohlendioxid, Broschüre der BASF AG, Ludwigshafen, Date unknown
[233] Beck, E.: UV-Härtung in neuen Dimensionen, Vortrag in Paderborn, 26.11.2002
[234] Beck, E.: UV-Lacke mit Höhensonne unter Kohlendioxidatmosphäre härten, besserlackieren, No. 19, 16.11.2001
[235] Biehler, M., Beck, E., Menzel, K., Titusson, S., Daiss, A., Soljamo, K., Fagerholm, K.: UV Becomes 3-Dimensional, BASF AG, Germany and Tikkurilla Coatings, Finland, RadTech Europe 2005, Conference & Exhibition
[236] Horschig, J.: 3D-Substrate mit UV-Strahlenhärten, besserlackieren, No. 16, 07.10.2005
[237] Meichsner, G.: Grundlagen der UV-Härtung, 3. Esslinger Oberflächenseminar, Date unknown
[238] Macosko, C. W., Miller, D. R.: Marcomolecules 1976, 9, 199
[239] Zosel A.: Lack- und Polymerfilme, Curt Vincentz Verlag, Hannover, 1996, p. 60–61
[240] Neumüller, O.-A.: Römpps Chemie-Lexikon, Francksche Verlagshandlung, V. Keller & Co., Stuttgart, 1981, 8. Edition, p. 1495
[241] UV coatings, guideline recipe by Bayer Material Science AG, Leverkusen
[242] UV coatings, guideline recipe by Cytec Surface Specialties Austria GmbH
[243] UV coatings, guideline recipe by BASF AG, Ludwigshafen
[244] Prieto, J.: UV-härtende Lacke – Möbelbeschichtung ohne Restemissionen? Dokumente zu Lacken und Farben No. 6, Deutsches Lackinstitut, 1999
[245] DE 102004007503 A, 3H-LACKE Lackfabrik Hammen, 2005
[246] Brenni, S.: Das Mono-Cure-Verfahren und die Möbellackierung in Norditalien, I-Lack, 57. Vol., 9/1989, p. 307–310
[247] Stolze, E.: Anlagenlösungen und Arbeitsprinzipien der italienischen Möbelindustrie, I-Lack, 56. Vol., 6/1988, p. 198–201
[248] Rauch Möbel: Homepage (nach URL: http://www.rauchmoebel.de/Historie.17.0.html, (14.05.2006))
[249] Ungricht, G.: Aspekte der Gravurtechnik, Vortrag auf dem Symposium der Fa. Bürkle, Freudenstadt, 2004
[250] unknown: Oberflächenmarkt in Bewegung, Verlag Frankstein, 2005
[251] Kipphahn, H.: Handbuch der Printmedien, SpringerVerlag Heidelberg, 2000
[252] http://www.bvd-online.de/branche/040204br.html, 14.05.2006
[253] http://www.sohnrey.de/druckseiten/index.html, 21.05.2006
[254] material + technik möbel, Mitteilung: Interprint startet eigene Lasergravur, Februar2006
[255] Ungricht: Homepage (nach URL: http://www.ungricht.de/public/industriebereiche.html?lang=de&id=14 (2005))
[256] Harm, D.: Druck-Sache – Direktdruck-Technik von Hymmen mit neuem Maschinenkonzept, Surface-Magazin, 2005, p. 44–47
[257] Schreck, T.: Tiefdruck-Zone – Bürkle setzt auf Direktdruck für Möbel- und Fußbodenoberflächen, Surface-Magazin, 2005, p. 36–38
[258] Vahlhaus, A.: Neu-Auflage – Indirect Print Laminate, Surface-Magazin, 2005, p. 40–42
[259] Barth, R.: IndirekterTiefdruck: Schreckgespenst oder „schlafender Hund"? mt International, März 2005, p. 40–44
[260] Voigt, J.: Direct Printing, Vortrag, RadTech Europa 2005, Conference & Exhibition, Barcelona
[261] Grahn, R.: UV-härtende und wasserlösliche Druckfarben, Vortrag auf dem Bürkle Symposium, Freudenstadt, 2004
[262] Vahlhaus, A.: The Renaissance of Wet Application: New Opportunities forthe Laminate Flooring Industry HPC (High Performance Coatings), RadTech Europa 2005, Conference & Exhibition, Barcelona
[263] Thole, V.: Leichte Werkstoffe aus nachwachsenden Rohstoffen, WKI Vortrag, Braunschweig, 2004
[264] unknown: Vacumat – Das innovative Lackiersystem für Profile, Kanten, Flächen. Technische Unterlage der Fa. Schiele Maschinenbau GmbH, Niederzissen
[265] unknown: Hohe Effizienz durch Flexibilität und Vielseitigkeit. Technische Unterlage der Fa. Schiele Maschinenbau GmbH, Niederzissen 1999
[266] Hochstein, D.: Allgemeine Information Vakuum Lackieranlagen. Technische Informationen, Fa. Giardina, Südlohn
[267] unknown: Die UV-Vakuum-Lackiertechnik hat jetzt begonnen. I-Lack 59, p. 115–117 (1991)
[268] unknown: Vakumat-Präsentation, Fa. Schiele Maschinen GmbH, Niederzissen, 2000
[269] Nebel, B., Wegener, G., Zimmer, B.: Ökobilanzierung Holzfußböden, Herausgeber der Studie: Holzforschung München – HFM, Technische Universität München, 2. überarbeitete Auflage, August 2002
[270] Informationen der FEP (Föderation der seuropäischen Parkett-Industrie-Verbände) auf dem 31. Europäischen Par kettkongress am 31.05.2006 in Wiesbaden

[271] Müller, M.: Optimized performances of UV-Curing parquet coatings based on unsaturated acrylic, Vortrag der Firma Bayer AG, 2001
[272] Philips, M., Decaux, M., De Waele, L.; Abriebbeständigkeit – Vergleich von Formulierungen und Testmethoden, UCB Radcure-Symposium, Tagungband 6, 12.–13.09.1995, Chiemsee
[273] Biesbrouck, K.: Hard or Flexible Coatings – Influence on Measured Performances, Tagungsband RadTech Europa 2005 in Barcelona
[274] Scheithauer, M.: Möglichkeiten der Herstellung von Fußbodenoberflächen mittels UV und ESH, Zeitschrift HK (Holz und Kunststoff), No. 6, 1999, p. 64–67
[275] Spang, H.: Sorgfalt und Vorsicht beim Umgang mit Ölen und Wachsen, Magazin Parkett, No. 5, 2002, p. 118
[276] Devantier, B.: Problematische Pauschalaussagen – Bewertung von ausgewählten Parkettbeschichtungen durch das IHD Dresden, Innenausbau-Magazin 1999, Beilage des Holzzentralblatt 9/1999, p. 20–22
[277] Emmler, R.: Bewertung der Verschleißfestigkeit und Langzeitbeständigkeit von Möbeloberflächen, Tätigkeitsbericht 2004, Institut für Holztechnologie Dresden (ihd), p. 28–29
[278] Emmler, R.: Auf das Trägermaterial kommt es an, Fachzeitschrift Boden-Wand-Decke, 5/1999, p. 58–63
[279] Emmler, R., Devantier, B., Macht, K.-H.: Verlegequalität stellt Weichen für Langlebigkeit bei Laminatböden, IHD-Dresden, Fachzeitschrift Laminat-Magazin, 1999, p. 29–32
[280] Emmler, R.: Anforderungsprofil an Holzlackierungen – Einteilung in Nutzungsklassen, Institut für Holztechnologie Dresden (ihd), 2003
[281] Prieto, J., Kiene, J.: Lehrbuch der Lacke und Beschichtungen, Kittel Vol. 9, p. 319–351, S. Hirzel Verlag Stuttgart, Leipzig, 2004
[282] Billiani, H.: Synthetische Polymer-Bindemittel in wässriger Lösung, Chapter 3.1, Kittel Bd. 3, Bindemittel für wasserverdünnbare Systeme, S. Hirzel Verlag Stuttgart – Leipzig, 2001
[283] Bieganska, B., Zubielewicz, M., Smieszek, E.: Anticorrosive WaterBorne Paints, Progress in organic coatings, No. 15, 33–56 (1987)
[284] Müller, B., Poth, U.: Coatings Formulation, 3rd Ed., page 177, Table III-1.2, Vincentz Network, Hannover, 2017, p. 163
[285] Albalat, J., Font, C., Pulido, A.: New ultraviolet curing acrylic dispersion forindustrial wood finishing, Cray Valley Ibérica, Spain, 2005
[286] Fischer, W., Lühmann, E., Rappen, D., Weikard, J.: UV-Curing High MolekularWeight Polyurethane Dispersions, Firmenunterlagen der BayerMaterial Science AG, Leverkusen
[287] Lange, H.: Wässrige Polyurethanacrylate als strahlenhärtende Lackrohstoffe, Farbe + Lack, 99. Vol., 7/1993, p. 597–601
[288] DE 4031732, BASF AG, 1992
[289] JP 02167328, Toyobo KK, 1990
[290] DE 2936039, Bayer AG, 1981
[291] DE 3900257, Bayer AG, 1990
[292] DE 4012339, Bayer AG, 1991
[293] US 4176099, Vianova Kunstharz AG, 1979
[294] US 6521702, Bayer AG, 2000
[295] Irle, C., Bayona, M.: Erfolgsstory vorgezeichnet, Holz- und Kunststoffverarbeitung (HK), 5/2006, p. 44–49
[296] Stoye D., Freitag, W.: Lackharze – Chemie, Eigenschaften und Anwendungen, Carl Hanser Verlag München – Wien, 1996
[297] Bayer, O.: Neuere Entwicklungen des Diisocyanat-Polyadditions-Verfahrens (Polyurethane), Fatipec-Kongress 1957, p. 11–18
[298] Schmitz, J.: Vom Standard Biuret zu speziellen Trimeren für Wasserlacke, High-Solid oder-schnelltrocknende PUR-Systeme, Vortrag am 21.03.2002 in Herford, Bezirksgruppe Ostwestfalen-Lippe des VILF
[299] Patent WO 0188006, Bayer AG, 2001
[300] Richtformulierung der Firma Bayer Material Science: Clearvarnish forspray, matt, IRL 5713, 2004
[301] Kubitza, W.: Wässrige Zweikomponenten-Polyurethan-Reaktivsysteme, Farbe + Lack, 97. Vol., 3/1991, p. 201–206
[302] Müller, B., Poth, U.: Lackformulierung und Lackrezeptur, Curt R. Vincentz Network, Hannover, 2003, p. 193–195
[303] Kahl, L., Bock, M., Jürgens, E., Laas, H. J.: Wässrige 2K-PUR-Klarlacke für die Automobilerstlackierung, Farbe + Lack, 102. Vol., 3/1996, p. 88
[304] Nawrath, T.: Abhängigkeit der Qualität von wässrigen 2K-Polyurethanen in Abhängigkeit von der anlagentechnisch vorgegebenen Mischungsgüte, Diplomarbeit, Universität-Gesamthochschule-Paderborn, 2004
[305] Kahl, L., Bock, M., Klinksiek, B.: Wässrige 2K-PUR-Klarlacke für die Autoerstlackierung, Teil II: Vermischungstechnologie, I-Lack, 64. Vol., 2/1996, p. 63–64
[306] Anlagenschema der Firma Venjakob Maschinenbau, Rheda-Wiedenbrück
[307] Prieto, J., Gehrmann, V.: Umsetzung der EU-VOC-Richtlinie aus Sicht der Holzlackhersteller, Informationsveranstaltung des Lackverbandes in Frankfurt, 1999
[308] Prieto, J., Kiene, J.: Chapter 9.2.5 Holzlackieren, Kittel–Lehrbuch der Lacke und Beschichtungen, Bd. 9: Verarbeitung von Lacken und Beschichtungsstoffen, 2004, p. 344–346
[309] Diekmann, M.: Industrielle Bearbeitungstechnik alternativerMöbeloberflächen. In: IBH-Tagung vom 10. Und 11. Mai 1996
[310] unknown: Moderne Möbellacke. Deutsches Lackinstitut, Frankfurt 1999

[311] Hägele, V.: Öle und Wachse zur Oberflächenbehandlung von Holz, Hrsg.: Landesverband Holz + Kunststoff, Baden-Württemberg und Landesgewerbeamt Baden-Württemberg, 2003

[312] Stetter, K.: Alternative Lacksysteme, WKI Bericht 31: Umweltfreundliche und emissionsarme Möbel, 1995, S.191–198

[313] „Einsatz von Ölen und Wachsen zu Oberflächenbeschichtung", Projektbericht im Rahmen des Gesamtprojektes „Nutzbarmachung des VOC-Minderungspotentials im Schreinerhandwerk, Umweltzentrum für Handwerk und Mittelstand e.V. im Auftrag des Ministeriums für Umwelt und VerkehrBaden-Württemberg, 2001

[314] Philipp, B., Lauter, N.: Kaltwachs auf Möbeloberflächen applizieren, besserlackieren, 2001, No. 1, p. 9

[315] Poth, U.: Chapter 1.4.1 Naturrohstoffe, Kittel – Lehrbuch der Lacke und Beschichtungen, Bd. 1: Geschichte, Grundlagen, Naturstoffe, anorganische Bindemittel, 1998, 153–204

[316] Penning, M.: Schellack – ein „nachwachsender" Rohstoff mit interessanten Eigenschaften und Anwendungen, Seifen – Öle – Fette – Wachse, 115, 1990, p. 221–224

[317] Nüßer, O.: Die Lösemittel-Probelmatik, Dokumente zu Lacke und Farben, No. 5, Deutsches Lackinstitut, 1998, p. 34–40

[318] Bauch, H: Elektrostatische Oberflächenbehandlung von Holzwerkstoffen mit thermo- und strahlenhärtenden Pulverlacken, Förderkennzeichen 01 RV 9612/9, BMBF gefördertes und DFO koordiniertes F&E-Programm zur Erweiterung der Anwendung von Pulverlacken, 1996–1999 Förderkennzeichen: 01 RV 9625/8

[319] Krüger, A., Schmich, P.: Einsatzmöglichkeiten und Bedingungen für die Pulverlack-Beschichtung von Holz und Holzwerkstoffen, Förderkennzeichen: 01 ZH 9309/1, BMBF gefördertes und DFO koordiniertes F&E-Programm zur Erweiterung der Anwendung von Pulverlacken 1994–1996, Förderkennzeichen 01 ZH 9307

[320] Krüger, A., Dreyer, C., Friebel, S.: Erarbeitung von Grundlagen für einen praxisgerechten Prozess zur Pulverbeschichtung von Holz und Holzwerkstoffen, Förderkennzeichen 01RV9611/6, 01RV9610/3, BMBF gefördertes und DFO koordiniertes F&E-Programm zur Erweiterung der Anwendung von Pulverlacken, 1996–1999 Förderkennzeichen: 01 RV 9625/8

[321] Friebel, S.: Wechselwirkungen zwischen Holzwerkstoff und Pulverlack, DFO-Berichtsband 42: Pulverbeschichtung von Holzwerkstoffen – ein innovatives Lackierkonzept, 1999, p. 21–28

[322] Bauch, H.: Elektrostatische Beschichtung von Holzwerkstoffen mit thermo- und strahlenhärtbaren Pulverlacken, DFO-Berichtsband 42: Pulverbeschichtung von Holzwerkstoffen – ein innovatives Lackierkonzept, 1999, p. 35–48

[323] Dreyer, C., Krüger, A.: Prozessablauf und Schichteigenschaften, DFO-Berichtsband 42: Pulverbeschichtung von Holzwerkstoffen – ein innovatives Lackierkonzept, 1999, p. 29–34

[324] Kiene, J., Vesper, H.: Powder Coating on Wood and Wood Composites, Woodcoatings – Challenges and solutions in the 21st Century, Den Haag 2000, Paper 15

[325] Spohnholz, R.: Rohstoffentwicklungen von Pulverlackharzen für die Holz- und Holzwerkstoffindustrie, DFO-Berichtsband 42: Pulverbeschichtung von Holzwerkstoffen – ein innovatives Lackierkonzept, Paderborn 1999, p. 5–12

[326] Bauch, H.: Elektrostatische Pulverlackierung von Holzwerkstoffen, Holz-Zentralblatt, 1996, No. 68/69

[327] Netuschil, P., Schmich, P.: Pulverlackbeschichtung von Holz- und Holzwerkstoffen – Ökologische Chancen und technische Möglichkeiten, Holz-Zentralblatt, 1996, No. 68/69

[328] Bauch, H.: Pulverbeschichtung von Holzwerkstoffen, Journal für Oberflächentechnik (JOT), Ausgabe 10, 1998

[329] Zimmermann, F.: personal message

[330] Piearce, C.: Powder Coating of Wood – Mythor Reality?, Woodcoatings – Challenges and solutions in the 21st Century, Den Haag 2000, Paper14

[331] Zimmermann, F.: Pulverlacke für Holzwerkstoffe, Thermoface und Pulverlack, Dresden 2003

[332] unknown: Pulverbeschichtung von MDF-Platten, Journal für Oberflächentechnik (JOF) 40 (6), 2000

[333] Kiene, J., Prieto, J.: Chapter 9.2.5 Holzlackieren, Kittel-Lehrbuch der Lacke und Beschichtungen, Bd. 9: Verarbeitung von Lacken und Beschichtungsstoffen, 2004, p. 344–346

[334] Sauter, W., Harsch, M., Mertins, L.: „Demonstrationsanlage zur umweltfreundlichen MDF-Pulverbeschichtung mit hohen Qualitätsanforderungen für die Möbelherstellung", Abschlußbericht, Fördernde Institution: Umweltbundesamt, Aktenzeichen 30441–1/22, Vorhaben No. 20055, November 2004

[335] Hauber, P.: Pulverbeschichten von MDF in der Praxis aus Sicht des Möbelherstellers, EPS – Der Pulvertreff 2003, Fachtagung überdas elektrostatische Pulverbeschichten, 2003

[336] Hauber, P.: Pulverbeschichten von MDF in der Praxis aus Sicht des Möbelherstellers, Thermoface und Pulverlack, Dresden 2003

[337] Mayer, C.: Pulverbeschichtung von MDF in der Praxis aus Sicht des Lohnbeschichters, Thermoface und Pulverlack, Dresden 2003

[338] Paauwe, J.: Low Temperature Curing Resins, Präsentation DSM Coatings Resins, 2006

[339] Bayards, R. A., Groen, J. H., Koldijk, F., Benkers, D.: Enlarging the curing window of powder coatings, European Coating Show 2003
[340] Buysens, K.: UV Curable Powder Coatings – A New Coating Technology for MDF, RadTech Europe 1999, Conference Proceedings
[341] Buysens, K.: UV Curable Powder Coatings for the Finishing of Wood Composites, Woodcoatings – Challenges and solutions in the 21st Century, 2000, Paper17
[342] Buysens, K.: UV Curable Powder Coatings on Wood – New Developments, RadTech Europe 2001, Conference Proceedings
[343] Buysens, K., Keil, E., Cremer, J.: „UV-härtbare Pulverlacke – Stand der Technik", Journal für Oberflächentechnik (JOT), 2001, Ausgabe 02
[344] Schmidt, T.: Pulverlackierung von Holzwerkstoffen, Thermoface und Pulverlack, Dresden 2003
[345] Groen, J. H., Bayard R., A., Kaptijn, W., H.: Application Properties of UV Curable Powder Coatings, RadTech Europe 2003, Conference Proceedings
[346] Maetens, D.: Optimal Performance of UV Powders on Different Substrates by Combining Polyesterand Epoxy Chemistry, RadTech Europe 2003, Conference Proceedings
[347] Fink, D., Buysens, K.: Strahlenhärtende Beschichtungen, Thermoface und Pulverlack, Dresden 2003
[348] Williams, F., Armengol, JM., Grau, E., Chambat, C., Laurens, C.: Urethane-Acrylated Resins for UV-Curable Powder Coatings: A Versatile Chemistry to Look forthe Excellence, Rad Tech Europe 2005, Conference Proceedings
[349] Patent: EP 0846034, DuPont Powder Coatings, 1996
[350] Patent: WO 02/100957, DuPont, 2002
[351] Patent: DE 198 35 867, BASF AG, 1998
[352] Patent: DE 198 35 917, BASF AG, 1998
[353] Patent: DE 199 03 725, BASF AG, 1999
[354] Patent: EP 585742, Hoechst, 1995
[355] Patent: EP 739922, UCB, 1995
[356] Patent: EP 1258296, (DSM), 2001
[357] Spyrou, E.: Storage Stable UV-Powder Coating Formulations, Rad Tech Europe 2005, Conference Proceedings
[358] Bender, J., Laver, H., Lehmann, K., Margraf, R., Schmid, O.: The Formulation of UV-Curable Powder Coatings, Rad Tech Europe 1999, Conference Proceedings
[359] Chamberlin, D. T., Laver, H. S.: UV-Curing and Formulation of Powder Coatings, Powder Coatings: What's Next, 1999, Conference Proceedings
[360] Jovanovic, Z., Lahaye, J., Laver, H., Megert, S., te Walvoort, C.: "UV-Curable Powder Coatings: A Guide to Succesful Formulation, Ciba-Broschüre
[361] Krug, D., Wolf, H.: Die pulverlackierbare MDF, Thermoface und Pulverlack, Dresden 2003
[362] Raatz, C.: Richtungsweisende Profilvergütung durch Thermoface, Thermoface und Pulverlack, Dresden 2003
[363] Fuchs, I.: Thermoglätten – ein ökonomisches und ökologisches Feinbearbeitungsverfahren, Thermoface und Pulverlack, Dresden 2003
[364] Rehm, K., Raatz, C.: Thermoglätten mit Zukunft – Staubfreie MDF-Feinbearbeitung für die Industrie, Holztechnologie 46 (2005)
[365] Strid, M., Bolm, H., Pilqvist, M.: „Continuation of a Succes Story with UV Powder", Rad Tech Europe 2003, Conference Proceedings
[366] Karlsson, L.: UV Powder on MDF for High Quality & Design, Rad Tech Europe 2003, Conference Proceedings
[367] Zimmermann, F.: personal message
[368] Timmermann, E.-H., Kleber, W.: Chapter 9.1.3 Pulverbeschichten, Kittel-Lehrbuch der Lacke und Beschichtungen, Bd 9: Verarbeitung von Lacken und Beschichtungsstoffen, p. 98–141, 2004
[369] Blatter, K., MacFadden, B., Strid, M., Niggemann, F.: UV – Pulverlacke erste Erfahrungen in der Anwendung, Journal für Oberflächentechnik (JOT), 1998, Ausgabe 02
[370] Heathcote, J.: UV-Powder Coating Process for MDF, Radtech Report, September/October 2001
[371] Bayards, R., Blatter, W., Concha, C., Griese, C., Shepherd, B., Sitzmann, E.: Test for Success – UV-Powder Coatings for MDF, Radtech Report, September/October 2001
[372] Mitchell, S.: Holz mit UV-härtenden Pulverlacken schnell und Kosten sparend beschichten, besser lackieren, 2002, No. 4
[373] Grimm, T.: Pulverauf MDF/Holz, Thermoface und Pulverlack, Dresden 2003
[374] Firmenunterlagen: Venjakob Maschinenbau
[375] Baroncini, B.: Pulverlackierbeschichtung auf MDF mittels horizontaler oder vertikaler Beschichtungsstraße? Thermoface und Pulverlack, Dresden 2003
[376] Santandrea, S.: „UV Powder Coating on MDF a Consolidating Technology", RadTech Europe 2005, Conference Proceedings
[377] Prieto, J.: personal message
[378] Roux, Marie-Lise: Finitions poudres: Les possibilités actuelles en ameublement, Brèves du CTBA, 2006
[379] URL: http://www.infolink.com.au/articles/AE/0C02F0AE.aspx
[380] Görmar, G.: Dekorfinishfolien aus Leipzig, Tagungsunterlagen, Workshop des ihd Dresden, Juni 2005
[381] Berger, E.: Endlosstrukturauf Folie, I-Lack, 63. Vol., 10/1995, p. 356–357
[382] Görma, G.: Dekorfinishfolien – ein Produkt mit Zukunft, Zeitschrift Holztechnologie, 47. Vol., Ausgabe No. 6, 2006, p. 43–47

[383] Hogg, J., Puusa, S.: Vorimprägnat – Chancen aus der Marktperspektive, Jaakko Pöyry Consulting, Arjo Wiggins Symposium, 10/2003

[384] Schwarz, G.: Aktuelle feste und flüssige Beschichtung von Holz und Holzwerkstoffen, Teil 1, source and date unknown

[385] Publication of Technocell Dekor: Einflussfaktoren auf die Nassdehnung von Dekorpapier, Herausgeber Frank Stein, Material and ideas forfuture living, Februar 2006

[386] Eibl, M.: Zellstofffasern aus einjährigen Pflanzen für Spezialpapiere, Frank Stein – Material and ideas forfuture living, Ausgabe 7, 2006, p. 45–50

[387] Erhard, K., Frohberg, K.: Verbesserung der Spaltfestigkeit von Papierdurch CMC-modifizierte Faserstoffe, PTS-Forschungsbericht, PTS Heidenau, 2005

[388] unknown: Vorimprägnat auf dem Vormarsch, Material+Technik (m+t), 06/2003, p. 34–36

[389] Beitrag in der Zeitschrift Material + Technik (m+t International), 07/2004, p. 44–45

[390] Patent WO200/074577, Ahlstrom Osnabrück GmbH, 2004

[391] Sczepan, R.: Melaminharzbeschichtungen, HK, 4/1997, p. 52–57

[392] unknown: Fortschritte in der Verarbeitung wirken auf Dekorpapierzurück, Euwid Holzspecial: Massivholz, Holzwerkstoffe, Oberflächen, 1997, p. 38–40

[393] Patent EP 0739435 B1, Arjo Wiggins SA, 1996

[394] Patent EP 0223922 A1, KaemmererGmbH, 1987

[395] Patent DE 19728250 A1, Koehler Decor GmbH & Co KG, 1999

[396] Patent DE 19758479 A1, Koehler Décor GmbH & Co KG, 1999

[397] Kazmeier, T.: Vorimprägnierte Dekorpapiere, Arjo Wiggins Forum Foil, 2003

[398] Janssen, W.: Das Vorimprägnat Teil 5, Zeitschrift Material + Technik (m+t), Juni 2004, p. 16–17

[399] Fuchs, R. A., Matscheko, H.: Bemerkungen zum Vorimprägnat, Zeitschrift Material+Technik (m+t), 06/2003, p. 17–18

[400] unknown: Dekorpapierbranche im Umbruch, Euwid Holzspezial: Massivholz, Holzwerkstoffe, Oberflächen, 1997, p. 42–44

[401] unknown: Formaldehyd, I-Lack, 53. Vol., 5/1985, p. 179

[402] Guideline recipe: Wasserverdünnbarer Lack für Holzdekorfolien, BASF AG, 8/1999

[403] Guideline recipe: 2-component acid curing fordecorative wood foil, No. VI.3.1, Worlée Chemie, 10/1995

[404] Preis, J.: Strahlenhärtende Foliensysteme, WKI-Bericht No. 34, Juni 1999, p. 171–188

[405] Bauch, H., Mehnert, R.: Verschleißfeste Lackierungen für Holzfußböden und Möbeloberflächen, besserlackieren, No. 13, August 2006, p. 11

[406] Abschlussbericht zum AiF-Forschungsprojekt 13730 BR: Entwicklung kratz- und abriebfester Beschichtungen für Holzoberflächen auf der Basis von lösemittelarmen Nanokomposit-Lacken, Mai 2006

[407] Bauch, H., Emmler, R.: Verschleißfeste Lackierungen mit UV-strahlenhärtenden Nano-Komposit-Systemen für Holzfußböden und Möbeloberflächen, 33. Internationales Rosenheimer Oberflächenseminar, Tagungsband LHK Rosenheim 11–12 Mai, 2006

[408] Böhme, P.: Dekorfolien – Herstellung, Verarbeitung und Eigenschaften von Dekorfolien für Möbel, VEB Fachbuchverlag Leipzig, 1988, 2. Edition

[409] Wüllenweber, K.-H.: Immer auf dem laufenden, Holz- und Kunststoffverarbeitung, No. 8, 1986, p. 62–64

[410] Ober, M. et. al: Oberflächentechnik, Ligna Plus, Neuheiten-Symposium, 05.05.2005

[411] Patent JP 61285

[412] Patent DE 2945217

[413] Patent US 3811915

[414] Schwarz, G.: Das Kalandrierverfahren – neue Beschichtungstechnik in der Holz- und Möbelindustrie, I-Lack, 52. Vol., 9/1984, p. 334–336

[415] Informationen der Firma Hymmen, Report, Vol. No. 5, June 1982

[416] Oberflächenbehandlung von Holz im Außenbereich, Merkheft 11, Deutsche Gesellschaft für Holzforschung, München 1998, 2. Edition

[417] Anforderungen an Beschichtungssysteme von Holzfenstern und Haustüren, VFF-Verband Fensterund Fassade Merkblatt H0.03 2004-04

[418] Neises, H.: Anwendungstechnische Eigenschaften von Alkydharzen in Abhängigkeit vom molekularen Aufbau, Farbe und Lack, 77. Vol., No. 9, 1971, p. 850–865

[419] Hubert, J.C. u.a.: Singlet Oxygen Drying of Alkyd Resins and Model Compounds, Journal of Coatings Technolgy, Vol. 69, No. 869, June 1977

[420] Muizebelt, W.J. a.o.: Oxidative Crosslinking of Alkyd Resins Studied with Mass Spectrometry and NMR Using Model Compounds, Journal of Coatings Technology, Vol. 70, No. 876, January 1998

[421] Goldschmidt, A., Streitberger H.-J.: BASF Handbook Coating Technology, Vincentz Network, Hanover, 3rd Edition 2018

[422] Valet, A., Braig A., Light Stabilizers for Coatings, 2nd Ed., Vincentz Network, Hanover, 2017

[423] Olsen, J. H.: Waterborne high-solids exterior-wood coatings, Surface Coatings International, Part A, Issue 08/2005, p. 317–322

[424] Gobec, M.: Modern resins forenvironmentally friendly architectural paints, 15th SLF Congress 1997 in Lillehammer

[425] Angelmayer, K., Merten G., Awad, R.: Hoechst Kunstharz Nachrichten 27, 06/91

[426] Désor, U., Krieger, S.: Waterborne acrylic dispersions, European Coatings Journal, 12, (1996), p. 920–923

[427] Kittel: Lehrbuch für Lacke und Beschichtungen, Vol. 3, Filmbildner für wässrige Systeme, 2. Edition, S. Hirzel Verlag, Stuttgart 2001

[428] Baumstark R., Tiarks, F.: Studies fora new generation of acrylic binders forexteriorwood coatings, Macromol. Symposium 187, 2002, S.177–186

[429] Zosel, A.; Ley, G.: Farbe und Lack 100 (1994) 239

[430] Zosel, A. et al.: Macromol. Chem. Macromol. Symp 35/36 (1990) 423

[431] Bufkin, B.G., Grawe, J. R.: J. Coat. Technol. 50 (1978) 41

[432] Ooka, M., Ozawa, H.: Prog. in Org. Coat. 23 (1994) 325–338

[433] Hora, G., Désor, U.: Improving wet adhesion of waterborne acrylic dispersions on wood by bond activation, Wood coatings cogress, Den Haag, Oktober 2002

[434] Kreis, Sherman: Development of a ureido functional monomerforpromoting wet adhesion in latex paints; Water-Borne and HigherSolids Coating Symp.; 02/1988; New Orleans, p. 222–234

[435] Singh, Chang; di Leone, Siesel: Novel wet adhesion monomers foruse in latex paints; 23. International Conference in Organic Coatings, Vouliagmeni, Athen 1997; p. 427–441

[436] Désor, U.; Krieger, S.: Wässrige Dispersionen für Holzlacke, Welt der Farben 7 und 8 (1997), S.6–12

[437] Hoffmann, H., Ulbricht; W.: Tenside Surfactants Detergents 24 (1987) 23

[438] B. Gerharz, H. J. Butt, B. Momper: morphology of heterogeneous latex particles investigated by atomic force microscopy, Prog. Colloid Polym. Sci. (1996) 100 p. 91–95

[439] Désor, U., Krieger, S., Apitz, G., Kuropka, R.: Waterborne acrylic dispersions forindustrial wood coatings, Surface coatings international, 82, (No. 10), 1999, S.485–496

[440] Wagner, O.: Farbe und Lack 97 (1991) 109

[441] Heussen, I.: Wässrige Lack- und Beschichtungssysteme: Polyurethandispersionen, Techn. Akademie Wuppertal, 1994

[442] Désor, U., Leon, B., Kuropka, R.: Comparison of waterborne acrylic and polyurethane dispersions and mixtures of both in 1-component clearvarnishes forwood, 25. Fatipec Kongress, Turin, Oktober 2000

[443] Friebel, Hennecke, Wittenberg, Overhagenböck, Goletz, Pohl: Industrial transparent UV-Coatings for outdoor wood applications, Fifth international wood coatings congress, Prague, 17–18 Ocotober 2006

[444] Wicks, Z.J.: Coatings Technol 12 (1986)

[445] Baumstark. R., Schwarz, M.: Dispersionen in Bautenfarben: Acrylatsysteme in Theorie und Praxis, Vincentz Network, Hannover, 2001, p. 192–212

[446] Gonzales-Gomes: Transparente Eisenoxide als UV-Absorberin Holzlasuren, Sachtleben Symposium, 08.06.2000

[447] Hein, J. T.: Holzschutz – Holzwerkstoffe erhalten und veredeln, Wegra Verlag

[448] Kastien, H.: Influence of the binderand additives on the weathering stability of colourless aqueous wood varnishes, Polymers Paint Colour Journal, 181, No. 4286 (1991), p. 366–368, 371

[449] Informationsdienst Holz: Anstriche für Holz und Holzwerkstoffe im Außenbereich, Arbeitsgemeinschaft Holz, Düsseldorf 12/1999

[450] Verband der Fenster und Fassadenhersteller e.V. (Hrsg.): Auswahl der Holzqualität für Fenster und Haustüren, VFF-Merkblatt HO 02; Frankfurt, 2003

[451] Hora, G., Belz, A.: Moderne Beschichtungssysteme für Holzfenster, Holz-Zentralblatt, 12.02.1999, p. 258–259

[452] Grüll, G: Oberflächenbeschichtungen für Fenster und Türen

[453] Verband der Fenster und Fassadenhersteller e.V. (Hrsg.): Auswahl der Holzqualität für Fenster und Haustüren, VFF-Merkblatt HO 02; Frankfurt, 2003

[454] Institut für Fenstertechnik e.V. (Hrsg.): Merkblatt Anstrichsysteme für Holzfenster, Fensterund Fassade No. 2, Rosenheim, 1991

[455] Verband der Fenster und Fassadenhersteller e.V. (Hrsg.): Klassifizierung von Beschichtungen für Holzfenster und -haustüren, VFF-Merkblatt HO 01; Frankfurt, 2004

[456] Verband der Fenster und Fassadenhersteller e.V. (Hrsg.): Runderneuerung von Kastenfenstern, VFF-Merkblatt HO 09; Frankfurt, 02/2003

[457] Verband der Fensterund Fassadenhersteller e.V. (Hrsg.): Holzarten für den Fensterbau, VFF-Merkblatt HO. 06; Frankfurt, 2004

[458] Tretter, A.: Holzlackschäden – Beschichtungsmängel an Fenstern: Erkennen Vermeiden Sanieren, DRW Verlag, Leinfelden-Echterdingen, 2004

[459] Initiative Pro Holzfenster e.V. (Hrsg.): Empfehlungen für die werkseitige Fertigbehandlung von Fenstern und Haustüren aus Holz, Mintraching, 2003

[460] Österreichisches Normeninstitut (Hrsg.): Ö Norm C 2350, Wien

[461] Bundesausschuss für Farbe und Sachwertschutz (BFS) (Hrsg.): Merkblatt No. 18: Beschichtungen auf maßhaltigen Außenbauteilen aus Holz, insbesondere Fenster und Außentüren

[462] Verband der Fenster und Fassadenhersteller e.V. (Hrsg.): Anforderungen an Beschichtungssysteme für die werkseitige Beschichtung von Holzfenstern und Haustüren, VFF-Merkblatt HO. 03; Frankfurt, April 2004

[463] Grüll, G.: Lufteinschlüsse in Holzfensterbeschichtungen; Holzforschung Austria, March 2004

[464] Meijer, M., Militz, H.: Adhesion of low VOC coatings on wood: a quantitative analysis, verfkroniek (1999) p. 25–30

[465] Schlatter, H.: Gesamt-Anlagen für die wirtschaftliche Holzfensterlackierung, (EISENMANN Holztechnik, Holzgerlingen), Workshop der REITER GmbH + Co. KG, Winnenden, vom 12.–14.01.2000

[466] Sikkens: Forum Holz Oberflächenbehandlung von A–Z, Akzo Nobel Deco GmbH, Köln, 2001

[467] ReiterOberflächentechnik: Holzfenster elektrostatisch beschichten, besserlackieren

[468] Hora, G.; Belz, A.: Ganzheitliche Flutlackierung zur Holzbeschichtung; Holz-Zentralblatt; No. 69; p. 980

[469] Initiative pro Holzfenstere.V. (Hrsg.): Empfehlungen für die werkseitige Fertigbehandlung von Fenstern und Haustüren aus Holz, Berlin (2003)

[470] Initiative pro Holzfenstere.V. (Hrsg.): Instandhaltung und Pfleganleitung, Berlin (2000)

[471] unknown: Variable Fensterlackierstrasse mit modernerAusstattung nutzen, besserlackieren, Mai 2003

[472] Tschorn, Ulrich: Fenstermarkt in Deutschland und Europa, Fenster-Türen-Treff 2016

[473] Selbstheilende Lacktechnologie von Adler: www.adler-lacke.com/de/newspresse, 19.04.2016

[474] Hochglanzbeschichtung in der Fensterprofilummantelung – ein neuer Trend?, Information aus gff-magazin.de 2017

[475] EGGER Holzwerkstoffe Brilon GmbH & Co. KG: Technical Information – MDF plate MBP-L powder coating: https://www.egger.com/shop/de_DE/Shop/Muster/Rohspanplatten/982594-MDF--Platte-MBP-L-Pulverbeschichtung-297-x-210-x-19/p/982594

[476] unknown (2015): CO_2-Fußabdruck verschiedener Lacksysteme, IGP Pulvertechnik AG, CH-9500 Will, 2015

[477] unknown (2015): Hochwertig, wirtschaftlich, umweltfreundlich, Veröffentlichung der Fa. Ramseier Woodcoat AG, CH-3608 Thun, 2015

[478] unknown (2016): Carbon footprint study for industrial coatings applied on a metal substrate focus on Powder Coatings, DSM Coating Resins, Zwolle, The Netherlands, 2016

[479] unknown (2016): Sustainability Report FY16, IKEA Group, Ingka Holding B. V., 2016

[480] unknown (2017): Decorative surfaces market – paper based, Pöyry, 2017

[481] unknown (2016): Sustainability Report FY16, IKEA Group, Ingka Holding B. V., 2016

[482] (2017): Trend Magazine 2017, Akzo Nobel Hilden GmbH, 2017

[483] (2015): Tätigkeitsbericht – Annual Report, Institut für Holztechnologie Dresden, 2015

[484] (2010): IHD-Güterichtlinie für Dekorfinishfolien, Fassung 01 (2010), Institut für Holztechnologie Dresden (2010)

[485] Wenker, E.: Comparing the autoxidative mechanism of Co-EH and FeONIX, Leiden Institute of Chemistry, 2010

[486] Konrad, M., Gol, F., Horakh, J.: Gut Holz ohne Cobalt, Farbe und Lack, Vincnetz Network, June 2015

[487] Gezici-Koç, Ö., et al.: In-depth study of drying solvent-borne alkyd coatings in presence of Mn- and Fe-based catalysts as cobalt alternatives, Materials Today Communications, 22–31, 2016

[488] Patent: WO2001064795A1, Steinert A., Borchers GmbH, 2001

[489] Grüll, G., Forsthuber, B., Tscherne, F.: Farblose Beschichtungen – Sonnenschutzmittel für Holz im Außenbereich? Wiener Holzschutztage, 2010

[490] unknown: Formulating with HALOX Tannin Stain Inhibitors for Improved Coating Performance, Presentation Company Halox

[491] Grüll G., Tscherne F., Forsthuber B.: Abschätzung der Lebensdauer von Holzbeschichtungen, Wiener Holzschutztag 2013

[492] Wenker, E.: Comparing the autoxidative mechanism of Co-EH and FeONIX, Leiden Institute of Chemistry, 2010

[493] TRGS 905: List of carcinogenic, germ cell mutagenic or reprotoxic substances, 08.06.2017

[494] Van Gilder, S.: Demystifying Decorative Foil Finishes, Surface and Panel 2012

[495] Painted, Printed and Foiled Wooden Surfaces; Economy and Technology, Swedish Paint and Printing Ink. Makers Association, 2011

[496] Forsthuber, B., Grüll, G., Tscherne, F.: Farblose Beschichtungen – Sonnenschutzmittel für Holz im Außenbereich? Holzforschung Austria 2010

4 Pre-treatment of wood and wood-based materials

The surface of wood or wood-based materials has to be pre-treated and properly prepared prior to the actual coating. Different procedures of pre-treatment can be applied depending on the type of wood and wood-based material as well as their intended application and requirements profile:

- Drying of sawn wood
- Grinding/smoothing/brushes
- Removal of resins and/or removing of glue from bleeding
- Bleaching
- Watering

The issue of drying of sawn wood and sawn wood-based materials should be left unaddressed in the scope of this book. It is applied to dry the wood to a set humidity, so that the wood achieves an appropriate and prescribed wood humidity during the processing.

High resinous woods have to be removed of resin prior to the coating since otherwise the resin later may bleed on the surface. The removing of resin can be done by saponification of the resins by means of a diluted alkaline solution, or the resins are dissolved by means of solvents on the surface and subsequently wiping-off of the solvent. After desiccation at the air, in both techniques one may grind using fine abrasive paper. In practice, solid woods such as pine are not removed of resin. Adhesives or glues from bleeding may cause disturbances in the pickling process as well as bright spots and adhesion problems. The removal of the residues always depends on the type of the adhesive or glue and is done by different chemical solutions which will not be described here.

4.1 Grinding

Due to the hygroscopic properties of the wood, in direct contact with a coating material the wood fibres suck up a coating material up to their fibre saturation limit. Thus, the wood fibres swell, place themselves on the surface like bristles and disturb the coating. Therefore, all wood-based materials should be grinded already prior to the actual coating in order to achieve a surface as smooth and even as possible. For this reason: the finer the pre-grinding, the lower is the risk of uprising wooden fibres, and the more paint-saving is the process of coating. The grinding also is applied for the calibration of substrates on the required dimensions (calibrated finish), the removal of scratch marks, splitters, glue residues and adhesive residues, other surface defects as well as for ensuring a sufficient adhesion between substrate and paint or between two coating layers (intermediate grinding). Therefore, a distinction is made between:

- Calibrated finish (coarse grinding of solid wood and wood-based materials, removal of adhesive residues, thickness grinding)
- Pre-grinding of wood (fine grinding of solid wood, veneered surfaces as well as wood-based panels)

- Preliminary and intermediate coating grinding (creation of a flat base area, removal of roughness, removal of accents (paint dust))
- Coating finish grinding (surface finish)

The grinding of wood and wood-based materials is a mechanical shaping process. Due to the cutting movement, the individual abrasive grits dive into the material and withdraw chips similar to a slicer blade. The chips accrue as grinding dust and have to be disposed. Since the geometry of the abrasive cuts is not precisely defined, the grinding also is termed as a cutting shaping processes with indefinite cutting geometry. Depending on the material and material geometry, one differentiates between [2]:
- Surface grinding: broadband grinding/narrow surface grinding
- Profile grinding
- Form small parts grinding

4.1.1 Abrasive and abrasive carrier

The most commonly applied abrasive material in the processing of wood and wood-based materials are abrasive on carriers according to DIN ISO 2976. Paper, tissue, as well as combinations of both are applied as carriers for the grinding of wood and wood-based materials. The applied abrasive papers usually have a mass of 180 to 300 g/m². The mechanical strength as well as the bending strength increase with the increasing mass of the paper. Abrasive papers are characterized by a low elongation, very regular surface and relatively low costs. A problem of the papers is that these are hygroscopic, can easily become warped at different exposures to humidity can warp and cannot function properly.

Fabric carriers are differentiated by their bending strength (flexibility). So-called X- and Y-carriers are applied for the grinding of wood and wood-based materials. Cotton or fabric carriers are applied as fibre materials. The mass of X fabric carriers is approximately 450 g/m², while the mass of Y fabric carriers is approximately 550 g/m². For special applications, especially heavy fabric carriers with a mass of more than 1,000 g/m² are applied. These fabric carriers have a very high tensile strength and bending stiffness. Fabric abrasives are characterized by a very high tear strength as well as tear propagation strength and are largely insensitive to fluctuations in humidity. Combinations of paper and fabrics are applied as abrasive carrier for the particularly strong mechanical requirements during grinding of chipboard panels and MDF.

The abrasive grit is scattered on the surface and fixed on the abrasive carrier with base attachment and covering attachment which consist of synthetic resins or glue in case of modern abrasives. A scattering of corundum (Al_2O_3) has been proven for the processing of solid wood, while a silicon carbide abrasive grain (SiC) has been proven for the coating grinding and the grinding of wood-based panels. Zirconium corundum is a special type of corundum. It is a microcrystalline mixed oxide of zirconium oxide (ZrO_2) 45 % aluminium oxide. This abrasive is characterized by a particularly high toughness. During wood processing, it is applied in high-performance grinding of solid wood and in particular in the grinding of parquets. Ceramic cutting grain materials are a more recent development. With "Trizact", the company 3M sells a new grinding material with a defined structure. As the company 3M claims [3], the structured surfaces of "Trizact" are based on a precise, three-dimensional arrangement of very small mineral particles. Continuously new surfaces of the grinding particles are released during the grinding process. This is related to a much longer lifetime of the grinding belt.

Table 4.1: Applications of abrasives on wood and coated wood according to[2]

Requirement (type of wood/coating system)	Carrier paper (grammage)	Abrasive parameter		
		Type of grain	Coating density	Corn size no.
Pine parallel to the fibre	D (146–185 g/m²)	Corundum (Al_2O_3)	Open	P 120 P 150
Pine vertical to the fibre	E (218–242 g/m²)	Compact grain (Al_2O_3)	Tight	P 220 P 240
Oak parallel to the fibre	F (300 g/m²)	Corundum (Al_2O_3)	Tight	P 120 P 150
Oak vertical to the fibre	Not mentioned	Zirconium corundum (ZrO_2/Al_2O_3)	Tight	P 220 P 240
Chipboard panel	Combination of fabric and paper	Silicon carbide	Open	P 80 (100) P 120
MDF	Combination of fabric and paper	Silicon carbide	Slightly open	P 400
Intermediate coating grinding	D, E	Ceramic	Dense	P 400
Smoothing of paraffin-containing polyester	C (114–126 g/m²)	Silicon carbide or ceramic	Dense	P 600–P800

The grain size as well as the scattering density are of particular importance are of particular importance. In terms of the grain size, a distinction is made between macro grains (grain size P 12 to P 220) and micro grains (grain size P 240 to P 1200). The smaller the grain, the greater the number P. Different spreading densities result in different types of scattering of abrasive grains. A distinction is made between slightly open, open, wide open as well as dense scattering (> 80 % of the area are covered with abrasive grains) in the areal coverage with abrasive grains. In most cases and particularly for the finer grain sizes, a dense, almost approximately 100 % comprehensive scattering is applied. An open, nearly 60 % comprehensive scattering is applied for the grinding of resin-rich wood-based materials in order to avoid an excessive heating of the belts. Table 4.1 exemplary illustrates different applications of abrasives according to specified requirements.

Electrical charges of several hundred kV may arise from the grinding of the poorly conductive material such as wood. These charges involve two disadvantages. On the one hand, these charges may result in spontaneous spark discharges, and on the other hand electrostatically charged grinding dust adheres on the abrasives as well as on machine components. As a result, the stability of the abrasive is reduced. An anti-static equipment of the grinding belts at the manufacturer strongly reduces the electrical charging of the abrasive belt.

4.1.2 Grinding procedures and grinding aggregates

In principle, different grinding procedures are applied for the machine processing in the industrial company, i.e. for the surface grinding:

- Longitudinal belt grinding
- Transversal belt grinding

- Brush grinding machines or smoothing machines
- Cylinder head grinding

The longitudinal belt grinding is the oldest belt grinding procedure in the processing of wood-based materials. During the longitudinal belt grinding, the feed rate performed by the work-piece is perpendicular to the movement of the grinding belt. As its name suggests, the special feature of this process is the very long belt with a length of 6 to 8 m – occasionally also of more than 8 m. The belt is pressed towards the surface which is to be grinded with suitable press-on elements. It is differentiated between two process variants: longitudinal belt grinding machine with a printing plate and longitudinal belt grinding machine with a printing bar. This procedure is applied in the manufacture of pianos as well as in the grinding of room doors. One speaks of transversal belt grinding, if the feed rate of the work-piece and the grinding direction of the grinding unit are parallel to each other.

The usual bandwidths are 600 to 1,500 mm whereby most frequently operational widths from 1,100 to 1,350 mm are applied. The transversal belt grinding is one of the most efficient grinding process for wood and wood-based materials. Fundamentally, the transversal belt grinding is performed in a straight course. If more intensive grinding operations are required, two or more grinding units are switched in series. A distinction is made depending on the shape and surface of contact between abrasive belt and work-piece:

- Transversal belt grinding with a cylindrical operational zone, also referred to as transversal belt grinding with contact roller
- Transversal belt grinding with flat operational zone, also referred to as transversal belt grinding with printing bar

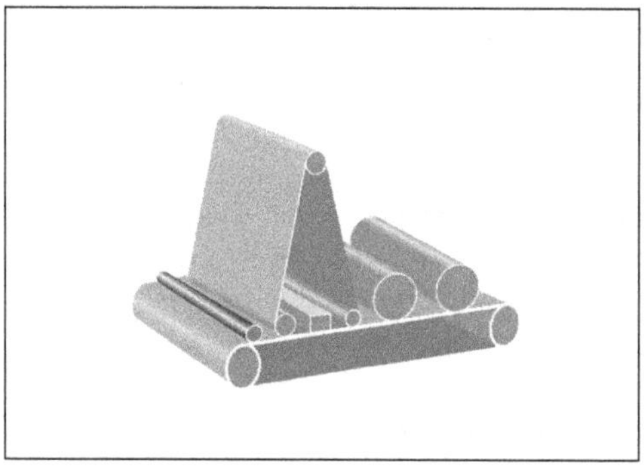

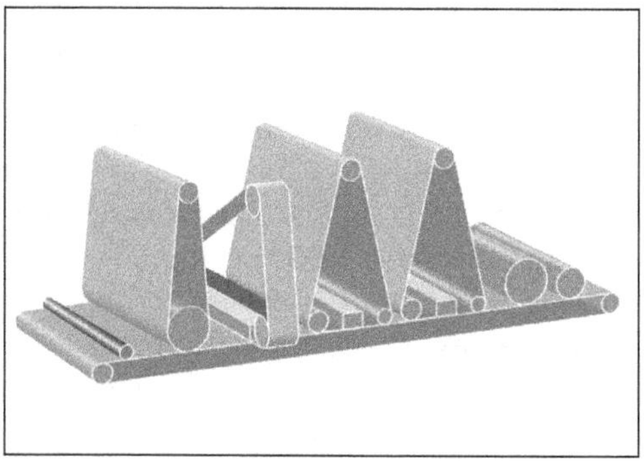

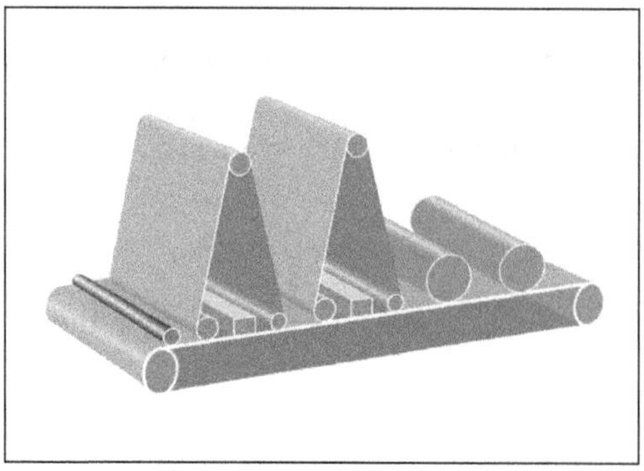

Figure 4.1 (left above): Contact roller and brushing roller aggregate for the calibration of industry parquet and solid wood [4]

Figure 4.2 (right above): Contact roller, transversal grinding unit, two longitudinal grinding units and brushing rollers for the calibration, cross grinding and finish grinding [4]

Figure 4.3 (left): Two longitudinal grinding units and two brushing units for the intermediate coating grinding for higher feed rates [4]

Source: Karl Heesemann Maschinenfabrik

Transversal belt grinding machines with printing bars exclusively are applied to improve the surface quality. However, transversal belt grinding machines with support rollers or contact rollers have to produce a flat surface and often are placed prior to machines with printing bars. This results in a classic combination of the transversal belt grinding process. The pre-grinding is performed with a cylindrical belt support, while the finish grinding is performed with a flat belt support. Depending on the task, i.e. depending on the thickness of the layer to be grinded as well as on the state of the grinded surface, also two grinding units are switched in series, whereby for example differently hard sheaths of support rollers or contact rollers, or different constructions of the supporting surface of printing bars determine the outcome. If solid wood panels or veneers are grinded, it is known that more material is grinded when grinding across the fibre when grinding parallel to the fibre under the same conditions. This fact is applied when cross grinding. In the process, the longitudinal belt grinding is combined with the transversal belt grinding.

A cross-grinding longitudinal belt as the first grinding element in a combination has the following advantages:
- The largest grinding performance is achieved across the grain
- The contaminants on the surface of the work-piece are reduced
- A greater flatness of surface is achieved
- The costs of a cross belt considerably are lower than for a transversal belt

The cross-cutting mainly is applied in the processing of surfaces in the manufacturing of furniture, doors and panels as well as in the interior construction. It does not matter if volume is grinded; what is important is to produce high-quality surfaces. If there is the requirement to calibrate the work-piece in addition to a regular surface grinding, then a hard contact roller as a first element of the grinding unit is applied. Various aggregates are successively installed in series in modern grinding machines whereby the abrasive belts may have different directions of rotation. The surface has to be purified subsequently to each performed grinding process since the pores are enriched with grinding dust. Brushing machines with suction of air applied in industrial manufacturing in order to remove the grinding dust.

If soft forming edges and profiles shall be smoothed, then special edge grinding machines and profile grinding machines are applied. Coating grinding disks with different grain sizes are applied for an intermediate grinding of pre-coated edges. Brush grinding machines are preferred for the intermediate grinding and smooth grinding in the frame editing. The Figures 4.1 and 4.2 illustrate possible arrangements of grinding machines for the calibration ("customer-oriented") of industrial parquets and solid wood. Figure 4.3 illustrates a possible arrangement of grinding machines for the coating intermediate grinding.

4.2 Smoothing process

4.2.1 Mechanical smoothing

The surfaces additionally can be smoothed after grinding or coating. In this way, a smooth grinding is performed with paraffin-containing polyester surfaces using brush rollers and a SiC paper with a granulation of P500 to P800. Following the same procedure, an intermediate grinding machine is applied for a UV curing oil for parquets.

4.2.2 Thermo smoothing

The principle of this procedure involves a pressing of wood fibres being exposed from the previous milling procedure into a plasticized subsurface by means of a bladeless tool being fitted to the contour of the milled profile under simultaneous impact of heat and pressure and subsequently fixed permanently. In Chapter 3.1.9, the process of thermo smoothing is described in detail.

4.3 Bleaching

Wood colours may vary as a result of exposure to light and especially by UV radiation in the sunlight. Thus, woods such as rosewood can be brighter, and other woods such as macore become darker. Wood surfaces that are different in colour may achieve a nearly uniform colour structure by bleaching. In addition, stains are removed, and areas are brightened areas. Bleaching is an oxidative process which is carried out only with hydrogen peroxide (H_2O_2) and small amounts of alkaline bleaching activators. Due to the oxidation of wood-based materials (wood dyes), the wood becomes brighter and more uniform in colour. Subsequently to the bleaching, the surface is treated with a stain in order to achieve the original or the desired colour of the wood.

A 35 % hydrogen peroxide solution and a bleach activator, usually a 10 % solution of ammonia, in the ratio of 1:5 or 1:10 has been proven successfully. While an odour nuisance due to the ammonia solution is to be expected in this treatment, however no salts on the surface remain in this bleaching process. It is a lengthy process that takes at least 16 hours at ambient temperature. The process can be accelerated by increasing the temperature. However, an installation of a drying equipment is essential. The relevant bleaching agents are applied on a surface by means of a sponge, a brush or spray gun. Subsequently these bleaching agents are dried for a period of time specified by the manufacturer specified time soak. The surface is washed with water or water vapour and dried for further processing.

4.4 References

[1] Argyropoulos, G. A.: Vol. 2: Schleifen plattenförmiger Werkstücke, Fachbuchreihe Holzbearbeitung, AFW Werbeagentur GmbH, Kassel, 1991

[2] Rothkamm, M.: Mechanische Bearbeitung des Holzes für Beschichtungen, Lackhandbuch Holz, DRW Verlag Weinbrenner GmbH & Co., Leinfelden Echterdingen, 2003, p. 77–108

[3] http://solutions.3m.com/portal/3M/de_DE/Abrasive-Systems/

[4] Figures provided by the company Karl Heesemann Maschinenfabrik GmbH & Co. KG, Bad Oeynhausen

5 Application process for the coating of wood

Several application processes are used for the coating of wood and wood-based materials due to the variety of requirements, due to the quality of the surface as well as due to the geometry of the parts to be coated. The following procedures will be applied in the industrial coating:
- Dipping
- Flowing
- Tumbling
- Spraying
- Curtain coating
- Rolling
- "Vacumat"

5.1 Dipping

The oldest and most simple application process for the coating of wood with a coating layer consists of a dipping of the wooden material into a container filled with a coating, to pull it out again, to drip off the excess coating and to dry. In comparison to the spraying procedure, the dipping procedure has a higher material yield and lower investment costs. The great advantage of the dipping procedure is the enhanced application efficiency of 85 to 99 %. Disadvantageous is the difficultly controllable thickness depending on the pre-treatment of the wood, moisture content and the residence time in the dipping container. The dipping process equally is suitable for the coating of wood-based materials as well as for the coating of larger single components such as wooden windows. Nowadays, the dipping process is applied in the coating of furniture components consisting of solid spruce or pine, wooden chairs, wicker works, wooden pencils and folding rules. In recent years, the dipping process grows in significance in the economic coating of inexpensive furniture components. The finish qualities achievable in the dipping process are not suitable for high-quality furniture components. Within the industrial dipping procedure, the work-pieces (such as windows) are dipped into a low-viscous coating by means of a cycle procedure or run-through procedure and pulled out. Subsequently to this, the work-pieces pass a drip-off line and are put into a drying oven. The drip-off zone often including a heated blow-off zone is of particular importance in order to avoid surface defects due to the formation of bubbles and coating sagging. In the case of construction units for outdoor elements, the dipping procedure is almost exclusively used for the impregnation and priming of windows and garden fences since the optical surface quality does not longer meet the current requirements. In terms of occupational hygiene and in order to avoid the risk of fire during the processing, exclusively one-component water-borne coatings are applied nowadays. The dipping pools should be equipped with an automatic coating circulation in order to avoid a possible settling of pigments and other components.

Table 5.1: Advantages and disadvantages of the flow coating [2]

Advantages	Disadvantages
– Possible for floatable objects	– Possible foam formation in the coating
– Risk of settling due to a rapid recirculation is lower than due to dipping	– High solvent losses result from the application of solvent-containing coatings
– More advantageous for large components, as these require larger dip tanks otherwise	– Due to qualitative reasons, the flow coating actually can be applied for the primer and intermediate coating of components
– Reduced application of coatings due to the recirculation of the coating	
– The coating application rate in the storage tank is lower in comparison to the spray application with the same number of pieces	

5.2 Flow coatings

In the flow coating process, the coating material is applied on the passing wood-based materials by means of directed spray nozzles. The excess coating material runs off, is collected in drip pans and fed back to the process after a multi-stage filtration process. The application efficiency amounts 85 to 99 %. Currently, the flow coating process mainly is applied for the intermediate coating, for the priming as well as for the intermediate coating of wooden windows and components for outdoor area depending on the quality of the coating material. The flow coating process is a very low-cost procedure. The industrial facilities consist of a closed flow-coat system, suspension system (such as B. Power & Free conveyer) and the control and regulation unit. The costs of acquisition for a complete fully automated system are between 20,000 and 100,000 Euro depending on the requirements. The essential advantages and disadvantages of the current flow coating or the applied coating systems, respectively, are presented in Table 5.1. Since 2005, the flow coating process successfully is applied for the industrial wood staining and priming of disassembled wooden chairs. Here, only water-borne wood stains and coating systems are applied.

5.3 Tumbling

The tumbling process has been proven for the coating of small parts (children's toys, wooden buttons, etc.). The small parts and the coating are put into the barrel. The coating is carried out in a rotating barrel, sheathed with a housing. More modern procedures facilitate a coating spraying by means of spraying guns during the movement of the barrel as well as a simultaneous drying of the small parts by supplying heated and filtered fresh air without adhesion. The incoming air can be heated up to a temperature of 150 °C. The small parts are kept in motion in the barrel as long as these are dry completely. Mainly water-borne coatings are applied. The barrel process also is characterized by a high application efficiency of 95 to 100 %.

5.4 Spraying

Due to its slight ease of handling and great flexibility, the spray application is essential for the wood coating and the most commonly applied procedure. Intricately shaped parts as well as a frequently requested variety of colour shades facilitate any other processing methods. Despite the drawbacks of sludge volumes due to coating losses when spraying, due to the high emissions of solvents and due to the enhanced energy consumption by exhaust air, spraying procedures and coating systems are optimised and adapted furthermore. Mainly the pneumatic atomisation (air pressure atomisation), the hydraulic atomisation (airless) as well as the hydro-pneumatic atomisation (airless with air support) are applied. In this atomisation process, the degree of the utilisation of coating materials amounts between 20 and 50 % depending on the geometry of the components [3]. When applying the low-pressure process (HVLP), the degree of utilisation of coating materials is increased by approximately 15 to 30 % in comparison to the compressed air atomisation in the high-pressure process. Even higher degrees of utilisation of the coating materials can be achieved by applying the electrostatic coating process. However, since the work-piece wood often does not have a sufficient electrical conductivity, the electrostatic procedures only are applied in special cases. The electrostatic coating of chairs and windows is known. The chairs are coated electrostatically in so-called omega loops by means of high speed rotational discs. In practice, however, the components have to be finished manually depending on the geometry of the components. The not always homogeneous wood moisture of ≥ 8 % as well as the continuous assurance of the ground contact during the coating process impede the further distribution. For years, different plant concepts successfully are applied in the industrial coating of wood-based materials [4]:

- Manual spray booths with turnstile or hanging of components, respectively
- Flow spray booths with overhead conveyors, ski conveyors or transport tables
- Surface spraying machine in continuous flow for flat components
- Surface spraying machine in a batch process for flat components (spraying robots)

The decision what plant concept is the best is made by the user depending on the geometry of the components, belt coverage density, number of coated wood substrates per day in square meters, number of colour shades, surface effect or surface quality, respectively, as well as the maximum costs per square metre for the coating.

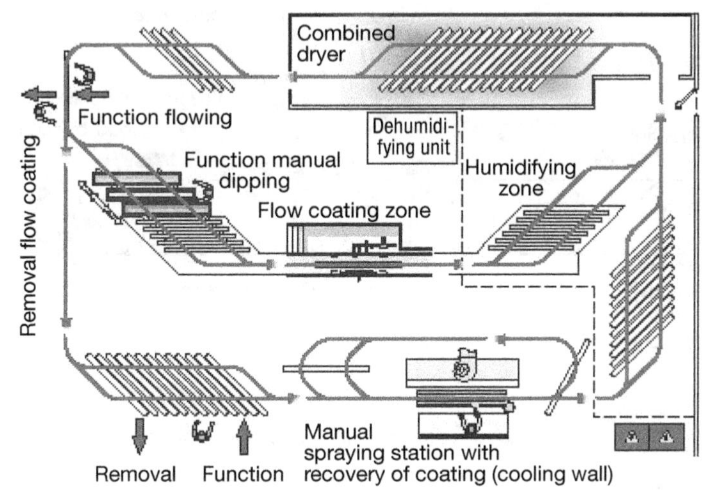

Figure 5-1: Combination method (dipping, flowing and spraying) for the coating of wooden windows [1]

5.5 Curtain coating

In the curtain coater process, almost flat and slightly shaped substrates pass through a coating curtain and receive a uniform coating application on one side. The procedure is very interesting since the application efficiency almost is 95 to 99 %. This procedure mainly is applied in order to achieve high-quality smooth topcoat surfaces in the front area of the furniture, in the interior design and in the industrial manufacturing of doors. In comparison to the rolling procedure where only flat work-pieces can be coated, the curtain coater is suitable for slightly curved or differently shaped substrates. A fundamental prerequisite is that all positions of the substrate surface can be achieved by the coating curtain. Transparent and pigmented 2C PUR coatings, UP coatings, UV/EB coatings and water-borne 1C coatings as well as water-borne UV coatings are applied. A curtain coater machine is equipped with a curtain coater head reaching over the entire width of the machine. On the underside of this curtain coater head there is a gap which is adjustable in its width [5]. The curtain coater coating leaves the curtain coater head through the gap as a homogeneously closed film in order to coat the surface of the substrate. The work-piece is transported under the curtain coater head by means of a conveyor belt. The coating which does not wet the work-piece runs back into the coating reservoir within a closed system over a trough of metal, so that there are only minimal coating losses. A conveyor pump again conveys the coating from the coating container into the curtain coater head (see Figure 5.2). The coating exactly can be adjusted to the applied quantity due to the adjustability of the curtain coater lips in the curtain coater head or by variation of the throughput speed of the work-piece. The curtain coater machines generally applied in the practice differ in the design of the curtain coater heads. Two variants are known [7]:

- Closed, universal curtain coater head
- Overflow curtain coater head

There are several settings at the closed curtain coater head:

- Pressureless curtain coater

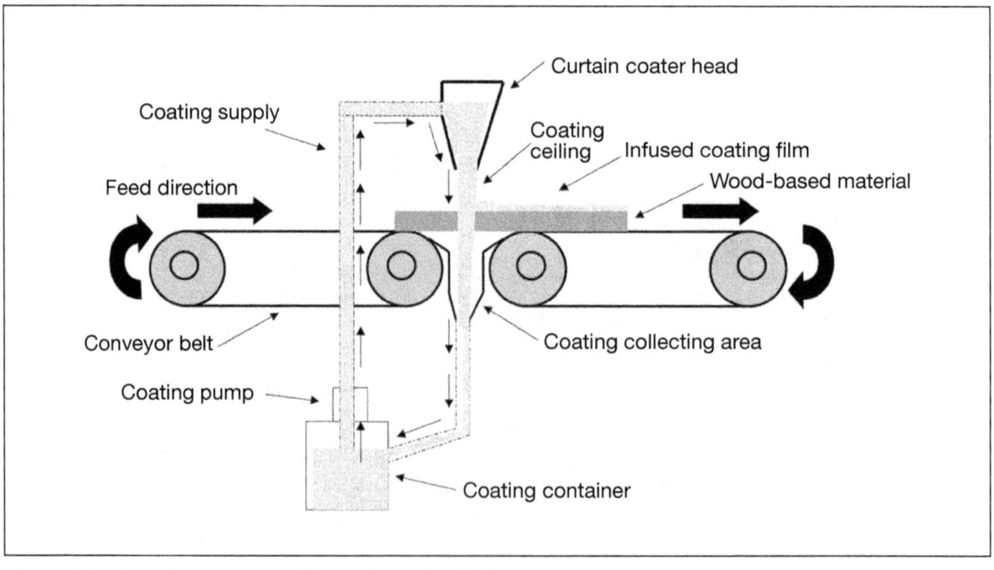

Figure 5.2: Curtain coater machine, schematically [8]

- The coating flows through a curtain coater gap, the coating application rate is controlled by the distance of the curtain coater lips

Curtain coater with low pressure drop (negative pressure curtain coater head)

The adjustment of a defined negative pressure in the curtain coater head facilitates a stripe-free realisation of low application amounts. The discharge velocity of the coating from the gap decreases. Thus, the gap width can be increased again.

Slight excess pressure by closing the overflow line

At the overflow system, the coating film takes advantage of the gravity in order to flow via a so-called dosing bar. This variant preferably is applied for the usage of pigmented curtain coater coatings as no pigments may deposit in the curtain coater gap. A newer possibility is the dosage of a metering roller. A metering roller in the curtain coater head conveys a certain amount of coating from a coating container supported by the rotational speed. The machine operator can adjust the required amount of coating quickly and very accurately by means of a speed regulation of the metering roller. The amount of coating on the metering roller is stripped off contactless by means of a drain-off lip and subsequently forms a curtain coater film. Such rolling heads/curtain coater heads do not work with a curtain coater gap. This eliminates the settling of pigments and impurities. The curtain coater units/rolling units particularly are suited for the foam-free and streak-free application of water-borne UV coatings. These technical variants facilitate the processing of coating systems with a discharge time of approximately 20 to 70 s in the 4 mm discharge cup. The application amounts are between 50 and 400 g/m^2. The curtain coater process is applied in a wide variety of process variants for the processing of unsaturated polyester coatings (UP coatings) in the piano industry. The following processes are known:

- Contact curtain coater process („active basic coat procedure")
- In a first step, a peroxide primer (hardener) is applied on the surface to be coated (65 to 80 g/m^2) in the curtain coater procedure. Subsequently, the primer is dried at ambient temperature or in a convection dryer at a low temperature. This is followed by an application of the UP curtain coater coating in the curtain coater process.

Double-head curtain coater procedure or (two-head curtain coater procedure)

A subset of the UP coated is mixed with an accelerator (metal salts) while the other subset is mixed with hardeners (peroxide). The type of coating with the accelerator is placed in the first head of the two-head curtain coater machine, while the type of coating with the hardener is placed in the second head of the two-head curtain coater machine. Both subsets are applied in a single operation. The application of the coating is repeated after a gelation time of approximately 15 to 20 minutes. The processing of thixotropic UP coatings (paraffin containing) which subsequently are wobbled tends to a certain scaredness or leather grain of the coating surface, respectively, in the double-head curtain coater procedure. The surface defects "scar" are a visual disturbance of the coating film surface according to the wobbling process and polishing process. The scaredness is visible under a certain viewing angle and by means of fluorescent tubes. A modification of the double-head curtain coater procedure, the so-called sandwich technique, permits high gloss surfaces which are similar to those of the UP coatings processed in spraying and do not exhibit a pronounced scaredness.

Sandwich process

The accelerated UP coating as well as the sandwich-hardener (peroxide solution in styrene-free UP resin) is applied wet-in-wet by means of a two-head curtain coater machine. The accelerated UP coating again is applied in a thick layer after a flash-off time of approximately 15 minutes. Normally, an additional finish coating layer is poured after a further gelation which merely serves as a grinding and polishing aid.

5.6 Rolling coater process

The rolling process exclusively is used for a single-sided or double-sided coating of flat wooden work-pieces. The coating is applied by means of a rotating rubber roller. The hardness of the used rubber rollers is 40 to 50 on average. The application dosing is performed by means of a chrome metering roller. The contact pressure or the gap width between the metering roller and application roller, respectively, influences the application amount. Other influencing factors are the coating viscosity and the rheological behaviour of the coating systems, speed of the conveyor belt in m/min, peripheral speed of the applicator roller and dosing roller. The rolling process mainly is applied for the application of approximately 100 % UV coating systems, wood stains, adhesion promoters and aqueous coating systems. The wet application amount per roller unit usually is between 4 and 120 g/m^2 for roller filling machines (Figure 5.3 and 5.4). During rolling, a distinction is made between the unidirectional principle and reverse procedure.

Unidirectional principle

The underlying unidirectional principle is that the applicator roller and the work-piece which is to be coated always run in the same direction (Figure 5.5). For today's coating applicator rollers, the dosing roller usually works with an autonomous drive motor. The advantage is that the dosing roller can be driven in line with the applicator roller as well as in opposite direction with the applicator roller. In addition, the circumferential speed between the applicator roller and metering roller can be varied. This reduces the material abrasion between the two rollers, and a smoothing effect of the coating surface is created. Table 5.2 illustrates the factors affecting the unidirectional principle (dosing roller in line with the ap-

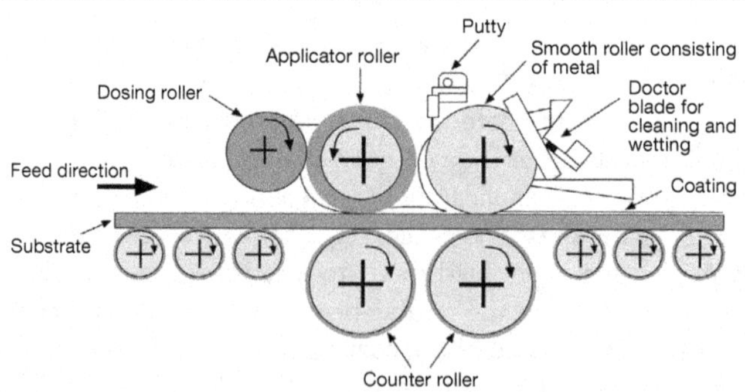

Figure 5.3: Schematic illustration of a heavy coating putty machine[8]

Table 5.2: Influencing parameters/effects for the unidirectional principle; dosing roller in line with the applicator roller[9]

Processing parameters	Impacts
Increase of the velocity of the dosing rollers	Increase of the application amount
Lowering of the velocity of the dosing rollers	Lowering of the application amount
Increase of the velocity of the applicator rollers	Increase of the application amount, structural effect, risk of chatter marks and inlet edges (applicator roller circumferential velocity maximally 1.5 m/min difference to the velocity of the conveyor belt)
Lowering of the velocity of the applicator rollers	"Coating effect"; the wave structure decreases
With a metal doctor blade or plastics doctor blade on the dosing roller and applicator roller	Smoother coating surface; lower material application; plastics scraper usually causes a lower formation of stripes compared to a metal scraper (very often, the dosing roller is operated without a scraper)
Lowering of the contact pressure of the rollers	Increase of the coating application amount

plicator roller). In opposite to an adjustment at which the dosing roller is driven in line with the applicator roller, the adjustment at which the dosing roller is driven in opposite direction with the applicator roller (also referred to as relative procedures) facilitates an optimum control of the application amount. This process variant results in significantly smoother surfaces of the rollers, since the merging of the dosing roller and the applicator roller does not split the coating film (Figure 5.6). The combination of one or more of such rollers in the area of topcoats, for example with and without UV gelation of the coating surface, facilitates a regulation and improvement of the surface smoothness, the coating tension and the gloss level of the resulting furniture surface (see Chapter 3.1.6). Table 5.3 compiles the parameters affecting the unidirectional principle for the dosing roller driven in opposite direction with the applicator roller.

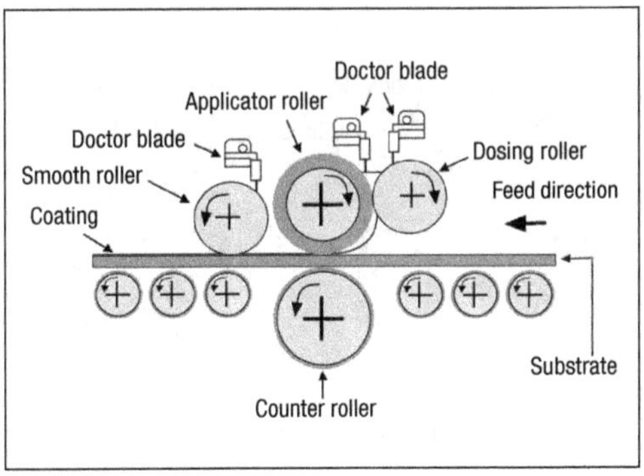

Figure 5.4: Schematic illustration of a lightweight levelling machine

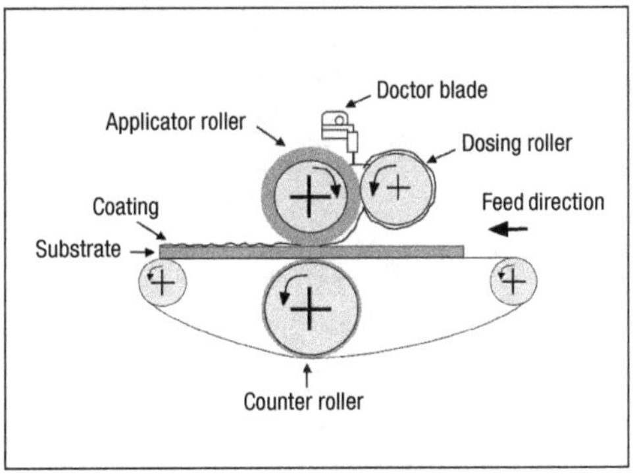

Figure 5.5: Rolling machine in unidirectional principle, dosing roller in line with the applicator roller[9]

Table 5.3: Influencing parameters/effect for the unidirectional principle; dosing roller in the opposite direction with the applicator roller[9]]

Processing parameters	**Impacts**
Increase of the velocity of the dosing rollers	Lowering of the application amount; finer structure; too fast = risk of stripe formation
Lowering of the velocity of the dosing rollers	Increase of the application amount; more structure
Increase of the velocity of the applicator rollers	Increase of the application amount; more structure; push effect; inlet edge
Lowering of the velocity of the applicator rollers	"Coating effect"; the wave structure decreases
With a metal doctor blade or plastics doctor blade on the dosing roller and applicator roller	Smoother coating surface; the plastics doctor blade causes fewer stripe formation compared to a metal doctor blade; (the dosing roller always is applied by means of a doctor blade)
Lowering of the contact pressure of the rollers	Increase of the application amount

Reverse procedure

Further efforts to reduce the existing rolling structure in the unidirectional procedure or relative procedure, respectively, led to a development of the reverse procedure. This rolling procedure is characterised by the fact that the applicator roller runs in the opposite direction of the transport of the work-piece. Thus, very smooth and almost structure-free coating surfaces are achieved since the splitting of the coating film between the applicator roll and the wooden substrate completely is avoided. Application amounts of more than 120 g/m^2 can be achieved. Today, the reverse procedure is practiced in combination with a rolling unit located upstream from the unidirectional process. In the unidirectional procedure, the wooden pores or the veneer joints, respectively, are filled. In the subsequent reverse procedure, the applied coating layer is smoothed optimally. Depending on the dosage and the circulation velocity, the final application amount is affected by the reverse unit. Following the coating procedure, the first reverse aggregates showed an enhanced coating application (coating bulge) such as in the production of interior doors and especially in the area of inlet edge and discharge edge of doors. Under plant engineering aspects, this fundamental problem could be solved elegantly

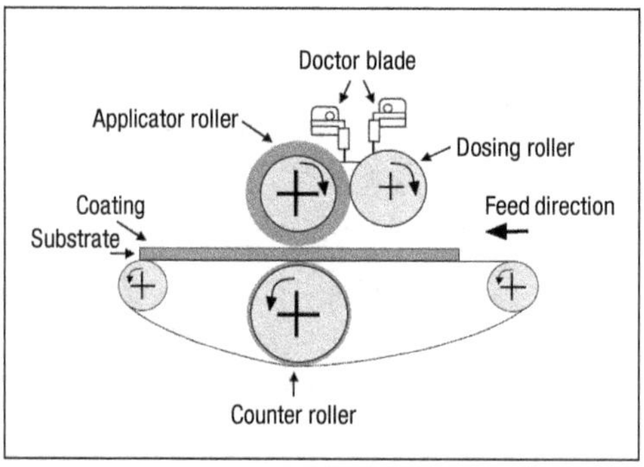

Figure 5.6: Rolling machine in unidirectional principle, dosing roller in opposite direction with the applicator roller[9]

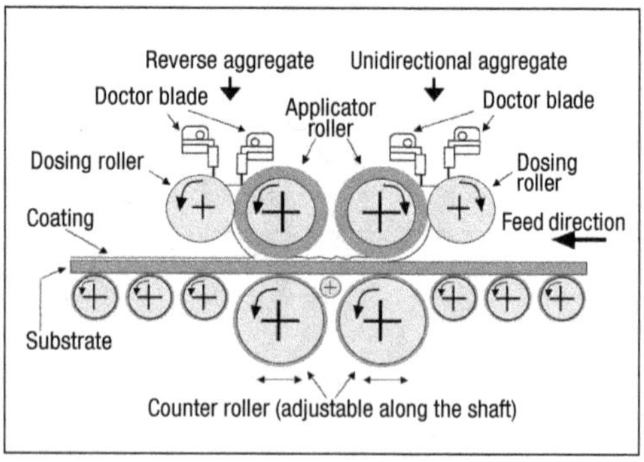

Figure 5.7: Combination rolling machine with unidirectional and reverse aggregate[8]

by means of a separate control of the rubberised bottom counter rollers (axis shift). Figure 5.7 illustrates a combination rolling machine with a unidirectional unit and reverse unit.

Grooved rubber roller („Laser rollers“)

Coating surfaces similar to curtain coatings can be achieved due to the utilisation of grooved rubber rollers in the UV topcoat unit. The applied rubber rollers are sorted according to the number of grooves per inch (25.4 mm). In the primer area, very often grooved rollers with a gear number of 64 and a gear number of approximately 80 to 84 for the UV topcoat application – depending on the manufacturer – are applied. A gear number of 64 means that there are 64 grooves per inch. Depending on the manufacturing process, there exist laser-cut or polished rubber rollers. Depending on the manufacturer, different grinding angles partly are applied, i.e. the rollers may exhibit different surface roughness. In extreme cases, this means that grooved rollers from various manufacturers can produce different results of rolling despite the same number of gears. For the application in the coating of furniture, the best shore hardness of the rollers is approximately 25 to 30. Very often, the topcoat application is applied in a combination of a smooth roller wet-in-wet with a grooved roller (gear number 80 to 84) with a wet application amount of approximately 23 to 25 g/m^2. In addition, combinations of two grooved rollers working wet-in-wet are applied for the application of high-gloss rolling topcoats. Here, the application amounts are approximately 30 g/m^2. The UV topcoat application with a grooved rubber roller on a polished UV primer leads to non-optimal results. A contact pressure auf 1.2 to 1.8 mm towards the work-piece is necessary in order to guarantee a good coating profile in the application of grooved rubber rollers (gear number 80, shore hardness approximately 25).

5.7 'Vacumat'

Aside from the more common application processes spraying, curtain coater and rolling for the wood coating with UV coatings of rods, bars, panels, profiles, edges, ribs and windows, the 'Vacumat' application increasingly gains in importance. The 'Vacumat' coating initiates not only a reduction of the organic solvent emissions, but also a saving of the injection overspray from the coating sludge. The application efficiency in the 'Vacumat' process is approximately 100 %. The first 'Vacumat' plants were put into operation nearly 35 years ago. A 'Vacumat' coating unit essentially consists of the following components:

- Coating conveying system (coating pump)
- Application chamber
- Vacuum pump
- Coating separation system

Functional principle

A coating pump (double diaphragm pump) is applied in order to pump the coating from a storage tank over a flowing frame on the work-piece in the application chamber of the 'Vacumat'. The coating is swirled up by an air suction (negative pressure) and sucked upwards. The substrate is transported through the resulting coating fog. A uniform coating is performed from all sides whereby the inlet and outlet of the application chamber corresponds to the cross section of the wood to be coated and facilitates an air circulation in order to produce the necessary vacuum in the entire machine (vacuum pump). The distance between the matrix and the work-piece has to be only a few millimetres. The air suction has a double function:

Table 5.4: Process parameter 'Vacumat'[9]

Parameter/determining parameters	Impacts
Feed rate: slower higher	Low application amount enhanced application amount
Vacuum setting: higher vacuum (lower pressure) lower vacuum (higher pressure)	Low application amount enhanced application amount
Margin of the matrices to the coated substrate	Inexact matrices lead to vacuum losses and irregular application amount. Margin maximal 1.3–1.5 mm
Soil/impurities	Surface defects, fluctuations of the gloss level, abrasive surface. Improvement of the suction in the grinding process: Application of the coating filter automatic system. Enclosure of the application heads and supply of filtered air
Coating temperature: too high temperature too low temperature	May lead to thickening of the UV coatings. Impact of gradation, UV reactivity, optional coating temperature is approximately 30–45 °C
Application of unsuitable pumping systems	May lead to foaming of the coatings: Double diaphragm pump with a gear transmission of 1:1 are recommended

at first, the air suction has to swirl up the coating colour in the swamp of the container. Furthermore, the air suction has to remove the excess coating at the discharge matrix in form of a laminar air flow and to repatriate into the application chamber, where it again is applied. The removal process causes a smoothing of the surface. In order to achieve an enhanced application efficiency, the swirled-up coating has to be separated from the air. This process occurs in the coating separation system. A separation of the coating material and the air is achieved by means of this process. Subsequently, the coating again accumulates in the coating tank.

Process parameter

The control of the application amount in a "Vacumat" can be influenced by means of the essential parameters shown in Table 5.4.

5.8 References

[1] Company brochure: Holzlackierung, Firma Eisenmann Maschinenbau, Böblingen

[2] Hora, G.: Flutlackieurng von Holzbauteilen, Fraunhofer-Institut für Holzforschung (WKI), Braunschweig, presentation on the occasion of the12th Holztechnischen Kolloquium in Braunschweig, 2004

[3] Prieto, J.: Holzlackieren. Published by DFO, DFO-Lehrgang Lackieren von Holz und Holzwerkstoffen, Braunschweig, 2001

[4] Hoffmann, U.: Jahrbuch für Lackierbetriebe, Vincentz Verlag, Hanover, 2000, p. 423–437

[5] Eichhorn, J.: Jahrbuch für Lackierbetriebe, Vincentz Verlag, Hanover, 2000, p. 195

[6] Frick, A.: Lackieren von Holz/Möbeloberflächen, VDI-Verlag, Düsseldorf, 1979, p. 49–60

[7] Rutkowski, R.: Glasurit Handbuch Lacke und Farben, Glasurit-Werke M. Winkelmann AG, Hamburg – Hiltrup – Berlin, 1969

[8] Unknown: Verniciatrici a Rulo, technical paper by Giardina

[9] Prieto, J., Kiene, J.: „ Lehrbuch der Lacke und Beschichtungen", Kittel Vol. 9, p. 319–351, S. Hirzel Verlag Stuttgart, Leipzig, 2004

[10] Unknown: "Vakumat" Presentation by Schiele Maschinen GmbH, Niederzissen, 2000

6 Recycling processes for coating systems

In recent years, better and better water-borne coatings with residual solvent levels between 0 and 15 % have been developed for the wood processing industry due to environmental requirements and the implementation of the solvents regulation (see Chapter 9.1). Moreover, the overspray of 1C water-borne coatings (air drying, UV curing) can be recovered in form of recycling coatings and processed again. This makes sense, especially when using application methods with high loss of coating spray. This implies not only a relief of the burden on the environment, but also a significant cost savings. The cost pressure on the woodworking industry has grown continuously in the past years. Thus, the application of in-house overspray is appropriate [1]. The following criteria should be met:

- The quality of the recovered coating has to be comparable to the quality of the fresh coating
- A sufficient amount of recyclable coatings in has to be processed
- As few as possible colour changes should be performed
- The plant engineering has to be adapted on the coating system and on the recovery process

The decision for an economically suitable process only can be made individually. Of course, an essential criterion is the consumption of coating. In the recycling of the overspray, it is mainly differentiated between:

- Indirect recovery of the coating from leaching water (ultrafiltration, coagulation)
- Direct recovery of coatings with collection tough
- Belt or scraper system
- Recovery of coatings with cooled collection areas
- Coating-in-coating cabin
- Combined procedures

Today, some of these techniques [1–7] are applied in the wood processing industry; equally for wood in the interior and exterior.

Belt or scraper system

The simplest procedure for a recovery of coatings is realised with drip tapes. Here, the coating overspray deposits on a circumferential band which also can be applied as a conveyor belt for the coating of flat parts (see Figure 6.1). A wiper belt with a V-shaped scraper merges the overspray in the middle of the wiper belt. The overspray is collected in a container and can be reused as a coating. This system is widely applied in the coating of flat wood substrates or wood-based materials.

Recovery of coatings with cooled collection areas

The recovery of coatings using cooled surfaces already is applied in many companies. The principle is that the resulting coating overspray is captured on a cooled wall or column. The humidity of the air condenses at the collection tough by cooling below the dew point of the water. This prevents a drying of the overspray. The overspray may run off or it is doctored. A plant concept is described in Chapter 3.2.3.2. The free design of the collecting area facilitates an adjustment even in existing systems. The low space requirement is an essential advantage.

Coating-in-coating cabin

The coating-in-coating process is based on an overspray recycling with a coating floated cabin. This means that the resulting coating overspray again is bound in the coating at the bulkhead which is sprinkled with the same material. After an adequate filtering, this material can be fed back into the coating process. In the work breaks, an air humidification plant is applied for the air drying of water-borne coatings in order to prevent the drying of the oversprays. When changing the coating, the circulation coating is pumped into a reservoir, and an automatic flushing of the plant with fully desalinated water is installed downstream [2]. Figure 6.2 illustrates the principle of the plant technology.

Figure 6.1: Belt purification with V belt scraper
Source: Venjakob Maschinenbuau

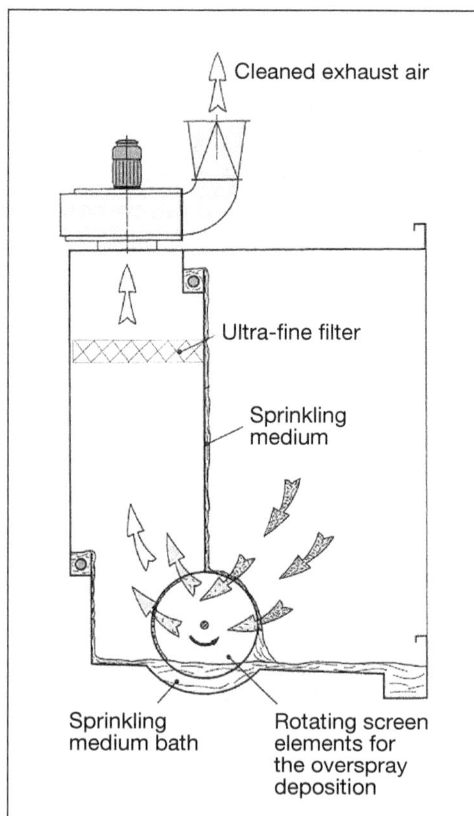

Figure 6.2: Principle of the coating-in-coating spray booth *Source: Rippert Anlagentechnik*

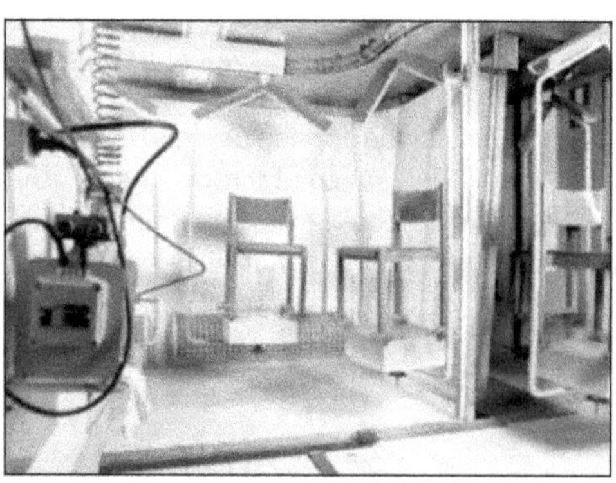

Figure 6.3: Application of the coating-in-coating spray booth for the coating of chairs
Source: Rippert Anlagentechnik

This procedure is suitable for small-scale plants as well as large-scale plants, if the minimum consumption of the coating is more than 50 kg per day. In this process, only one shade of colour can be recovered in pure. All other shades of colour can be recycled as mixed colour. As an example, a plant for the coating of chairs is illustrated in Figure 6.3.

In recent years, also combinations of recycling methods appear on the market (Inlac, Relac, etc.). Before a plant operator opts for a coating recycling technology, the following aspects should be considered for a detailed planning of a plant:

- Variety of the colour shades of the coating plant and incidence of a colour change
- Stability and strength of the coating in the cabin (mechanical stress of the coating, shear of the coating, temperature)
- Drying rate of the collected overspray
- Cleanliness and varietal purity of the incidental coating overspray
- Operational mode of the processes
- Economic efficiency calculation

6.1 References

[1] Prieto, J., Kiene, J.: Holzlackieren, Chapter 9.2.5, Lehrbuch der Lacke und Beschichtungen, Kittel Vol. 9 Verarbeitung von Lacken und Beschichtungsstoffen, S. Hirzel Verlag Stuttgart – Leipzig, 2004

[2] Ondratschek, D.: Rückgewinnungstechnik für Spritzlacke, Chapter 9.1.2.3.2, Kittel Vol. 9 Verarbeitung von Lacken und Beschichtungsstoffen, S. Hirzel Verlag Stuttgart – Leipzig, 2004

[3] Unknown, Overspray-Recycling von Wasserlack im Betrieb – Welche Verfahren gibt es?, Journal für Oberflächenbehandlung (JOT), p. 45 (1998)

[4] Eberhard, G.: Recycling von wasserbasierenden Lacken – Stand der Technik und neue Entwicklungen. In: Proceedings CD 15th World Interfinish Congress and Exhibition, Garmisch-Partenkirchen 2000

[5] W. Heine: Lackrückgewinnung wässriger und lösemittelfreier, flüssiger Beschichtungsstoffe, In: Proceedings CD 15th World Interfinish Congress and Exhibition, Garmisch-Partenkirchen 2000

[6] Acker, G., Holzfensterlackierung in Möbelqualität, Journal für Oberflächentechnik (JOT), S. 32 (1999)

7 Drying and curing processes

The process of film forming is a material exchange process by submitting organic solvents and/or water in combination with chemical reactions, if curing coating systems are applied. These processes are accelerated or initiated by supplying energy in form of heat or electromagnetic radiation. The application of drying and curing procedure widely depends on the customer-specific requirements such as ecological and economic aspects[1]. Due to the wide range of component geometries as well as technologies for wood coating, there exist no ready-made patent solution for the planning of drying systems and curing plants. This chapter discusses the most important drying and curing procedures in the field of coating of wood and wood-based materials. Figure 7.1 generally illustrates the fundamental influencing factors for the drying process of solvent- or water-borne coating systems. The drying process generally depends on a variety of different parameters such as suction behaviour of the underground, wet application amount, substrate temperature, air temperature, air humidity, air velocity in m/s as well as the applied coating system. Table 7.1 illustrates the costs per square metre [€/m²] depending on the production volume with different drying procedures. The costs per square metre inevitably decreases with increasing production volume. Aqueous coating systems have been used for drying tests showing that there are no significant differences in the processing costs between the drying techniques at an output of 140 m²/h. In comparison to glass or metal, timber is a significantly more inhomogeneous and especially a hydrophilic substrate. Water-borne coatings are the most widely used alternative to solvent-borne systems in a wide range of area. However, the application of water-borne coatings to wood poses various problems (Table 7.2) which require a complex approach to the coating process. The exposure time of the water to the surface can be reduced by means of forced drying methods. This results in a lower swelling as well as in a reduced polishing effort and enables a faster subsequent processing. The wood moisture as well as the roughness of the wood surface influence the penetration of the water into the surface[20]:

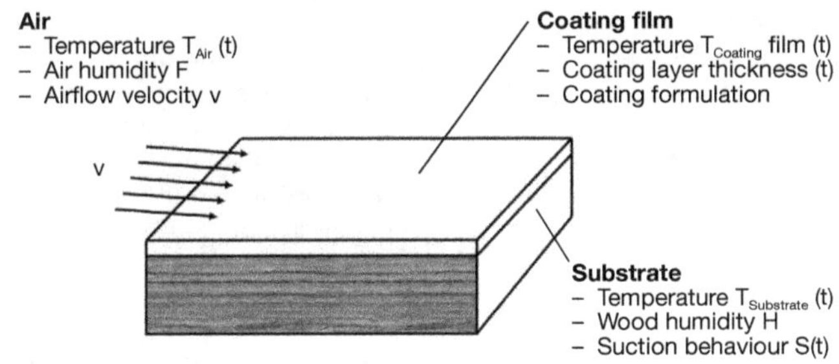

Figure 7.1: Influencing factors for the drying and curing of coating systems[2]

1 According to DIN 55945, the term coating drying describes the passing of the paint or coating, respectively, from a liquid state in a solid state. In the coating technology, the physical drying generally describes the evaporation of solvents, additives and other auxiliary materials without chemical reactions. The curing describes the linkage (cross-linking) of resins due to chemical reactions, whereby the cross-linking specifically means the formation of a three-dimensional network.

Table 7.1: Challenge for the coating of wood with water-borne coating systems [20]

Challenges in the use of water-borne coatings	**Challenges on wooden substrates**
High surface tension of water → film forming agents often necessary → foam formation	**Wood swelling leads to** → increased expense to grinding the base coat → poor penetration and adhesive properties on porous surfaces → discolouration and adhesion problems on knots (pine)
High boiling point of water → longer drying time → high energy and space (drying units) requirements → corrosion of metal components and transfer pipes	**Wood ingredients (e.g. in the case of woods rich in tanning agents such as oak or larch) can react with aqueous alkaline coatings and lead to discolouration of the coated surface** → greenish discoloration of oak → yellowish dscolouration of pine
General → equipment must be adapted (stainless steel or plastic) → tools must be cleaned immediately → cured water-borne coatings are very difficult to remove → high paint finish quality require an intensive surface-cleaning → free from dirt, dust, oily and greasy residues → high-gloss surfaces are difficult to achieve → costs per litre higher than conventional coatings (same solid content) (compensation throught lower application amount)	**Low transparency and lower grain accentuation of the wood substrate** → dark woods (cherry wood, Mahogani, walnut) should be stained before coating to get a stronger intensification
	Resin galls (tropical wood) → this leads to adhesion and wetting problems

Table 7.2: Advantages and disadvantages of air drying [20]

Advantages	**Disadvantages**
– Very low purchase costs for trolleys – Comparatively low energy consumption (for the environment of the trolley) – Often optically good surfaces due to gentle drying	– Strong dependence on external climatic influences (temperature, humidity) – Very long drying times – Comparatively large space requirement – Lower resistances of the coated surfaces, lower cross-linking – Greater swelling of the wood surface leads to bigger grinding effort

- Freshly grinded surfaces incorporate more water than deposited surfaces
- Finer grinded surfaces (from 280 grit) incorporate less water than coarsely grinded surfaces
- The optimum moisture content of the wood is 8 to 12 %; moisture content lower than 8 % may result in problems in the film formation by increased roughness.

Meanwhile, a lot of drying methods using different principles are available on the market. However, it is difficult for the user to choose the optimal technology from the technologies offered [20].

7.1 Air drying

The simplest form is the drying in a dust-free environment in air without additional drying units. The air circulation in the room takes over the removal of moisture. In accordance with the climatic conditions, the water escapes sometimes faster or slower (usually long drying times), but very gentle, so that surfaces with a good appearance and good tactile properties thus can be realized [20]. A disadvantage is the risk of dirt accumulation due to the relatively long drying time. For lower production volumes and simple applications, the air drying is an economical alternative with lowest investment costs especially for physically drying coating systems unless there is sufficient room for the drying (see Table 7.2).

7.2 Convection drying

If the heat is transmitted by air to the coating surface/substrate, then this is referred to as convection drying. In practice, jet dryers as well as circulating-air dryers are applied. The energy required for the heating of air is obtain from the waste utilisation (combustion of wood) in form of saturated steam or hot water. As for reasons of explosion hazard, the legislator has set uniform regulations for the concentration of circulating-air in dryers [3] due to the regulations BGV D25 and BGV D24 issued by the professional associations. The regulation BGV D25 prescribes a concentration of 0.8 vol. % for flammable organic solvents in the air of the oven air as an upper limit. The required amounts of exhaust air and fresh air may vary depending on the coating system and solvent content. The drying is performed at a circulating-air temperature between 25 and 80 °C depending on the temperature sensitivity of the substrates (such as the resistance to edge gluing). The drying of coated hardboard panels applies convection temperatures of approximately 120 to 140 °C. Generally, convection dryers require a relatively large space requirement due to the drying time of coating systems. It must be kept in mind, that only 2 to 5 % of the totally applied energy in drying tunnels effectively are applied for the coating drying. The heat losses are very large.

Jet dryer

Jet dryers directly blow hot air from numerous round nozzles or slotted nozzle vertically or diagonally on the substrates which are moved alongside. Due to extremely high air velocities, such dryers achieve a very rapid heat transfer and enormously shorten the drying time of the coating systems. The air velocity amounts 15 to 30 m/s. In order to achieve an optimal heat convection, nozzle dryers often are subdivided into several, independently separated temperature zones according to temperature and air speed. Jet dryers are applied for example for a fast drying of wood stains and aqueous rolling primers. The convection drying by means of heat radiation can be supported with the help of integrated infrared lamps. Thus, the drying times can be further reduced. In many applications, a convection dryer usually can be combined like the jet dryer with a pre-dryer for faster drying.

Circulating-air dryer

Circulating-air dryers widely are applied in the furniture industry. The air is passed on the substrates to be dried in a counter-flow procedure or cross-flow procedure, respectively. The dryers often are operated with air velocities of approximately 0.5 to 10 m/s. In comparison to the jet dryer, the circulating-air dryers considerably are more gentle for the coating systems. So-called flat belt dryers with belt transport, cross-rod transport or rack trolley dryer

Table 7.3: Costs per square metre (Euro/m²) depending on the coated production volume [1]

Production volume	Circulating-air dryer	Circulating-air dryer with dehumidified air ("Hydrex")	Microwave	IR emitter unit/ circulating-air ("OIR dryer")
[m²/h]	[Euro/m²]	[Euro/m²]	[Euro/m²]	[Euro/m²]
10	2.384	2.653	2.714	2.362
30	0.795	0.884	0.905	0.787
50	0.477	0.531	0.543	0.472
70	0.341	0.379	0.388	0.337
100	0.238	0.265	0.271	0.236
120	0.199	0.221	0.226	0.197
140	0.170	0.190	0.194	0.169

are available for flat components. When coatings have to be dried gently, rack trolleys with 12 to 14 floors are available which are transported cyclically or continuously by means of a tunnel dryer. The drying times may amount between 20 min up to 6 hours at a temperature of 25 to 45 °C. An ascending and descending temperature profile often is realised in such tunnel dryers.

Vertical dryer

A vertical dryer consists of pallets, whose number and dimensions each are designed to the customer specifications. The components are transported into the dryer after application of the coating. The components stop on a pallet which subsequently lifts. The following pallet then is available for the next incoming components. For example, a vertical drying may consist of four drying zones each equipped with 24 pallets, for example. If the pallet has achieved the top position in the first drying zone, this pallet is transferred transversely into the second zone. In this second zone, the pallet is transported downwards where it is transferred into the third zone, and so forth. Figure 7.4 illustrates a vertical dryer. The vertical dryer consists of four independently ventilated drying zones with 96 pallets, for example. The first zone is applied as an evaporation zone and settling zone, while the second and third zone are applied as a drying zone. The fourth zone is applied as a cooling zone. The required amount of air is 25,000 to 30,000 m³/h. The air in the evaporation zone always has to be dissipated completely. In the other zones (drying and cooling zone), about 30 parts of fresh air are mixed with 70 parts of exhaust air. The drying times appear as follows:

- 12 min evaporation time (25 °C circulating-air temperature)
- 24 min forced drying (35 to 50 °C circulating-air temperature)
- 12 min cooling (25 °C circulating-air temperature)

Vertical dryers generally are suitable for the drying of solvent-containing and water-borne systems. Frequently, nozzle dryers and circulating-air dryers are equipped with short-wave infrared lamps in the last third of the plant. Thus, the drying of the water-borne coatings are accelerated.

In order to increase further the efficiency of the circulating-air dryers during the drying of water-borne coatings, the applied air additionally is dehumidified. Thus, the air may absorb more moisture per unit of time, and the drying process is shortened by approximately 20 to 40 % in comparison to conventional systems. There are two fundamental methods for the removal of moisture from the air:

- Cooling of the air with elimination of water by means of a cooling unit
- Dehumidification of the air by means adsorption agents.

Condensation of air with removal of water

In 1973, *Hellmann* has developed the "Hygrex" drying process. The dehumidification process also is referred to as "Hygrex" (Venjakob) and "Dry-air method" (Rahman). In the furniture industry, the conventional circulating-air dryers are operated with air temperatures between +40 and +120 °C. The air enriched with water vapour is discharged and balanced with outside air. Due to the additional installation of a coldness condenser, the moist process air is dehumidified at a temperature of -10 to -15 °C on a cooling surface and subsequently fed back to the drying process with an absolute humidity of 1 to 4 g/m³ and with a temperature of 30 to 40 °C. A pure circulating-air circulation is possible by means of a condensation of the humidity of the air. The heat of evaporation is released during the condensation free again and

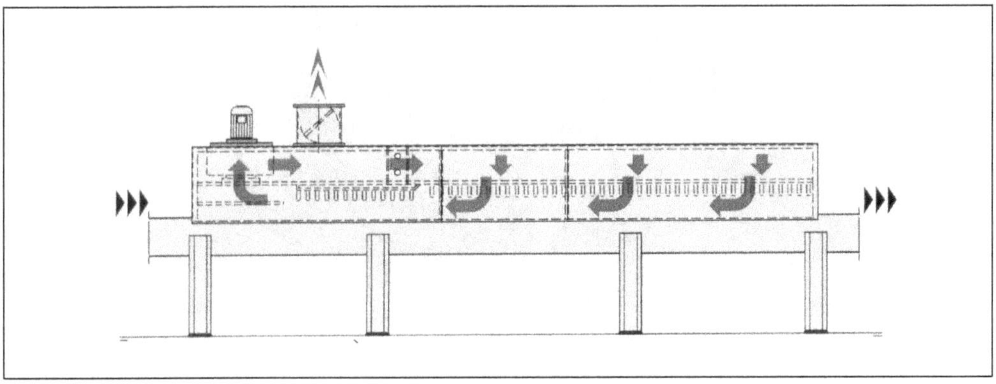

Figure 7.2: Schematic illustration of a jet dryer with round nozzles[4]

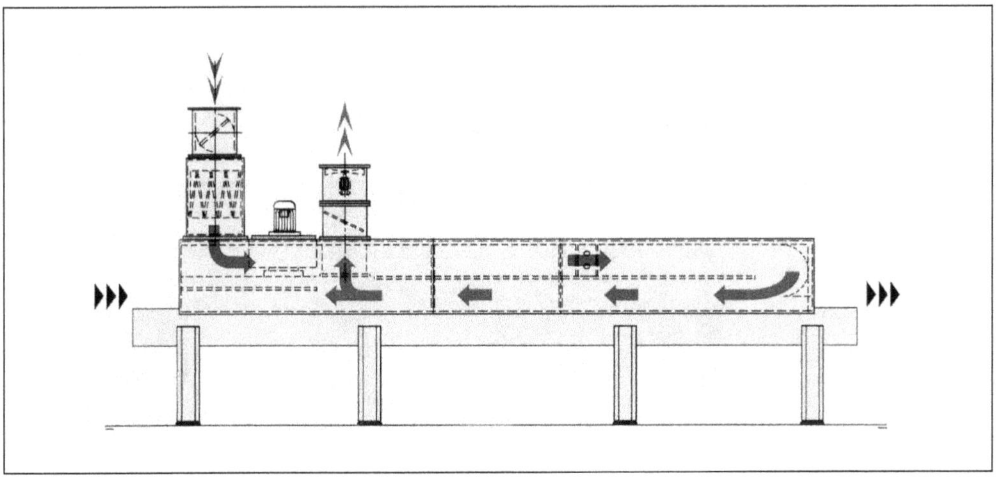

Figure 7.3: Schematic illustration of a conventional circulating-air dryer[4]

passes back into the drying cycle via a heat exchanger. Thus, the process air can be applied more economically. Systematic investigations have shown that water-borne UV coatings significantly dry faster by means of this drying concept in comparison to conventional circulating-air dryers. Additionally, the grain rise is marginal. Drying times of <10 minutes are available with UV water-borne coatings and wet application amounts of approximately 80 to 90 g/m^2. In order to prevent surface defects by rapid drying of water-borne coatings, the relative humidity should be not less than 20 %. Practical values are between 20 and 30 %. Refrigerant dryers cannot be applied in such a broad range of temperature (30 to 40 °C). Otherwise, the refrigerant dryers have the advantage of a very economic application under certain conditions. The disadvantage of a low dehumidification capacity per unit volume is compensated by an enhanced air throughput.

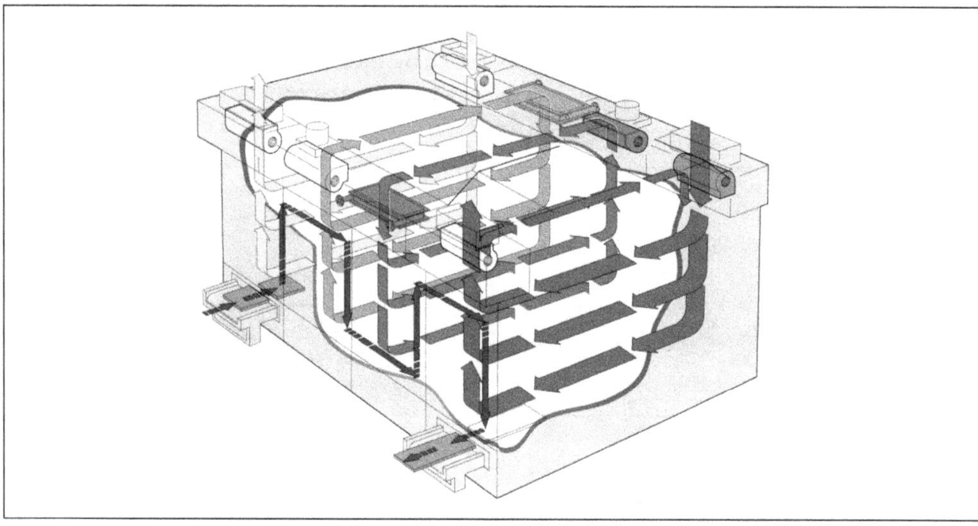

Figure 7.4: Schematic illustration of a vertical dryer[5]

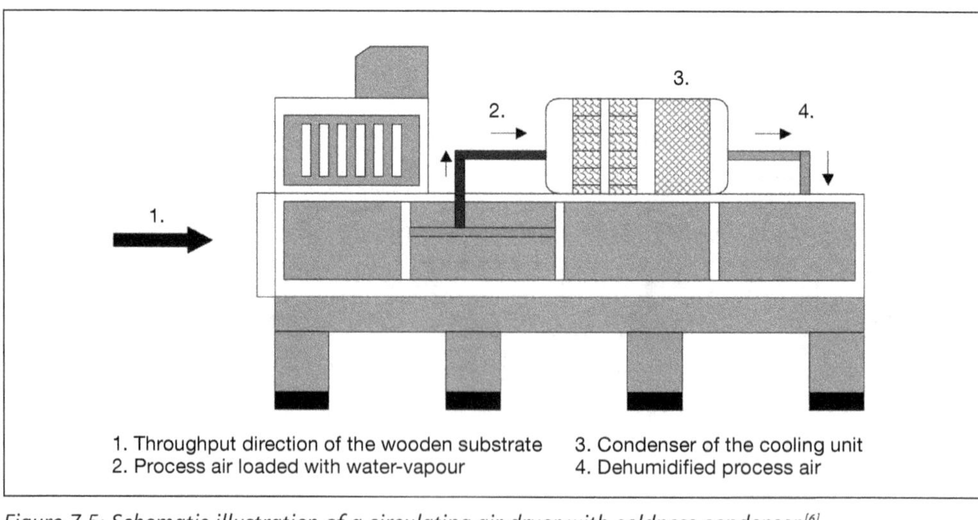

Figure 7.5: Schematic illustration of a circulating-air dryer with coldness condenser[6]

Dehumidification of air with adsorption agents ("sorption dryer") [7]

The air to be dehumidified continuously passes a rotating rotor (drying wheel). The rotor is loaded with sorbents[2] such as lithium chloride or non-crystalline silica gel (titanium silica gel). The rotor consists of a process air compartment (270°) and a regeneration compartment (90°). The air to be dehumidified (process air) flows through the process air compartment of the rotor which has an axially arranged honeycomb structure. The sorbent absorbs the moisture from the process air, and the dehumidified air is available at the outlet for dry air. Simultaneously, a separate air flow (regeneration air) is heated and passed through the regeneration compartment. Then, the heated air is dissipated by means of a ventilator for the regeneration air. The rotor moves (about 8 revolutions per hour) in order to continuously regenerate the compartments which have absorbed moisture. Simultaneously, a regenerated compartment of the rotor continuously is supplied to the process air compartment. Thus, the dehumidification of the air can be performed continuously. Overall, it can be said that the convection drying (circulating-air/nozzle dryer) is a very effective technique for the drying of water-borne coatings. The air velocity in m/s, the humidity of the dryer air and the temperature are crucial for the drying rate. The convection drying can be applied more effectively by means of dehumidification techniques. The dehumidification always guarantees a constant relatively humidity in the drying room. Thus, this system is regardless of the ambient conditions and can be applied very flexible as well as with an enhanced process reliability. Depending on the manufacturer, those dryers also are called "Hydrex" dryer, dry air dryer or "Hydrex" dryer and often are used with jet dryers [21].

7.3 Heat radiation drying (IR radiation)

The transmission of heat by means of thermal radiation (electromagnetic waves) such as infrared radiation is referred to as radiation drying. If infrared radiation meets matter, an interaction takes place due to reflection, absorption and scattering. This cause the molecules to vibrate resulting in heat generation. The applied coating materials exhibit absorption bands of considerable strength in the wavelength range of the total infrared radiation so that an enhanced heat transfer due to broad emission spectra of the lamps is achieved. With 0.8 to 6.0 μm, the optimum wavelengths for the heat transfer lie in the long-wavelength range of the visible light. The infrared radiation transmits large amounts of energy in a short time. The result is a quick, targeted and non-contact heating of the layer to be dried with a high heat transfer performance in comparison to conventional convection dryers. The full power of the infrared radiation is immediately available after switching on the IR unit. Thus, the heat specifically can be applied by pushing a button. The laws of physics systematically are presented in the BASF Handbook [3]. Depending on the wavelength spectrum, a distinction is made between short-wave, medium-wave and long-wave infrared lamps for practical application. The infrared radiation can be divided into several areas:

- NIR (near infrared): from 0.72 to 1.20 μm
- Short-wave infrared: 1.20 to 2.00 μm

2 Sorption is the generic term for the accumulation of a substance or particles, respectively, in a phase boundary region. Sorption mainly takes place at the interface between the liquid and solid phase as well as between the gaseous and solid phase. Sorption also may occur at the interface between the liquid and gas phase. The sorbent is the sorbing solid phase or boundary, respectively. The substance which is to be adsorbed is referred to as sorptive, while the accumulated or embedded substance is referred to as sorbate

- Medium-wave infrared: 2.00 to 3.00 µm
- Long-wave infrared: 3.00 to 1,000 µm

By applying infrared lamps (for example short-wave infrared lamp) it always has to be considered that a colour shade dependent heating of the coating will take place. Water, for example, best absorbs the radiation at a wavelength of 3.4 µm. In order to achieve reproducible results, different energy intensities per colour shade have to be adjusted. This does not occur frequently in the everyday practice. For safety reasons, fresh air volumes have to be supplied when applying solvent-borne coating systems. Thus, the sink rate of the air does not descent to zero in the area of IR lamps, but amounts approximately 0.5 to 2 m/s. In principle, IR lamps should be used in combination with convection dryers in order to achieve a more efficient drying of water-borne coatings (removal of the water vapour). Water vapour has its strongest absorption in the wavelength range of approximately 2 to 3 µm. For this reason, medium-wave infrared lamps are applied for the drying of water-borne coatings. Different IR lamp systems such as "thermal reactors" have been established in the market.

Thermal reactors (IR-radiator)

Thermal reactors (such as "Sunkiss") generate an electromagnetic radiation due to a catalytic oxidation of natural gas or propane gas [8]. The generated range of wavelengths is between 2 and 8 µm and thus in the range of infrared radiation. The intensity of the radiation can be regulated by the temperature of the lamp (300 to 800 °C). Compared to conventional gas lamps, thermal reactors do not generate an open flame. The object is to dry the coating film from inside to outside to dry uniformly. In order to provide the oxygen being necessary for the oxidation, fresh air which is introduced by ventilation continuously is blown onto the surface of the heater. Since the combustion air also is heated and therefore causes a slight

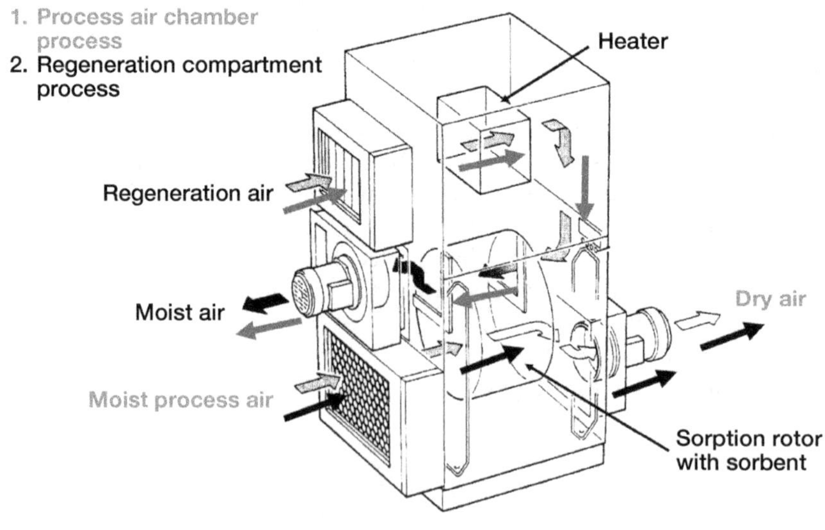

Figure 7.6: Functional principle of an air dehumidifier from Munter [7]

effect of convection in the dryer, the releasing convection heat supports additionally the drying of water-borne coatings in the shadow area of the infrared radiation. The thermal reactor already is applied for the drying of solvent-borne and water-borne coating systems in the coating of furniture, stairs and doors [9].

7.4 High frequency drying and microwave drying

Here too, the basic idea is to dry the coating layers of aqueous products evenly and more quickly than with convection drying from the inside to the outside. During the microwave drying, the heat is generated by direct conversion of electromagnetic energy into the kinetic energy of the molecules within the layer of wet coating. Microwaves are high-frequency elec-

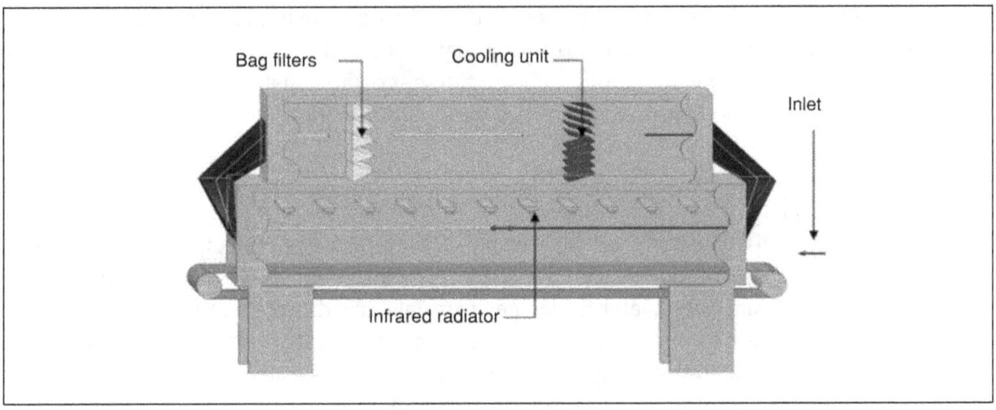

Figure 7.7: Scheme of a dryer with optimised infrared radiation [21]

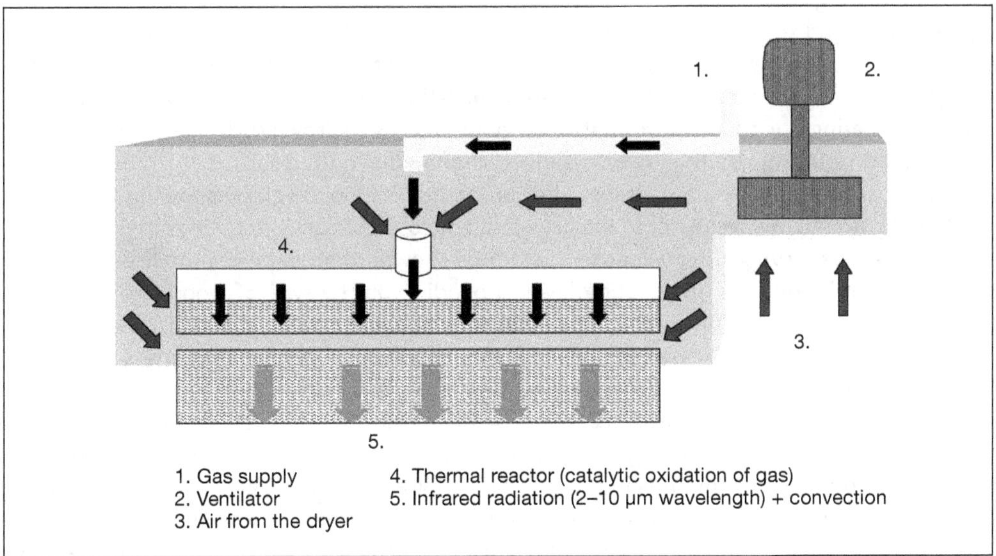

Figure 7.8: Functional principle of the thermal reactor [8]

tromagnetic waves in a frequency range of 300 MHz to 300 GHz. Microwaves are produced in a magnetic field tube (magnetron). The best known and most widely used frequency is 2.45 GHz which corresponds to a wavelength of 0.12 m. The microwave heating belongs to the field of dielectric heating. This term describes the heat generation in electrically non-conductive or only weakly conductive materials by exposure to a high-frequency electrical alternating field [10]. The drying with high-frequency electromagnetic waves is divided as follows:
- High frequency drying in the frequency range: 0.3 to 300 MHz
- Microwave drying in the frequency range: 300 MHz to 25 GHz

High frequency drying

In the lower frequency range (0.3 to 300 MHz) it is referred to as the high frequency heating or capacitor field heating. With certain exceptions, the high frequency drying has not prevailed for the drying of water-borne coatings. The high frequency drying (27 MHz) is applied for the drying of UV water-borne spraying coatings in the industrial coating of stair steps. Here, the flash-off times of water are specified to approximately 90 seconds. Then, the material can be curing with UV lamps immediately. For the homogeneous dissipation of the process air loaded with water vapour, convection heat additionally is required. The application of homogeneous wood-based materials generally is critical for the reproducible uniform drying.

Microwave drying

Only coating materials can be dried by means of high-frequency electromagnetic waves which are able to transform the electromagnetic energy into heat. Water-borne coatings particularly are suitable, since the water molecules form electric dipoles aligning themselves in the electrical field. The high-frequency alternating field stimulates these electric dipoles of the water to rotational vibration. High frequency energy is absorbed due to intermolecular friction and converted into heat energy for drying [10]. Also, polar solvent such as isopropyl alcohol exhibit a very good absorption across the whole considered spectrum of frequency. Metal-containing substrates are excluded since metals act as an antenna due to the free-moving electrons, and since an excessive heating occurs due to the alternating frequency. In this context, the wood moisture content of common wood is of importance, since their water content (6 to 12 weight percent) also is excited to vibration due to the microwave resulting in a substrate heating due to an appropriate energy input. The microwave drying of coated, laminated wood resulting in a boiling of the glue components in the past.

Swaboda presents fundamental studies on neutral supports (glass) showing that a pretreatment with microwaves prior to the drying of aqueous UV curable coatings in the jet dryer reduces the total drying time. Wooden substrates showed that water is mobilised from the support and migrates into the coating layer depending on the type of wood, moisture content and supplied power of microwave radiation which in turn can affect the resulting surface quality [11]. This is expressed for example in blistering and greater roughness of the surface. In addition, an increase in the average roughness of wood such as solid oak and pine was detected after the microwave treatment. Economically, the sole application of the microwave technology is not interesting for the drying of water-borne coatings, since at least a retention period of several minutes in the dryer is required until the water will be transported out. The application of an enhanced power of microwave radiation accelerates the evaporation of the water, albeit with the disadvantage of a strong deformation and drying of the solid wood. Nowadays, the microwave drying is applied in the pre-drying (flash-off zone) of 2C UV curing water-borne coatings such as on foil-coated chipboard panels or MDF panels. The retention period in the microwave amounts 1 to 3 minutes. The installed power of such microwave dryers is 20 KW. According to *Baumgärtner*, the greatest losses arise in the implementation of the electric current in the microwave energy. Under favourable conditions, a degree of efficiency of up to 50 % can be assumed [12]. The air circulation in the microwave tunnel is approximately 5,000 m^3/h. Subsequently, the main drying is done with an air circulating dryer and/or jet dryer. Only afterwards, the surfaces are UV cured. The entire drying process (microwave, air-circulating dryer and jet dryer) takes approximately 6 to 8 minutes for flat components and 10 to 12 minutes for profiled kitchen fronts at a wet application amount of approximately 130 to 140 g/m^2. According to the current state of the art, microwave dryers are not applicable universally. Further still on-going investigations will help to get a better understanding of the microwave drying under industrial conditions. The cost-benefit ratio should be considered always prior to the application of the microwave drying. A finer calibration of microwave power would be very helpful for a further distribution of the microwave drying technology in order to perform an adjustment depending on the substrate/coating system. The efficiency of the dryer is affected by the field strength of the microwave field, the frequency and the corresponding loss value of the coating system. The field is continuously adjustable in the performance spectrum of the magnetron. The radiation shielding is governed by the regulation IEC 705. According to this, the limit of the maximum radiation in a distance of 5 cm has to be maintained to 2.5 mW/cm. The market also offers variable microwave devices with an integrated jet drying, a much greater flexibility and range of applications (also for high-gloss water-borne coatings).

7.5 UV curing

UV radiation is a part of the spectrum of the electromagnetic waves and joins the short-wave part of the visible light (violet). According to DIN 5031 part 7, rays in the wavelength range from 100 to 380 nm ultraviolet are referred to as ultraviolet UV rays or shortly UV rays. The UV range is divided into the following sections:

- UV-C 100 to 280 nm
- UV-B 280 to 315 nm
- UV-A 315 to 380 nm
- UV-V 380 to 450 nm

A UV curing system essentially consists of the UV rays, the reflector system, the electric units (supply and control), the transport unit, the air suction unit and a cooling unit.

UV lamps

In addition to the electrode-free microwave-excited UV lamps, in the furniture industry mainly mercury low-pressure lamps as well as mercury high-pressure lamps mainly are applied. These are gas discharge lamps. In a quartz tube (quartz exhibits only a little absorption in the UV range), the filling material (mercury, noble gas and halogen additives) is excited at pressures between 1 and 10 bar for the emission of radiation. Since the discharge performs in the saturated pressure of mercury vapour, the gas discharge lamp has to be applied under optimal operating temperatures in order to achieve a good degree of efficiency. In this process, the mercury high-pressure lamps emit a characteristic spectrum with the main lines 254, 302, 313, 366, 405, 436 nm (see Figure 7.9). The radiation of this wavelength range is very energetic in order to cleave photoinitiators to divide and thus to start the radical polymerisation. The power input of the spotlight often is between 80 to 160 W/cm. The average lifetime of the applied lamps amounts 1,500 to 2,000 hours until the performance is decreased to 80 %. Usually, lamps with discharge lengths of 50 mm up to 2,500 mm are manufactured. Here, the discharge lamp is defined as the distance between the two electrodes. Depending on the lamp output, the diameter of the lamp varies between 19 and 28 mm. In the energy balance of the electrode-excited lamps, the UV radiation accounts for only 25 to 30 % of the UV-C, UV-B and UV-A radiation, while further 10 to 15 % are due to the visible range of light and approximately 50 to 60 % due to infrared radiation. However, microwave-excited lamps consist of only 35 to 40 % of infrared radiation. With about 36 %, the energy balance in the UV range is slightly higher than with electrode-excited mercury lamps. The specific lamp performance [W/cm] has become established for the performance comparison of UV lamps. It has to be considered that the specific lamp power cannot make a statement about the intensity and energy density of the UV radiation arriving at the substrate, since the following information cannot be considered:

- Reflector geometry
- Focussing strength (diameter of the lamp bulb and reflector)
- Distance of the lamp to the substrate
- UV-spectrum
- Feed rate of the UV line

Metal halide-doped mercury lamps

The spectral distribution can be influenced by adding metal halides such as gallium iodide or iron iodide to the lamp filler material. The emission spectrum of a gallium-doped lamp moves to the long-wave visible range of light with the main emission lines at 403, 410 and 417 nm (see Figure 7.9). These lamps especially are suitable for the curing of white pigmented coating systems (see Chapter 3.1.6). The guaranteed lifetime of gallium-doped lamps amount 1,000 to 1,200 hours. Subsequently, the lamps should be replaced.

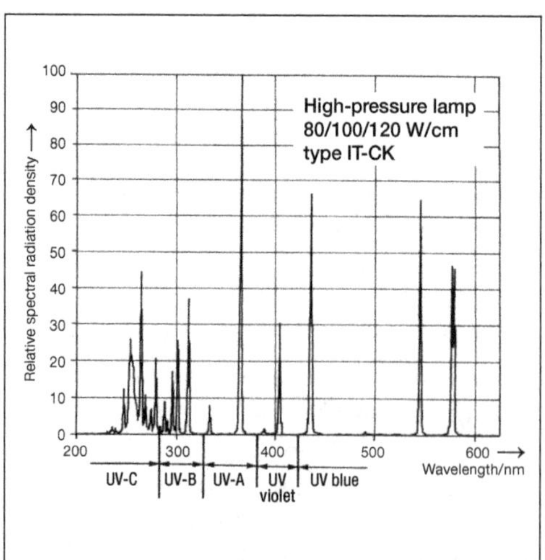

Figure 7.9: Spectrum of a mercury high-pressure lamp Source: IST Metz GmbH

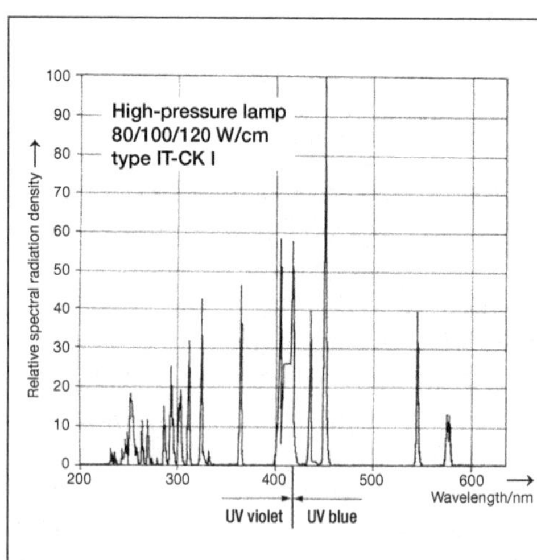

Figure 7.10: Spectrum of a gallium-doped mercury high-pressure lamp Source: IST Metz GmbH

Special lamps

The mercury low-pressure lamps also referred to as super actinic fluorescent lamps (type TL 05 and TL 03, Philips) are applied for the gelation of styrene-containing UV coatings. The focus of the radiation of the lamp TL 05 still is in the UV-A range, while the focus of the radiation of the lamp TL 03 is in the visible range. The fluorescent lamps have a very low output, so that a hardening of white pigmented UV coatings only with fluorescent lamps is not possible.

Ozone

Due to the UV radiation emitted by mercury high-pressure lamps, ozone is formed from oxygen in the ambient air. This led to the development of modified UV lamps where the disturbing development of ozone (formation of ozone at 187.6 nm) is switched off by special quartz materials with filter effect for the UV-C range. The odour threshold for ozone is 0.01 ml/m³ (ppm), the MAC value amounts 0.1 ml/m³ (ppm). The UV lamps could not prevail in many applications, since the absence of the short-wave component often resulted in an insufficient curing result. During operation, the surface of the aforementioned lamps achieves temperatures between 900 and 950 °C. Higher temperatures may cause a deformation of the UV lamp and an uneven distribution of the mercury. For this reason, always an air cooling or a water cooling is applied which prevents the heating-up of the environment as well as the lamp and dissipates the disturbing ozone.

UV reflectors

Reflectors collect and/or reflect the UV light in order to focus the UV radiation on the substrate. In principle, two main types are applied. The elliptical reflector is the mostly applied type. It focuses on a focal line in which an optimum of UV radiation is present. The distance has to be taken into account exactly. The parabolic reflector reflects the radiation parallel, a focal line does not arise. Furthermore, mixing types of the above-mentioned reflectors are applied. Due to its low reflection behaviour (efficiency > 90 %), pure aluminium is applied as reflection material. For its protection, it is coated with a layer of quartz.

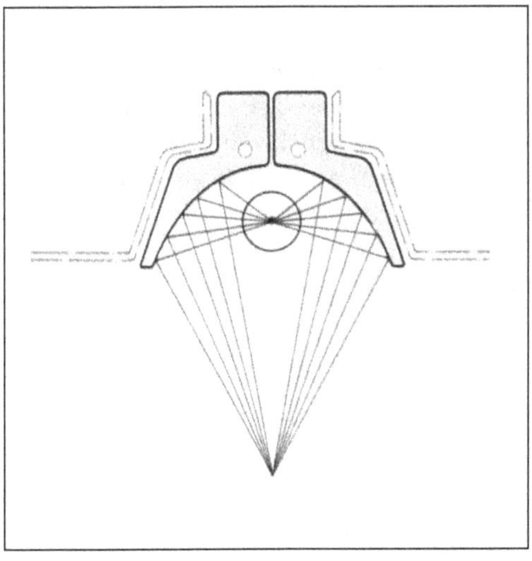

Figure 7.11: Illustration of a focussing reflector Source: IST Metz GmbH

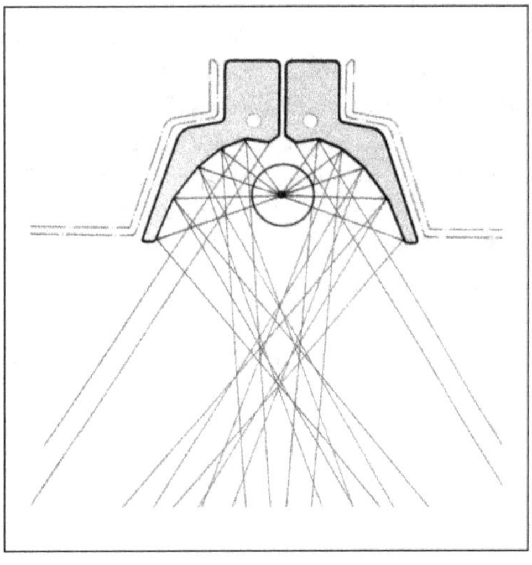

Figure 7.12: Illustration of a focussing and scattering reflector Source: IST Merz GmbH

Irradiation (energy dose)/irradiance (UV intensity)

The feed rates of the coating processes in the wood coating amount between 10 and100 m/min. The energy required for the curing of UV coatings is done by a series connection of several UV lamps. Since the output emitted by the UV lamp is not equal to the intensity available on the surface of the substrate, the UV systems are characterized by two quantities which can be measured with a UV photometer. These are the irradiation (energy dose) and the irradiance (UV intensity). Generally, both quantities depend on the wavelength. The size units E and H refer to a specific wavelength range and are referred to as spectral irradiance (H) and spectral radiation (E) means. Different values for E and H are received for different spectral distributions of radiation.

Irradiation (energy dose)

The dose (irradiation) is the total radiation energy, which affects the object while passing the UV dryer. It is specified in Joule per cm^2. The intensity of the total radiation energy is associated with the time. Thus, the dose depends on the belt speed of the UV system.

irradiation (E) [$W*s/cm^2 = J/cm^2$] = intensity * time

Irradiance (UV intensity)

The term intensity describes by definition the emission of the radiation from a radiation source. In the practice of the UV lamps, this term has become established for the irradiation impinging on the component. The intensity is the maximal radiation output which reaches the surface of the substrate. It is specified in Watt per cm^2 or mW/cm^2. The intensity is characteristic of the applied type of lamp and reflector and is independent of the feed rate of the coating line. The irradiance is an important characteristic value for the chemical cross-linking and the setting of the gloss level of the UV coatings.

UV measuring instruments

When measuring radiation physical quantities, an optical signal is transformed into an electrical signal. The measuring instruments available on the market are referred to as photometers. A proven method is the measurement of the integral irradiance and/or the integral radiation in the UV dryer. Since there is no standardisation of the measuring instruments, the measured values are not comparable. Even instruments of the same type are not always with each other. It is important to specify the device type and serial number. Different radiation profiles are achieved for measuring the irradiance along the operational width of the UV lamp. A decline of irradiance of about 20 to 30 % is measurable in the electrode area. Fundamentally, a UV measuring device should record the following data:

- Specification of the radiation (energy dose) in J/cm^2 or mJ/cm^2
- Maximal irradiance (UV intensity) in W/cm^2 or mW/cm^2
- Graphic illustration of the irradiation profile
- Measurement in different wavelength ranges (UV-C, -A, -B, -V)

The compliance of the system parameters and the maintenance of UV dryers is inevitable for a reproducible surface quality and to avoid residual emissions (see Chapter 9.2). In the practice, it is noted rather frequently, that the UV lamps and reflectors are not maintained regularly. However, contaminated reflectors reduce the emitted UV radiation by 30 to 50 % to the substrate which may cause sticky coating surfaces in extreme cases. In addition, cleavage products at an extremely low concentration of the photoinitiator may lead to odour problems, since a too low irradiance leads to an increased formation of cleavage products. A checklist for UV systems has to include the following measures, for example [9]:

- At least once in a month, reflectors and UV lamps are to be purified with a lint-free, alcohol-soaked cotton cloth
- Weekly inspection of the UV lamps with a UV measuring instrument
- Verification of the hours of operation, power consumption, irradiation and irradiance
- Inspection of the UV lamp cooling by means of the measurement of the temperature in the exhaust air of the UV dryer
- Ensuring the continuous cooling in order to prevent a turbidity in the quartz tube and thus a loss of the irradiance

7.6 Electron beam curing

7.6.1 Introduction

The historical background of the electron beam curing technology (EB) already is described in Chapter 3.1.6. With a few exceptions, this technology has not found any industrial application on a large scale in the area of the coating of wood and wood-based materials. Thus, this chapter only reports on the essential aspects of the EB technology.

7.6.2 Mechanism of the EB technology

Electron beams as well as UV rays may excite or ionize molecules. In contrast to the UV radiation which can be considered as "pure" electromagnetic waves, electron beams also exhibit a corpuscular character (particle character). Even if the energy density of the electron beams is lower than that of the pure corpuscular radiation, the electron radiation is a highly ionising radiation and also may penetrate into the substrate to be coated. Its penetration capability depends on the acceleration voltage (see below) [13–15]. Electron beams do not require any photoinitiators and photosensitizers for the generation of initiator radicals. These additives can be dispensed with in the production of electron beam curable coatings. Since the reaction initiator is formed within fractions of a second, the curing process occurs very quickly. This is an important economic factor for the application of electron beam curing. In principle, there are three primary processes:

Ionisation

The accelerated electrons may knock out a bonding electron (Equation I), whereas the ionized molecule (cation) dissociates in a free radical and a radical ion (Equation II).

$$AB \rightarrow AB^{+} + e \qquad \text{Equation I}$$

$$AB^{+} \rightarrow A\cdot + \cdot B^{+} \qquad \text{Equation II}$$

If the free electron has a sufficiently high energy, it once again may ionise a molecule.

Excitation

An excitation may occur (Equation III), if the energy of the knocked-out electron is too low in order to ionize another molecule. As shown above, the dissociation of the excited molecule leads to two radicals (Equation IV).

$AB \rightarrow AB^*$ Equation III

$AB^* \rightarrow A\cdot + \cdot B$ Equation IV

Electron capture

Electrons with even lower energy can be captured by molecules thus forming an anion (Equation V). This can decompose again into two radicals (Equation VI).

$AB + e \rightarrow AB^-$ Equation V

$AB- \rightarrow A\cdot + \cdot B^-$ Equation VI

As it was at the UV curing process, the radicals produced in the primary processes lead to radical chain reactions and thus to a polymer network in a fraction of a second.

7.6.3 Generation of electron beams

Electron beams are produced in so-called electron accelerators in vacuum (at a pressure of about 10^{-2} to 10^{-4} Pa). By heating a hot cathode – usually a tungsten cathode – those electrons which are movable easily in the metal composite are emitted and accelerated to the anode. Simplistically, the industrially applied electron accelerators can be compared with a television tube. However, while the electrons in the television tube only excite certain luminescent materials, an EB system should send as many electrons as possible from the vacuum on a work-piece. The radiation output of such a plant is several orders of magnitude greater in comparison to the radiation output of a television tube. In general, acceleration voltages of 150 to 300 kV are realised, and rarely up to 500 kV. The empirical values [15] for industrial accelerator voltages are:

- 130 to 150 kV for thin films in the printing ink sector or silicone release materials
- 165 to 180 kV for furniture foils or pressure sensitive adhesives
- 180 to 250 kV for panels, parquet, strips

The generated electrons leave the electron accelerator through a thin metal window that mostly consists of a titanium foil with a thickness of 12 to 15 µm. The window has to be so stable that a sufficient vacuum can be maintained. It also has to be sufficiently permeable for the accelerated electrons. Approximately 20 to 25 % of the electron energy are absorbed in the metal foil. Since the escaping electrons strongly heat up the titanium foil, the titanium foil has to be cooled sufficiently in order to avoid a melting. This can be done by means of water cooling as well as by an intensive air-jet ventilation. Nowadays, the lifetime of such a titanium foil is more than 2,000 hours. The electrons that penetrate the titanium foil are slowed down by interactions with the air or material. The electrons only have a defined penetration depth. In this case, the following arises:

Backscattered electrons

These electrons effectuate a curing of the coating on the lateral surfaces and edges especially when irradiating moulded parts.

Secondary electrons

The cross section of the secondary electrons is large enough to ionize molecules and to generate radicals. The fast primary electrons as well as backscattered electrons do not cause chemical reactions. The fast electrons only are required for the production of electrons outside of the vacuum.

X-ray Bremsstrahlung ('braking radiation')

According to the laws of electrodynamics, the deceleration of primary electrons leads to an emission of a very short-wave electromagnetic radiation. The thus produced radiation is referred to as X-ray bremsstrahlung. The hardness or the range of x-rays depends on the energy of the primary electrons. The faster the primary electron, the harder the resulting X-ray bremsstrahlung. However, the energy of the X-ray bremsstrahlung never can be greater than the energy of the primary electrons. For security reasons, the electron accelerator as well as the irradiation room have to be shielded against the release of these X-rays.

7.6.4 Process parameter

Penetration depth

The acceleration voltage determines the penetration depth of electrons and thus the maximal thickness of the coating layer to be cured. The production of usable layers with a thickness of approximately 80 to 400 µm requires usual voltages of 150 to 300 kV. The usable penetration depth is defined as the penetration depth which is necessary in order to secure a full curing of the coating. A dosage drop in the coating to a value of 50 % of the maximum dosage is admitted in practice. Figure 7.13 illustrates the dependence of the usable penetration depth of the acceleration voltage. The absorption of electrons in the emission window as well as in the air gap between the titanium window and object surface has to be considered. The range of the electrons thus reduces to approximately 117µm.

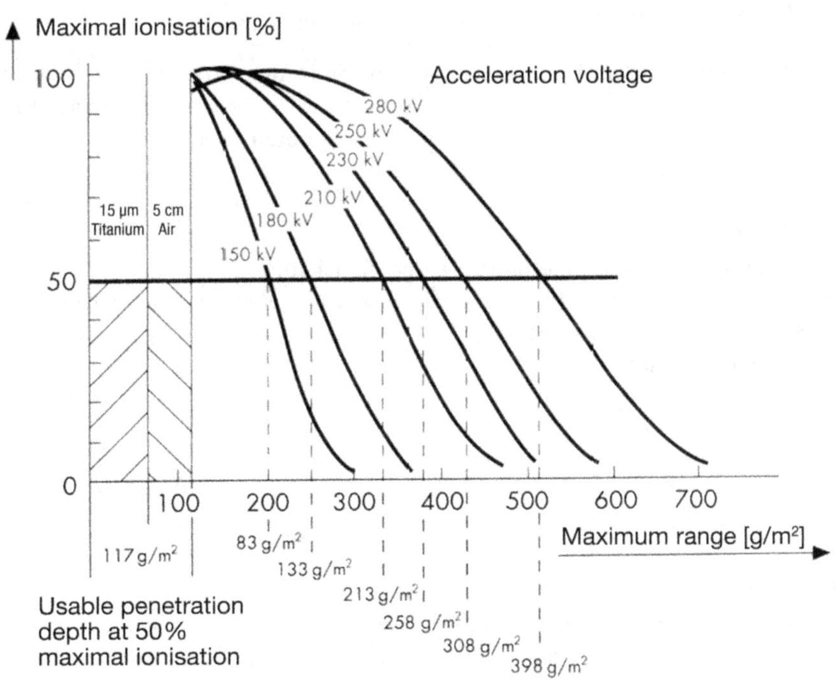

Figure 7.13: Absorption curves of the electrons depending on the accelerator voltage Source [13]

Beam output

The beam output is defined as the product of acceleration voltage and electron current. It is specified in Watt (W). Together with the operational width, the beam output determines the achievable throughput of the radiation facility.

Radiation dosage

The dosage of the electron beam absorbed by the object is directly proportional to the number of electrons which leave the accelerator (Equation VII).

$$D \sim I_0 \qquad \text{Equation VII}$$

where D = dosage and I_0 = electron current in the object

The radiation dosage (Equation VIII) is the energy absorbed per unit mass of the irradiated material. The valid unit for the radiation dosage is Gray [Gy] (Equation IX).

$$D = J/kg \ (J = Ws) \qquad \text{Equation VIII}$$

$$1 \text{ Gray [Gy]} = 1 \text{ J/kg} \qquad \text{Equation IX}$$

The coating has to be exposed to the radiation as long as it has absorbed the required dosage of radiation. Thus, the radiation dosage is a measure for the rate of curing. An even distribution of the dosage across the operational width is important for a high-quality surface. By means of simple measures, the operator of a facility has to be able to establish his products with an enhanced surface quality. There is a simple relationship between the dosage (D), the electron current (I) and the line velocity. The following linear relationship arises for the applied dosage:

$$D = k\,(I/v) \qquad \text{Equation X}$$

where v = line velocity and k = k-factor

The factor k contains the acceleration voltage, the deflection width and all device constants. These are varied only rarely. In general, the line velocity as well as the dosage at the change of the coating are varied. It is quite simply possible to obtain a product with consistent quality by applying the Equation X.

Types of accelerator

Different types of electron accelerators are distinguished:
- Devices with spot cathode and magnetically deflected electron beam (scanning system)
- Devices with linear cathode system
- UV lamp with grid scattering
- UV lamp without grid scattering
- Devices with large-area cathode

Within the scope of this chapter, only the multistage electron accelerator which works according to the principle of scanning is reviewed.

Multistage electron accelerator

Electron accelerators of this type of construction exhibit acceleration voltages from 150 to 280 kV. In this device, the electron accelerators are complied with electrical currents of 20 to 200 mA. The simple design combined with self-monitoring microprocessors facilitate the operator of this device to change the cathode and electron discharge window quickly and economically without the help of the manufacturer. The dosage distribution across the entire object width is better than ± 3 % for the deflection frequencies of more than 800 Hz. The accelerator is cooled by water and convection by means of supporting structures at the exit window.

In the multistage electron accelerator (see Figure 7.14), the high-voltage line between the hot cathode and anode is divided in several stages. A potential difference between these stages amounts 60 kV. In this "cascade accelerators", high voltages may lead to electrical flashovers in the individual stages as well as cascade-like discharges inside or outside of the accelerator tube. Sulphur hexafluoride (SF_6) is applied as a compressed gas for the insulation of the high-voltage line.

Energy consumption of the electron beam facilities

Since many parameters play a role in a coating process, and since each company calculates its costs of coating in a different way, only essential technical data are specified at this point. All electron accelerators are subject to a wear at the cathode and electron discharge window. Depending on the stress of the electron emitter, the cathode has to be replaced after 300 to 1,000 hours. The cost will be approximately 100 Euro for a replacement time of about 15 minutes including vacuum pumps and reconnecting of the high-voltage. Generally, a titanium foil lasts for over 2,000 hours at full-load operation. Depending on the operational width of a lamp, the costs of a titanium foil will be between 100 and 500 Euro at a changeover time of 1 hour including vacuum pumps and reconnecting of the high-voltage.

Inertisation

The polymerisation reactions can be disturbed by oxygen radicals significantly. The possibility of a curing process at such a high level of ionisation density that the polymerisation reaction is completed prior to the oxygen diffusion also is not given in the process of electron beam curing. Thus, the quality of a radiation-cured surface mainly depends on the measures of inertisation. The inertisation commonly used today is between 200 and 500 ppm. I.e., at least the chamber should be flushed seven times with inert gas in order to get a flawless surface. Experiences for complex components showed a far greater flush volume. The inertisation during the electron beam curing of coatings causes the largest part of operating costs during the operation of an electron beam curing plant. The plant manufacturer does everything in order to construct the coating curing zone with relatively small amounts of inert gas consumption.

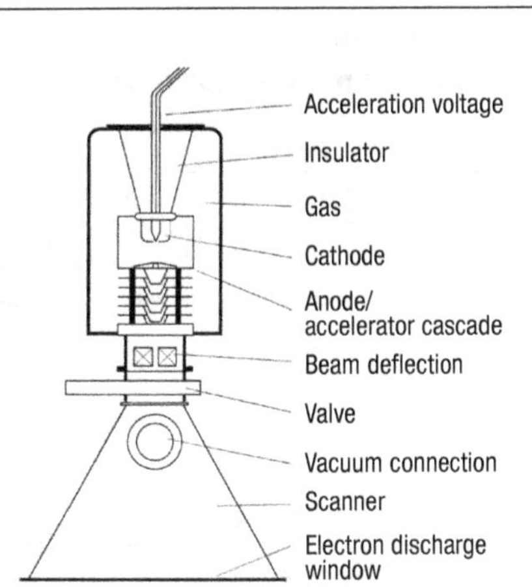

Figure 7.14: Multistage electron accelerator
Source [14]

Pigmented systems

Generally, pigmented systems can be cured by means of the EB technology. Especially in the thick layer range, pigmented systems cause significantly fewer difficulties than the UV curing pigmented systems, since pigmented systems do not require photoinitiators for the curing, and electron rays penetrate deeply into the interior of the coating. However, there are investigations [16] which point out that the impact of organic coloured pigments on the curing process by means of electron beams is not negligible.

7.6.5 Plant concepts

There are some applications [13–19] of the EB technology in the field of coating of wood as well as wood-based materials, and some procedures are converted by means of UV curing coating systems or still are converted in future. The following examples ought to be mentioned here.
- Finishing coating of doors (e.g. Svedex, Varsseveld, NL) [17]
- Wood cement panels (e.g. Falco, Szombathely, HU)
- Finishing coating of front parts of furnitures (PPG, Saultain, F)
- Impregnation and finish coating at laminates (HPL, CPL) [19]

7.6.6 Advantages and disadvantages of the EB technology

The electron beam curing offers several advantages compared to other techniques. The essential advantages are listed here:
- Solvent-free system
- Products are stackable immediately
- Very high throughput capacities are possible
- Thick layers are curable
- Cold process
- All pigmentations are translucent.

Especially the comparison of the UV technology with the EB technology is interesting. In the light of the current state of knowledge, the EB technology has the following advantages compared to the UV technology:
- Film properties (degree of cross-linking, mechanical resistance, etc.)
- Curing of thick-layer, pigmented coatings within the shortest time
- Adhesion
- Storage stability
- Coating of porous substrates

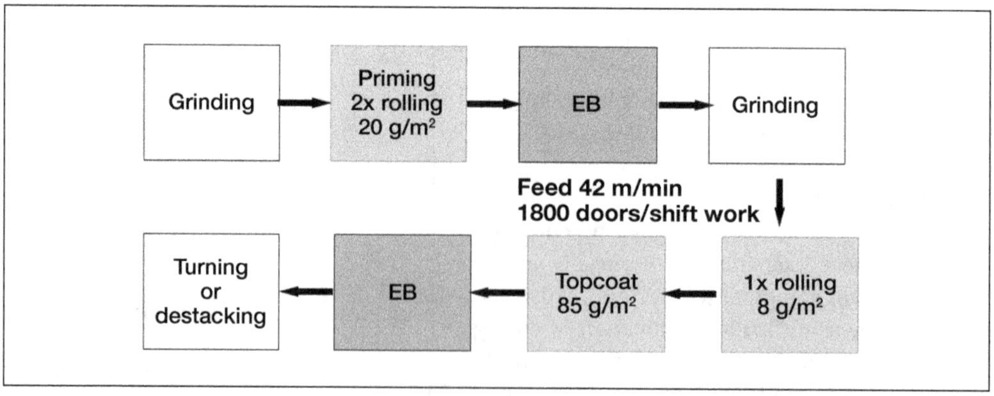

Figure 7.15: Coating of doors by means of EB technology Source [17]

- Rate of curing
- Curing temperature

The UV technology has the following advantages in comparison to the EB technology:
- Investment costs
- Space requirement
- Coating of three-dimensional objects
- Coating of objects sensitive to radiation
- Oxygen inhibition
- No radiation protection advisor necessary

The high investment costs are the main disadvantage of the EB technology. Furthermore, the UV technology has undergone a considerable boost to innovation (see Chapter 3.1.6) in recent years, so that a cost-benefit analysis usually only favours the EB procedure in the wood finishing industry when the processing of larger quantities of flat surfaces (more than 2.5 million m^2/a) is made.

7.7 References

[1] Hruschka, R., Schneider, M.: Trocknung von Wasserlacken auf Holzwerkstoffoberflächen, HOLZ, 1/2001, p. 37

[2] Dauth, C.: Effektive Trocknungstechnologien für die Möbelindustrie, besser lackieren, Vol. 13, 08/2006

[3] Goldschmidt, A., Streitberger, H.-J.: BASF Handbook, Vincentz Network, Hanover, 2007

[4] Technical documentation by company Barberán SA, Barcelona, Spain

[5] Technical documentation by company Cefla Finishing, Imola, Italy

[6] Technical documentation by company Venjakob Maschinenbau, Rheda-Wiedenbrück, Germany

[7] Technical documentation by company Munters: Luftentfeuchter Baureihe MX (B)/MXT (B), Munters Europe AB, 1998

[8] Wuchter, J.: Neues Trocknungsverfahren mit Thermoreaktoren, I-Lack, Vol. 64, 6/1996, p. 331–333

[9] Prieto, J.: Die ökoeffiziente Beschichtung mit UV-härtenden Lacksystemen, Dokumente zu Lacken und Farben, No. 6, Deutsches Lackinstitut, Frankfurt, 1999, p. 20–30

[10] Schleusener, J.: Mikrowellen zum Trocknen haben Zukunft, Farbe & Lack, Vol. 105, 7/1999, p. 111–112

[11] Swaboda, C.: Combination of microwave drying and UV-curing for waterborne lacquers on wood surfaces, Presentation at RadTech 2005, Barcelona

[12] Baumgärtner, O.: Lacktrocknung mit Mikrowellen – Ergebnisse von Grundlagenversuchen, IPA Stuttgart, date unknown

[13] Garatt, P. G.: Strahlenhärtung, aus der Reihe: Die Technologie des Beschichtens, Curt. R. Vincentz Verlag, Hannover, 1996

[14] Kiene, J.: Grundlagen der Elektronenstrahlhärtung, Manuskript des DFO-Lehrganges: Strahlungshärtung von Lacken, Düsseldorf, 2001

[15] Holl, P.: Stand der Elektronenstrahlhärtung und Kostensituation für Holzoberflächen, I-Lack, Nr. 12, 1995, p. 420–423

[16] Häring, E., Ahlers, K., Holtmann, G.: Die Buntpigmentierung elektronenstrahlhärtbarer Lacke

[17] Lobert, M.: Elektronenhärtbare Systeme, published in HK No. 12, 1995

[18] Unknown: Elektronenstrahlhärtung spart Energie und Material, I-Lack, No. 5, 1981, p. 175–177

[19] Patent: EP 0166153, Hoechst AG

[20] Emmler, R., Swaboda, C.: Leitfaden zum Einsatz forcierter Trocknungsverfahren für die Applikation von Wasserlacken auf Holz- und Holzwerkstoffen, Forschungsprojekt 11639/01 (AiF/BMWI/DFO), 2004, ihd Dresden

[21] Ondratschek, D., Schneider, M., Vogelsang, H.: Forcierung des Wasserlackeinsates durch neue Trocknungsverfahren, Universität Stuttgart, 2001

8 Discolouring of wood

Wood is a non-homogeneous natural product which fully develops its structure, brilliance and colouration by means of surface coating. Apart from the optical aspects, a coating has to preserve the wood against the stress of the everyday life such as dirt effects, household chemicals and mechanical influences. In addition, the coating system should protect the wood from premature discolouration caused by an exposure to sunlight and heat. However, without UV protection due to a surface treatment, most types of wood only are slightly colour stable. The colour changes due to the influence of light are to be considered above all in furniture and interior construction as well as in the area of the wooden floor. These can be weakened, retarded or largely avoided by appropriate surface treatments (wood pickling, glazes, clear coats). Thus, woods such as oak only coated with clear coats become gradually yellowish-brown. Ash trees become yellow while mahogany and cherry tree become red. While maple, birch, pine and spruce become brownish, walnut trees become brighter and wengé strawy. The reddish tone of alder and steamed beech turned to the yellowish [6–9]. It is also known, that discolorations in the wood not only are initiated by direct exposure to sunlight but also by complex biochemical and chemical reactions. For beech trees for example, these reactions also may occur in the living tree (appearance of red corns) during storage as well as during treatment and processing of the wood (for example stain formation during steaming) [1]. In many other cases, an exact separation of discolorations which are not caused by UV light is not always possible since these discolorations also can be caused by combined reactions. The impact of sunlight and diffuse light in interior spaces contributes to colour changes. The degree of the occurring colour changes depends not only on the time but also on the location of the furniture. Compared to the light exposure, the exposure of heat often is of secondary importance. However, it can have an impact if the furniture is near heating elements or exposed to higher ambient temperatures [7].

Wood consists of different fibre elements whereby the respective wood element consists of a different arrangement of the individual fibres. When wood is exposed to sunlight and, above all, to the UV radiation of the sun, wood components and mainly the lignin (polyaromatic phenolic structure) are degraded on the surface. This results in a discolouration. If the wood surface also is exposed directly to outdoor weathering, the water-soluble degradation products of the lignin are washed out whereby the photo-chemically stable silvery-white cellulose is left. However, the wood moistening due to dew and rain results in a population by dark-coloured mould fungi as well as to an entry of dust particles so that the surface turns grey to black over time [8 14].

For example, the lignin of coniferous woods predominantly consists of coniferyl units (approximately 90 %). The lignin formation occurs in a free-radical formation whereby this free-radical formation is performed enzymatically. Otherwise, the further reaction of this free-radical formation is not performed enzymatically. Thus, the composition and proportions of the individual building blocks are highly variable; a direct connection according to an always the same pattern does not exist. The aromatic structure of lignin mainly contains four different functional groups whose relative number per aromatic unit varies depending on the type of wood. While monomeric phenol hardly absorbs near-ultraviolet light above a wavelength of 300 nm, conjugated phenols exhibit a strong absorption between 300 nm and

330 nm. Such biphenyl-like structures occur in native lignin and cause the formation of light-induced phenoxyl radicals and ketyl radicals (phenoxyl mechanism and ketyl mechanism) [17].

After sun exposure, the natural coating of light woods such as male, pine or spruce with those colourless wood coatings which do not have a special UV protection point out severe yellowish-brown discolorations. This is attributable to the occurrence that colourless clear coat let pass the UV-radiation whereby this UV protection modifies photo-chemically the wood component lignin. The various lignin structures absorb the light at different UV ranges. Chromophoric groups (chromophores) are formed in the lignin whereby these chromophoric groups selectively absorb within the visible light wave spectrum (380 to 780 nm) [2].

The UV-B rays extend in a wavelength range of 280 to 315 nm and have the shortest wavelength of the sunlight occurring on the Earth's surface. These UV-B rays are responsible for the most destructions in the coating film. Furthermore, the UV-B rays cause the sunburn well known to us. The window glass as it is also found in our living rooms lets pass only the UV-A component and the visible light in the wavelength range of 380 to 780 nm. The transmitted UV-A rays in the wavelength range of 315 to 400 nm still are sufficient to destroy the lignin of the wood. Due to its aromatic structure, the lignin absorbs UV light and is subject to a rapid degradation by photo-oxidation when exposed to UV-light [3, 4]. The photochemical modification of lignin results in modified properties which make their presence felt by means of cracking, yellowing or changes in the colour shade as well as an enhanced sensitivity to water [3].

For many years, the application of micronized iron oxide pigments are the state of the art for wood coatings in order to delay or to suppress the discolouration of light-coloured wood. The iron oxide pigments have been highly successful in the UV protection for glazes in the outdoor area. However, iron oxide pigments have the disadvantage that the paint film is coated by the pigment in a natural coating. However, this is not desired with natural coatings.

Additionally, inorganic, nanoscale UV absorbers such as ceria and zinc oxide, as well as white pigments can be used for an active UV protection. The white pigments have the disadvantage that not always homogeneous and clear coat films may be obtained depending on the application process, the layer thickness and pigment concentration. This can mean that there is no uniform UV protection against discolouration at positions which are not sufficient covered with the white pigment. Today, the stabilisation of high-quality coating systems with UV absorbers and sterically hindered amines (HALS)[1] are used. Different class of absorption agents such as oxalanilides, hydroxy benzophenones, hydroxy phenyl-s-triazine and hydroxy phenylbenztriazoles are applied. An appropriate light stabilizer has to be selected according to the following criteria:

- Effective UV absorption with a good protection for the underground against the photochemical degradation phenomena in the particularly critical wavelength range of 280 to 400 nm
- Easy incorporation into coating systems
- Any interference with the properties of the wood coating

In our view, in the coatings of furniture and construction elements actually there exists no satisfactory UV stabilizers or combinations which can be used in different paint technologies universally. This is understood as an application of an active component in both the cellulose nitrate coating, 2C polyurethane coating as well as in water-borne coatings.

1 HALS = Hindered Amine Light Stabilizers

However, in recent years active ingredient combinations have been developed that effectively are used in various wood coating systems for the UV protection of light wood. When using light stabilizers, an optimum dosage has to be determined in order to ensure their effectiveness (filter effect). Apart from the absorption characteristics, the effectiveness of UV absorbers primarily is determined from Lambert-Beer's Law [5].

$$E = \log \frac{I_0}{I} = \varepsilon \cdot c \cdot d$$

E = extinction
I_0 = intensity of the incident light
I = intensity of the outgoing light
ε = extinction coefficient [L/(mole*cm)]
c = concentration [mole/L]
d = film thickness of the coating [cm]

Chemical group	Hardwood lignin	Softwood lignin
Methoxyl	139–158	92–97
Phenolic hydroxyl	10–15	15–30
Benzylic hydroxyl	40–50	30–40
Carbonyl	10–15	10–15

Figure 8.1: Functional groups in lignin among hard- and softwoods (number per 100 phenyl units)

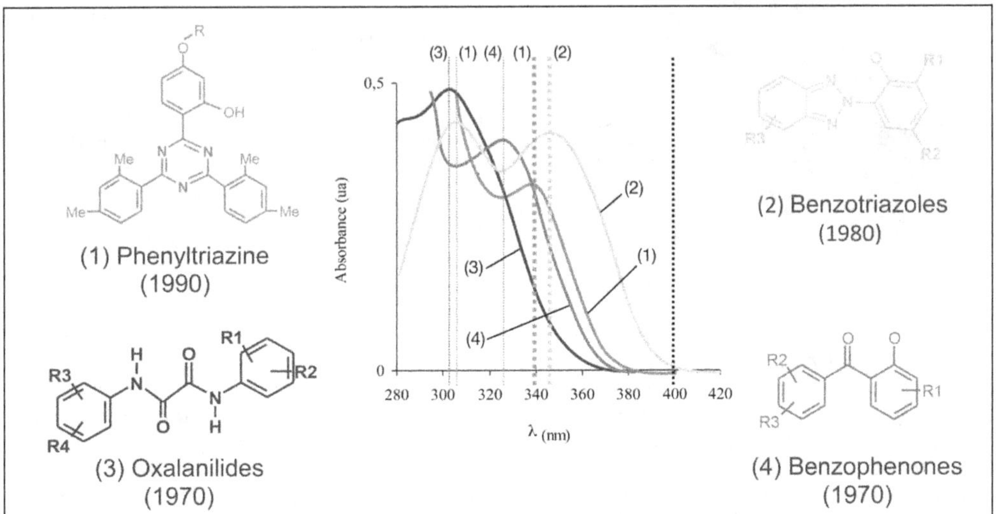

Figure 8.2: Important classes of UV absorbers [17]

The higher the extinction, the more UV radiation is filtered out. The extinction E also depends on the extinction coefficient ε, on the concentration c of the light stabilizer in the coating as well as on the thickness d of the transparent coating film. One may increase the concentration c of the light stabilizer or the coating film thickness d in order to enhance the protective effect.

Up to now, it is still not succeeded to protect woods such as cherry, walnut, oak, ash or beech effectively against discolouration by means of classical light stabilisers or combinations with HALS components while maintaining the natural wood character. Despite otherwise statements by leading manufacturers of coatings, today it has not been tackled successfully to protect light woods such as maple, spruce and pine effectively against UV radiation. Here, 2C polyurethane coatings and water-borne coatings are applied. However, the application of light stabilizers in UV curing spray coatings and rolling coatings is limited since the UV curing does not permit an effective concentration under industrial conditions. However, the application of light stabilizers in UV curing spray coatings and rolling coatings greatly is limited since the UV curing does not permit an effective concentration under industrial conditions. Even more difficult is the protection of the colour shade of tropical and dark woods which often contain red pigments. These woods cannot be effectively protected against colour modifications to yellowish brown. The red colorant is degraded by the UV radiation.

Dark natural and thermally modified timber (TMT)

These woods exhibit a pronounced tendency to a light-induced discolouration. This is often the result of bleaching ($\Delta L^* > 0$) and yellowing ($\Delta b^* > 0$) in the case of selected woods such as cherry tree or hemlock as well as a result of darkening ($\Delta L^* > 0$) of exposed native or transparently coated wood surfaces[13]. Figure 8.3 illustrates examples of unexposed and exposed

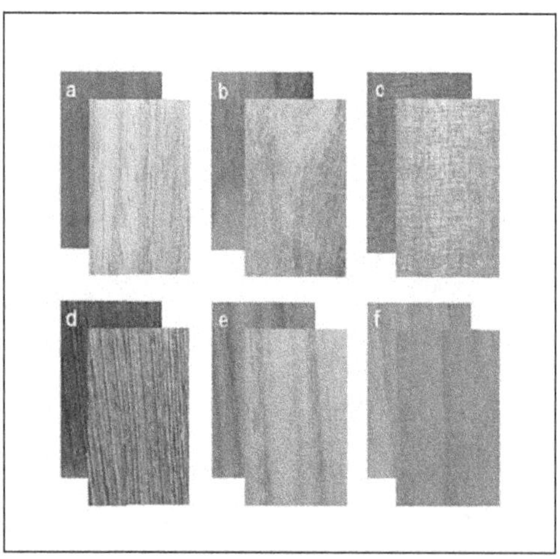

Figure 8.3: Left: Transparently coated, non-stabilised wood (a: African Padouk, b: East Indian Rosewood, c: African mahogany, d: Wengé, e: American Walnut, f: American bird cherry) before (background) and after 42 days of free exposure to light behind window glass (foreground) according to ISO 877-2 (2009); right: corresponding CIE-Lab colour parameters[13]

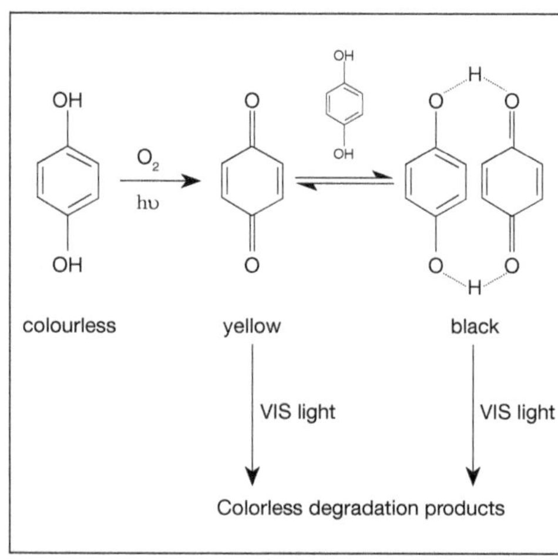

Figure 8.4: Photo-oxidation of a hydroquinone (I) to a yellow-coloured p-quinone (II); Interaction of II with I results to dark coloured charge-transfer complexes (III) (darkening); Degradation/polymerisation of II and III to colourless reaction products by means of visible light (photobleaching); modified by Furman & Lonsky, 1988a, 1988b)

surfaces of different woods as well as modifications of the associated CIE-Lab colour parameters (ISO 1664-4) [10–12].

The colour changes are based on photo-oxidatively induced formation of chromophoric structures (e.g. yellow or deep dark coloured quinones (II) and charge transfer complexes (III) which are formed from colourless phenolic structures (I) of the lignin or accessory wood components (e.g., tanning agents)) or result from a photochemical degradation or polymerisation of these to form colourless reaction products by visible light (Figure 8.2). Darkening and bleaching out of wood surfaces may therefore proceed in parallel or with a time offset and are specific to the type of wood [13, 14, 17].

Due to their absorption behaviour, on the market light-protection additives (organic and nanoscale inorganic UV absorbers, free radical scavengers: HALS compounds) are designed especially for the colour stabilisation of light woods [13, 14] and have a very low effect or even result in an increased bleaching out of the wood surface in comparison to untreated dark woods [13 15].

In cooperation with the Institut für Holztechnologie GmbH in Dresden as well as industry partners, new, light-stabilizing wood impregnations which can be adapted to the requirements of different light-sensitive dark woods have been developed. Especially a depth effect of the novel light protection additives has been identified in this research project by providing impregnating coating systems [16]. The fact that the light-induced discolouration of dark wood species and TMT also is caused by light of the visible region (Δ > 400 nm), and that the associated colour changes are caused by wood's own extractives were also taken into account (Figure 8.5) [6, 7]. The chromophores responsible for this which were identified for example in the wood extractives of macassar, walnut tree, dalbergia nigra and padouk are methyl juglon (I), Juglon (II), 4-hydroxy alberdigion and santalin A (IV) [16, 17].

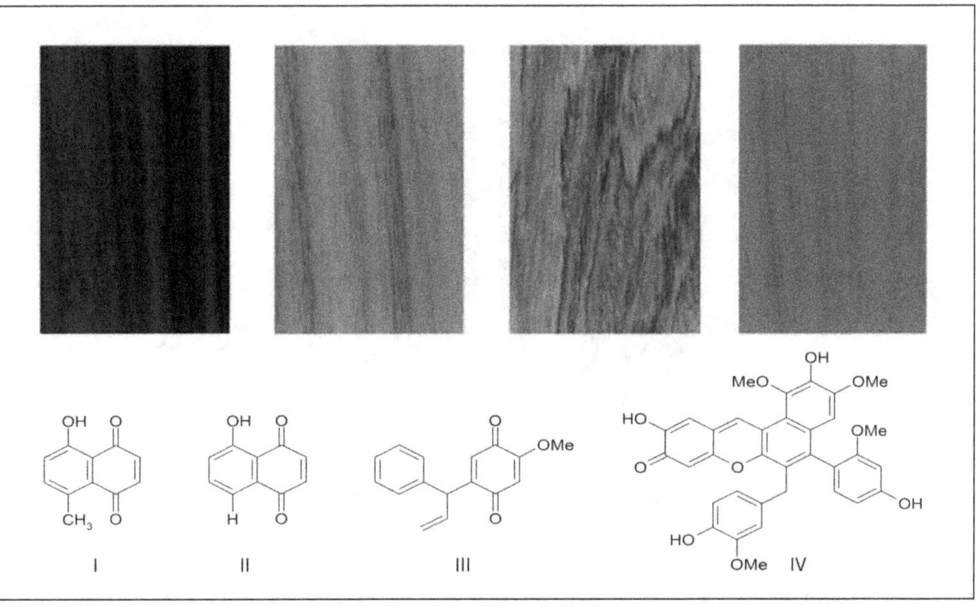

Figure 8.5: Wood textures (above) and quinoid substances identified in the extractives of the respective wood species: Macassar ebony (methyljuglone I), Eastern black walnut (juglone II), Brazilian rosewood (4-hydroxydalbergione III) and Padouk (santalin A IV) *Source:* [15]

Table 8.1: Groups of wood species and effective light protection impregnations

Composition of light protection	Wood species
LS-1	TMT (spruce, beech, ash tree), wengè, East Indian palisander
LS-2	American walnut tree, beech, African padouk
LS-3	Mahagoni, African padouk
LS-4	Beech, American bird cherry

The focus of the investigations was the development of aqueous, polymer-based, low-viscosity formulations with an impregnating character serving as carriers for a new group of active substances with stabilizing effect for light-sensitive and responsible for the inherent colouring of the wood. Alkyd resins and polyester resins were used as binders. The application was performed by means of spreading, rolling and spraying depending on the adjusted viscosity. In order to ensure the acceptance of stabilising formulations, it was considered that the initial colour and texture of the treated wood surfaces should be retained largely. The efficacy of the stabilizers was determined by means of xenon arc irradiation (ISO 11341) and natural exposure of impregnated and coated wood behind window glass (ISO 877-2) [10–12]. A significant improvement in the light stability of the wood species tested could be achieved by applying stabilising impregnations (Figure 8.6). the effect is based on the stabilisation or immobilisation, respectively, of light-sensitive, chromophoric wood ingredients [13]. The innovative concepts of light protection enable the application of dark wood surfaces in the area of high-quality furniture manufacture as well as in the area of high-quality of the interior construction. Furthermore, these innovative concepts also enable the application of dark parquet floors with significantly improved colour stability.

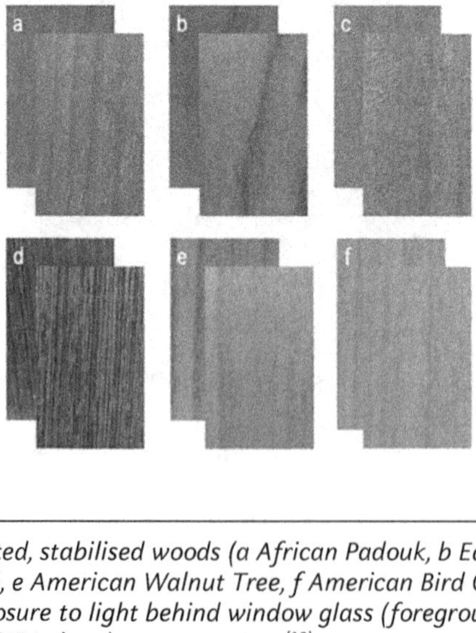

Figure 8.6: Transparently coated, stabilised woods (a African Padouk, b East Indian Rosewood, c African Mahagony, d Wengé, e American Walnut Tree, f American Bird Cherry) before (background) and after 42 days of free exposure to light behind window glass (foreground) according to ISO 877-2 (2009); right: corresponding CIE-Lab colour parameters [13]

In so far, a wood intrinsic impact of the novel light protection systems could be demonstrated which necessitates a matching of light protection impregnations to the wood species or group of wood species, respectively, to be protected (Table 8.1).

A particularly stabilising effect of the light protection impregnation on all CIE-Lab colour components could be demonstrated for the wood species of East Indian Rosewood and Teak (Figure 8.4, right). In particular, photo-bleaching (ΔL^*) and yellowing (Δb^*) could be significantly reduced in comparison to the unprotected surfaces (Figure 8.1). Likewise, a good efficacy was achieved in the particularly light-sensitive woods Padouk, Wengé and walnut tree, whereby a photo-stabilisation of the flavonoid santalin A (Figure 8.3) being responsible for the red colouration of Padouk remained problematic. This photo-stabilisation only could be achieved insufficiently ($\Delta a^* = -18.5$). It has also been found that selected formulations are suitable for the colour stabilisation of native woods, in particular for beech and to a limited extent also for oak.

8.1 References

[1] Prieto, J., Tran, P-H.: Helle Hölzer im Möbel- und Türenbau vor UV-Strahlen schützen, besser lackieren, No. 19, 15.11.2002

[2] Prieto, J., Tran, P-H.: Helle Hölzer im Möbel- und Türenbau vor UV-Strahlen schützen, besser lackieren, Nr. 20, 06.12.2002

[3] Hayoz, P., Peter W., Rogez D.: Farbstabilisierung und Lichtschutz bei naturbelassenen und gebeizten Hölzern für den Innenbereich, Farbe + Lack, Vol. 109, 7/2003, p. 26–32

[4] Devantier, B., Emmler, R.: Das Farbänderungsverhalten beschichteter Holzoberflächen, Farbe + Lack, 103 Jahrgang, 4/1997, p. 194–203

[5] Valet, A., Braig, A.: Light Stabilizers for Coatings, Vincentz Network, Hanover, 2017

[6] Baar, J.; d'Amico, S.; Wimmer, R. ProLigno 2013, 9, p. 581-584

[7] Beyer, M.; Müller, M. (2010). Light protection of TMT and other dark wood species for interior application. In Proceedings of the European Furniture Coatings Conference. March 02–03, Berlin, Germany, pp 143–152

[8] Burtin, P.; Jay-Allemand, Ch.; Charpentier, J.-P.; Janin, G. Trees 1998, 12, p. 258–264, Forsthuber, B.; Grüll, G. Polym. Degrad. Stab. 2010, 95, p. 746–755

[9] Furman, G. S., Lonsky, W. F. W. J. Wood Chem. Technol. 1988, 8, p. 165–189 Furman, G. S., Lonsky, W. F. W. J. Wood Chem. Technol. 1988, 8, p. 191–208, Hrutfiord, B.F.; Luthi, R.; Hanover, K.F. J. Wood Chem. Technol. 1985, p. 451–460

[10] ISO 877-2 (2009): Plastics – Methods of exposure to solar radiation – Part 2: Direct weathering and exposure behind window glass

[11] ISO 11341 (2004): Paints and varnishes – Artificial weathering and exposure to artificial radiation – Exposure to filtered xenon-arc radiation

[12] ISO 1664-4(2008): Colorimetry – Part 4: CIE 1976 $L^*a^*b^*$ Colour space Mayer, I.; Koch, G.; Puls, J. Holzforschung 2006, 60, p. 589–594

[13] Prieto, J.; Beyer, M.; Passauer, L. (2014) Novel color stabilization concepts for decorative surfaces made of dark wood and TMT. In Proceedings of the 2nd European Technical Coatings Congress, September 02–05, Cologne, Germany, pp 36–37

[14] Proft, B., Mocken, J. (2010) White pigments, UV absorbers and inorganic additives for furniture coatings. In Proceedings of the European Furniture Coatings Conference, March 02–03, pp 131–142

[15] Rowe, J.W. (Ed.) Natural Products of Woody Plants, Springer, Berlin, 1989 Schaller, C.; Rogez, D. J. Coat. Technol. Res. 2007, 4, p. 401–409

[16] Weichelt, F.; Beyer, M.; Emmler, R.; Flyunt, R.; Beyer, E.; Buchmeister, M. Macromol. Symp. 2011, 301, p. 23–30

[17] Prieto, J.: Untersuchungen zu Lichtschutzlösungen für dunkle Hölzer, DFO Tag der Holzbeschichtung am 18.03.2014 in Bad Salzuflen

9 VOC emissions

In the course of environmental protection legislation, the coating industry and its customers had to face considerable challenges in recent years. In the past, problem-free processability and price were the most important criteria for the development of new coating systems. Since the end of the 1990s, particularly in Europe, ecological requirements have been an important feature. A cutting edge is the so-called ‚solvent regulation'. Before describing in detail, the essential requirements and the resulting measures for coating manufacturers and coating processors especially for wood processing companies, the authors would like to point out that environmentally sound technologies and economic success are compatible with each other if one considers the following definition of an environmentally compatible painting process:

The environmentally compatible painting process is the result of the desire to coat as much surface as possible with the least possible use of resources (coating material, chemicals, water, energy, etc.), with a high-quality level and as short a time as possible.

Unfortunately, the term ‚solvent' and the term ‚VOC' could not be agreed on a uniform definition throughout the world and Europe. At least, one agrees the translation of the term VOC = volatile organic compound. The term „volatile" implies that the substances belonging to the group of VOCs evaporate rapidly (volatilize) due to their high vapour pressure or low boiling point. According to WHO[1], volatile organic compounds are classified according to their boiling point or volatility as follows:

- Very volatile organic compounds (VVOC), boiling range: < 0 to 50 ... 100 °C
- Volatile Organic Compounds (VOC), boiling range: between 50 ... 100 °C and 240 ... 260 °C
- Semi Volatile Organic Compounds (SVOC), boiling range between 240 ... 260 °C and up to 380 ... 400 °C

Even definitions of the term 'vapour pressure' are also common. Accordingly, data on the emission of VOCs can only be evaluated if, in addition to the specification, the definition used is also mentioned. The definition according to DIN ISO 11890-1, 2 of 2007 [1, 2] is the most general definition and therefore the least vulnerable. Thereafter, VOCs are generally any organic liquid and/or any organic solid which evaporates by itself under the prevailing ambient conditions (temperature/pressure) (ISO 4618-1). In the Chapters 9.1 and 9.2 different definitions for the term VOC are applied.

9.1 VOC emissions during processing in installations

9.1.1 Solvents Regulation

At the beginning of the 1990s, the discussion on a directive on the reduction of VOC emissions, which was introduced in 1999 with the ‚Council Directive 1999/13/EU of 11 March 1999 on the limitation of emissions of volatile organic compounds used in certain activities

1 WHO: World Health Organisation

Jorge Prieto, Jürgen Kiene: Wood Coatings

and in certain activities Plants resulting from the use of organic solvents' [3]. The Directive obliges the Member States to adopt the laws, regulations and administrative provisions necessary to comply with the EU Directive. In Germany, this was done with the 'Ordinance on the Implementation of Directive 1999/13/EU on the Limitation of Emissions of Volatile Organic Compounds' [4] (31st BImSchV[2]) of 21st August 2001 (Federal Law Gazette 1, 2180). A revised version was adopted, state of the art (paper by Verband der deutschen Lack- und Druckfarbenindustrie e.V. (Coatings and Air Pollution Control)), references to the state of the art in the 31st Ordinance on the Federal Immission Control Act. Similar implementations of the directive took place in other EU Member States.

The purpose of the directive or regulation is to reduce the emissions of volatile organic compounds (VOCs). Among these, a broad spectrum of various organic substances is included, for example organic solvents, benzine fumes, hydrocarbons e.g. In exhaust gases etc. A volatile organic compound in the sense of the solvent regulation is, according to § 2 No. 11 in conjunction with No. 25 *'means a compound containing at least carbon and one or more of the elements hydrogen, halogen, oxygen, sulphur, phosphorus, silicon or nitrogen or a plurality thereof, excluding carbon oxides and inorganic carbonates and bicarbonates obtained from 293.15 Kelvin (20 °C) Has a vapor pressure of 0.01 kilopascal or more or has corresponding volatility under the respective conditions of use'* [4].

Volatile organic compounds contribute decisively to the formation of ground-level ozone in the summer. In the case of intense sunlight, the ozone formation increases markedly and leads to the known problem of summer smog. Emissions from solvent applications have been consistently on a level of over 1,000 kilotons per year for years. After the use of solvents, a number of plant main groups are particularly relevant in Germany: painting, printing and

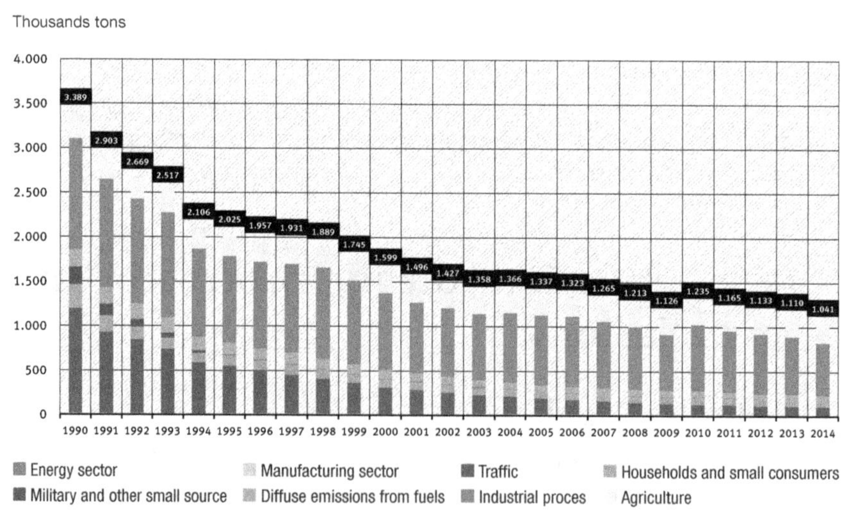

Figure 9.1: Emissions of volatile organic compounds without methane (NMVOC) by source category Source: Umweltbundesamt, Germany [41]

2 Federal Emission Protection Directive

surface cleaning systems. Approx. at the beginning of the century, 60 % of the solute emissions were from paint applications.

From 1990 to 2014, NMVOC emissions have been reduced from 3.4 million tons (million tons) to 1.04 million tons, which is more than two-thirds (see Figure 9.1). The decline can be explained primarily by the development of emissions from road traffic as well as by solvent applications in the industrial and commercial sectors.

The NMVOC emissions caused by the use of paints and cleaning agents have been halved since 1990, with a lower solvent content and the partial conversion to water-borne systems, especially in paint shops, printing plants and metalworking and processing plants.

Even products outside of plants contribute significantly to emissions at approximately 300 kilotons per year. In December 2004, the 'Decopaint Directive', the first regulation to limit the solvent content in products, was submitted to this area and transposed into national law by the 'Solvent-borne Paint and Coating Regulation' (ChemVOCFarbV)[3]. It sets requirements for the reduction of solvent content in building paints in two stages (2007 and 2010).

9.1.1.1 Contents and Requirements of the Solvent Regulation

The ,Ordinance on the Implementation of Directive 1999/13/EC on the limitation of emissions of volatile organic compounds' essentially consists of two parts. First, the 2nd Ordinance on the Federal Immission Control Act was amended, and secondly, the Ordinance on the Limitation of Emissions of Volatile Organic Compounds in the Use of Solvents in Specific Installations, 31st BImSchV or also the Solvents Ordinance. The main points in the amended 2nd BImSchV are as follows:

- This Regulation shall not apply to installations pursuant to the Second Ordinance on the Implementation of the BImSchG, in which organic solvents containing light volatile halogenated organic compounds are used.
- Installations that are subject to authorization under the terms of an immission protection law are also within the scope of the 2nd BImSchV.
- Surface cleansing has provided limited possibilities for the use of substances other than the substances Tetrachloroethene (Per), trichloroethene (Tri) and dichloromethane, which have so far only been permitted.
- The emission limit for dichloromethane has been reduced from 50 mg/m^3 to 20 mg/m^3.

In addition, the 2nd BImSchV provides an obligation for plant operators to make a notification to the authority if the requirements of the regulation are not complied with. According to Article 3 of the German Solvents Ordinance, a number of requirements are set for all plants. To be mentioned first is the substitution requirement for carcinogenic, mutagenic or reproductive toxic substances taking into account the utility and proportionality. This applies to substances or mixtures which are marked with the hazard warnings H340, H350, H350i, H360D or H360F. This requirement is already known from other areas, such as occupational health and safety, hazardous substances legislation. In addition, measures are to be taken to minimize emissions when the plant is started. Measures to reduce emissions during the transfer of solvents with a boiling point above 100 °C with a volume of 100 tons per year shall be taken. In addition, operators are required to notify the authorities if the requirements of the

3 Ordinance on the regulation of chemicals to limit the emissions of volatile organic compounds (VOC) by restricting the marketing of solvent-containing paints and coatings (ChemVOCFarbV), published on 22.12. 2004 (BGBl.I, No. 70, p. 3508)

Table 9.1: Installations within the scope of the Regulation (Annex II) for the coating of wood and wood materials

Activity and designation of the installation	Threshold value for the solvent consumption (t/a)
9. Coating of wood and wood-based materials	
9.1 Installations for the coating of wood or wood based materials with a solvent consumption up to 15 t/a	5
9.2 Installations for the coating of wood or wood-based materials with a solvent consumption of > 15 t/a	15

Table 9.2: Requirements for installations of No. 9.2 of Annex III.1: Thermal after burning (TAB)

Emission limit values for exhaust gases	VOC in the exhaust gas (mg C/Nm³) coating and drying	Diffuse emissions (% of the applied solvent content)	Coating particle [mg/m³] spray zone
> 15–25 t/a	100	25	3
> 25 t/a	50 20 with TAB[1)]	20	3

1) Thermal after burning (TAB); mg C/Nm³ = milligram carbon per standard cubic metre

Solvent Ordinance are not complied with. For substances with high environmental relevance or high potential hazards, emission limit values have been set for all plants. Emission measurements shall be carried out in accordance with §3 (No. 4) if, according to the state of the art, exhaust gas purification is necessary to comply with the limit values.

9.1.1.2 Special requirements for the coating of wood and wood-based materials[4]

Two criteria are decisive for the question whether an installation is within the scope of the solvent regulation. Firstly, activities must be carried out according to Annex II (see Table 9.1), and second, the solvent consumption must be above the threshold set out in Annex I when exercising this activity. Under the solvent consumption (§ 2, No. 19) is *'the total amount of organic solvents used in an installation per calendar year or within any twelve-month period, minus all VOCs returned for recycling'*. The term solvent also includes detergents, dispersants, preservatives, plasticizers or viscosity or surface tensioning agents.

In No. 9 of the Annex II, the activity of the wood coating is defined as follows: *'Any activity in which one layer is applied to surfaces of wood or wood by single or multiple application.'* The threshold is 5 tons per annum of solvent consumption. Under this threshold, the Solvent Ordinance is not relevant for the coating of wood and wood materials.

Pursuant to section 5 sentence 2 of the Solvents Ordinance, installations according to the Federal Pollution Control Act must be approved or notified to the competent authority if they are not subject to authorization – as is the case in most cases. Annex III of the Solvents Ordinance contains the specific requirements for individual plants. Here, specific limit values of the solvent emissions are defined for the areas of activity as well as for the coating of wood and wood materials. Depending on the type of emissions, these limit values are defined differently:

4 As this book does not deal with every detail, such as the different interpretations of the licensing authorities, the authors would particularly like to point out the literature on this chapter.

- Fanned emissions (as concentration values). These are all exhaust gases which are released via a chimney or other exhaust gas lines.
- Diffuse emissions (as a percentage of solvents used). These are unencumbered exhaust fumes, such as emissions from doors, windows or ventilation shafts. Untreated exhaust fumes shall be assigned to diffuse emissions for certain Annex III installations.
- Total emissions (mass flow relative to a product such as a coated area) or as a percentage of solvents used.

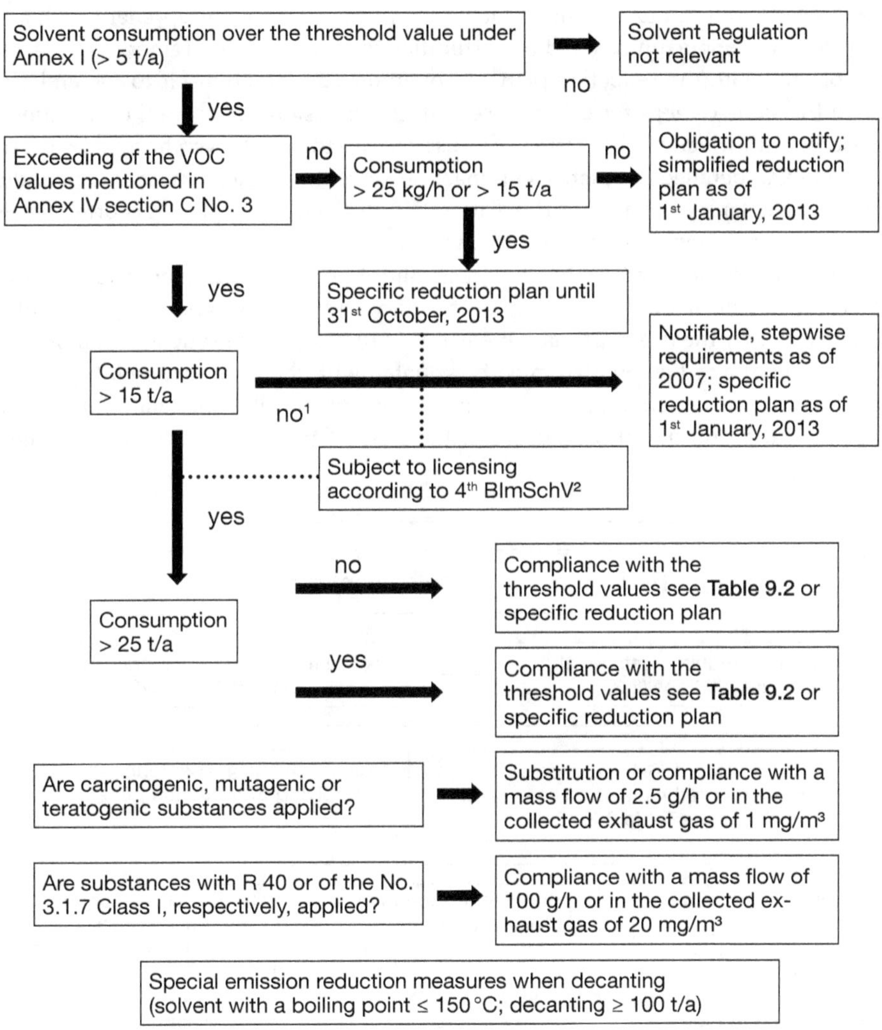

1 For installations with a consumption between 5–15 t/a, a need to licensing may except in exceptional circumstances if the consumption capacity exceeds 25 kg/h

2 In addition to the requirements of the solvent regulation, installations subject to licensing have to comply with the requirements of the TA Air provided that it does not concern to the requirements for VOC limits in these systems.

Figure 9.2: Test scheme of the Solvent Regulation for the coating of wood and wood-based materials

Volatile organic compounds contained in captured untreated exhaust gases are diffuse emissions when coating wood and wood materials. Table 9.2 schematically shows the plant-specific emission limit values of the Solvent Ordinance for the coating of wood and wood materials. They only apply to installations of no. 9.2 according to Annex III.

However, the compliance with the emission limit values, which have been verified by measurement, is an opportunity to meet the specific requirements of the regulation. The operator of an installation may also decide to achieve the required emission reduction by means of a reduction plan according to Annex IV. A reduction plan can, for example, provide for the reduction of emissions by changing the paint technology and the solvent content in the paints used. The total emissions may not be greater than for the application of the emission limit values set out in Annex III. The reduction plan provides the entrepreneur with a multitude of individual opportunities and opportunities to meet the requirements of the regulation on solvent. With the reduction plan, the operator commits himself to the emission reduction at the same level as for compliance with the emission values. In the reduction plan, the measures with which the limitation of emissions are to be achieved have to be presented in a concrete and detailed manner. The impact on solvent emissions must also be indicated in the reduction plan. As a basis for the elaboration of a reduction plan, it is often necessary to first establish a solvent balance (see below).

The plant operator therefore has the possibility to comply with plant-specific limit values (Annex III) by means of an 'end of pipe' technique (exhaust gas purification plant) or to reduce emissions by the use of primary measures in the form of a reduction plan (Annex IV). The application of a reduction plan must be equated with the introduction into the production-integrated environmental protection. In the case of installations falling under the No. 9.2 Annex III, afterburning may be useful if the state of the technology does not allow the

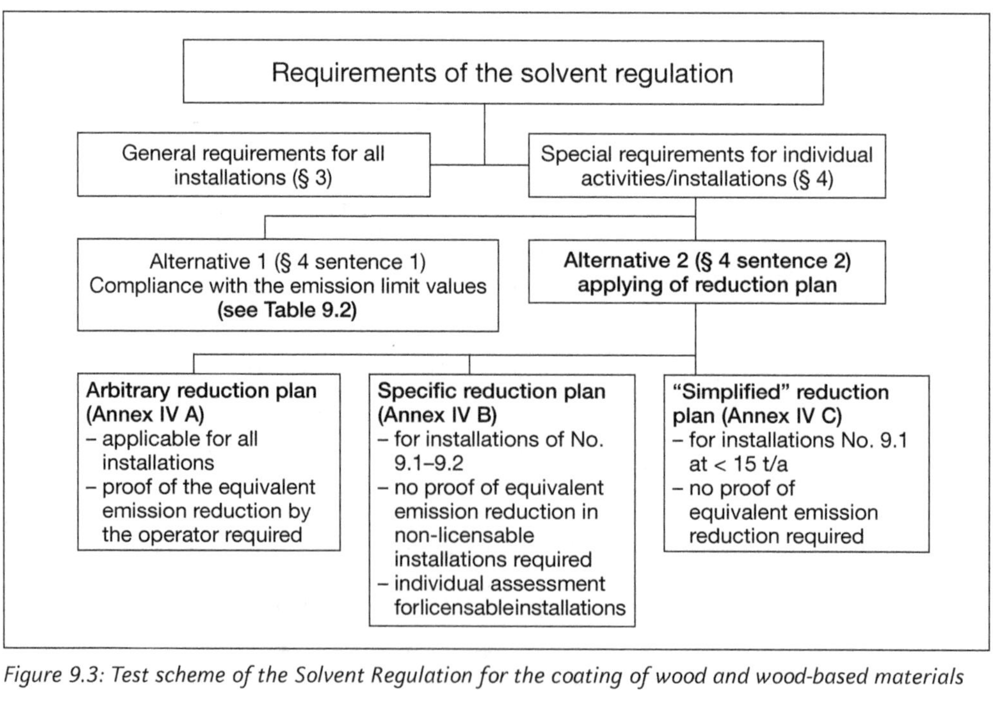

Figure 9.3: Test scheme of the Solvent Regulation for the coating of wood and wood-based materials

production of sufficient surface quality by the use of low-solvent or free-coating systems. Such afterburning plants are expected when the solvent content of the exhaust air is so high that the afterburning plant can be operated auto-thermally. With a solvent content of > 2 g/m^3, it can be ensured in any case that auto-thermal operation can be guaranteed. Frequently, however, it is not the case in practice. Figures 9.2 and 9.3 show schematically the most important requirements of the Solvents Ordinance for the coating of wood and wood materials.

According to Annex III, No. 9.1. of the 31st BImSchV, the following requirements are applied to plants of No. 9.1 of Annex I with an annual solvent consumption of up to 15 tons:

- Reduction of VOC emissions through the use of low-solvent feedstocks according to the state of the art
- Since 01st November, 2007: Determination of VOC emissions at least once a year by means of a solvent balance according to Annex V
- Since 01st January, 2013: Application of a reduction plan according to Annex IV

Reduction plan

Three variants of the reduction plan are possible according to the regulation and are sketched in Figure 9.3.

- For arbitrary reduction plan', proof must be provided that an emission reduction is achieved at least at the same level (Annex IV A). Anything that contributes to an emission reduction can be used. The equivalence shall be indicated to the authority.
- The 'specific reduction plan' has been developed for coating plants and assumes a reduction of at least the same level as for the application of the limit value rule in Annex II. A target emission calculated according to a given method must be complied with (Annex IV B).
- In the 'simplified reduction plan', plant operators are obligated to use substances with a given low VOC value or solvent content (Annex IVC). The same emission reduction as for the application of the limit value rule in Annex II is also assumed.

For new installations, the reduction plan must be submitted to the authority before commissioning. Only the specific and the simplified reduction plan is discussed below.

Specific reduction schedule

The specific reduction plan referred to in Annex IV B must comply with certain target emissions. The following parameters are used for the calculation of the mission:

- The amount of solid (kg/a) used in the coating period during the reporting period
- The multiplication factor for the determination of a yearly reference emission
- The percentage for the determination of the mission

Table 9.3 lists the data for the calculation.

The target emission is calculated according to the following formula:

Target emission =
Solid (kg/a) x Multiplication factor x Percentage of the target emissions

The solid content can be e.g. the safety data sheets or other product information. The mass of the solids can be calculated from the amounts of the substances used and the solids contents. The reference emission is the product of the mass of the substances and the multiplication

Table 9.3: Determination of the target emissions according to Annex IV B: Calculation factors; percentage is given by the sum of the respective limit values for diffuse emissions and a defined value for the proportion of VOC emissions recorded but not deposited

No.	Solvent consumption (t/a)	Multiplication factor for determining the annual reference emission	Percentage for determining the target emission
9.1	> 5–15	4	(25 + 15) %
9.2	> 15–25	3[1)]	(25 + 15) %
9.2	> 25	3[1)]	(20 + 15) %

1) With an order efficiency > 85 % the multiplication factor 4 can be used. Determination of the target emissions according to Annex IV B: Calculation factors; The percentage is given by the sum of the respective limit values for diffuse emissions and a defined value for the proportion of the VOC emissions detected but not deposited

factor. It represents a fictitious total emission which would be released from an installation by using conventional solvent-containing coating materials without the use of pollution-reducing measures. It is calculated according to:

Reference emission = Solid (kg/a) x Multiplication factor

The requirements are met if the solvent balance is the result of emissions that are smaller than the target emissions. For installations requiring approval pursuant to the Federal Emission Control Act, § 5 (1) No. 2 BImschG must also be observed. This may require further action.

Simplified reduction plan

The simplified reduction plan is a simplified proof for adherence to the target emissions of the specific reduction plan, which cannot be used for elaborate solvent balancing. It applies to the coating of wood and wood materials for plants in the threshold range of 5 to 15 tons (Annex I, point 9.1). Operators wishing to apply the simplified reduction plan in their plants must submit a binding declaration to the competent authority on the use of solvent-free or free-flowing substances. Since January 1, 2013, the simplified reduction plan must be adhered to. For the following substances with a very low solvent content, a binding declaration must be given to the competent authority when applying the simplified reduction plan for the above plants:

- Only use of coating materials with a VOC < 250 g/l when coating flat and flat surfaces
- Only use of coating materials with a VOC value < 450 g/l for coatings of other surfaces
- Only use of aqueous pickling with a VOC value of 300 g/l

Simplified proof is the safest and most cost-effective way to meet the requirements of 31st BImSchV, as no annual solvent balance and no equivalency proof is required. Furthermore, it is the least expensive for the authority and for the authority. The VOC content of coating materials must be calculated according to the following formula:

$$\text{VOC value (g/L)} = (100 - \text{nfa} - \text{mw}) \cdot s \cdot 10$$

where nfa = non-volatile content, mw = mass fraction of the water in percent, s = density of the coating material

Solvent balance

The solvent balance is a new planning and verification instrument in the German immission control law, which was introduced by the German Solvents Ordinance.

The solvent balance serves:

- The determination of the solvent consumption in the test whether threshold values are exceeded
- As proof of compliance with the limit values for diffuse emissions and / or total emissions
- As proof of compliance with the reduction plan

Annex V of the Regulation describes the general principles of solvent release. Due to the complexity of solvent preparation, the authors would like to elaborate on the further literature references [5–7] and here only important aspects. Figure 9.4 shows the possible inputs and outputs of solvents in a fictive installation.

All relevant input and output streams of solvents of an installation are considered in the solvent balance. Input currents are marked with 'I' and output currents are marked with 'O' and a number. The currents, indicated by dashed arrows and cursively labelled, are among the diffuse emissions. As proof of the compliance with No. 9.2, the diffuse emissions are determined according to Annex III (see Table 9.2). In principle, two methods are possible for the determination of the emissions as the sum of the diffuse and enclosed emissions. In the direct method, diffuse emissions, such as solvent emissions in wastewater, as impurities and residues in the product, are summed up as uncontrolled emissions and solvent emissions

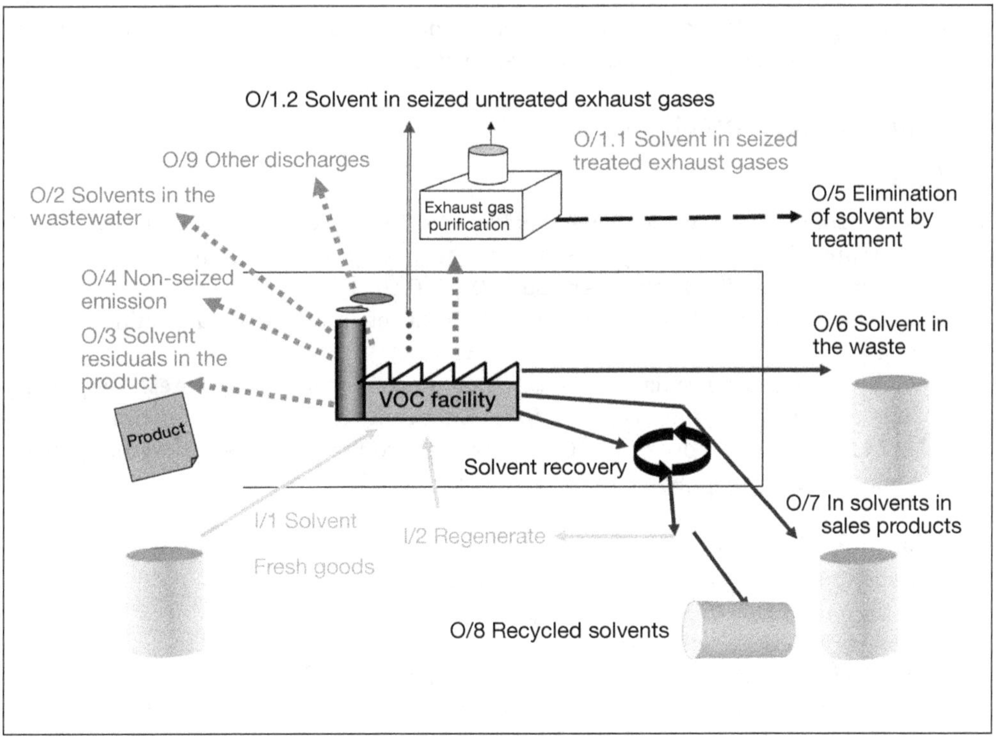

Figure 9.4: Presentation of the input flows and output flows Source: Federal Environmental Agency and Ökopol

from other routes. Means of choice will usually be the indirect method. The diffuse emissions are determined in which solvent streams which are relatively easy to determine are removed from the solvent used. These are the solvents in the exhaust gas, solvents that are destroyed in the exhaust gas treatment, solvents in the waste, solvents as product constituents and also recovered solvents. Untreated exhaust fumes can be attributed to diffuse emissions in a range of plants. Of the solvents used, e.g. In the case of paint production, deduct the quantities, leave the plant in the product and in the waste and, if necessary, be destroyed in the exhaust gas purification. The remainder corresponds to the diffuse emissions.

These solvent balances must, in principle, be compiled for all plants and represent a considerable effort. When e.g. smaller companies do not use external help, 1 to 2 working days for a knowledgeable employee should be expected. To determine the solvent consumption and thus to determine whether a plant is above the threshold of 5 tons per year solvent consumption, each plant operator should carry out a solvent assessment. The following points must be observed:

- The determination of the solvents used can be determined, for example, by means of the purchase orders or the delivery notes and changes in the stock level. The solvent content in preparations can be taken from the safety data sheets or requested from the manufacturer.
- The number of loose parts of all process steps, secondary equipment or parts of this activity must be summed up
- If solvents are reused, these quantities must be subtracted from the solvent inlet. Such solubilized solvents are to be recorded and documented in detail.
- The solvent consumption is to be compared with the threshold in Annex I

Solvents in secondary devices and in upstream or downstream process steps are easily ignored. Solvents in adjuvants are often underestimated in their significance.

Examples of good practice

Example 1

A furniture manufacturer [5] is mentioned here as a first example for the reduction of the solvent consumption in wood coating. This is a coating line for chairs with the following data:

- Coating line for chairs: 60 chairs per hour, 105,000 chairs per year
- Coating sludge: approx. 12,000 kg/a (50 to 60 % of the used amount of coating)
- Process:
  - water-borne powder stain, 15,750 kg/a, solvent content = 0 %, high-pressure spraying
  - primer with CN coating, 14,700 kg/a, solvent content = 70 %, dipping process
  - UV topcoat (water-borne), 10,500 kg/a, solvent content = 8 %, electrostatic spraying
- Consumption: coating 40,950 kg/a, solvent (11,130 kg/a)

The facility is subject to notification because the solvent consumption is > 5 t/a.

The operator has to perform a balancing of solvents every always upon expiry of three years. As from the year 2013, a reduction plan has to be prepared.

The following reductions were made:

- Topcoat facilities with recovery of the coating (paint-in-paint cabin)
- Installation of a circular conveyor with chair carriers which can be rotated by 4 x 90° respectively
- Primer coat with a water-dilutable 1C water-borne coating

Table 9.4: Comparison 2C PU coatings/1C UV water-borne coatings for the coating of stairs

	2C PU coating	**1C UV water-borne coating**
Solid content (nfA)	26 %	38 %
Solvent content	74 %	3 %
Water content	-	59 %
Application amount per m² inclusive loss of coatings	350 g (250 g)	220 g (200 g)
Solvent application per m²	259 g	6.6 g
Dry film thickness [µm]	65	76
Total coating throughput period of a stair step	12 hours	2 hours
Application efficiency	55 %	90 %
Recovery of the coating (recycling capability)	-	++
Potlife	Yes	No
Visual evaluation of the surface	++	++
Chemical resistance DIN 68861 Part 1 1B	Yes	Yes
Abrasion resistance	++	++

Table 9.5: Total amount of coating per year for 60,000 m² surface to be coated

	2C PU coating	**1C UV water-borne coating**
Total amount of coating	21,000 kg	13,200 kg
Total amount of solvent	15,540 kg	396 kg
Total amount of wastes	12,600 kg*1	1,980 kg1)

1) Total coatings quantity for 60000 m² of surface to be coated per year

- Process:
  - Wood stain: 15,750 kg/a, (solvent content = 9 %, solvent based water stain), High-pressure spraying
  - Primer coat with 1C water-borne coating, 14,700 kg/a, solvent content = 9 %, dipping process
  - UV topcoat (water-borne), 10,500 kg/a, solvent content = 8 %, electrostatic spraying
- Consumption: coating 40,950 kg/a, solvent (3,580.5 kg/a)

The solvent consumption is declined to 3.6 t/a. Thus, the facility is no longer in the scope of the regulation and therefore not subject to notification. Furthermore, the solvent vapours at immersion pools as well as wastes could be reduced.

Example 2:

For the abrasion resistant and robust coating of stair steps very often quickly drying 2C PU single-layer coatings are applied in spraying processes. The processing solid content of these coating systems usually amounts 25 to 30 %. The solvent content ensures an optimal running and elegant coating surface. The chemical reaction (polyaddition) of the base coat with the hardener results in chemical resistant surfaces in accordance with DIN 68861 Part 1 1B.

As an environmentally friendly and low-solvent alternative for solvent-borne 2C PU spray coatings, especially the 1C UV water-borne coatings are increasingly important. A major advantage of UV water-borne coatings in comparison to conventional 2C PU spray coatings is that the wood substrates can be stacked or are packable after evaporation of the water and the chemical UV curing. The abrasion resistance and the visual appearance of the coating surface of the UV water-borne coatings is absolutely comparable with the 2C PU coatings (see Table 9.4).

The application is carried out by means of a surface spraying machines with a belt conveyor with automatic doctoring for the overspray (plant concept, see Chapter 3.1.7). Nearly 20 to 25 % of the thus recovered overspray is added to the UV water-borne coating which exclusively is used as a primer. The application efficiency is increased to about 90 to 95 % by means of the recovery. Thus, the disposal costs for coating sludge can be saved [8]. As shown in the example (see Table 9.5), the procedure is subject to the solvent regulation (solvent consumption > 15 t/a) for application of 2C PU coatings and a surface of 60.000 m² to be coated. By using surface spraying machines and 1C UV water-borne coatings, this facility is not subject to the solvent regulation (solvent consumption < 5 t/a).

The advantages of the UV-water-spraying process are the following:

- Significantly lower amounts of coating are required (approx. 38 % less coating)
- Reduction of the total amount of solvent by approx. 97 %
- The amount of coating wastes is reduced by 84 %
- No deterioration of the technical properties of the coatings
- The processing times significantly are reduced by the UV technology

More practical examples can be taken from further literature.

9.2 Residual emissions after drying and curing of coated wood and wood-based materials

In recent years, the theme ‚residual emissions from coated wood and wood-based materials' enormously has gained great importance in the population in the age of environmental and health consciousness. Not only the quality of the air contributes to this generally, but also the modified lifestyles of the people in the industrial society. Evaluations of the Federal Environmental Association show that in Germany adults with an age between 25 and 69 years reside approx. 20 hours per day indoors.

A human being spends an average 80 to 90 % of his life time in interior spaces. The fresh air supply in interior spaces was reduced significantly by means of the improved thermal insulation of buildings after the oil crises of the 1970ies and by means of the application of more tight windows [9, 10]. This has also led to further problem areas in the hygiene of interior spaces. The mould problems or fogging effects (black dwellings) are examples of.

In the case of unfavourable ventilation conditions, various volatile compounds from building materials, insulation materials, furniture, decorative wall cladding and ceiling claddings can naturally also accumulate in the interior air. Up to a few years ago, furniture still continued to produce large amounts of formaldehyde, which in combination with the above-mentioned modifications present a greater health risk than more than 60 years ago. The reason for these emissions were wood materials, adhesives and coating materials. The odour threshold for formaldehyde is 0.12 ppm (160 μg/m³ air). However, odour sensitive people can already perceive formaldehyde from 0.05 ppm (60 μg/m³). Due to the toxicity of the formaldehyde (see Chapter 2.5.3), the formaldehyde release of furniture has been investigated extensively in recent years, significantly reducing the yield potential of the wood materials, adhesives and coating materials used [11–13].

In 1986, the formaldehyde concentration in the room air was regulated by law and anchored in the Ordinance on Hazardous Substances (GefStoffV). Since this, wood materials have been prohibited in furniture which results in a concentration of more than 0.1 ml/m³ (= 0.1 ppm) in a defined test chamber. This has been fixed since 1993 also in the Chemical Prohibition Ordinance (ChemVerbotsV). E1 chipboard panels adhere to these conditions. By the legal regulation, the formaldehyde release of wood materials for furniture has been reduced to such an extent that formaldehyde emissions from furniture have now been shifted into the background. Large furniture houses have also gained experience with formaldehyde emissions. After having reported high formaldehyde emissions, mainly from the adhesives used in the chipboard in the 1980s, the magazine "Der Stern" claimed that the popular "Billy shelf" gave formaldehyde and that when buying this shelf the consumer would play "Russian roulette" with his health [14]. In the autumn of 1993, the company decided not to use any acid-curing coatings for furniture coating. Today the Billy shelves are coated with environmentally friendly water-curing and UV-curing roller paints.

However, furniture can still be a significant source of emissions in interior spaces. What was formerly highlighted as a quality feature and made the owner's pride, today a highly sensitized public generally defends itself as a deficiency. The smellable spectrum of the residual emissions of factory-fresh products is therefore not only a nuisance but also a health risk. It is now assumed that humans can distinguish about 4,000 to 10,000 odours (see Table 9.6).

9.2.1 Residual emissions from furniture surfaces

Since about 1989, intensive care has been devoted to residual effects from coated furniture and wood surfaces [16, 17]. For example, numerous institutes are working to develop an evaluation concept of ingredients in indoor air and house dust. A much debated proposal was published by the AGÖF (Arbeitsgemeinschaft ökologischer Forschungsinstitute e.V.). The AGÖF orientation values for volatile air constituents are statistically derived. The orientation values are based on the results of the investigations carried out by the involved AGÖF member institutes over a period of 10 years. They represent measured values of more than 2,000 room air measurements and more than 3,500 house dust analyses. Another evaluation concept is based on toxicological tests, which are often based on animal experiments. Both methods have their advantages and disadvantages. In the future one will have to wait and see which evaluation methods will be implemented in Europe and worldwide. Table 9.7 shows the average concentrations of some organic substances in indoor spaces from 1993.

The emission test chamber method (DIN EN ISO 16000-9) and emission test cell methods (DIN EN ISO 16000-10) were developed in order to systematically record and standardise the residual emissions. Sampling, storage of the samples and preparation of the test pieces are standardized in DIN EN ISO 16000-11. Figure 9.5 schematically shows an emission test chamber [20]. The wood substrates to be tested are pre-conditioned after 23 °C,

Table 9.6 Examples of odours distinguishable by human beings [15]

Woody, resinous	According to cork	According to smoke
According to turpentine oil	Rancid	According to vinegar
According to bark, birch bark	Oily, fatty	Fruity (citrus fruit)
According to oak wood	According to coating, painting	According to eucalyptus

Table 9.7: Average concentrations of some organic substances in indoor rooms from 1993 [18, 19]

Substance	**Concentration [µg/m³]**
Formaldehyde	58.3
n-Hexane	9.5
Cyclohexane	8.0
Benzene	9.0
Toluene	78.0
Ethylbenzene	10.0
m,p-Xxylene	22.0
Styrene	2.0
1,1,1-Trichloroethane	8.0
α-Pinene	10.0
n-Butylacetate	6.2
n-Hexanal	1.5
n-Butanol	1.3

50 % relative atmospheric humidity and 1/hr air exchange in a climate chamber. Thereafter, the samples are placed in the test chamber and, after 24 hours of equilibration[5], the chamber concentration is measured to the final determination after 28 days. The 28 test days were chosen forthe reason that, statistically speaking, this period goes by from the production of a piece of furniture to delivery to the end customer. Analytical detection is by means of gas chromatography and mass spectrometry. For the emissions to be determined in the test chamber, the following definitions apply to DIN ISO 16000-6:

- VOC: all individual substances in the retention range[6] C_6 to C_{16}
- TVOC: sum of all individual substances ≥ 5 µg/m³ in the retention range C_6 to C_{16}
- SVOC: all individual substances in the retention area> C_{16} to C_{22}
- SVOC: sum of all individual substances ≥ 5 µg/m³ in the retention range> C_{16} to C_{22}

The interior air area is hitherto largely not legally regulated, that is, there are no comprehensive requirements in the form of laws, regulations or other legal binding regulations. This applies not only to the German but also to the international area. In recent years, the smell of new furniture is a frequent reason for complaint. Due to ever shorter delivery times and partly air-tight packaging films, the ventilation phase of the furniture frequently only takes place at the end-user premises.

The smell of new furniture can be composed of [22–25]:

- Emissions from wood and wood materials (for example formaldehyde, terpenes, see Chapter 2.4)

5 Equilibration: to balance

6 Retention time: time that the molecules of a substance need to pass through the chromatographic column (residence time) from injection to detection

- Organic solvents from adhesives, cleaners, coating materials, plasticizers
- Cleavage products of photoinitiators and residual monomers from UV coatings
- Residual monomers of unsaturated polyester resins (e.g., styrene)
- degradation products from oxidatively curing oils (e.g. aldehydes)

For some years, one has been increasingly concerned with solvent emissions, which are released during the life of furniture, in order to be able to evaluate and limit residual emissions more thoroughly. Investigations at the Braunschweig Wilhelm-Klauditz-Institut from 1993 to 1995 have shown that mainly organic solvents cause the residual emissions in the first weeks after completion of the furniture, depending on the products and processing conditions used. As shown in Figure 9.6, the solvent release takes place in the emission test chamber method as a typical decay function [26]. The following factors of the coating and drying process influence the residual emissions of furniture:
- Highly absorbent substrates
- Substrate temperature (cold or hot)
- Order quantity / layer thickness
- Drying parameters
- Air speed
- Temperature in the dryer
- Duration of drying
- Relative humidity

The basis for the award of the RAL-UZ 38 eco-label for flat, low-emission products (for example, coated doors, panels, laminate flooring, prefabricated floors, etc.) is shown in Table 9.8. After the 28th day, the residual end-of-life value must be ≤ 300 µg/m³ organic compounds (boiling point 50 to 250 °C) and ≤ 100 µg/m³ (boiling point > 250 °C). Three-dimensional components, e.g. cabinet furniture may have a residual emission of ≤ 600 µg/m³. In order to be able to carry out the restoration tests as economically as possible, a preliminary test of the furniture components in the emission test chamber is carried out after 48 hours. If the TVOC value (total volatile organic compounds) is ≤ 1.200 µg/m³, the test is terminated. It is known

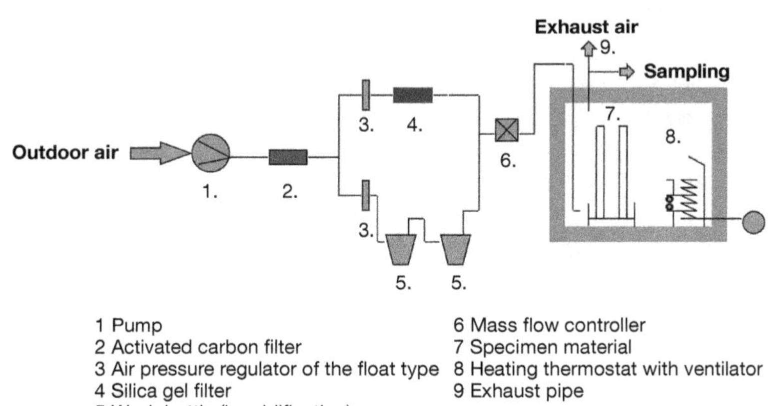

Figure 9.5: Schematic representation of an emission test chamber [18, 21]

Table 9.8: Award requirement RAL-UZ 38 – emission values for planar, flat products (for examples doors, panels, laminate floorings, prefabricated parquet) [28

Substance	Initial value (24 + 2 h)	Final value (28th day)
Formaldehyde	-	0.05 ppm
Organic compounds boiling point 50 up to 250 °C	-	300 µg/m³
Organic compounds boiling point > 250 °C	-	100 µg/m³
CMT substance	> 1 µg/m³	< 1 µg/m³

from practical experiments that after a test period of 28 days the TVOC value is ≤ 600 µg/m³. In the event that the residual value after 48 hours is more than 1200 µg/m³, the test is carried out until the 28th day.

9.2.1.1 AgBB scheme

In a concerted action, the Regional and Federal Authorities set up the AgBB (German Committee for the Health-Related Evaluation of Building Products) in 1997 in order to arrange uniform assessment criteria for volatile organic pollutants (VOCs) and their summary accumulation (TVOCs) in indoor spaces, thus ensuring the health of the building users.

Under the title „Controlled Indoor Air“, associations and institutes are given the task of using the guideline values or the assessment rules of the AgBB. The current AgBB scheme is intended to help assess health evaluations of residual emissions from construction products [29]. It was published in 2001 and revised in 2015 (see Figure 9.7). The German Institute for Structural Engineering (DIBt) has adopted this assessment scheme for the approval of floor coverings such as wood floors [29]. It is clearly foreseeable that the assessment of emissions according to the AgBB scheme should also be extended in the field of furniture products. In the AgBB scheme, not only sum values of the residual emissions from the chamber test procedure are considered, but also individual material evaluations are carried out. The test of the residual emissions is carried out in the test chamber after 28 days, as per DIN EN ISO 16000-9. The individual VOC values are then derived using so-called NIK values (lowest

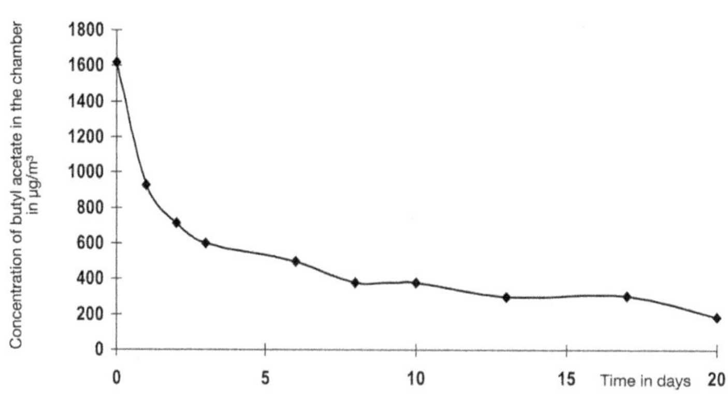

Figure 9.6: Decay curve of butyl acetate using the example of a cellulosic lattice coating in the emission test chamber method according to DIN EN ISO 16000-9 [27]

concentration of interest), which are based on toxicological data, such as the MAK[7] values, taking into account safety factors (factor 100 to 1,000). The NIK evaluation is based on the dimensionless R-value. Substances which at present do not have a toxicological evaluation, i.e. have no NIK classification, are also limited in their permissible concentration. For deepening the AgBB scheme refer to further literature [29, 30].

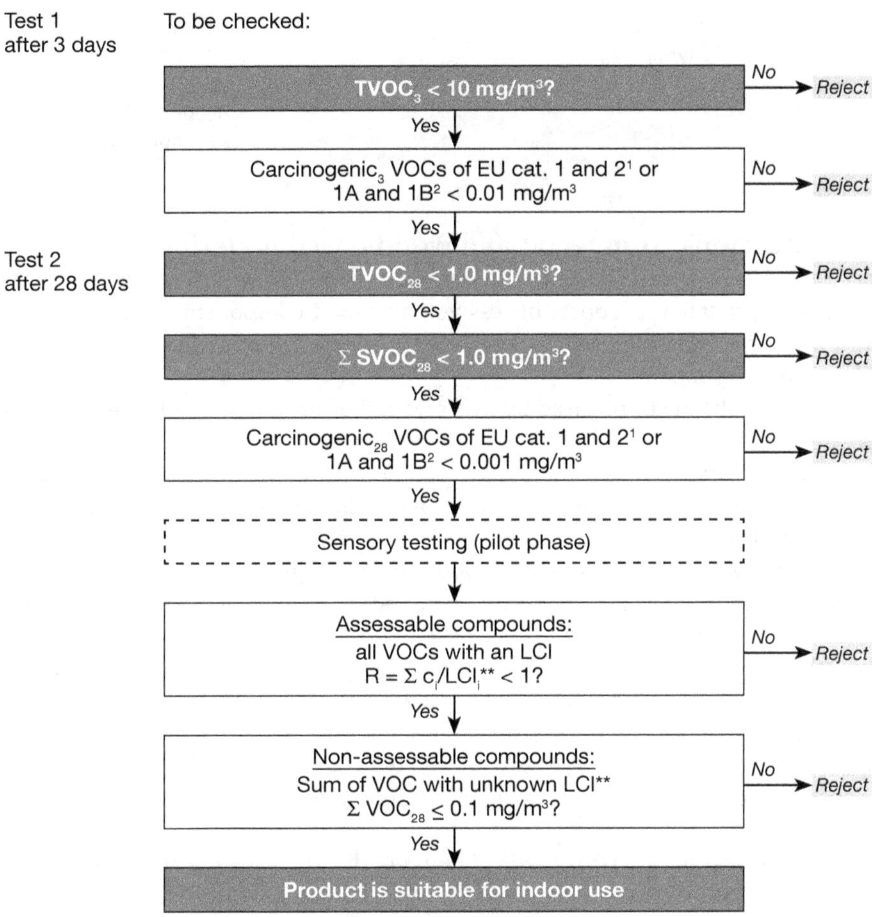

Footnotes
* VOC, TVOC: Retention range C_6–C_{16}, SVOC: Retention range C_{16}–C_{22}
** LCI: Lowest Concentration of Interest (German: NIK)
European Emission Test Standard prEN ISO 16000-9 to -11
1 Classification according to Directive 67/548/EEC Appendix I respectively Regulation (EC) No 1272/2008 Appendix VI Table 3.2
2 Classification according to Regulation (EC) No 1272/2008 Appendix VI Table 3.1

Figure 9.7: AgBB scheme for the health assessment of construction product emissions [30]

7 MAK value: indicates the maximum permissible concentration of a substance as gas, steam or suspended matter in the (respiratory) air at the workplace where no health damage is to be expected, even if the concentration is generally 8 hours a day, maximum 40 (42) hours a week. (outdated but still in use)

Table 9.9: Volatile components from wood and wood-based materials [15]

Volatile wood ingredients		**reactive released VOC**	
Monoterpenes	– α-Pinene – β-Pinene – 3-Carene – Limonene – β-Phellandrene – γ-Terpene – p-Cumene – Terpinolene – β-Myrcene	Aldehydes	– Formaldehyde – Acetaldehyde – Propanal – Pentanal – Hexanal – Heptanal – Octanal – Nonanal – Furfural
		Acids	– Formic acid – Acetic acid – Hexanoic acid

9.2.1.2 Emissions from wood and wood-based materials

Untreated wood products can contribute to the re-emission behaviour due to volatile wood components or degradation products of less volatile wood components (see Chapter 2.4). The volatile organic compounds (VOC) from wood products can be divided into two groups:

- Volatile wood components
- Volatile organic substances released by hydrolytic, thermal and/or oxidative reactions.

It is known that coniferous cones contain between 0.5 and 2 % monoterpenes, such as α- and β-pinene, 3-carene and limonene. In Table 9.9, volatile components of wood and wood materials are exemplified.

The relevant VOCs made of wood and wood materials (monoterpenes, acetic acid and higher aldehydes) are natural products with comparatively low acute toxicity. The question of limitation and reduction is therefore not based on the health risk caused by such emissions, but rather on the general demand of residential hygiene for clean, unstressed interior air and is also to be assessed from the point of view of odour perceptions. In the case of the monoterpenes, only the wood products of conifers are affected, with the emissions decreasing as the wood has been altered by processing and digestion. Figure 9.8 shows the very high terpene concentration in a schoolroom after laying a hardwood floor and sealing with a natural resin coating [29].

Particle boards and fibre boards made of conifers therefore give relatively low amounts of terpenes, while OSB (orientated strand or structural boards), plywood, glued timber and solid wood have higher values for the same raw material base. The OSB boards produced from relatively large chips have a typical odour pattern, especially after the hot pressing. In addition to monoterpenes, the odour-intensive components such as pentanal, hexanal and hexanoic acid are responsible. Figure 9.9 shows the decay curve of hexanal from an OSB plate [29]. Monoterpenes are particularly abundant in various species of fruit. They are particularly important for aromas. For example, menthol from peppermint, carvone from caraway oil and lemon from citrus oil belong to this group. As can be seen in Figure 9.10, very high monoterpene concentrations (about 700 μg/m^3) are still detectable in a white-coated furniture component made of pinewood, even after 3 days in the chamber test. The reactive VOC's (aldehydes, carboxylic acids) are emitted, in particular, by thermal or hydrolytic processing of wood products (for example, wood). This also applies to lentils. Studies on wood materials

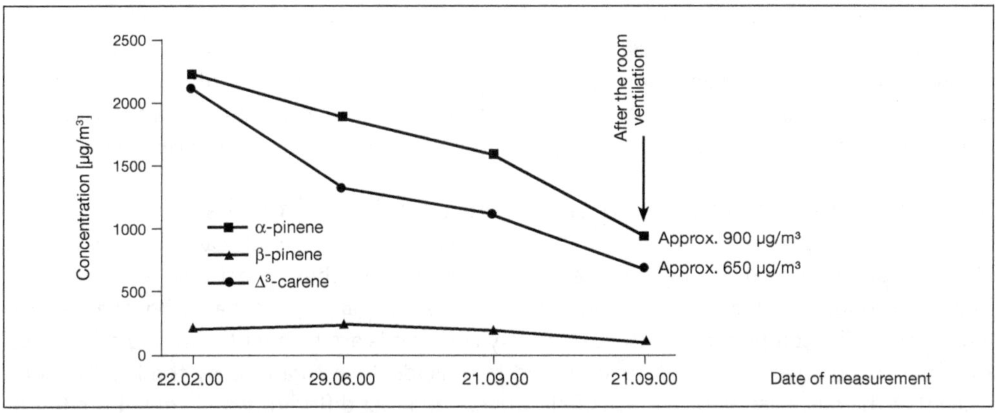

Figure 9.8: Terpene concentrations in a classroom with surface-treated softwood parquet

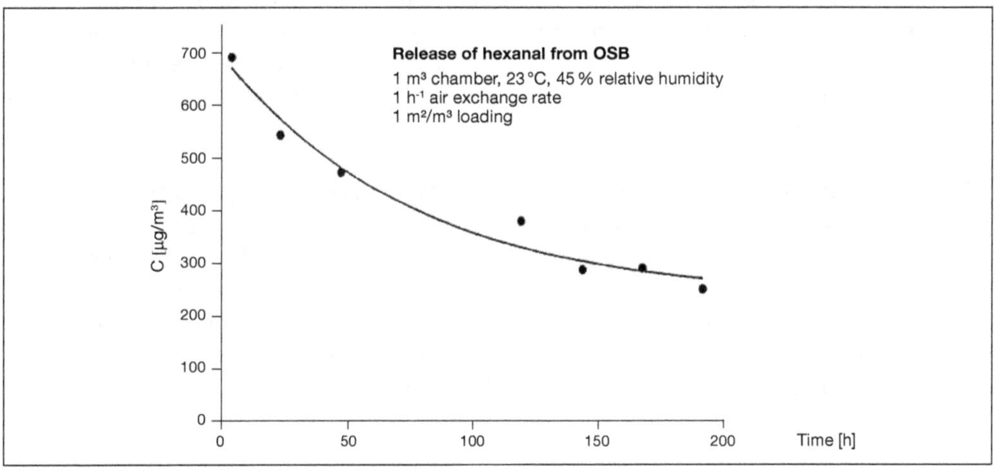

Figure 9.9: Release of hexanal from an OSB board

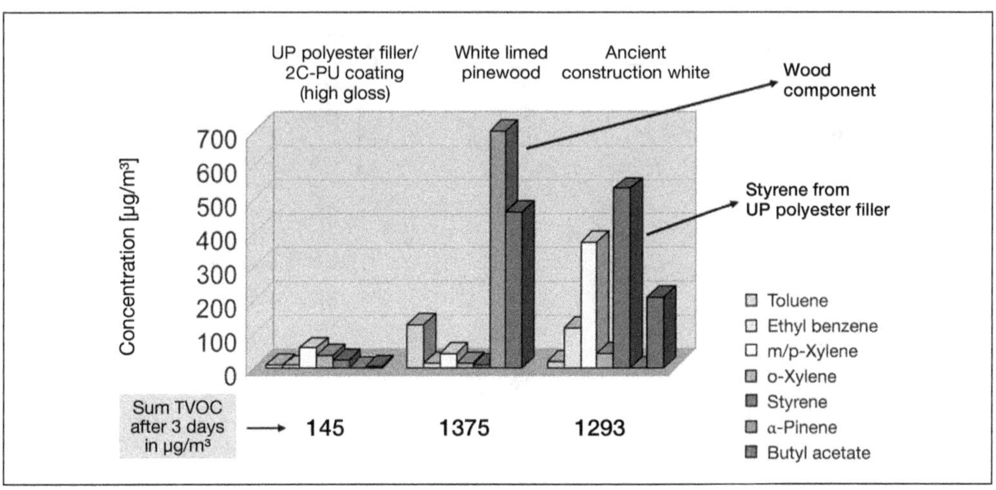

Figure 9.10: Measurements of residual emissions (TVOC) of different wood coating systems after 3 days

have shown that relatively high acetic acid emissions are found, especially in particle boards bonded with alkaline phenolic resins [15]. In the thermal treatment of wood (> 60 °C), small amounts of formaldehyde emissions are detected. Frequently, in the case of residual emission measurements, wood contents are incorrectly interpreted as constituents of wood coatings. The wood contents should be marked separately in the presentation of the residual values.

9.2.1.3 Residual emissions from oxidatively curing oils

Oxidatively curing oils contain unsaturated fatty acids, which can still react chemically even after the purchase and during the use of the furniture. In these reactions, low-molecular cleavage products such as organic acids or aldehydes (propanal, pentanal or hexanal) are often emitted. The components of the oxidatively curing oils are frequently linseed oil, linseed stand oil, wood oil and oils with unsaturated fatty acids. Investigations by the IHD Dresden show that the oils, despite of the use of siccatives can react differently and contribute to different aldehyde emissions [31]. For example, linseed oil products reach the highest aldehyde doses of 700 to 800 mg/kg within two to three days. Particularly odour-intensive compounds emit from wood oil or surfaces coated with wood oil. On the other hand, olive oil-based products have lower emissions (200 to 250 mg/kg). Siccative oil formulations lead to higher aldehyde emissions. In general, coating with oxidatively curing oils almost always leads to a higher odour intensity than uncoated samples [31].

9.2.1.4 Residual emissions from UV-curing coating systems

The emission behaviour of the different coating technologies is primarily dependent on the solvent content of the formulations and on the process parameters such as drying and curing. Efforts to increase the solids content are conceivable, but technically difficult to realise, taking into account customer requirements and processes. The increased use of low-solvent and free UV-curing coating systems significantly contributes to the reduction of residual emissions from furniture surfaces [32]. Previous studies have shown, however, that cleavage products can be formed from the photoinitiators by the UV radiation in small amounts, which

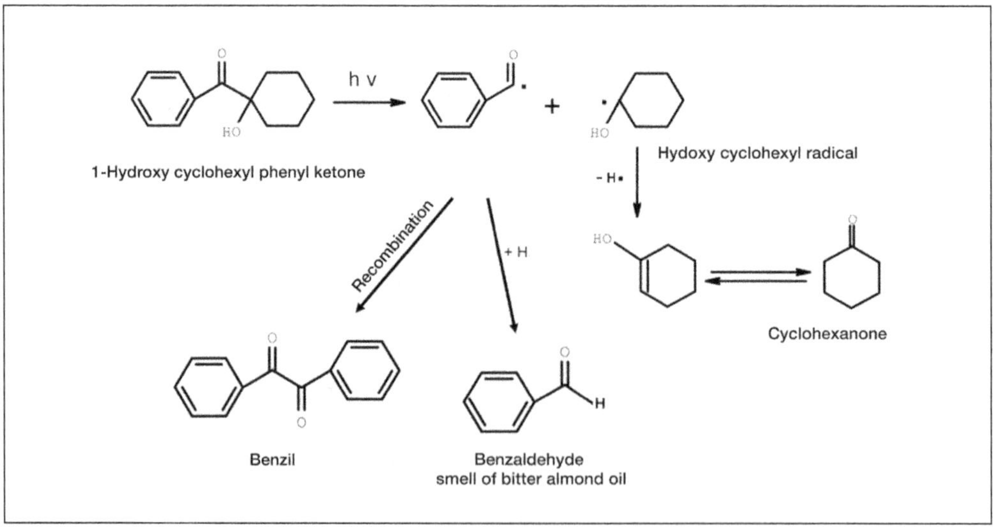

Figure 9.11: Potential cleavage products of 1-hydroxy cyclohexyl phenyl ketone

contribute to odour evolutions and residual emissions. Since hardly any systematic work on the subject of residual emissions from UV coatings was available, a research project on the residual emissions behaviour was carried out from 1996 to 1998 at the Wilhelm-Klauditz-Institute for Wood Research (WKI), in cooperation with the VdL coating association and paint and raw material manufacturers of UV coatings [33, 34]. The results show that the main components emitted are mainly the following substances:

- Solvent
- Photoinitiators and photoinitiator fractions
- Residual monomers
- Other substances

Compared to residual solvent emissions, the smell of minute cleavage product quantities from the photoinitiator decreases only slowly. As a rule, fission product concentrations are significantly below 100 µg/m³ in the emission test chamber after 28 days. In the Figure 9.11 and 9.12, 1-hydroxycyclohexyl-phenyl-ketone and 2-hydroxy-2-methyl-1-phenylpropan-1-one possible cleavage products are shown schematically for the photoinitiators. As shown in the reaction scheme in Figure 9.11 for 1-hydroxycyclohexylphenyl ketone, cyclohexanone is also formed from the hydroxycyclohexyl radical, hydrogen abstraction and keto-enol tautomerism [35]. A by-product of the α-cleavage of 2-hydroxy-2-methyl-1-phenyl-propan-1-one is acetone formed from the 2-hydroxypropyl radical. The odour-intensive cleavage product produced by both photoinitiators is benzaldehyde [36]. When using optimized UV coating formulations and observing the process parameters determined in the pre-field, the user is usually expected to experience no-smelling, odours development [23]. In practice, there are different efforts to further reduce possible smells of UV coatings.

Optimization of UV coatings and photoinitiators

In the course of investigations, water-borne and solvent-thinnable UV coatings with different photoinitiators and variation of the process parameters were tested [21, 23, 27, 37, 38, 39].

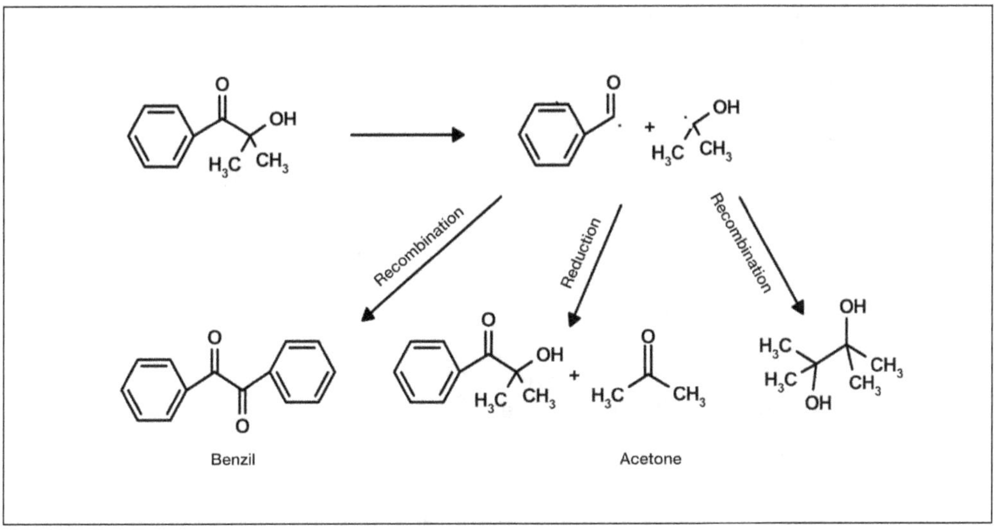

Figure 9.12: Potential cleavage products of 2-hydroxy-2-methyl-1-phenyl-propane-1-ol

Impact of different photoinitiators on the behaviour of residual emissions

Various photoinitiators were tested in UV coating systems in order to determine whether the released and odourily detectable fission products from photoinitiators can be further reduced. Figure 9.13 illustrates the residual values of three different photoinitiated UV roller paints. Experiments PI I and PI III show very small amounts of cleavage products, such as benzaldehyde and cyclohexanone. In PI II no fission products or the photoinitiator could be detected. The experiments show that by careful selection of photoinitiators the cleavage products and thus frequently associated odour development can be further reduced.

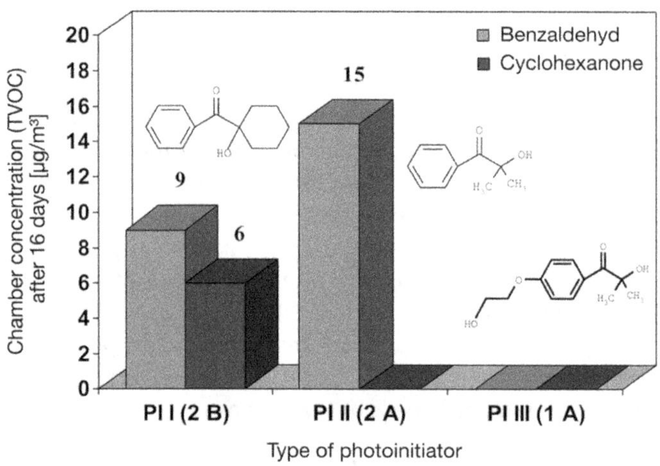

Figure 9.13: Behaviour of the residual emission of different types of photoinitiators

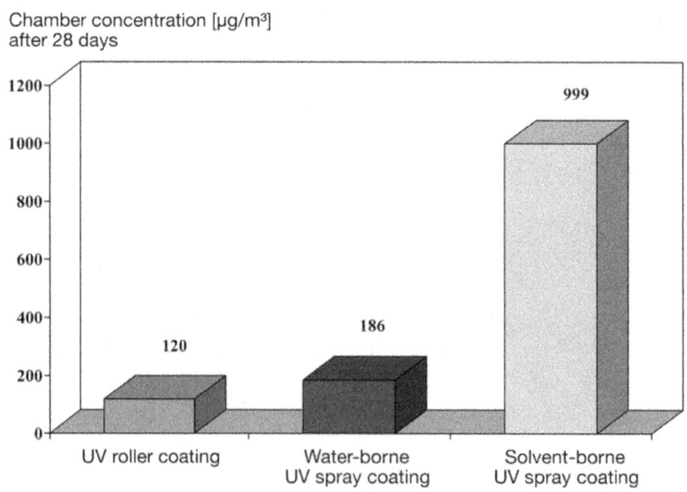

Figure 9.14: Behaviour of the residual emission of white pigmented UV coatings

Behaviour of the residual emission of white pigmented UV coatings

The influence of the pigmentation on the UV curing or the behaviour of the residual emission behaviour of white pigmented UV coating systems was examined. A white pigmented UV roller coating (approx. 100 % non-volatile), UV water-borne and a solvent-borne UV spray coating were tested. Figure 9.14 shows the resorption behaviour (TVOC) of the coating systems in the test chamber after 28 days. The pigmented samples contained a combination of 1-hydroxycyclohexyl-phenyl ketone and bis (2,4,6-trimethylbenzoyl) phenylphosphine oxide (BAPO II). Small amounts of benzaldehyde and cyclohexanone were identified in the chamber air. The 2,4,6-trimethylbenzaldehyde released from bis (2,4,6-trimethylbenzoyl) phenylphosphine oxide could be found on the material surface, but not in the chamber air [40]. The maximum concentration for the cleavage products from the photoinitiators was below 100 µg/m³ after 24 hours. After 28 days in the chamber test, the UV roller coating set-up showed a residual emission value of approx. 120 µg/m³ and the aqueous UV spray coating a value of 186 µg/m³. The residual value of 999 µg/m³ after 28 days for the solvent-containing UV spray coating was noticeable. The major constituents were butyl acetate and ethyl acetate. The systematic experiments show that the pigmented aqueous UV-spraying and UV-roller coating system has a low residual emission behaviour after an optimised selection of photoinitiators after 28 days. Due to the high content of organic solvents, only the solvent-borne UV spray coating build-up result in high residual values after 28 days.

Impact of the application procedure on the behaviour of residual emission of different UV clear coats

Various UV clear coats were applied on beech-veneered chipboard in roll, roll/pour, roll/injection and spraying as base coat and topcoat. Figure 9.15 shows the TVOC values after the 28th day in the emission test chamber. The minimum residual values are the areas (< 50 µg/m³) rolled with UV base and UV topcoat, followed by the rolling/curtain coater process (water-

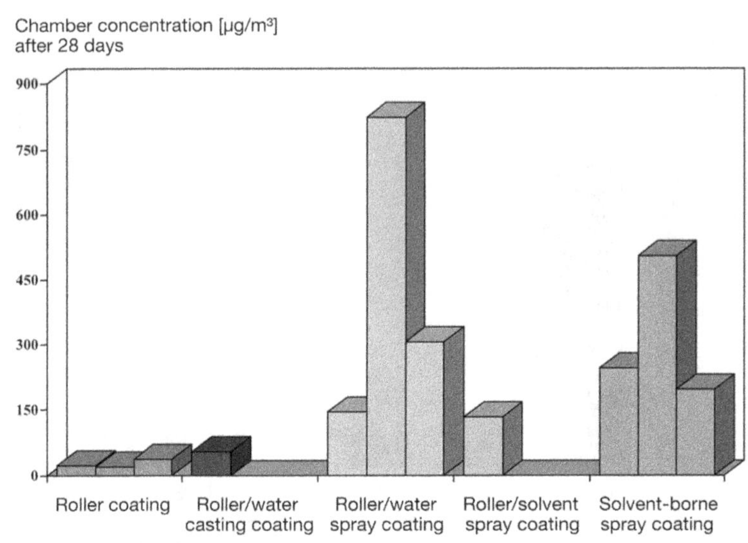

Figure 9.15: Impact of the application process on the residual emission behaviour of different clear coats

borne UV curtain coating, <50 µg/m³) and the rolling/spraying process (solvent-containing UV spray coating). The UV base (roller) variant in combination with the water-borne UV spray coating results in the highest residual emissions within the test series. The residual emissions are mainly caused by organic solvents (e.g., butyl glycol, ethyl diglycol) required for the film formation of water-borne paints. These are less volatile than organic solvents, e.g. ethyl acetate and butyl acetate which are used for the formulation of conventional UV spray paints. Although only 3 % of ethyl glycol was present in the water-borne UV spray coating used, the low volatility resulted in a slower evaporation from the coating film. For this reason, it is particularly important to better coordinate the drying parameters in advance of the introduction of new production processes with regard to the resorption behaviour (for example, longer drying times).

Impact of the UV dosage on the behaviour of residual emissions wof UV coatings

By changing the curing parameters (energy dose), the behaviour of the residual emission is influenced significantly. For this purpose, a UV roller assembly was applied to a beech-veneered chipboard and cured with a low, optimal (practically determined) and strongly increased UV dose. Figure 9.16 illustrates the residual emissions from UV roller coatings as a function of the energy dose during the UV curing. As expected, the area with the lowest UV dose initially shows the highest residual values for cleavage products from the photoinitiator. The coated surfaces which were cured with an optimal and greatly increased UV dose, showed very low residual values at a comparable level after 28 days. Although the residual values are at a very low level, the surface cured with the lowest UV dose contained the highest initial concentration of cleavage products. UV curing with insufficient imaging UV dose results in a higher concentration of cleavage products and thus to an intense smell formation, which can be perceived by the end user in new furniture.

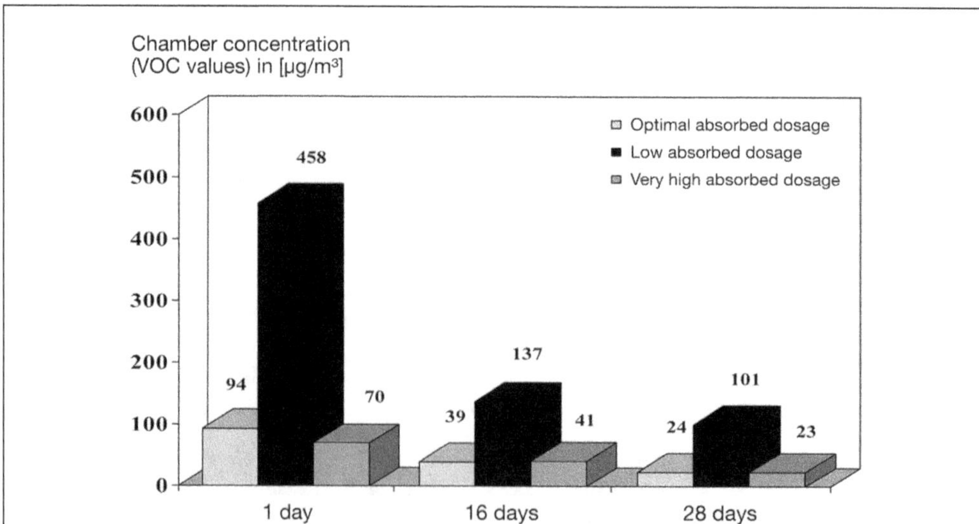

Figure 9.16: Residual emissions from UV roller coatings depending on the absorbed dose within the UV curing [27]

UV curing under inert gas conditions

The UV curing of UV coatings under inert gas is an elegant method to eliminate the interfering influence of the oxygen during the curing process. Under these conditions, significantly less photoinitiator is required for the curing of the UV coatings or less secondary reactions of the photoinitiator occur. In addition, the feed rate of the operated coating systems can be increased significantly without impairing the curing. The furniture industry has not yet used this technology, since the conversion of the UV curing units, the UV flat strata and the use of inert gas requires higher investment. In addition, there is the difficulty of uniformly coating the different component geometries with nitrogen. The UV curing under inert gas is frequently used in web-shaped substrates (for example for the coating of decorative film).

9.3 References

[1] Scheithauer, M.: Messen und Bewerten von Lösemittel-Emissionen bei der Verarbeitung von Lacken, Informationstagung des Verbandes der Lackindustrie e.V.: Lösemittel-Reduzierung bei der Möbelfertigung, 1998

[2] DIN ISO 11890: Bestimmung an flüchtigen organischen Verbindungen (VOC-Gehalt), Teil 1: Differenzierverfahren, Teil 2: Gaschromatographisches Verfahren

[3] Richtlinie 1999/13/EG des Rates vom 11. März 1999 über die Begrenzung von Emissionen flüchtiger organischer Verbindungen, die bei bestimmten Tätigkeiten und in bestimmten Anlagen bei der Verwendung organischer Lösungsmittel entstehen (ABl. EG Nr. L 85 p.1)

[4] Verordnung zur Umsetzung der Richtlinie 1999/13/EG über die Begrenzung von Emissionen flüchtiger organischer Verbindungen – 31. BImSchV, 21, August 2001 (BGBl. 1 p. 2180)

[5] Hackmack, U., Mahrwald, B.: Die Lösemittelverordnung – Einführung und Vorschläge zur Umsetzung in die Praxis, Eigenforschungsvorhaben des Umweltbundesamtes – Luftreinhaltung, Forschungsbericht 50044301, 2002

[6] Vollzug der 31. BImSchV bei der Holzbeschichtung, Fachtagung am 21, June 2005, Herausgeber: Bayerisches Landesamt für Umweltschutz, Augsburg, 2005

[7] Vesper, H., Schrübbers, H.: Entwicklung und Erprobung eines modellhaften Beratungskonzeptes zur Unterstützung des Technologietransfers in kleinen und mittelständischen Unternehmen mit dem Ziel der Einführung umweltgerechter Lackierverfahren, durchgeführt von: bregau (bremer gesellschaft für angewandte umwelttechnologie), Deutsche Forschungsgesellschaft für Oberflächenbehandlung e.V. (DFO) gefördert vom BMBF über die DLR, Förderkennzeichen 01RC0088 – Teil 1, 01RC0089 – Teil 2, 2002

[8] Prieto, J., Gehrmann, V.: Die praktische Anwendung emissionsarmer, wasserverdünnbarer, strahlenhärtender Lacksysteme für die Treppenstufenlackierung, Presentation Verband der deutschen Lackindustrie e. V., 1999

[9] Erhorn, H., Weiland, T.: Wirkt sich die Wohnsituation in der Bundesrepublik Deutschland auf die Anforderungen an den Mindestluftwechsel aus? Bauphysik, Nr. 8, 1986, p. 85–88

[10] Salthammer, T., Marutzky, R.: Emissionen organischer Verbindungen aus Möbeloberflächen, published in Holz-Zentralblatt, No. 144/95, p. 2405 und No. 6/96 p. 57–58

[11] Marutzky, R.: Prüfmethoden als Grundlagen der Formaldehydabgabeminderung bei Möbeln, Holztechnologie 28, 1987, Nr. 6, p. 301–304

[12] Marutzky, R.: Zur Herstellung formaldehydarmer Möbel aus furnierten Holzwerkstoffen, Holz- und Kunststoffverarbeitung (HK), No. 23, 1988, p. 272–275

[13] Marutzky, R., Schriever, E.: Emissionen aus Werkstoffen für Möbel und Inneneinrichtungen von Wohnwagen, congress paper, Wilhelm-Klauditz-Institut (WKI) „Möbel und Formaldehyd", 1990, p. 55–82

[14] Jungbluth, R.: Die 11 Geheimnisse des Ikea-Erfolges, Campus Verlag GmbH Frankfurt/Main, 2006, p. 256–257

[15] Salthammer, T, Marutzky, R., Aehlig, K., Scheithauer, M: Sensorische Prüfung von Produkten für den Innenraum, WKI report No. 38, 02/2003

[16] Fischer, M., Böhm, E.: Erkennung und Bewertung von Schadstoffemissionen aus Möbellacken, Erich Schmidt Verlag, 1994

[17] Unknown: Vertrauensplus für neue Möbel, Material + Technik, 01/1996, p. 19–20

[18] Salthammer, T.: Luftverunreinigende organische Substanzen in Innenräumen, Chemie in unserer Zeit, Vol. 28, No. 6, 1994, p. 280–289

[19] Bundesgesundheitsblatt No. 3, Bewertung der Luftqualität in Innenräumen. Berlin, 1993, p. 117–118
[20] Jann, O., Wilke, O., Brödner, D.: Procedure for the determination and limitation of VOC-Emissions from furnitures and coated wood based products, International Journal of Indoor Environment and Health, 1997
[21] Prieto, J., Schulte, A.: Moderne UV-härtbare Möbellacke, I-Lack, Vol. 67, 10/1999
[22] Marutzky, R.: Umweltfreundliche und Emissionsarme Möbel, congress paper, Wilhelm-Klauditz-Institut (WKI), No. 31, 1995, p. 1–8
[23] Prieto, J.: UV-härtende Lacke – Möbelbeschichtung ohne Restemissionen? Dokumente zu Lacken und Farben, No. 6, 1999, p. 20–30
[24] Salthammer, T, Schwarz, A., Fuhrmann, F.: Emission of reactive compounds and secondary products from wood-based furniture coatings, Atmospheric Environment, No. 33, 1999, p. 75–84
[25] Scheithauer, M.: Emissionen auf der Spur, Holz- und Kunststoffverarbeitung (HK), No. 11, 1998, p. 82–85
[26] Unknown: Holzlacke setzen auf Gütesiegel, Deutsches Lackinstitut, Lack im Gespräch, No. 44, 1996
[27] Prieto, J.: Home comforts, Polymer Paint Colour Journal (PPCJ), Vol. 191, No. 4442, July 2001, p. 15–17
[28] Plehn, W.: Was können emissionsarme Baustoffe und Möbel leisten? – Beispiele aus Produktprüfungen und Auswahlkriterien, Präsentation des Umwelt Bundesamt (UBA), date unknown
[29] Wensing, M., Schulz, N., Salthammer, T.: Kontrollierte Innenraumluft, Farbe und Lacke, Vol. 112, No. 2, 2006, p. 61–64
[30] Unknown: Vorgehensweise bei der gesundheitlichen Bewertung der Emissionen von flüchtigen organischen Verbindungen (VOC und SVOC) aus Bauprodukten, Ausschuss zur gesundheitlichen Bewertung von Bauprodukten, 09/2005
[31] Broege, M., Aehlig, K.: Öle und Wachse auf Massivholz – Ursachen für die Bildung geruchsintensiver Verbindungen, Möbeltage in Dresden, 2002, p. 73–89
[32] Unknown: Möbelemissionen realistisch betrachtet: I-Lack, Vol. 64, 05/1996
[33] Salthammer, T.: Emissionsarme UV-Systeme für Holzoberflächen, WKI report, No. 34, 06/1999
[34] Unknown: Umweltfreundliche und emissionsarme Möbel, I-Lack, Vol. 64, 03/1996, p. 137–139
[35] Dietliker, K.: Photoinitiators for free radical and cationic polymerisation, in P. K. T. Oldring: Chemistry & Technology of UV & EB formulation for coatings, inks and paints, Vol. 3, SITA Technology Ltd., London, 1991
[36] Salthammer, T.: Release of Photoinitiator Fragments from UV-Cured Furniture Coatings, Journal of Coatings Technology, Vol. 68, No. 856, 05/1996, p. 41–47
[37] Salthammer, T., Hofmockel, U., Lokai, M., Prieto, J., Hansemann, W.: UV-curable furniture coatings with low emissions, RadCureLetter, 6/1999, p. 92–95
[38] Salthammer, T., Bednarek, M., Fuhrmann, F.: Effect of climatic parameters on the release of VOC from UV-cured furniture coatings, Proceedings of the 8th International Conference on Indoor Air and Climate, Vol. 5, 1999, p. 99–104
[39] Salthammer, T.: Holzlacke für Innenräume – Restemissionen lassen sich minimieren, published in Farbe und Lack, No. 10, 1997, p. 142–150
[40] Salthammer, T.: Emissionsprüfungen zur Optimierung von UV-härtenden Möbellacken, published in Farbe und Lack, Vol. 107, No. 2, 2001
[41] http://www.umweltbundesamt.de/daten/luftbelastung/luftschadstoff-emissionen-in-deutschland/emission-fluechtiger-organischer-verbindungen-ohne

10 Trends

Trends are undoubtedly of great importance, and for this reason cannot be missing at the end of a book. A trend is nothing more than an action of change or a process of change. In order to be a step ahead of the competition, it is important to recognize social and technological trends at an early stage and to implement them in innovative products, technologies, processes or services. It is necessary to identify emerging markets, to develop strategies for the allocation of emerging gaps in the market and to achieve this with a wealth of ideas. Factors such as flexibility, customer orientation, quality or perfectly functioning technology have long been assumed by the customer. The rapid response to dynamic, fast evolving markets and products is an important issue, especially in design-oriented furniture and interior design, which is strongly influenced by fashion trends. The trend monitoring is less a case of predicting the next seasonal colour. Rather, it is a question of recognising current developments and developing a feeling for what remains important in the long term. The future certainly lies in sustainable and ecological materials.

Globalisation of the markets and to ensuring the supply of raw materials

The globalisation of the markets and in addition the focus of the wood coating producers have clearly been felt over the last 10 years. The strong growth of wood coatings is mainly caused by Asia and South America. By strong environmental requirements and the closure of non-environmentally friendly production facilities in China, the demand for water-borne and UV-curing paints should be further promoted. At the same time, the control of the entire industry by the environmental authorities has increased considerably. Since the commencement of the new environmental law as of January 1, 2015, responsible persons have been prosecuted for serious violations of the environmental protection in the company. For the globally active paint and coating manufacturers, the availability and securing of the raw material supply is a primary goal. Due to the further well-evolving markets in Asia and their demand for raw materials, commodity prices will continue to rise in the medium to long term and certain raw materials are becoming increasingly scarce, while at the same time a monopolisation of the raw material producers can be observed. As a result of the commencement of REACH[1] in Europe, some raw materials will soon be unavailable as the registration/ approval costs are extremely high.

Trends within the individual coating technologies

Technologies are evaluated primarily for their efficiency and cost-effectiveness and must have a higher degree of automation in the future. Above all, higher speeds and operational widths of 2.40 m for the roll coating of backings, parquet, laminate or other flat substrates are to be expected. In this way, feed rates of 100 to 120 m/min are required for the flat line flooring of laminates and MDF rear panels ("backings") and are already partially implemented in order to remain competitive as a manufacturer. For the decorative finishing of pre-

1 REACH stands for **R**egistration, **E**valuation and **A**uthorization of **Ch**emicals. REACH prescribes for the first time the testing of chemicals for their environmental and health effects and concerns substances produced or imported in the EU from a quantity of 1 tons per year.

impregnates, the electron beam curing paints are continuously gaining market significance, since very fast processes and very good chemical and mechanical properties can be achieved with this. However, the water-borne acid-curing film topcoats also continue to grow in quantity.

Promoted by the European-wide Solvent Regulation (see Chapter 9), water-borne coating systems are gaining in importance for the spray application. Water-borne 1C UV and dual cure systems will further replace solvent-containing 2-component PU and acid-curing paints. For the substitution of e.g. cellulose nitrate coatings one-component and self-crosslinking water-borne coatings are still in operation. In the furniture coating as well as in the floor coating oxidative curing oils on the basis of renewable raw materials are continuing to advance. The powder coating (low-temperature powder) will also gain market share for technically feasible applications (e.g., IKEA).

Trend to high-gloss surfaces

The global trend for high-gloss interior finishes, furniture segments and flooring segments has steadily increased in recent years. The classic spraying process with two-component PU coatings followed by fine grinding and polishing is becoming increasingly important as it is very time-consuming and cost-intensive. The global trend towards high-gloss coatings on unpolished, melamine resin-coated wood-based panels represents new demands on processes and products. The perfect high-gloss finish is dependent on cost-effectiveness and process-engineering feasibility. Depending on the requirements, the following rolling processes or combination processes (rolling/curtain coater) are available with their advantages and disadvantages:

- "Melacoat" process (Oskar Nolte Lacke)
- "Hot-Coatings" (Kleiberit)
- "Fusion-Coatings" (Henkel)

Trend towards super mat coating surfaces

In addition to high-gloss surfaces, super mat coating surfaces are also gaining in importance for several years. There are three processes in the market that are gradually becoming established.

- "Calander Coating Inert Technology" (e.g., Hymmen) → setting of the gloss level by means of the microstructure of a plastic film
- "Excimer-Inert Method" (IOT-Leipzig) → setting of the gloss level via the physical microfolding after irradiation with an excimer lamp (172 nm)
- "Opal-Soft" Coating process (Oskar Nolte Lacke)

Trend towards sustainability and naturalness

The design of the personal residential area is gaining in importance. The own four walls are a place of retreat and an oasis of well-being – comfort and cosiness are the main focus. This trend has been around for a long time, and it has been enduring. The trend towards a healthy and sustainable way of life has been felt for years. This trend has a name and already has its own community: LOHAS (Lifestyle of health and sustainability). "LOHAS" is translatable with "Orientation of the way of life to health and sustainability." Permanent, elegant furnishings with practical handling accompany the inhabitants in their leisure time. The harmony with nature increasingly is emphasized – natural materials and objects consisting of organic materials are very popular. A trend in this area is the "raw wood effect", which makes the coating both haptically and visually barely recognisable, while at the same time people liberate from too much decoration, opulence and striking design. The "quiet" luxury is preferred by high-quality woods or perfectly planned furnishing solutions.

Trends by end user requirements

The emphasis here is on the desire for more individuality in the design and colour of the piece of furniture. The desire for functionality will also set new trends.
For example, desires for anti-bacterial equipment of kitchen furniture surfaces, easy-to-clean surfaces or after writable and wipeable surfaces are observed.

Functional coating surfaces

This trend is closely linked to consumer preferences. Thus, e.g. improved performance properties or further functionalities of surfaces can be still expected from nanoscale particles. Examples are:

- Scratch resistant surfaces
- Anti-bacterial surfaces (nano-scale silver)
- Dust-repellent surfaces due to electrically conductive wooden floors
- Anti-slip equipment
- Self-healing window coatings due to the installation of microcapsules
- Haptic surfaces (e.g., leather effect)

Trends in application

The furniture production of the future is characterised by individuality. The trend towards cost-effective serial production in lot size 1 poses new challenges for all industry players.
In furniture production, 3D printing will continue to grow steadily. Materials such as plastic or a mixture of wood fibres and polymers are used. The digital print will grow steadily for individualized solutions in the coating of paper and wood and is an alternative to indirect gravure printing. The digital printing provides an economic opportunity to react flexibly and quickly to short-term trends. Apart from the aspect of the short-term nature, digital printing technology, above all, provides planning security, given that the minimum purchase quantity of a rotogravure decor paper is one ton, or approximately 1,300 sheets.

Authors

Dr. Jürgen Kiene studied Chemistry at the University of Braunschweig, Germany and he received his PhD at the University of Paderborn, Germany in 1996. After an additional study of Chemistry and Technology of Coatings at the University of Paderborn he worked at the Fraunhofer Institute for Surface Engineering and Thin Films (IST) as a science assistant between 1996 and 1999. Amongst others he supervised the technical committee Coating of Wood and Wood-based materials at Deutsche Forschungsgesellschaft für Oberflächenbehandlung (DFO) in Dusseldorf, Germany. Furthermore, he was lecturer and head of the seminar Coating of Wood and Wood-based materials at DFO. From 2004 to 2006, he was the Technical Director of the China Paint Manufacturing (Shenzhen) Co., Ltd. Since 2006 he is working for Helcotec Chemie und Technik GmbH & Co. KG. in Mönchengladbach, Germany and is Technical Director and authorised officer since 2016.

Dipl.-Chem.-Ing. Jorge Prieto started his career in 1984 when he trained as a lacquer laboratory technician at BASF Coatings in Münster, Germany. From 1990 to 1994, he studied Chemistry and Technology of the Coatings Material at the University of Paderborn, Germany. In 1994 he joined BASF Coatings, Münster and was responsible as the head of laboratories for research and technical department wood coatings in Italy, Germany and Netherland. From 2002 to 2007 he was Technical Director and member of the company management at 3H-Lacke (Herford, Germany). In 2008 he founded 3P Coating Consulting Company in Lünen, Germany. During this time, he served worldwide various clients from the coatings and raw material industry. He has been the Managing Director at Oskar Nolte GmbH (industrial wood coatings) since January 2016.

Index

A

B

C

D

E

F

G

H

N

O

P

Q

R

S

T

U

V

W

X

Y

Z

Symboles

www.ingramcontent.com/pod-product-compliance
Lightning Source LLC
LaVergne TN
LVHW080339011026
849161LV00009B/2126
9783866306080